Embedded Systems Design

By the same author

VMEbus: a practical companion

Newnes UNIX™ Pocket Book

Microprocessor architectures: RISC, CISC and DSP

Effective PC networking

PowerPC: a practical companion

The PowerPC Programming Pocket Book

The PC and MAC handbook

The Newnes Windows NT Pocket Book

Multimedia Communications

Essential Linux

Migrating to Windows NT

All books published by Butterworth-Heinemann

About the author:

Through his work with Motorola Semiconductors, the author has been involved in the design and development of microprocessor-based systems since 1982. These designs have included VMEbus systems, microcontrollers, IBM PCs, Apple Macintoshes, and both CISC- and RISC-based multiprocessor systems, while using operating systems as varied as MS-DOS, UNIX, Macintosh OS and real-time kernels.

An avid user of computer systems, he has had over 60 articles and papers published in the electronics press, as well as several books.

Embedded Systems Design

Second edition

Steve Heath

Newnes

OXFORD AMSTERDAM BOSTON LONDON NEW YORK
PARIS SAN DIEGO SAN FRANCISCO SINGAPORE SYDNEY TOKYO

Newnes
An imprint of Elsevier
Linacre House, Jordan Hill, Oxford OX2 8DP
30 Corporate Drive, Burlington, MA 01803

First published 1997
Reprinted 2000, 2001
Second edition 2003
Reprinted 2005

British Library Cataloguing in Publication Data
A catalogue record for this book is available from the British Library

Library of Congress Cataloguing in Publication Data
A catalogue record for this book is available from the Library of Congress

ISBN 0 7506 5546 1

For information on all Newnes publications visit our website at www.newnespress.com

Typeset by Steve Heath
Printed and bound in Great Britain by MPG Books Ltd, Bodmin, Cornwall

Contents

Preface

The term embedded systems design covers a very wide range of microprocessor designs and does not simply start and end with a simple microcontroller. It can be a PC running software other than Windows and word processing software. It can be a sophisticated multiprocessor design using the fastest processors on the market today.

The common thread to embedded systems design is an understanding of the interaction that the various components within the system have with each other. It is important to understand how the hardware works and the restraints that using a certain peripheral may have on the rest of the system. It is essential to know how to develop the software for such systems and the effect that different hardware designs can have on the software and vice versa. It is this system design knowledge that has been captured in this book as a series of tutorials on the various aspects of embedded systems design.

Chapter 1 defines what is meant by the term and in essence defines the scope of the rest of the book. The second chapter provides a set of tutorials on processor architectures explaining the different philosophies that were used in their design and creation. It covers many of the common processor architectures ranging from 8 bit microcontrollers through CISC and RISC processors and finally ending with digital signal processors and includes information on the ARM processor family.

The third chapter discusses different memory types and their uses. This has been expanded in this edition to cover caches in more detail and the challenges associated with them for embedded design. The next chapter goes through basic peripherals such as parallel and serial ports along with timers and DMA controllers. This theme is continued in the following chapter which covers analogue to digital conversion and basic power control.

Interrupts are covered in great detail in the sixth chapter because they are so essential to any embedded design. The different types that are available and their associated software routines are described with several examples of how to use them and, perhaps more importantly, how not to use them.

The theme of software is continued in the next two chapters which cover real-time operating systems and software development. Again, these have a tremendous effect on embedded designs but whose design implications are often not well understood or explained. Chapter 9 discusses debugging and emulation techniques.

The remaining five chapters are dedicated to design examples covering buffer and data structures, memory and processor performance trade-offs and techniques, software design examples including using a real-time operating system to create state machines and finally a couple of design examples. In this edition, an example real-time system design is described that uses a non-real-time system to create an embedded system. The C source code is provided so that it can be run and experimented with on a PC running MS-DOS.

Steve Heath

Acknowledgements

By the nature of this book, many hardware and software products are identified by their tradenames. In these cases, these designations are claimed as legally protected trademarks by the companies that make these products. It is not the author's nor the publisher's intention to use these names generically, and the reader is cautioned to investigate a trademark before using it as a generic term, rather than a reference to a specific product to which it is attached.

Many of the techniques within this book can destroy data and such techniques must be used with extreme caution. Again, neither author nor publisher assume any responsibility or liability for their use or any results.

While the information contained in this book has been carefully checked for accuracy, the author assumes no responsibility or liability for its use, or any infringement of patents or other rights of third parties which would result.

As technical characteristics are subject to rapid change, the data contained are presented for guidance and education only. For exact detail, consult the relevant standard or manufacturers' data and specification.

1 What is an embedded system?

Whenever the word microprocessor is mentioned, it conjures up a picture of a desktop or laptop PC running an application such as a word processor or a spreadsheet. While this is a popular application for microprocessors, it is not the only one and the fact is most people use them indirectly in common objects and appliances without realising it. Without the microprocessor, these products would not be as sophisticated or cheap as they are today.

The embedding of microprocessors into equipment and consumer appliances started before the appearance of the PC and consumes the majority of microprocessors that are made today. In this way, embedded microprocessors are more deeply ingrained into everyday life than any other electronic circuit that is made. A large car may have over 50 microprocessors controlling functions such as the engine through engine management systems, brakes with electronic anti-lock brakes, transmission with traction control and electronically controlled gearboxes, safety with airbag systems, electric windows, air-conditioning and so on. With a well-equipped car, nearly every aspect has some form of electronic control associated with it and thus a need for a microprocessor within an embedded system.

A washing machine may have a microcontroller that contains the different washing programs, provides the power control for the various motors and pumps and even controls the display that tells you how the wash cycles are proceeding.

Mobile phones contain more processing power than a desktop processor of a few years ago. Many toys contain microprocessors and there are even kitchen appliances such as bread machines that use microprocessor-based control systems. The word control is very apt for embedded systems because in virtually every embedded system application, the goal is to control an aspect of a physical system such as temperature, motion, and so on using a variety of inputs. With the recent advent of the digital age replacing many of the analogue technologies in the consumer world, the dominance of the embedded system is ever greater. Each digital consumer device such as a digital camera, DVD or MP3 player all depend on an embedded system to realise the system. As a result, the skills behind embedded systems design are as diverse as the systems that have been built although they share a common heritage.

What is an embedded system?

There are many definitions for this but the best way to define it is to describe it in terms of what it is not and with examples of how it is used.

An embedded system is a microprocessor-based system that is built to control a function or range of functions and is not designed to be programmed by the end user in the same way that a PC is. Yes, a user can make choices concerning functionality but cannot change the functionality of the system by adding/replacing software. With a PC, this is exactly what a user can do: one minute the PC is a word processor and the next it's a games machine simply by changing the software. An embedded system is designed to perform one particular task albeit with choices and different options. The last point is important because it differentiates itself from the world of the PC where the end user does reprogram it whenever a different software package is bought and run. However, PCs have provided an easily accessible source of hardware and software for embedded systems and it should be no surprise that they form the basis of many embedded systems. To reflect this, a very detailed design example is included at the end of this book that uses a PC in this way to build a sophisticated data logging system for a race car.

If this need to control the physical world is so great, what is so special about embedded systems that has led to the widespread use of microprocessors? There are several major reasons and these have increased over the years as the technology has progressed and developed.

Replacement for discrete logic-based circuits

The microprocessor came about almost by accident as a programmable replacement for calculator chips in the 1970s. Up to this point, most control systems using digital logic were implemented using individual logic integrated circuits to create the design and as more functionality became available, the number of chips was reduced.

This was the original reason for a replacement for digital systems constructed from logic circuits. The microprocessor was originally developed to replace a mass of logic that was used to create the first electronic calculators in the early 1970s. For example, the early calculators were made from discrete logic chips and many hundreds were needed just to create a simple four function calculator. As the integrated circuit developed, the individual logic functions were integrated to create higher level functions. Instead of creating an adder from individual logic gates, a complete adder could be bought in one package. It was not long before complete calculators were integrated onto a single chip. This enabled them to be built at a very low cost compared to the original machines but any changes or improvements required that a new chip be developed. The answer was to build a chip that had some form of programmable capability within it. Why not build a chip that took data in, processed it and sent it out again? In this way, instead of creating new functions by analysing the gate level logic and modifying it — a very time-consuming process — new products could be created by changing the program code that processed the information. Thus the microprocessor was born.

Provide functional upgrades

In the same way that the need to develop new calculator chips faster and with less cost prompted the development of the first microprocessors, the need to add or remove functionality from embedded system designs is even more important. With much of the system's functionality encapsulated in the software that runs in the system, it is possible to change and upgrade systems by changing the software while keeping the hardware the same. This reduces the cost of production even lower because many different systems can share the same hardware base.

In some cases, this process is not possible or worthwhile but allows the manufacturer to develop new products far quicker and faster. Examples of this include timers and control panels for domestic appliances such as VCRs and televisions.

In other cases, the system can be upgraded to improve functionality. This is frequently done with machine tools, telephone switchboards and so on. The key here is that the ability to add functionality now no longer depends on changing the hardware but can be done by simply changing the software. If the system is connected to a communications link such as a telephone or PC network, then the upgrade can be done remotely without having to physically send out an engineer or technician.

Provide easy maintenance upgrades

The same mechanism that allows new functionality to be added through reprogramming is also beneficial in allowing bugs to be solved through changing software. Again it can reduce the need for expensive repairs and modifications to the hardware.

Improves mechanical performance

For any electromechanical system, the ability to offer a finer degree of control is important. It can prevent excessive mechanical wear, better control and diagnostics and, in some cases, actually compensate for mechanical wear and tear. A good example of this is the engine management system. Here, an embedded microprocessor controls the fuel mixture and ignition for the engine and will alter the parameters and timing depending on inputs from the engine such as temperature, the accelerator position and so on. In this way, the engine is controlled far more efficiently and can be configured for different environments like power, torque, fuel efficiency and so on. As the engine components wear, it can even adjust the parameters to compensate accordingly or if they are dramatically out of spec, flag up the error to the driver or indicate that servicing is needed.

This level of control is demonstrated by the market in 'chipped' engine management units where third party companies modify the software within the control unit to provide more power or torque. The differences can range from 10% to nearly 50% for some turbo charged engines! All this from simply changing a few bytes. Needless to say, this practice may invalidate any guarantee from the manufacturer and may unduly stress and limit the engine's mechanical life. In some cases, it may even infringe the original manufacturer's intellectual property rights.

Protection of intellectual property

To retain a competitive edge, it is important to keep the design knowledge within the company and prevent others from understanding exactly what makes a product function. This knowledge, often referred to as IPR (intellectual property rights), becomes all important as markets become more competitive. With a design that is completely hardware based and built from off-the-shelf components, it can be difficult to protect the IPR that was used in its design. All that is needed to do is to take the product, identify the chips and how they are connected by tracing the tracks on the circuit board. Some companies actually grind the part numbers off the integrated circuits to make it harder to reverse engineer in this way.

With an embedded system, the hardware can be identified but the software that really supplies the system's functionality can be hidden and more difficult to analyse. With self-contained microcontrollers, all that is visible is a plastic package with a few connections to the outside world. The software is already burnt into the on-chip memory and is effectively impossible to access. As a result, the IPR is much more secure and protected.

Replacement for analogue circuits

The movement away from the analogue domain towards digital processing has gathered pace recently with the advent of high performance and low cost processing.

To understand the advantages behind digital signal processing, consider a simple analogue filter. The analogue implementation is extremely simple compared to its digital equivalent. The analogue filter works by varying the gain of the operational amplifier which is determined by the relationship between r_i and r_f.

In a system with no frequency component, the capacitor c_i plays no part as its impedance is far greater than that of r_f. As the frequency component increases, the capacitor impedance decreases until it is about equal with r_f where the effect will be to reduce the gain of the system. As a result, the amplifier acts as a low pass filter where high frequencies will be filtered out. The equation shows the relationship where $j\omega$ is the frequency component. These filters are easy to design and are cheap to build. By making the CR (capacitor-resistor) network more complex, different filters can be designed.

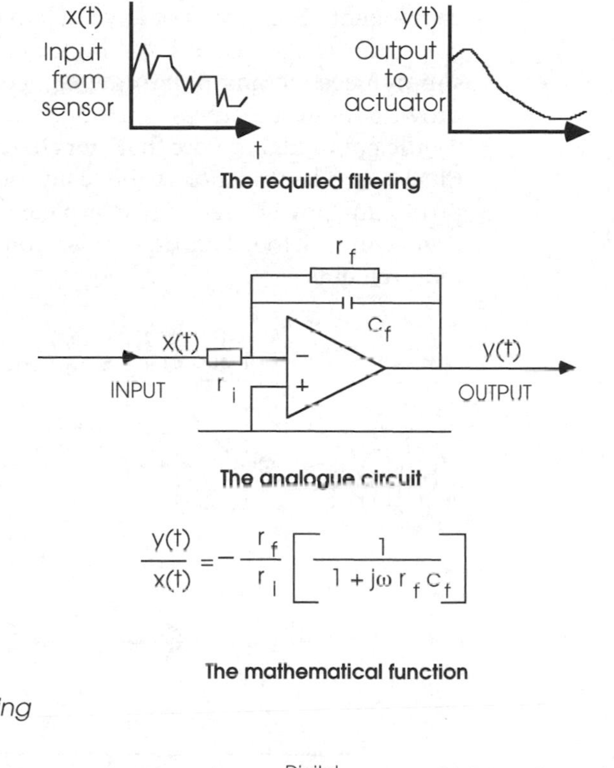

Analogue signal processing

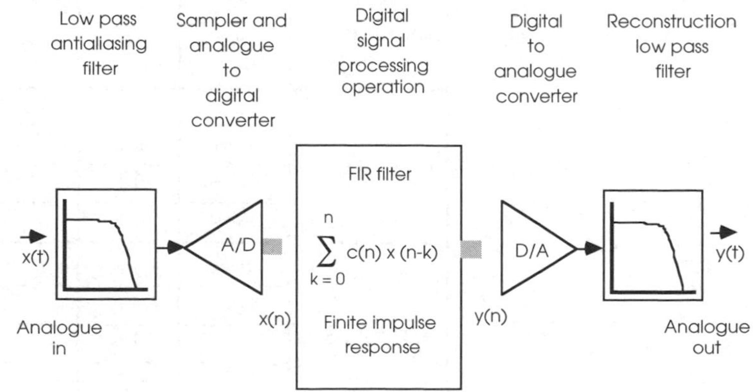

Digital signal processing (DSP)

The digital equivalent is more complex requiring several electronic stages to convert the data, process it and reconstitute the data. The equation appears to be more involved, comprising of a summation of a range of calculations using sample data multiplied by a constant term. These constants take the place of the CR components in the analogue system and will define the filter's transfer function. With digital designs, it is the tables of coefficients that are dynamically modified to create the different filter characteristics.

Given the complexity of digital processing, why then use it? The advantages are many. Digital processing does not suffer from component ageing, drift or any adjustments which can plague an analogue design. They have high noise immunity and power supply rejection and due to the embedded processor can easily provide self-test features. The ability to dynamically modify the coefficients and therefore the filter characteristics allows complex filters and other functions to be easily implemented. However, the processing power needed to complete the 'multiply–accumulate' processing of the data does pose some interesting processing requirements.

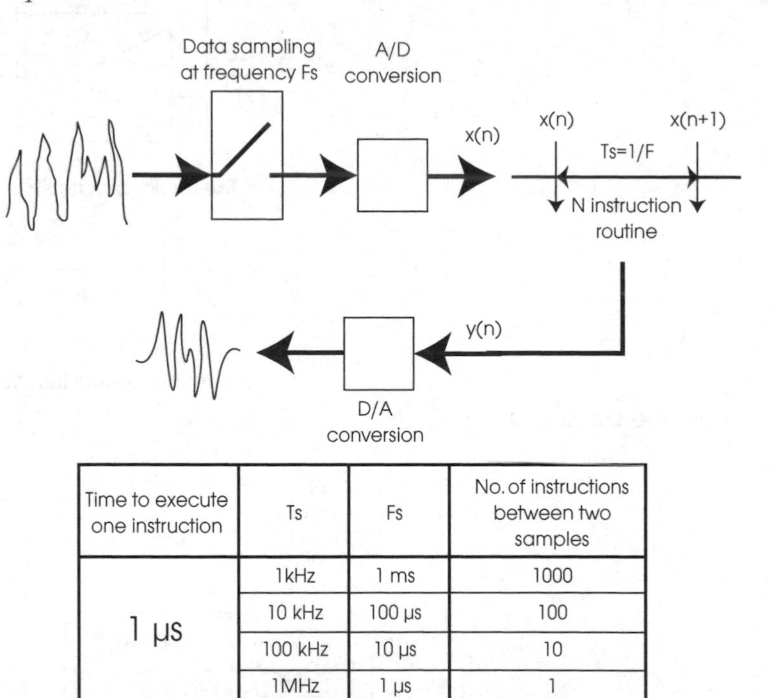

Time to execute one instruction	Ts	Fs	No. of instructions between two samples
1 μs	1kHz	1 ms	1000
	10 kHz	100 μs	100
	100 kHz	10 μs	10
	1MHz	1 μs	1
100 μs	1kHz	1 ms	10000
	10 kHz	100 μs	1000
	100 kHz	10 μs	100
	1MHz	1 μs	10

DSP processing requirements

The diagram shows the problem. An analogue signal is sampled at a frequency f_S and is converted by the A/D converter. This frequency will be first determined by the speed of this conversion. At every period, t_S, there will be a new sample to process using *N* instructions. The table shows the relationship between sampling speed, the number of instructions and the instruction execution time. It shows that the faster the sampling frequency, the more processing power is needed. To achieve the 1 MHz frequency, a 10 MIPS processor is needed whose instruction set is powerful enough to complete the processing in under 10 instructions. This analysis does not take into account A / D conversion delays. For DSP algorithms, the sampling speed is usually twice the frequency of the highest frequency signal being processed: in this case the 1 MHz sample rate would be adequate for signals up to 500 kHz.

One major difference between analogue and digital filters is the accuracy and resolution that they offer. Analogue signals may have definite limits in their range, but have infinite values between that range. Digital signal processors are forced to represent these infinite variations within a finite number of steps determined by the number of bits in the word. With an 8 bit word, the increases are in steps of 1 / 256 of the range. With a 16 bit word, such steps are in 1 / 65536 and so on. Depicted graphically as shown, a 16 bit word would enable a low pass filter with a roll-off of about 90 dB. A 24 bit word would allow about 120 dB roll-off to be achieved.

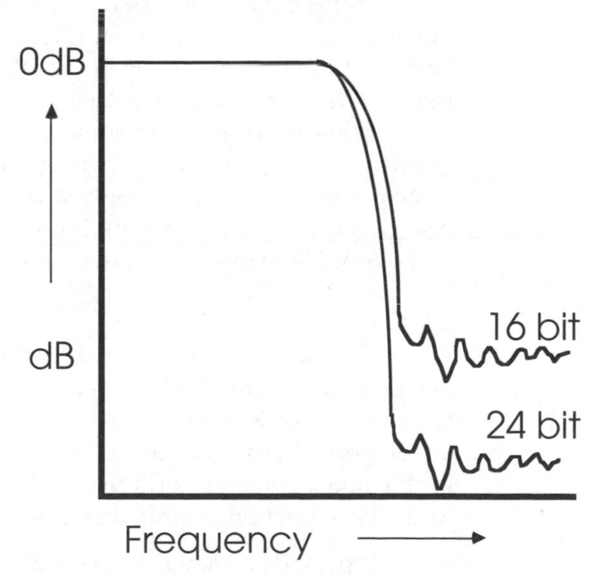

Word size and cutoff frequencies

DSP can be performed by ordinary microprocessors, although their more general-purpose nature often limits performance and the frequency response. However, with responses of only a few hundred Hertz, even simple microcontrollers can perform such tasks. As silicon technology improved, special building blocks appeared allowing digital signal processors to be developed, but their implementation was often geared to a hardware approach rather than designing a specific processor architecture for the job. It is now common for processors to claim DSP support through enhanced multiply–accumulate operations or through special accelerators. It is clear though, that as general purpose processing increases in capability, what was once the sole province of a DSP can now be achieved by a general purpose processor.

Inside the embedded system

Processor

The main criteria for the processor is: can it provide the processing power needed to perform the tasks within the system? This seems obvious but it frequently occurs that the tasks are either underestimated in terms of their size and/or complexity or that creeping elegance expands the specification to beyond the processor's capability.

In many cases, these types of problems are compounded by the performance measurement used to judge the processor. Benchmarks may not be representative of the type of work that the system is doing. They may execute completely out of cache memory and thus give an artificially high performance level which the final system cannot meet because its software does not fit in the cache. The software overheads for high level languages, operating systems and interrupts may be higher than expected. These are all issues that can turn a paper design into failed reality.

While processor performance is essential and forms the first gating criterion, there are others such as cost — this should be system cost and not just the cost of the processor in isolation, power consumption, software tools and component availability and so on. These topics are discussed in more detail in Chapter 2.

Memory

Memory is an important part of any embedded system design and is heavily influenced by the software design, and in turn may dictate how the software is designed, written and developed. These topics will be addressed in more detail later on in this book. As a way of introduction, memory essentially performs two functions within an embedded system:

- It provides storage for the software that it will run

  At a minimum, this will take the form of some non-volatile memory that retains its contents when power is removed. This can be on-chip read only memory (ROM) or external EPROM. The software that it contains might be the complete program or an initialisation routine that obtains the full software from another source within or outside of the system. This initialisation routine is often referred to as a bootstrap program or routine. PC boards that have embedded processors will often start up using software stored in an onboard EPROM and then wait for the full software to be downloaded from the PC across the PC expansion bus.

- It provides storage for data such as program variables and intermediate results, status information and any other data that might be created throughout the operation

  Software needs some memory to store variables and to manage software structures such as stacks. The amount of memory that is needed for variables is frequently less than that needed for the actual program. With RAM being more expensive than ROM and non-volatile, many embedded systems and in particular, microcontrollers, have small amounts of RAM compared to the ROM that is available for the program. As a result, the software that is written for such systems often has to be written to minimise RAM usage so that it will fit within the memory resources placed upon the design. This will often mean the use of compilers that produce ROMable code that does not rely on being resident in RAM to execute. This is discussed in more detail in Chapter 3.

Peripherals

An embedded system has to communicate with the outside world and this is done by peripherals. Input peripherals are usually associated with sensors that measure the external environment and thus effectively control the output operations that the embedded system performs. In this way, an embedded system can be modelled on a three-stage pipeline where data and information input into the first stage of the pipeline, the second stage processes it before the third stage outputs data.

If this model is then applied to a motor controller, the inputs would be the motor's actual speed and power consumption, and the speed required by the operator. The outputs would be a pulse width modulated waveform that controls the power to the motor and hence the speed and an output to a control panel showing the current speed. The middle stage would be the software that processed the inputs and adjusts the outputs to achieve the required engine speed. The main types of peripherals that are used include:

- Binary outputs

  These are simple external pins whose logic state can be controlled by the processor to either be a logic zero (off) or a logic one (on). They can be used individually or grouped together to create parallel ports where a group of bits can be input or output simultaneously.

- Serial outputs

  These are interfaces that send or receive data using one or two pins in a serial mode. They are less complex to connect but are more complicated to program. A parallel port looks very similar to a memory location and is easier to visualise and thus use. A serial port has to have data loaded into a register and then a start command issued. The data may also be augmented with additional information as required by the protocol.

- Analogue values

  While processors operate in the digital domain, the natural world does not and tends to orientate to analogue values. As a result, interfaces between the system and the external environment need to be converted from analogue to digital and vice versa.
- Displays

  Displays are becoming important and can vary from simple LEDs and seven segment displays to small alpha-numeric LCD panels.
- Time derived outputs

  Timers and counters are probably the most commonly used functions within an embedded system.

Software

The software components within an embedded system often encompasses the technology that adds value to the system and defines what it does and how well it does it. The software can consist of several different components:

- Initialisation and configuration
- Operating system or run-time environment
- The applications software itself
- Error handling
- Debug and maintenance support.

Algorithms

Algorithms are the key constituents of the software that makes an embedded system behave in the way that it does. They can range from mathematical processing through to models of the external environment which are used to interpret information from external sensors and thus generate control signals. With the digital technology in use today such as MP3 and DVD players, the algorithms that digitally encode the analogue data are defined by standards bodies.

While this standardisation could mean that the importance of selecting an algorithm is far less than it might be thought, the reality is far different. The focus on getting the right implementation is important since, for example, it may allow the same function to be executed on cheaper hardware. As most embedded systems are designed to be commercially successful, this selection process is very important. Defining and implementing the correct algorithm is a critical operation and is described through several examples in this book.

Examples

This section will go through some example embedded systems and briefly outline the type of functionality that each offers.

Microcontroller

Microcontrollers can be considered as self-contained systems with a processor, memory and peripherals so that in many cases all that is needed to use them within an embedded system is to add software. The processors are usually based on 8 bit stack-based architectures such as the MC6800 family. There are 4 bit versions available such as the National COP series which further reduce the processing power and reduce cost even further. These are limited in their functionality but their low cost has meant that they are used in many obscure applications. Microcontrollers are usually available in several forms:

- Devices for prototyping or low volume production runs

  These devices use non-volatile memory to allow the software to be downloaded and returned in the device. UV erasable EPROM used to be the favourite but EEPROM is also gaining favour. Some microcontrollers used a special package with a piggyback socket on top of the package to allow an external EPROM to be plugged in for prototyping. This memory technology replaces the ROM on the chip allowing software to be downloaded and debugged. The device can be reprogrammed as needed until the software reaches its final release version.

  The use of non-volatile memory also makes these devices suitable for low volume production runs or where the software may need customisation and thus preventing moving to a ROMed version.

  These devices are sometimes referred to as umbrella devices with a single device capable of providing prototyping support for a range of other controllers in the family.

- Devices for low to medium volume production runs

  In the mid-1980s, a derivative of the prototype device appeared on the market called the one time programmable or OTP. These devices use EPROM instead of the ROM but instead of using the ceramic package with a window to allow the device to be erased, it was packaged in a cheaper plastic pack and thus was only capable of programming a single time — hence the name. These devices are cheaper than the prototype versions but still have the programming disadvantage. However, their lower cost has made them a suitable alternative to producing a ROM device. For low to medium production quantities, they are cost effective and offer the ability to customise software as necessary.

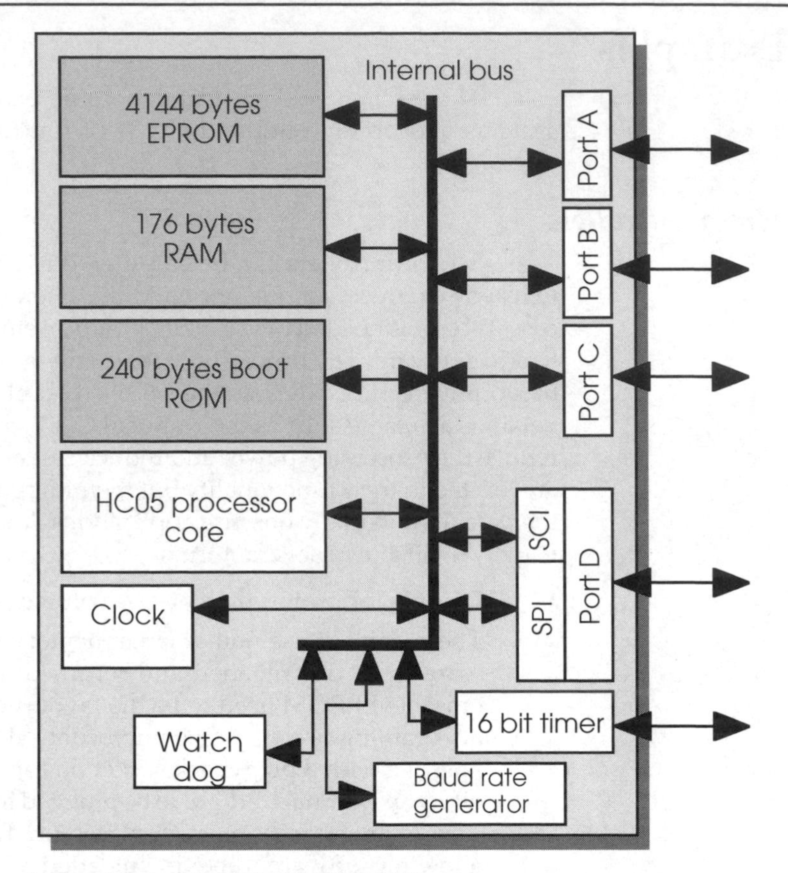

Example microcontroller (Motorola MC68HC705C4A)

- Devices for high volume production runs

  For high volumes, microcontrollers can be built already programmed with software in the ROM. To do this a customer supplies the software to the manufacturer who then creates the masks necessary to create the ROM in the device. This process is normally done on partly processed silicon wafers to reduce the turnaround time. The advantage for the customer is that the costs are much lower than using prototyping or OTP parts and there is no programming time or overhead involved. The downside is that there is usually a minimum order based on the number of chips that a wafer batch can produce and an upfront mask charge. The other major point is that once in ROM, the software cannot be changed and therefore customisation or bug fixing would have to wait until the next order or involve scrapping all the devices that have been made. It is possible to offer some customisation by including different software modules and selecting the required ones on the basis of a value read into the device from an external port but this does consume memory which can increase the costs. Some controllers can provide some RAM that can be used to patch the ROM without the need for a new mask set.

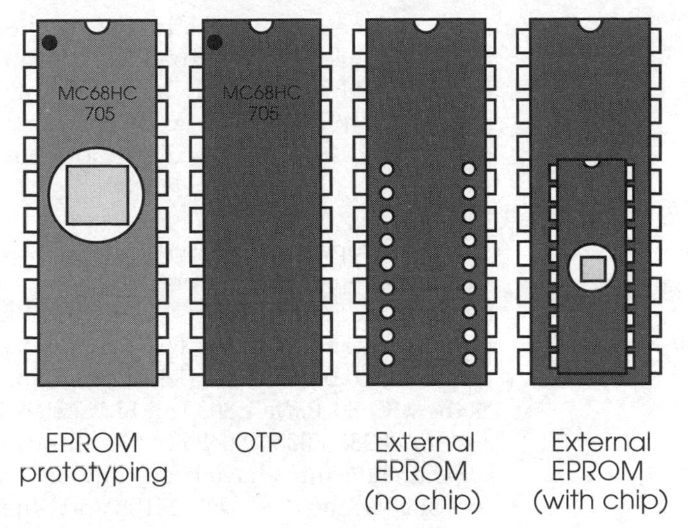

Prototype microcontrollers

Expanded microcontroller

The choice of memory sizes and partitioning is usually a major consideration. Some applications require more memory or peripherals than are available on a standard part. Most microcontroller families have parts that support external expansion and have an external memory and/or I/O bus which can allow the designer to put almost any configuration together. This is often done by using a parallel port as the interface instead of general-purpose I/O. Many of the higher performance microcontrollers are adopting this approach.

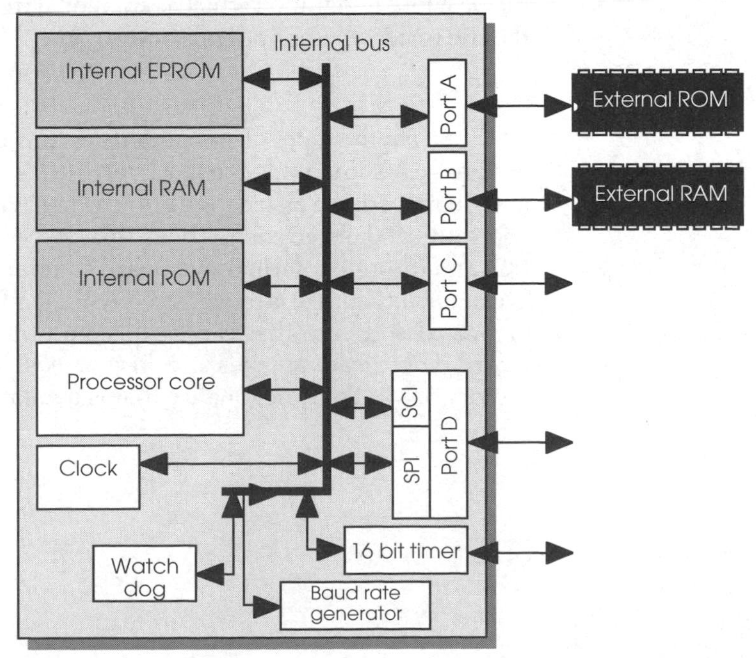

An expanded microcontroller

In the example shown on the previous page, the microcontroller has an expanded mode that allows the parallel ports A and B to be used as byte wide interfaces to external RAM and ROM. In this type of configuration, some microcontrollers disable access to the internal memory while others still allow it.

Microprocessor based

Microprocessor-based embedded systems originally took existing general-purpose processors such as the MC6800 and 8080 devices and constructed systems around them using external peripherals and memory. The use of processors in the PC market continued to provide a series of faster and faster processors such as the MC68020, MC68030 and MC68040 devices from Motorola and the 80286, 80386, 80486 and Pentium devices from Intel. These CISC architectures have been complemented with RISC processors such as the PowerPC, MIPS and others. These systems offer more performance than is usually available from a traditional microcontroller.

However, this is beginning to change. There has been the development of integrated microprocessors where the processor is combined with peripherals such as parallel and serial ports, DMA controllers and interface logic to create devices that are more suitable for embedded systems by reducing the hardware design task and costs. As a result, there has been almost a parallel development of these integrated processors along with the desktop processors. Typically, the integrated processor will use a processor generation that is one behind the current generation. The reason is dependent on silicon technology and cost. By using the previous generation which is smaller, it frees up silicon area on the die to add the peripherals and so on.

Board based

So far, the types of embedded systems that we have considered have assumed that the hardware needs to be designed, built and debugged. An alternative is to use hardware that has already been built and tested such as board-based systems as provided by PCs and through international board standards such as VMEbus. The main advantage is the reduced work load and the availability of ported software that can simply be utilised with very little effort. The disadvantages are higher cost and in some cases restrictions in the functionality that is available.

2 Embedded processors

The development of processors for embedded system design has essentially followed the development of microprocessors as a whole. The processor development has provided the processing heart for architecture which combined with the right software and hardware peripherals has become an embedded design. With the advent of better fabrication technology supporting higher transistor counts and lower power dissipation, the processor core has been integrated with peripherals and memory to provide standalone microcontrollers or integrated processors that only need the addition of external memory to provide a complete hardware system suitable for embedded design. The scope of this chapter is to explain the strengths and weaknesses of various architectures to provide a good understanding of the trade-offs involved in choosing and exploiting a processor family.

There are essentially four basic architecture types which are usually defined as 8 bit accumulator, 16/32 bit complex instruction set computers (CISC), reduced instruction set computer (RISC) architectures and digital signal processors (DSP). Their development or to be more accurate, their availability to embedded system designers is chronological and tends to follow the same type of pattern as shown in the graph.

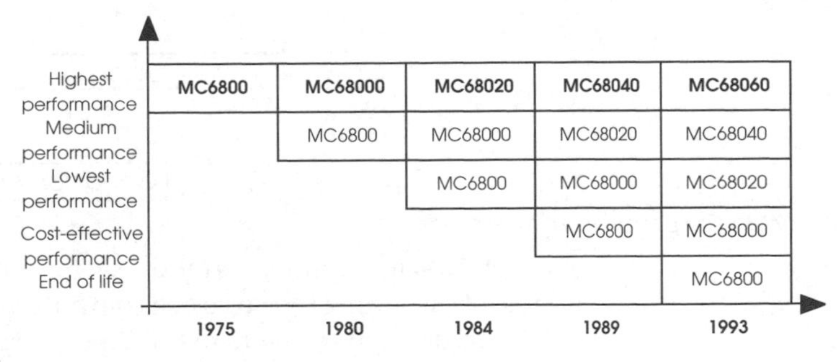

Processor life history

However, it should be remembered that in parallel with this life cycle, processor architectures are being moved into microcontroller and integrated processor devices so that the end of life really refers to the discontinuance of the architecture as a separate CPU plus external memory and peripherals product. The MC6800 processor is no longer used in discrete designs but there are over 200 MC6801/6805 and 68HC11 derivatives that essentially use the same basic architecture and instruction set.

8 bit accumulator processors

This category of processor first appeared in the mid-1970s as the first microprocessors. Devices such as the 8080 from Intel and the MC6800 from Motorola started the microprocessor revolution. They provided about 1 MIP of performance and were at their introduction the fastest processors available.

Register models

The programmer has a very simple register model for this type of processor. The model for the Motorola MC6800 8 bit processor is shown as an example but it is very representative of the many processors that appeared (and subsequently vanished). It has two 8 bit accumulators used for storing data and performing arithmetic operations. The program counter is 16 bits in size and two further 16 bit registers are provided for stack manipulations and address indexing.

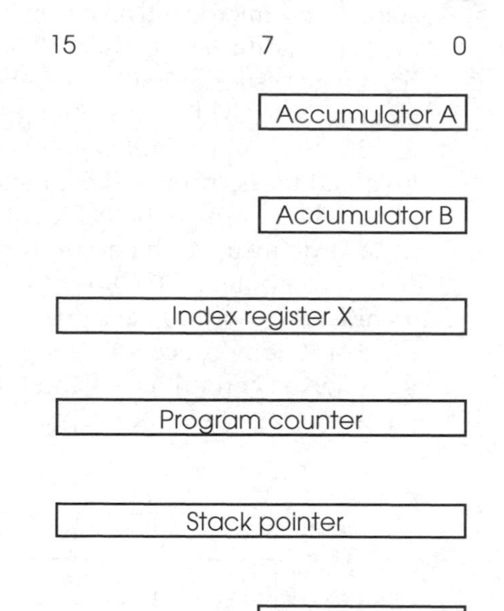

The MC6800 programmer's model

On first inspection, the model seems quite primitive and not capable of providing the basis of a computer system. There do not seem to be enough registers to hold data, let alone manipulate it! Comparing this with the register laden RISC architectures that feature today, this is a valid conclusion. What is often forgotten is that many of the instructions, such as logical operations, can operate on direct memory using the index register to act as pointer. This removes the need to bring data into the processor at the expense of extra memory cycles and the need for additional or wider registers. The main area within memory that is used for data storage is known as the stack. It is normally accessed using a special register that indexes into the area called the stack pointer.

This is used to provide local data storage for programs and to store information for the processor such as return addresses for subroutine jumps and interrupts.

The stack pointer provides additional storage for the programmer: it is used to store data like return addresses for subroutine calls and provides additional variable storage using a PUSH/POP mechanism. Data is PUSHed onto the stack to store it, and POPed off to retrieve it. Providing the programmer can track where the data resides in these stack frames, it offers a good replacement for the missing registers.

8 bit data restrictions

An 8 bit data value can provide an unsigned resolution of only 256 bits, which makes it unsuitable for applications where a higher resolution is needed. In these cases, such as financial, arithmetic, high precision servo control systems, the obvious solution is to increase the data size to 16 bits. This would give a resolution of 65536 — an obvious improvement. This may be acceptable for a control system but is still not good enough for a data processing program, where a 32 bit data value may have to be defined to provide sufficient integer range. While there is no difficulty with storing 8, 16, 32 or even 64 bits in external memory, even though this requires multiple bus accesses, it does prevent the direct manipulation of data through the instruction set.

However, due to the register model, data larger than 8 bits cannot use the standard arithmetic instructions applicable to 8 bit data stored in the accumulator. This means that even a simple 16 bit addition or multiplication has to be carried out as a series of instructions using the 8 bit model. This reduces the overall efficiency of the architecture.

The code example is a routine for performing a simple 16 bit multiplication. It takes two unsigned 16 bit numbers and produces a 16 bit product. If the product is larger than 16 bits, only the least significant 16 bits are retained. The first eight or so instructions simply create a temporary storage area on the stack for the multiplicand, multiplier, return address and loop counter. Compared to internal register storage, storing data in stack frames is not as efficient due the increased external memory access.

Accessing external data consumes machine cycles which could be used to process data. Without suitable registers and the 16 bit wide accumulator, all this information must be stored externally on the stack. The algorithm used simply performs a succession of arithmetic shifts on each half of the multiplicand stored in the A and B accumulators. Once this is complete, the 16 bit result is split between the two accumulators and the temporary storage cleared off the stack. The operation takes at least 29 instructions to perform with the actual execution time totally dependant on the values being multiplied together. For comparison, most 16/32 bit processors such as the MC68000 and 80x86 families can perform the same operation with a single instruction!

```
MULT16          LDX    #5                CLEAR WORKING REGISTERS
                CLR    A
LP1             STA    A       U-1,X
                DEX
                BNE            LP1
                LDX    #16               INITIAL SHIFT COUNTER
LP2     LDA     A      Y+1               GET Y(LSBIT)
                AND    A       #1
                TAB                      SAVE Y(LSBIT) IN ACCB
                EOR    A       FF        CHECK TO SEE IF YOU ADD
                BEQ    SHIFT             OR SUBTRACT
                TST    B
                BEQ            ADD
                LDA    A       U+1
                LDA    B       U
                SUB    A       XX+1
                SBC    B       XX
                STA    A       U+1
                STA    B       U
                BRA    SHIFT             NOW GOTO SHIFT ROUTINE
ADD             LDA    A       U+1
                LDA    B       U
                ADD    A       XX+1
                ADC    B       XX
                STA    A       U+1
                STA    B       U
SHIFT           CLR            FF        SHIFT ROUTINE
                ROR            Y
                ROR            Y+1
                ROL            FF
                ASR            U
                ROR            U+1
                ROR            U+2
                ROR            U+3
                DEX
                BNE            LP2
                RTS                               FINISH SUBROUTINE
                END
```

M6800 code for a 16 bit by 16 bit multiply

Addressing memory

When the first 8 bit microprocessors appeared during the middle to late 1970s, memory was expensive and only available in very small sizes: 256 bytes up to 1 kilobyte. Applications were small, partly due to their implementation in assembler rather than a high level language, and therefore the addressing range of 64 kilobytes offered by the 16 bit address seemed extraordinarily large. It was unlikely to be exceeded. As the use of these early microprocessors became more widespread, applications started to grow in size and the use of operating systems like CP/M and high level languages increased memory requirements until the address range started to limit applications. Various techniques like bank switching and program overlays were developed to help.

System integrity

Another disadvantage with this type of architecture is its unpredictability in handling error conditions. A bug in a software application could corrupt the whole system, causing a system to either crash, hang up or, even worse, perform some unforeseen operations. The reasons are quite simple: there is no partitioning between data and programs within the architecture. An application can update a data structure using a corrupt index pointer which overwrites a part of its program.

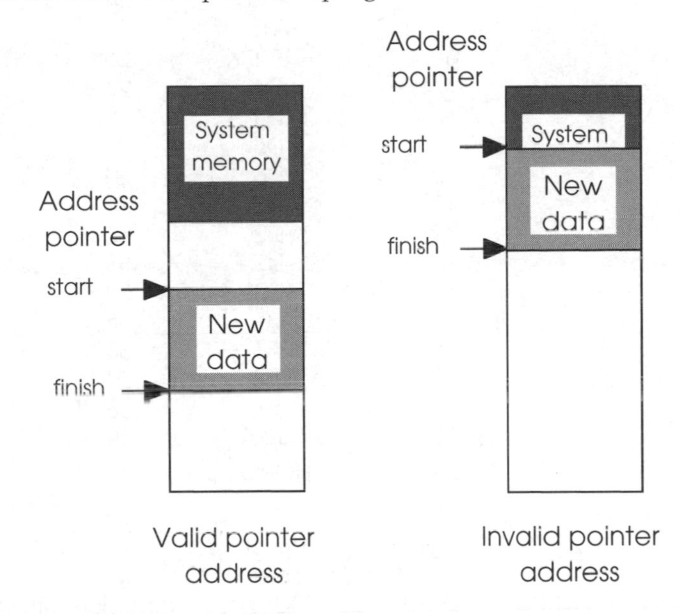

System corruption via an invalid pointer

Data are simply bytes of information which can be interpreted as instruction codes. The processor calls a subroutine within this area, starts to execute the data as code and suddenly the whole system starts performing erratically! On some machines, certain undocumented code sequences could put the processor in a test mode and start cycling through the address ranges etc. These attributes restricted their use to non-critical applications.

Example 8 bit architectures

Z80

The Z80 microprocessor is an 8 bit CPU with a 16 bit address bus capable of direct access to 64k of memory space. It was designed by Zilog and rapidly gained a lot of interest. The Z80 was based on the Intel 8080 but has an extended instruction set and many hardware improvements. It can run 8080 code if needed by its support of the 8080 instruction set. The instruction set is essential based around an 8 bit op code giving a maximum of 256 instructions. The 158 instructions that are specified — the others are reserved — include 78 instructions from the 8080. The instruction set supports the use of extension bytes to encode additional information. In terms of processing power, it offered about 1 MIP at 4 MHz clock speed with a minimum instruction time of 1 μs and a maximum instruction time of 5.75 μs.

Pin	Signal	Pin	Signal
1	A11	21	RD
2	A12	22	WR
3	A13	23	BUSAK
4	A14	24	WAIT
5	A15	25	BUSRQ
6	CLOCK	26	RESET
7	D4	27	M1
8	D3	28	RFSH
9	D5	29	GND
10	D6	30	A0
11	Vcc	31	A1
12	D2	32	A2
13	D7	33	A3
14	D0	34	A4
15	D1	35	A5
16	INT	36	A6
17	NMI	37	A7
18	HALT	38	A8
19	MREQ	39	A9
20	IORQ	40	A10

The Z80 signals

Signal	Description
A0 - A15	Address bus output tri-state
D0 - D7	Data bus bidirectional tri-state
CLOCK	CPU clock input
RFSH	Dynamic memory refresh output
HALT	CPU halt status output
RESET	Reset input
INT	Interrupt request input (active low)
NMI	Non-maskable interrupt input (active low)
BUSRQ	Bus request input (active low)
BUSAK	Bus acknowledge output (active low)
WAIT	Wait request input (active low)
RD, WR	Read and write signals
IORQ	I/O operation status output
MREQ	Memory refresh output
M1	Output pulse on instruction fetch cycle
Vcc	+5 volts
GND	0 volts

The Z80 pinout descriptions

The programming model includes an accumulator and six 8 bit registers that can be paired together to create three 16 bit registers. In addition to the general registers, a stack pointer, program counter, and two index (memory pointers) registers are provided. It uses external RAM for its stack. While not as powerful today as a PowerPC or Pentium, it was in its time a very powerful processor and was used in many of the early home computers such as the Amstrad CPC series. It was also used in many embedded designs partly because of its improved performance and also for its built-in refresh circuitry for DRAMs. This circuitry greatly simplified the external glue logic that was needed with DRAMs.

The Z80 was originally packaged in a 40 pin DIP package and ran at 2.5 and 4 MHz. Since then other packages and speeds have become available including low power CMOS versions — the original was made in NMOS and dissipated about 1 watt. Zilog now use the processor as a core within its range of Z800 microcontrollers with various configurations of on-chip RAM and EPROM.

Z80 programming model

The Z80 programming model essential consists of a set of 8 bit registers which can be paired together to create 16 bit versions for use as data storage or address pointers. There are two register sets within the model: the main and alternate. Only one set can be used at any one time and the switch and data transfer is performed by the EXX instruction. The registers in the alternate set are designated by a ′ suffix.

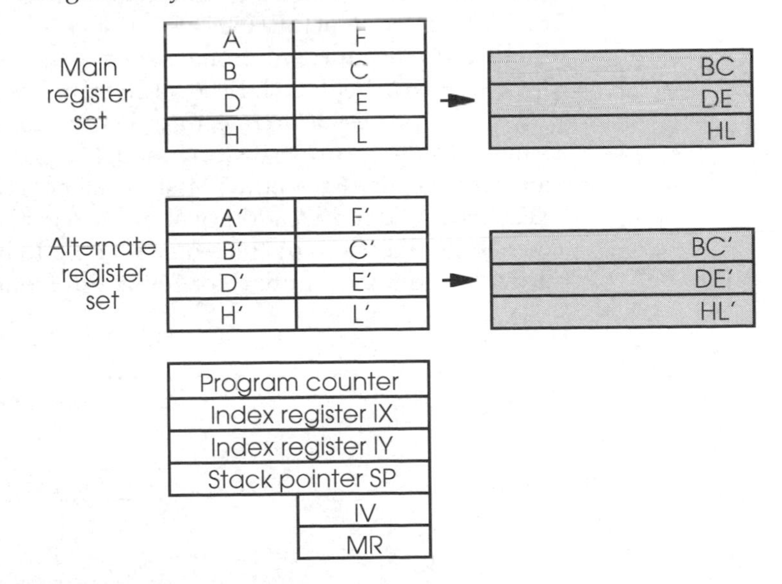

The Z80 programming model

The model has an 8 bit accumulator A and a flags register known as F. This contains the status information such as carry, zero, sign and overflow. This register is also known as PSW (program status word) in some documentation. Registers B, C, D, E, H and L are 8 bit general-purpose registers that can be paired to create 16 registers known as BC, DE and HL. The remaining registers are the program counter PC, two index registers IX and IY and a stack pointer SP. All these four registers are 16 bits in size and can access the whole 64 kbytes of external memory that the

Z80 can access. There are two additional registers IV and MR which are the interrupt vector and the memory refresh registers. The IV register is used in the interrupt handling mode 2 to point to the required software routine to process the interrupt. In mode 1, the interrupt vector is supplied via the external data bus. The memory refresh register is used to control the on-chip DRAM refresh circuitry.

Unlike the MC6800, the Z80 does not use memory mapped I/O and instead uses the idea of ports, just like the 8080. The lower 8 bits of the address bus are used along with the IORQ signal to access any external peripherals. The IORQ signal is used to differentiate the access from a normal memory cycle. These I/O accesses are similar from a hardware perspective to a memory cycle but only occur when an I/O port instruction (IN, OUT) is executed. In some respects, this is similar to the RISC idea of load and store instructions to bring information into the processor, process it and then write out the data. This system gives 255 ports and is usually sufficient for most embedded designs.

MC6800

The MC6800 was introduced in the mid-1970s by Motorola and is as an architecture the basis of several hundred derivative processors and microcontrollers such as the MC6809, MC6801, MC68HC05, MC68HC11, MC68HC08 families.

The processor architecture is 8 bits and uses a 64 kbyte memory map. Its programming model uses two 8 bit accumulators and a single 16 bit index register. Later derivatives such as the MC68HC11 added an additional index register and allowed the two accumulators to be treated as a single 16 bit accumulator to provide additional support for 16 bit arithmetic.

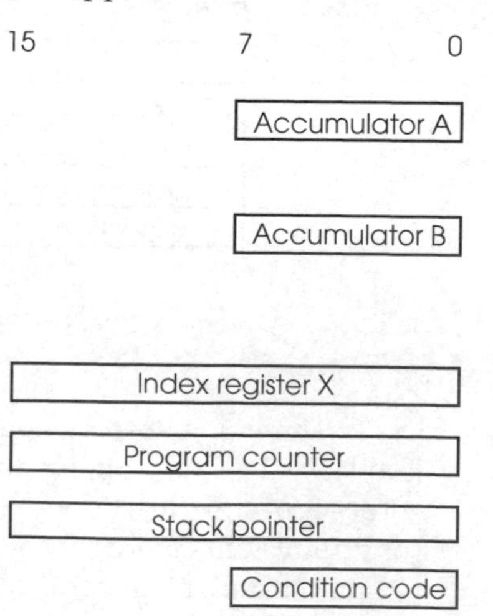

The MC6800 programmer's model

Its external bus was synchronous with separate address and data ports and the device operated at either 1, 1.5 or 2 MHz. The instruction set was essentially based around an 8 bit instruction with extensions for immediate values, address offsets and so on. It supported both non-maskable and software interrupts.

These type of processors have largely been replaced today by the microcontroller versions which have the same or advanced processor architectures and instruction sets but have the added advantage of glueless interfaces to memory and peripherals incorporated onto the chip itself. Discrete processors are still used but these tend to be the higher performance devices such as the MC68000 and 80x86 processors. But even with these faster and higher performance devices, the same trend of moving to integrated microcontroller type of devices is being followed as even higher performance processors such as RISC devices become available.

Microcontrollers

The previous section has described the 8 bit processors. While most of the original devices are no longer available, their architectures live on in the form of microcontrollers. These devices do not need much processing power — although this is now undergoing a radical change as will be explained later — but instead have become a complete integrated computer system by integrating the processor, memory and peripherals onto a single chip.

MC68HC05

The MC68HC05 is microcontroller family from Motorola that uses an 8 bit accumulator-based architecture as its processor core. This is very similar to that of the MC6800 except that it only has a single accumulator.

It uses memory mapping to access any on-chip peripherals and has a 13 bit program counter and effectively a 6 bit stack pointer. These reduced size registers — with many other 8 bit processors such as the Z80/8080 or MC6800, they are 16 bits is size — are used to reduce the complexity of the design. The microcontroller uses on-chip memory and therefore it does not make sense to define registers that can address memory that doesn't exist on the chip. The MC68HC05 family is designed for low cost applications where superfluous hardware is removed to reduce the die size, its power consumption and cost. As a result, the stack pointer points to the start of the on-chip RAM and can only use 64 bytes, and the program counter is reduced to 13 bits.

MC68HC11

The MC68HC11 is a powerful 8 bit data, 16 bit address microcontroller from Motorola that was on its introduction one of the most powerful and flexible microcontrollers available. It was originally designed in conjunction with General Motors for use within engine management systems. As a result, its initial versions had built-in EEPROM/OTPROM, RAM, digital I/O, timers,

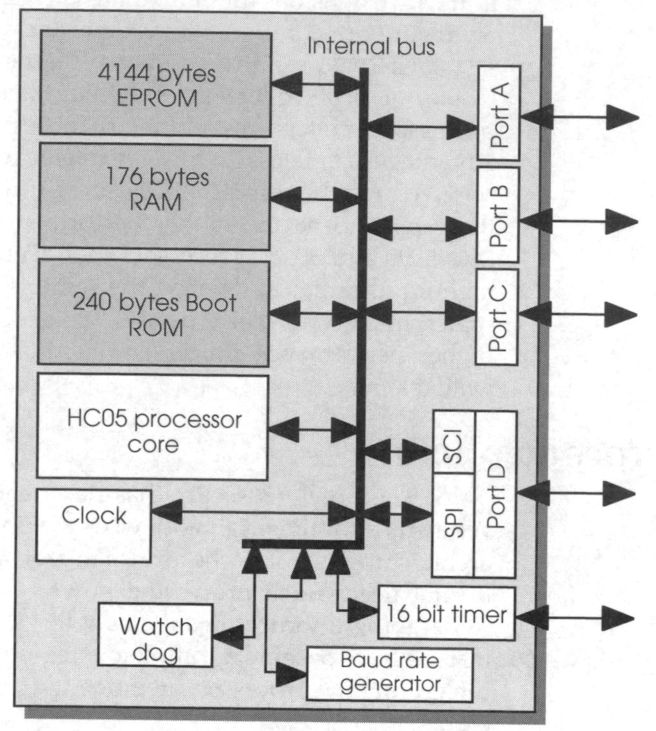

Example microcontroller (Motorola MC68HC705C4A)

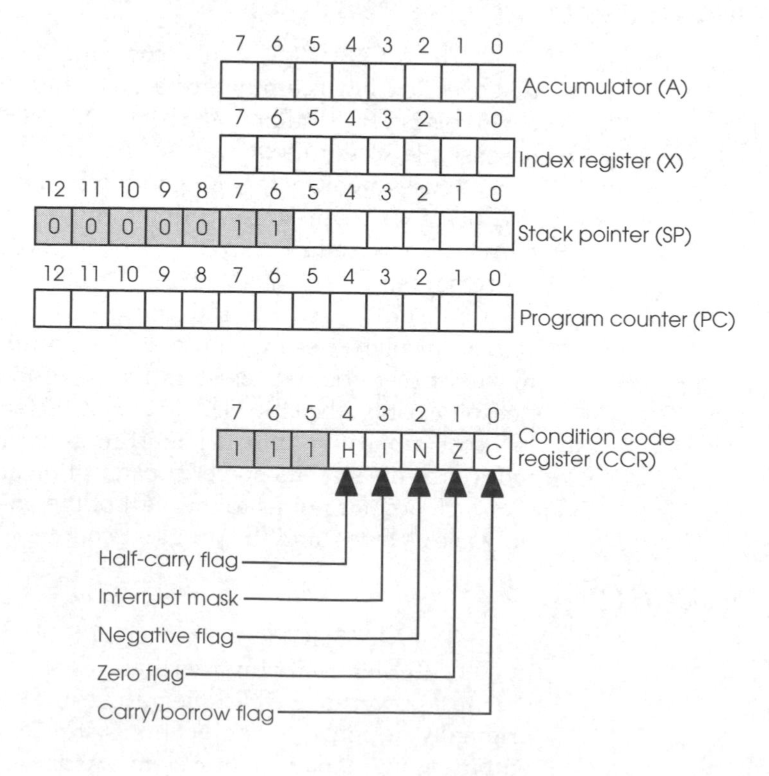

68HC05 programming model

8 channel 8 bit A/D converter, PWM generator, and synchronous and asynchronous communications channels (RS232 and SPI). Its current consumption is low with a typical value of less than 10 mA.

Architecture

The basic processor architecture is similar to that of the 6800 and has two 8 bit accumulators referred to as registers A and B. They can be concatenated to provide a 16 bit double accumulator called register D. In addition, there are two 16 bit index registers X and Y to provide indexing to anywhere within its 64 kbyte memory map.

Through its 16 bit accumulator, the instruction set can support several 16 bit commands such as add, subtract, shift and 16 by 16 division. Multiplies are limited to 8 bit values.

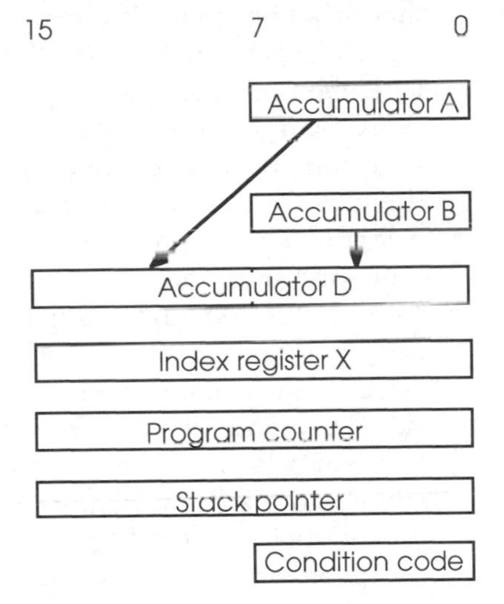

MC68HC11 programming model

Data processors

Processors like the 8080 and the MC6800 provided the computing power for many early desktop computers and their successors have continued to power the desktop PC. As a result, it should not be surprising that they have also provided the processing power for more powerful systems where a microcontroller cannot provide either the processing power or the correct number or type of peripherals. They have also provided the processor cores for more integrated chips which form the next category of embedded systems.

Complex instructions, microcode and nanocode

With the initial development of the microprocessor concentrated on the 8 bit model, it was becoming clear that larger data sizes, address space and more complex instructions were needed. The larger data size was needed to help support higher precision arithmetic. The increased address space was needed to support bigger blocks of memory for larger programs. The complex instruction was needed to help reduce the amount of memory required to store the program by increasing the instruction efficiency: the more complex the instruction, the less needed for a particular function and therefore the less memory that the system needed. It should be remembered that it was not until recently that memory has become so cheap.

The instruction format consists of an op code followed by a source effective address and a destination effective address. To provide sufficient coding bits, the op code is 16 bits in size with further 16 bit operand extensions for offsets and absolute addresses. Internally, the instruction does not operate directly on the internal resources, but is decoded to a sequence of microcode instructions, which in turn calls a sequence of nanocode commands which controls the sequencers and arithmetic logic units (ALU). This is analogous to the many macro subroutines used by assembler programmers to provide higher level 'pseudo' instructions. On the MC68000, microcoding and nanocoding allow instructions to share common lower level routines, thus reducing the hardware needed and allowing full testing and emulation prior to fabrication. Neither the microcode nor the nanocode sequences are available to the programmer.

These sequences, together with the sophisticated address calculations necessary for some modes, often take more clock cycles than are consumed in fetching instructions and their associated operands from external memory. This multi-level decoding automatically lends itself to a pipelined approach which also allows a prefetch mechanism to be employed.

Pipelining works by splitting the instruction fetch, decode and execution into independent stages: as an instruction goes through each stage, the next instruction follows it without waiting for it to completely finish. If the instruction fetch is included within the pipeline, the next instruction can be read from memory, while the preceding instruction is still being executed as shown.

The only disadvantage with pipelining concerns pipeline stalls. These are caused when any stage within the pipeline cannot complete its allotted task at the same time as its peers. This can occur when wait states are inserted into external memory accesses, instructions use iterative techniques or there is a change in program flow.

With iterative delays, commonly used in multiply and divide instructions and complex address calculations, the only possible solutions are to provide additional hardware support, add more stages to the pipeline, or simply suffer the delays on the grounds that the performance is still better than anything else! Additional hardware support may or may not be within a designer's real estate budget (real estate refers to the silicon die area, and directly to the number of transistors available). Adding stages also consumes real estate and increases pipeline stall delays when branching. This concern becomes less of an issue with the current very small gate sizes that are available but the problem of pipeline stalls and delays is still a major issue. It is true to say that pipeline lengths have increased to gain higher speeds by reducing the amount of work done in each stage. However, this has been coupled with an expansion in the hardware needed to overcome some of the disadvantages. These trade-offs are as relevant today as they were five or ten years ago.

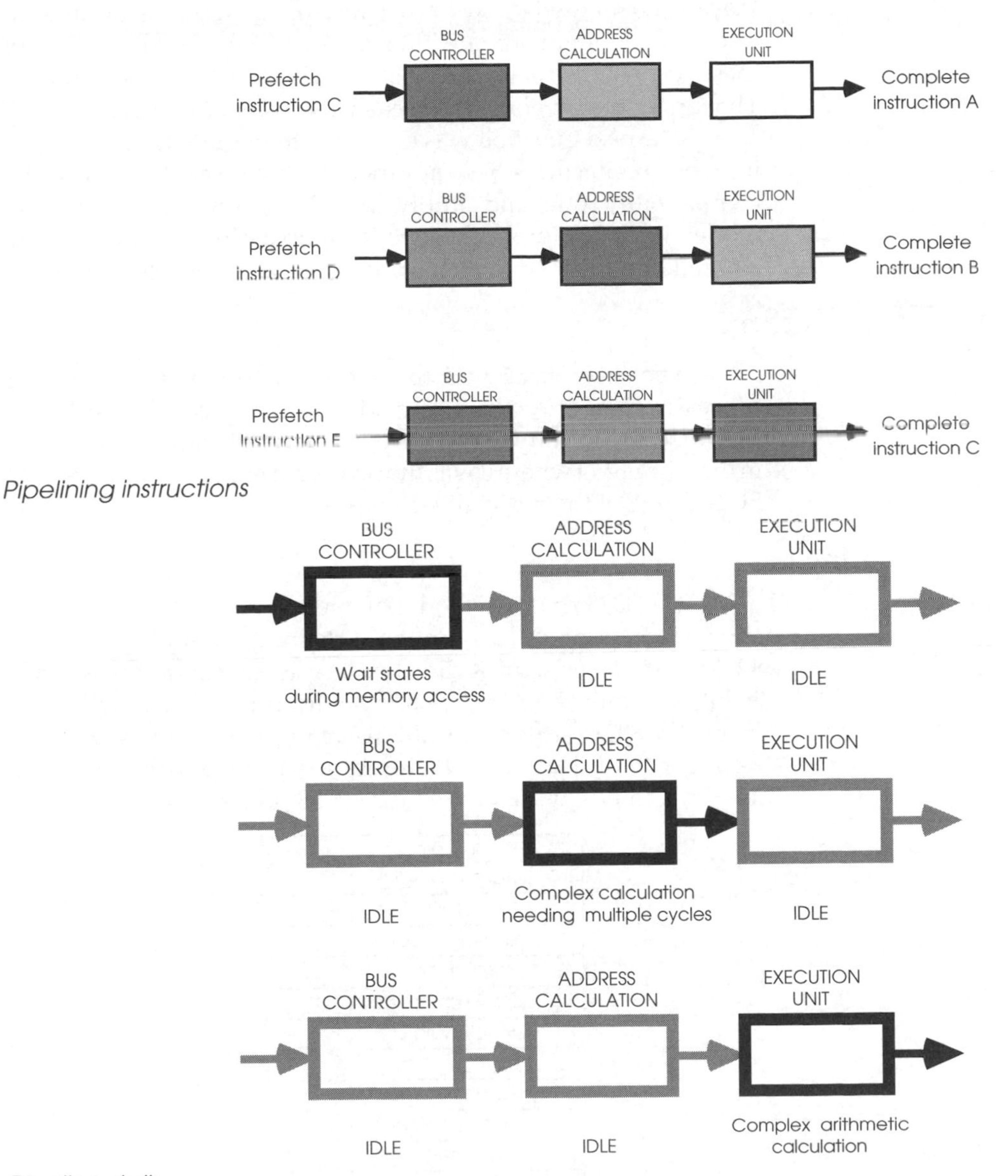

Pipelining instructions

Pipeline stalls

The main culprits are program branching and similar operations. The problem is caused by the decision whether to take the branch or not being reached late in the pipeline, i.e. after the next instruction has been prefetched. If the branch is not taken, this instruction is valid and execution can carry on. If the branch is taken, the instruction is not valid and the whole pipeline must be flushed and reloaded. This causes additional memory cycles before the processor can continue. The delay is dependent on the number of stages, hence the potential difficulty in increasing the number of stages to reduce iterative delays. This interrelation of engineering trade-offs is a common theme within microprocessor architectures. Similar problems can occur for any change of flow: they are not limited to just branch instructions and can occur with interrupts, jumps, software interrupts etc. With the large usage of these types of instructions, it is essential to minimise these delays. The longer the pipeline, the greater the potential delay.

The next question was over how to migrate from the existing 8 bit architectures. Two approaches were used: Intel chose the compatibility route and simply extended the 8080 programming model, while Motorola chose to develop a different architecture altogether which would carry it into the 32 bit processor world.

INTEL 80286

The Intel 80286 was the successor to the 8086 and 8088 processors and offered a larger addressing space while still preserving compatibility with its predecessors. Its initial success was in the PC market where it was the processor engine behind the IBM PC AT and all the derivative clones.

Architecture

The 80286 has two modes of operation known as real mode and protected mode: real mode describes its emulation of the 8086/8088 processor including limiting its external address bus to 20 bits to mimic the 8086/8088 1 Mbyte address space. In its real mode, the 80286 adds some additional registers to allow access to its larger 16 Mbyte external address space, while still preserving its compatibility with the 8086 and 8088 processors.

Accumulator	AX
Base register	BX
Counter register	CX
Data register	DX
Source index	SI
Destination index	DI
Stack pointer	SP
Base pointer	BP
Code segment	CS
Data segment	DS
Stack segment	SS
Extra segment	ES
Instruction pointer	IP
Status flags	FL
15	0

Intel 80286 processor register set

The register set comprises four general-purpose 16 bit registers (AX, BX, CX and DX) and four segment address registers (CS, DS, SS and ES) and a 16 bit program counter. The general-purpose registers — AX, BX, CX, and DX — can be accessed as two 8 bit registers by changing the X suffix to either H or L. In this way, each half of register AX can be accessed as AH or AL and so on for the other three registers.

These registers form a set that is the same as that of an 8086. However, when the processor is switched into its protected mode, the register set is expanded and includes two index registers (DI and SI) and a base pointer register. These additions allow the 80286 to support a simple virtual memory scheme.

Within the IBM PC environment, the 8086 and 8088 processors can access beyond the 1 Mbyte address space by using paging and special hardware to simulate the missing address lines. This additional memory is known as expanded memory. This non-linear memory mapping can pose problems when used in an embedded space where a large linear memory structure is needed, but these restrictions can be overcome as will be shown in later design examples.

Interrupt facilities

The 80286 can handle 256 different exceptions and the vectors for these are held in a vector table. The vector table's construction is different depending on the processor's operating mode. In the real mode, each vector consists of two 16 bit words that contain the interrupt pointer and code segment address so that the associated interrupt routine can be located and executed. In the protected mode of operation each entry is 8 bytes long.

Vector	Function
0	Divide error
1	Debug exception
2	Non-masked interrupt NMI
3	One byte interrupt INT
4	Interrupt on overflow INTO
S	Array bounds check BOUND
6	Invalid opcode
7	Device not available
8	Double fault
9	Coprocessor segment overrun
10	Invalid TSS
11	Segment not present
12	Stack fault
13	General protection fault
14	Page fault
15	Reserved
16	Coprocessor error
17-32	Reserved
33-255	INT *n* trap instructions

The interrupt vectors and their allocation

Instruction set

The instruction set for the 80286 follows the same pattern as that for the Intel 8086 and programs written for the 8086 are compatible with the 80286 processor.

80287 floating point support

The 80286 can also be used with the 80287 floating point coprocessor to provide acceleration for floating point calculations. If the device is not present, it is possible to emulate the floating point operations in software, but at a far lower performance.

Feature comparison

Feature	8086	8088	80286
Address bus	20 bit	20 bit	24 bit
Data bus	16 bit	8 bit	16 bit
FPU present	No	No	No
Memory management	No	No	Yes
Cache on-chip	No	No	No
Branch acceleration	No	No	No
TLB support	No	No	No
Superscalar	No	No	No
Frequency (MHz)	5,8,10	5,8,10	6,8,10,12
Average cycles/Inst.	12	12	4.9
Frequency of FPU	=CPU	=CPU	2/3 CPU
Frequency	3X	3X	2X
Address range	1 Mbytes	1 Mbytes	16 Mbytes
Frequency scalability	No	No	No
Voltage	5 v	5 v	5 v

Intel 8086, 8088 and 80286 processors

INTEL 80386DX

The 80386 processor was introduced in 1987 as the first 32 bit member of the family. It has 32 bit registers and both 32 bit data and address buses. It is software compatible with the previous generations through the preservation of the older register set within the 80386's newer extended register model and through a special 8086 emulation mode where the 80386 behaves like a very fast 8086. The processor has an on-chip paging memory management unit which can be used to support multitasking and demand paging virtual memory schemes if required.

Architecture

The 80386 has eight general-purpose 32 bit registers EAX, EBX, ECX, EDX, ESI, EDI, EBP and ESP. These general-purpose registers are used for storing either data or addresses. To ensure compatibility with the earlier 8086 processor, the lower half of each register can be accessed as a 16-bit register (AX, BX, CX, DX, SI, DI, BP and SP). The AX, BX, CX and DX registers can be also accessed as 8 bit registers by changing the X suffix for either H or L thus creating the 8088 registers AH, AL, BH, BL and so on.

To generate a 32 bit physical address, six segment registers (CS, SS, DS, ES, FS, GS) are used with addresses from the general registers or instruction pointer. The code segment (CS) is used with the instruction pointer to create the addresses used for instruction fetches and any stack access uses the SS register. The remaining segment registers are used for data addresses.

Each segment register has an associated descriptor register which is used to program and control the on-chip memory management unit. These descriptor registers — controlled by the operating system and not normally accessible to the application programmer — hold the base address, segment limit and various attribute bits that describe the segment's properties.

The 80386 can run in three different modes: the real mode, where the size of each segment is limited to 64 kbytes, just like the 8088 and 8086; a protected mode, where the largest segment size is increased to 4 Gbytes; and a special version of the protected mode that creates multiple virtual 8086 processor environments.

The 32 bit flag register contains the normal carry zero, auxiliary carry, parity, sign and overflow flags. The resume flag is used with the trap 1 flag during debug operations to stop and start the processor. The remaining flags are used for system control to select virtual mode, nested task operation and input/output privilege level.

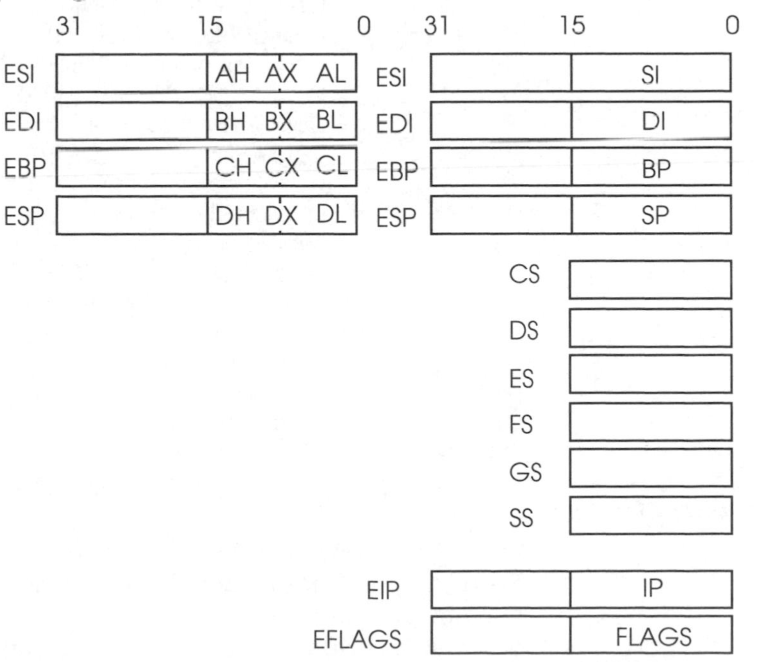

Intel 80386 register set

For external input and output, a separate peripheral address facility is available similar to that found on the 8086. As an alternative, memory mapping is also supported (like the M68000 family) where the peripheral is located within the main memory map.

Interrupt facilities

The 80386 has two external interrupt signals which can be used to allow external devices to interrupt the processor. The INTR input generates a maskable interrupt while the NMI generates a non-maskable interrupt and naturally has the higher priority of the two.

During an interrupt cycle, the processor carries out two interrupt acknowledge bus cycles and reads an 8 bit vector number on D0–D7 during the second cycle. This vector number is then used to locate, within the vector table, the address of the corresponding interrupt service routine. The NMI interrupt is automatically assigned the vector number of 2.

Software interrupts can be generated by executing the INT *n* instruction where *n* is the vector number for the interrupt. The vector table consists of 4 byte entries for each vector and starts at memory location 0 when the processor is running in the real mode. In the protected mode, each vector is 8 bytes long. The vector table is very similar to that of the 80286.

Vector	Function
0	Divide error
1	Debug exception
2	Non-masked interrupt NMI
3	One byte interrupt INT
4	Interrupt on overflow INTO
S	Array bounds check BOUND
6	Invalid opcode
7	Device not available
8	Double fault
9	Coprocessor segment overrun
10	Invalid TSS
11	Segment not present
12	Stack fault
13	General protection fault
14	Page fault
15	Reserved
16	Coprocessor error
17-32	Reserved
33-255	INT n trap instructions

Instruction set

The 80386 instruction set is essentially a superset of the 8086 instruction set. The format follows the dyadic approach and uses two operands as sources with one of them also duplicating as a destination. Arithmetic and other similar operations thus follow the A+B=B type of format (like the M68000). When the processor is operating in the real mode — like an 8086 processor — its instruction set, data types and register model is essentially restricted to a that of the 8086. In its protected mode, the full 80386 instruction set, data types and register model becomes available. Supported data types include bits, bit fields, bytes, words (16 bits), long words (32 bits) and quad words (64 bits). Data can be signed or unsigned binary, packed or unpacked BCD, character bytes and strings. In addition, there is a further group of instructions that can be used when the CPU is running in protected mode only. They provide access to the memory management and control registers. Typically, they are not available to the user programmer and are left to the operating system to use.

LSL	Load segment limit
LTR	Load task register
SGDT	Store global descriptor table
SIDT	Store interrupt descriptor table
STR	Store task register
SLDT	Store local descriptor table
SMSW	Store machine status word
VERR	Verify segment for reading
VERW	Verify segment for writing

Addressing modes provided are:

Register direct	(Register contains operand)
Immediate	(Instruction contains data)
Displacement	(8/16 bits)
Base address	(Uses BX or BP register)
Index	(Uses DI or SI register)

80387 floating point coprocessor

The 80386 can also be used with the 80387 floating point coprocessor to provide acceleration for floating point calculations. If the device is not present, it is possible to emulate the floating point operations in software, but at a far lower performance.

Feature comparison

There is a derivative of the 80386DX called the 80386SX which provides a lower cost device while retaining the same architecture. To reduce the cost, it uses an external 16 bit data bus and a 24 bit memory bus. The SX device is not pin compatible with the DX device. These slight differences can cause quite different levels of performance which can mean the difference between performing a function of not.

In addition, Intel have produced an 80386SL device for portable PCs which incorporates a power control module that provides support for efficient power conservation.

Although Intel designed the 80386 series, the processor has been successfully cloned by other manufacturers (both technically and legally) such as AMD, SGS Thomson, and Cyrix. Their versions are available at far higher clock speeds than the Intel originals and many PCs are now using them. These are now available with all the peripherals needed to create a PC AT clone integrated on the chip and these are extensively used to create embedded systems based on the PC architecture.

Feature	i386SX	i386DX	i386SL
Address bus	24 bit	32 bit	24 bit
Data bus	16 bit	32 bit	16 bit
FPU present	No	No	No
Memory management	Yes	Yes	Yes
Cache on-chip	No	No	Control
Branch acceleration	No	No	No
TLB support	No	No	No
Superscalar	No	No	No
Frequency (MHz)	16,20,25,33	16,20,25,33	16,20,25
Average cycles/Inst.	4.9	4.9	<4.9
Frequency of FPU	=CPU	=CPU	=CPU
Address range	16 Mbytes	4 Gbytes	16 Mbytes
Frequency scalability	No	No	No
Transistors	275000	275000	855000
Voltage	5 v	5 v	3 v or 5 v
System management	No	No	Yes

Intel i386 feature comparison

INTEL 80486

The Intel 80486 processor is essentially an enhanced 80386. It has a similar instruction set and register model but to dismiss it as simply a go-faster 80386 would be ignoring the other features that it uses to improve performance.

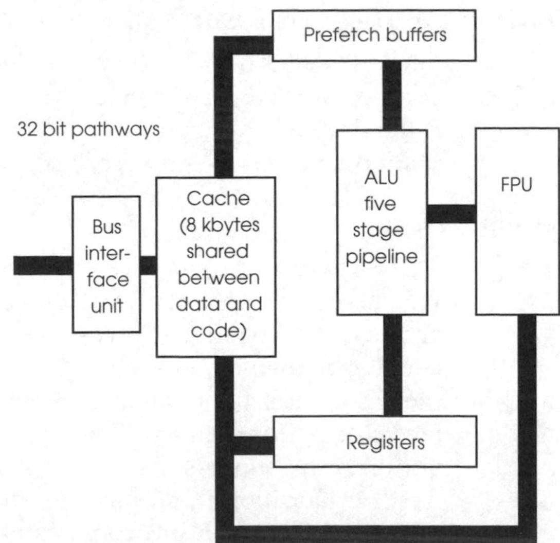

The 80486 internal architecture

Like the MC68040, it is a CISC processor that can execute instructions in a single cycle. This is done by pipelining the instruction flow so that address calculations and so on are performed as the instruction proceeds down the line. Although the pipeline may take several cycles, an instruction can potentially be started and completed on every clock edge, thus achieving the single cycle performance.

To provide instruction and data to the pipeline, the 80486 has an internal unified cache to contain both data and instructions.

This removes the dependency of the processor on faster external memory to maintain sufficient data flow to allow the processor to continue executing instead of stalling. The 80486 also integrates a 80387 compatible fast floating point unit and thus does not need an external coprocessor.

Instruction set

The instruction set is essentially the same as the 80386 but there are some additional instructions available when running in protected mode to control the memory management and floating point units.

Intel 486SX and overdrive processors

The 80486 is available in several different versions which offer different facilities. The 486SX is like the 80386SX, a stripped down version of the full DX processor with the floating point unit removed but with the normal 32 bit external data and address buses. The DX2 versions are the clock doubled versions which run the internal processor at twice the external bus speed. This allows a 50 MHz DX2 processor to work in a 25 MHz board design, and opens the way to retrospective upgrades — known as the overdrive philosophy — where a user simply replaces a 25 MHz 486SX with a DX to get floating point support or a DX2 to get the FPU and theoretically twice the performance. Such upgrades need to be carefully considered: removing devices that do not have a zero insertion force socket can be tricky at best and wreck the board at worst. Similarly, the additional heat and power dissipation has also to be taken into consideration. While some early PC designs had difficulties in these areas, the overdrive option has now become a standard PC option.

The DX2 typically gives about 1.6 to 1.8 performance improvement depending on the operations that are being carried out. Internal processing gains the most from the DX2 approach while memory-intensive operations are frequently limited by the external board design. Intel have also released a DX4 version which offers internal CPU speeds of 75 and 100 MHz.

For embedded system designers, these overdrive chips can be a gift from heaven in that they allow the hardware performance of a system to be upgraded by simply swapping the processor. As the speed clocking is an internal operation, the external hardware timing can remain as is without affecting the design. It should be stated that getting the performance budget right in the first place is always preferable, but having the option of getting more performance without changing the complete hardware is always useful as a backup plan. It should be remembered that this solution will only address CPU performance issues and not problems caused by external memory access delays or I/O speed problems. This approach can be used with many other processors where a pin compatible but faster CPU is available.

Feature	i486DX2-40	i486DX2-50	i486DX2-66
Address bus	32 bit	32 bit	32 bit
Data bus	32 bit	32 bit	32 bit
FPU present	Yes	Yes	Yes
Memory management	Yes	Yes	Yes
Cache on-chip	8K unified	8K unified	8K unified
Branch acceleration	No	No	No
TLB support	No	No	No
Superscalar	No	No	No
Frequency (MHz)	40	50	66
Average cycles/Inst.	1.03	1.03	1.03
Frequency of FPU	=CPU	=CPU	=CPU
Upgradable	Yes	Yes	Yes
Address range	4 Gbytes	4 Gbytes	4 Gbytes
Frequency scalability	No	No	No
Transistors	1.2 million	1.2 million	1.2 million
Voltage	5 and 3	5 and 3	5

Feature	i486SX	i486DX	i486DX-50
Address bus	32 bit	32 bit	32 bit
Data bus	32 bit	32 bit	32 bit
FPU present	No	Yes	Yes
Memory management	Yes	Yes	Yes
Cache on-chip	8K unified	8K unified	8K unified
Branch acceleration	No	No	No
TLB support	No	No	No
Superscalar	No	No	No
Frequency (MHz)	16,20,25,33	25,33	50
Average cycles/Inst.	1.03	1.03	1.03
Frequency of FPU	N/A	=CPU	=CPU
Upgradable	Yes	Yes	No
Address range	4 Gbytes	4 Gbytes	4 Gbytes
Frequency scalability	No	No	No
Transistors	1.2 million	1.2 million	1.2 million
Voltage	5 and 3	5 and 3	5
System management	No	No	No

Intel i486 feature comparison

Intel Pentium

The Pentium is essentially an enhanced 80486 from a programming model. It uses virtually the same programming model and instruction set — although there are some new additions.

The most noticeable enhancement is its ability to operate as a superscalar processor and execute two instructions per clock. To do this it has incorporated many new features that were not present on the 80486.

As the internal architecture diagram shows, the device has two five-stage pipelines that allow the joint execution of two integer instructions provided that they are simple enough not to use microcode or have data dependencies. This restriction is not that great a problem as many compilers have now started to concentrate on the simpler instructions within CISN instruction sets to improve their performance.

To maintain the throughput, the unified cache that appeared on the 80486 has been dropped in favour of two separate 8 kbyte caches: one for data and one for code. These caches are fed by an external 64 bit wide burst mode type data bus. The caches also now support write-back MESI policies instead of the less efficient write-through design.

Branches are accelerated using a branch target cache and work in conjunction with the code cache and prefetch buffers. The instruction set now supports an 8 byte compare and exchange instruction and a special processor identification instruction. The cache coherency support also has some new instructions to allow programmer's control of the MESI coherency policy.

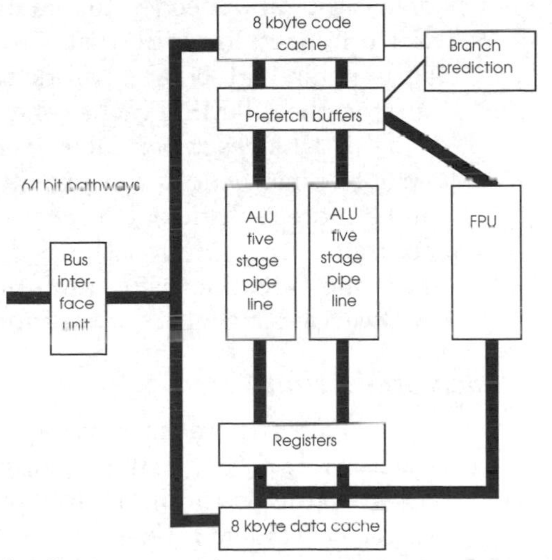

The Intel Pentium internal architecture

The Pentium has an additional control register and system management mode register that first appeared on the 80386SL which provides intelligent power control.

Feature	Pentium
Address bus	32 bit
Data bus	64 bit
FPU present	Yes
Memory management	Yes
Cache on-chip	Two 8 kbyte caches (data and code)
Branch acceleration	Yes — branch target cache
Cache coherency	MESI protocol
TLB support	Yes
Superscalar	Yes (2)
Frequency (MHz)	60, 66, 75, 100
Average cycle/Inst.	0.5
Frequency of FPU	=CPU
Address range	4 Gbytes
Frequency scalability	No
Transistors	3.21 million
Voltage	5 and 3
System management	Yes

Intel Pentium feature comparison

Pentium Pro

The Pentium Pro processor is the name for Intel's successor to the Pentium and was previously referred to as the P6 processor. The processor was Intel's answer to the RISC threat posed by the DEC Alpha and Motorola PowerPC chips. It embraces many techniques that are used to achieve superscalar performance that have appeared previously on RISC processors such as the MPC604 PowerPC chip. It was unique in that the device actually consisted of two separate die within a single ceramic pin grid array: one die is the processor with its level one cache on-chip and the second die is the level 2 cache which is needed to maintain the instruction and data throughput needed to maintain the level of performance. It was originally introduced at 133 MHz and gained some acceptance within high-end PC/workstation applications. Both caches support the full MESI cache coherency protocol.

It achieves superscalar operation by being able to execute up to five instructions concurrently although this is a peak figure and a more realistic one to two instructions per clock is a more accurate average figure. This is done through a technique called dynamic execution using multiple branch prediction, dataflow analysis and speculative execution.

Multiple branch prediction

This is where the processor predicts where a branch instruction is likely to change the program flow and continues execution based on this assumption, until proven or more accurately, until correctly evaluated. This removes any delay providing the branch prediction was correct and speeds up branch execution.

Data flow analysis

This is an out of order execution and involves analysing the code and determining if there are any data dependencies. If there are, this will stall the superscalar execution until these are resolved. For example, the processor cannot execute two consecutive integer instructions if the second instruction depends on the result of its predecessor. By internally reordering the instructions, such delays can be removed and thus faster execution can be restored. This can cause problems with embedded systems in that the reordering means that the CPU will not necessarily execute the code in programme order but in a logical functional sequence. To do this, the processor needs to understand any restrictions that may affect its reordering. This information is frequently provided by splitting the memory map into areas, with each area having its own attributes. These functions are described in Chapter 3.

Speculative execution

Speculative execution is where instructions are executed speculatively, usually following predicted branch paths in the code until the true and correct path can be determined. If the processor has speculated correctly, then performance has been gained. If not, the speculative results are discarded and the processor continues down the correct path. If more correct speculation is achieved than incorrect, the processor performance increases.

It is fair to say that the processor has not been as successful as the Pentium. This is because faster Pentium designs, especially those from Cyrix, outperformed it and were considerably cheaper. The final problem it had was that it was not optimised for 16 bit software such as MS-DOS and Windows 3.x applications and required 32 bit software to really achieve its performance. The delay in the market in getting 32 bit software — Windows 95 was almost 18 months late and this stalled the market considerably — did not help its cause, and the part is now overshadowed by the faster Pentium parts and the Pentium II.

The MMX instructions

The MMX instructions or multimedia extensions as they have also been referred to were introduced to the Pentium processor to provide better support for multimedia software running on a PC using the Intel architecture. Despite some over-exaggerated claims of 400% improvement, the additional instructions do provide additional support for programmers to improve their code. About 50 instructions have been added that use the SIMD (single instruction, multiple data) concept to process several pieces of data with a single instruction, instead of the normal single piece of data.

To do this, the eight floating point registers can be used to support MMX instructions or floating point. These registers are 80 bits wide and in their MMX mode, only 64 bits are used. This is enough to store a new data type known as the packed operand. This supports eight packed bytes, four packed 16 bit words, two packed 32 bit double words, or a single 64 bit quad word. This is extremely useful for data manipulation where pixels can be packed into the floating point register and manipulated simultaneously.

The beauty of this technique is that the basic architecture and register set does not change. The floating point registers will be saved on a context switch anyway, irrespective of whether they are storing MMX packed data or traditional floating point values. This is where one of the problems lies. A program can really only use floating point or MMX instructions. It cannot mix them without clearing the registers or saving the contents. This is because the floating point and MMX instructions share the same registers.

This has led to problems with some software and the discovery of some bugs in the silicon (run a multimedia application and then watch Excel get all the financial calculations wrong). There are fixes available and this problem has been resolved. However, the success of MMX does seem to be dependent on factors other than the technology and the MMX suffix has become a requirement. If a PC doesn't have MMX, it is no good for multimedia. For embedded system, this statement is not valid and its use is not as obligatory as it might seem.

What is interesting is that MMX processors also have other improvements to help the general processor performance and so it can be a little difficult to see how much MMX can actually help. The second point is that many RISC processors, especially the PowerPC as used in the Apple Macintosh, can beat an MMX processor running the same multimedia application. The reason is simple. Many of the instructions and data manipulation that MMX brings, these processors have had as standard. They may not have packed data, but they don't have to remember if they used a floating point instruction recently and should now save the registers before using an MMX instruction. What seems to be an elegant solution does have some drawbacks.

The Pentium II

The Pentium II was the next generation Intel processor and uses a module based technology and a PCB connector to provide the connection to a Intel designed motherboard. It no longer uses a chip package and is only available as a module. Essentially, a redesigned and improved Pentium Pro core with larger caches, it was the fastest Intel processor available until the Pentium III appeared. It is clear that Intel is focusing on the PC market with its 80x86 architecture and this does raise the question the suitability of these processors to be used in embedded systems. With the subsequent Pentium III and Pentium IV processors requiring specialised motherboard support, their suitability for embedded designs is limited to completely built boards and hardwares. The other problem with these types of architectures is that the integrated caches and other techniques they use to get the processing speed mean that the processor becomes more statistical in nature and it becomes difficult to predict how long it will take to do a task. This is another topic we will come back to in later chapters.

Motorola MC68000

The MC68000 was a complete design from scratch with the emphasis on providing an architecture that looked forward without the restrictions of remaining compatible with past designs. Unlike the Intel approach of taking an 8 bit architecture and developing it further and further, Motorola's approach was to design a 16/32 bit processor whose architecture was more forward looking.

The only support for the old MC6800 family was a hardware interface to allow the new processor to use the existing M6800 peripherals while new M68000 parts were being designed.

Its design took many of the then current mini and mainframe computer architectural concepts and developed them using VLSI silicon technology. The programmer's register model shows how dramatic the change was. Gone are the dedicated 8 and 16 bit registers to be replaced by two groups of eight data registers and eight address registers. All these registers and the program counter are 32 bits wide.

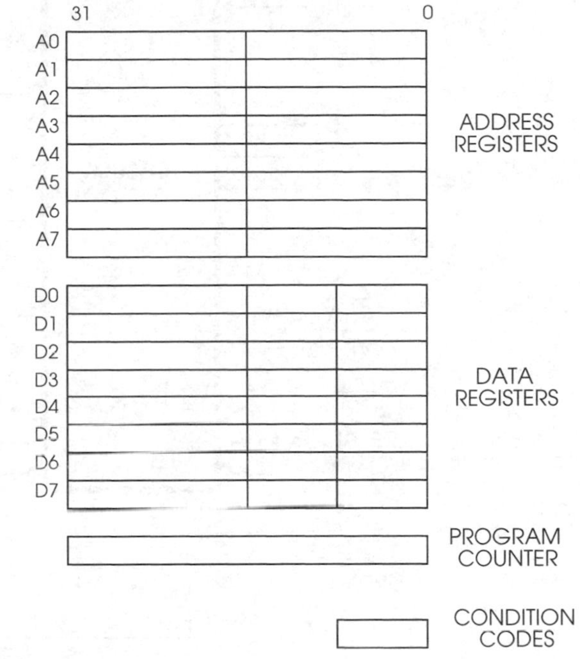

The MC68000 USER programmer's model

The MC68000 hardware

Address bus

The address bus (signals A1 – A23) is non-multiplexed and 24 bits wide, giving a single linear addressing space of 16 Mbytes. A0 is not brought out directly but is internally decoded to generate upper and lower data strobes. This allows the processor to access either or both the upper and lower bytes that comprise the 16 bit data bus.

Data bus

The data bus, D0 – D15, is also non-multiplexed and provides a 16 bit wide data path to external memory and peripherals. The processor can use data in either byte, word (16 bit) or long word (32 bit) values. Both word and long word data is stored on the appropriate boundary, while bytes can be stored anywhere. The diagram shows how these data quantities are stored. All addresses specify the byte at the start of the quantity.

If an instruction needs 32 bits of data to be accessed in external memory, this is performed as two successive 16 bit accesses automatically. Instructions and operands are always 16 bits in size and accessed on word boundaries. Attempts to access instructions, operands, words or long words on odd byte boundaries cause an internal 'address' error.

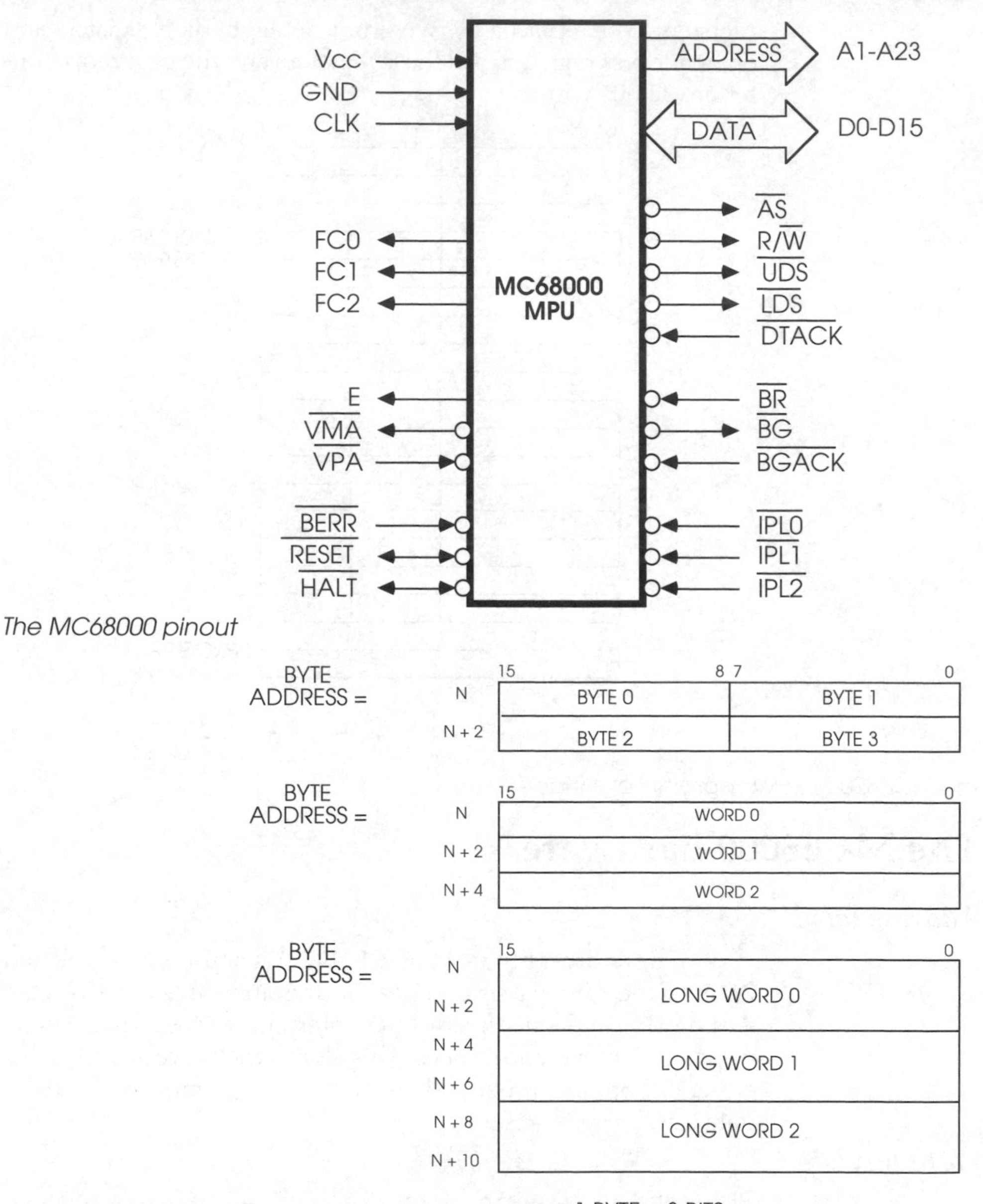

The MC68000 pinout

MC68000 data organisation

Function codes

The function codes, FC0–FC2, provide extra information describing what type of bus cycle is occurring. These codes and their meanings are shown in the table below. They appear at the same time as the address bus data and indicate program / data and supervisor / user accesses. In addition, when all three signals are asserted, the present cycle is an interrupt acknowledgement, where an interrupt vector is passed to the processor. Many designers use these codes to provide hardware partitioning.

Interrupts

Seven interrupt levels are supported and are encoded on to three interrupt pins IP0–IP2. With all three signals high, no external interrupt is requested. With all three asserted, a non-maskable level 7 interrupt is generated. Levels 1–6, generated by other combinations, can be internally masked by writing to the appropriate bits within the status register.

The interrupt cycle is started by a peripheral generating an interrupt. This is usually encoded using a 148 priority encoder. The appropriate code sequence is generated and drives the interrupt pins. The processor samples the levels and requires the levels to remain constant to be recognised. It is recommended that the interrupt level remains asserted until its interrupt acknowledgement cycle commences to ensure recognition. Once the processor has recognised the interrupt, it waits until the current instruction has been completed and starts an interrupt acknowledgement cycle. This starts an external bus cycle with all three function codes driven high to indicate an interrupt acknowledgement cycle.

Function code			Reference class
FC0	FC1	FC2	
0	0	0	Reserved
0	0	1	User data
0	1	0	User program
0	1	1	Reserved (I/O space)
1	0	0	Reserved
1	0	1	Supervisor data
1	1	0	Supervisor program
1	1	1	CPU space/interrupt ack

The MC68000 function codes and their meanings

The interrupt level being acknowledged is placed on address bus bits A1–A3 to allow external circuitry to identify which level is being acknowledged. This is essential when one or more interrupt requests are pending. The system now has a choice over which way it will respond:

- If the peripheral can generate an 8 bit vector number, this is placed on the lower byte of the address bus and DTACK* asserted. The vector number is read and the cycle completed. This vector number then selects the address and subsequent software handler from the vector table.
- If the peripheral cannot generate a vector, it can assert VPA* and the processor will terminate the cycle using the M6800 interface. It will select the specific interrupt vector allocated to the specific interrupt level. This method is called autovectoring.

To prevent an interrupt request generating multiple acknowledgements, the internal interrupt mask is raised to the interrupt level, effectively masking any further requests. Only if a higher level interrupt occurs will the processor nest its interrupt service routines. The interrupt service routine must clear the interrupt source and thus remove the request before returning to normal execution. If another interrupt is pending from a different source, it will be recognised and cause another acknowledgement to occur.

Error recovery and control signals

There are three signals associated with error control and recovery. The bus error BERR*, HALT* and RESET* signals can provide information or be used as inputs to start recovery procedures in case of system problems.

The BERR* signal is the counterpart of DTACK*. It is used during a bus cycle to indicate an error condition that may arise through parity errors or accessing non-existent memory. If BERR* is asserted on its own, the processor halts normal processing and goes to a special bus error software handler. If HALT* is asserted at the same time, it is possible to rerun the bus cycle. BERR* is removed followed by HALT* one clock later, after which the previous cycle is rerun automatically. This is useful to screen out transient errors. Many designs use external hardware to force a rerun automatically but will cause a full bus error if an error occurs during the rerun.

Without such a signal, the only recourse is to complete the transfer, generate an immediate non-maskable interrupt and let a software handler attempt to sort out the mess! Often the only way out is to reset the system or shut it down. This makes the system extremely intolerant of signal noise and other such transient errors.

The RESET* and HALT* signals are driven low at power-up to force the MC68000 into its power-up sequence. The operation takes about 100 ms, after which the signals are negated and the processor accesses the RESET vector at location 0 in memory to fetch its stack pointer and program counter from the two long words stored there.

Motorola MC68020

The MC68020 was launched in April 1984 as the '32 bit performance standard' and in those days its performance was simply staggering — 8 million instructions per second peak with 2–3 million sustained when running at 16 MHz clock speed. It was a true 32 bit processor with 32 bit wide external data and address buses as shown. It supported all the features and functions of the MC68000 and MC68010, and it executed M68000 USER binary code without modification (but faster!).

- Virtual memory and instruction continuation were supported. This is explained in Chapter 7 on interrupts.
- The bus and control signals were similar to that of its M68000 predecessors, offering an asynchronous memory interface but with a three–cycle operation (instead of four) and dynamic bus sizing which allowed the processor to talk to 8, 16 and 32 bit processors.
- Additional coprocessors could be added to provide such facilities as floating point arithmetic and memory management, which used this bus to provide a sophisticated communications interface.
- The instruction set was enhanced with more data types, addressing modes and instructions.
- Bit field data and its manipulation was supported, along with packed and unpacked BCD (binary coded decimal) formats. An instruction cache and a barrel shifter to perform high speed shift operations were incorporated on-chip to provide support for these functions.

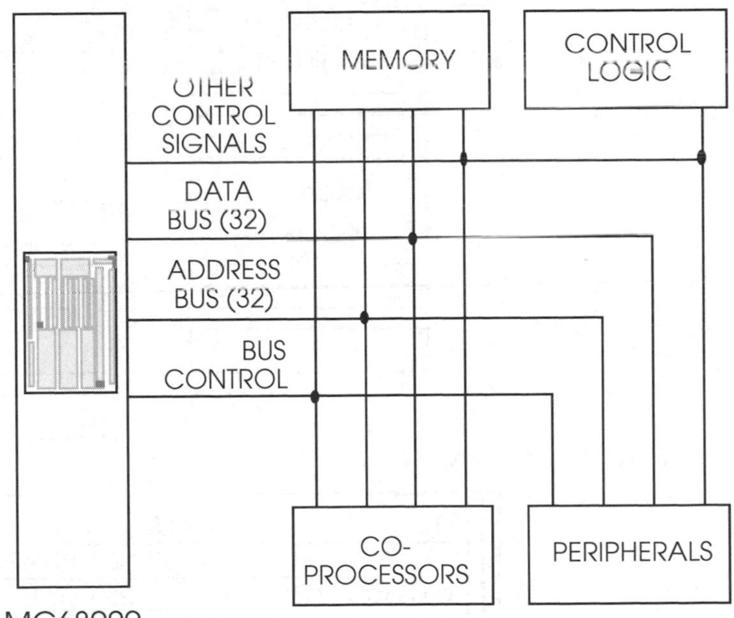

A simple MC68020 system

The actual pipeline used within the design is quite sophisticated. It is a four–stage pipe with stage A consisting of an instruction router which accepts data from either the external bus controller or the internal cache. As the instruction is processed down the pipeline, the intermediate data can either cause micro and nanocode sequences to be generated to control the execution unit or, in the case of simpler instructions, the data itself can be passed directly into the execution unit with the subsequent speed improvements.

The programmer's model

The programmer's USER model is exactly the same as for the MC68000, MC68010 and MC68008. It has the same eight data and eight address 32 bit register organisation. The SUPERVISOR mode is a superset of its predecessors. It has all the registers found in its predecessors plus another three. Two registers are associated with controlling the instruction cache, while the third provides the master stack pointer.

The supervisor uses either its master stack pointer or interrupt stack pointer, depending on the exception cause and the status of the M bit in the status register. If this bit is clear, all stack operations default to the A7´ stack pointer. If it is set, interrupt stack frames are stored using the interrupt stack pointer while other operations use the master pointer. This effectively allows the system to maintain two separate stacks. While primarily for operating system support, this extra register can be used for high reliability designs.

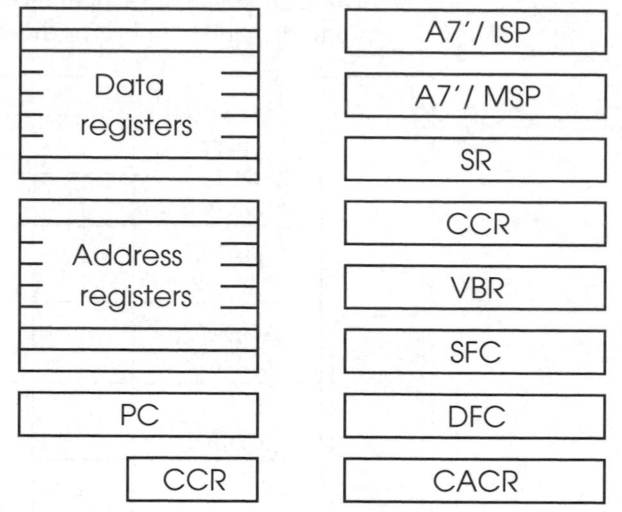

The MC68040 programming model

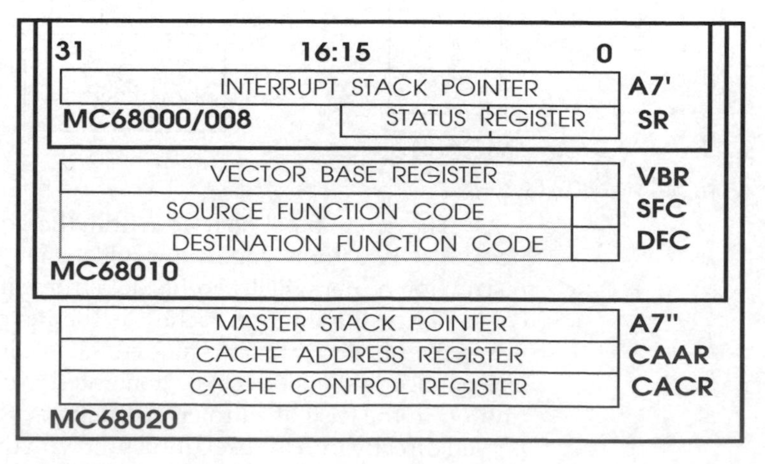

The M68020 SUPERVISOR programming model

The MC68020 instruction set is a superset of the MC68000/ MC68010 sets. The main difference is the inclusion of floating point and coprocessor instructions, together with a set to manipulate bit field data. The instructions to perform a trap on condition operation, a compare and swap operation and a 'call-return from module' structure were also included. Other differences were the addition of 32 bit displacements for the LINK and Bcc (branch on condition) instructions, full 32 bit arithmetic with 32 or 64 bit results as appropriate and extended bounds checking for the CHK (check) and CMP (compare) instructions.

The bit field instructions were included to provide additional support for applications where data does not conveniently fall into a byte organisation. Telecommunications and graphics both manipulate data in odd sizes — serial data can often be 5, 6 or 7 bits in size and graphics pixels (i.e. each individual dot that makes a picture on a display) vary in size, depending on how many colours, grey scales or attributes are being depicted.

Mode	Description
Dn	Data Register Direct
An	Address Register Direct
(An)+	Address Reg. Indirect w/ Post-Increment
-(An)	Address Reg. Indirect w/ Pre-Decrement
d(An)	Displaced Address Register Indirect
d(An,Rx)	Indexed, Displaced Address Reg.
d(PC)	Program Counter Relative
d(PC,Rx)	Indexed Program Counter Relative
#xxxxxxxx	Immediate
$xxxx	Absolute Short
$xxxxxxxx	Absolute Long
MC68000/008/010	
(bd,An,Xn.SIZE*SCALE)	Register Indirect
	Memory Indirect
((bd,An,Xn.SIZE*SCALE),od)	Pre-Indexed
((bd,An),Xn.SIZE*SCALE,od)	Post-Indexed
	Program Counter Memory Indirect
((bd,PC,Xn.SIZE*SCALE),od)	Pre-Indexed
((bd,PC),Xn.SIZE*SCALE,od)	Post-Indexed
MC68020/MC68030	

The MC68020 addressing modes

The addressing modes were extended from the basic M68000 modes, with memory indirection and scaling. Memory indirection allowed the contents of a memory location to be used within an effective address calculation rather than its absolute address. The scaling was a simple multiplier value 1, 2, 4 or 8 in magnitude, which multiplied (scaled) an index register. This allowed large data elements within data structures to be easily accessed without having to perform the scaling calculations prior to the access. These new modes were so complex that even the differentiation between data and address registers was greatly reduced: with the MC68020, it is possible to use data registers as additional address registers. In practice, there are over 50 variations available to the programmer to apply to the 16 registers.

The new CAS and CAS2 'compare and swap' instructions provided an elegant solution to linked list updating within a multiprocessor system. A linked list is a series of data lists linked together by storing the address of the next list in the chain in the preceding chain. To add or delete a list simply involves modifying these addresses.

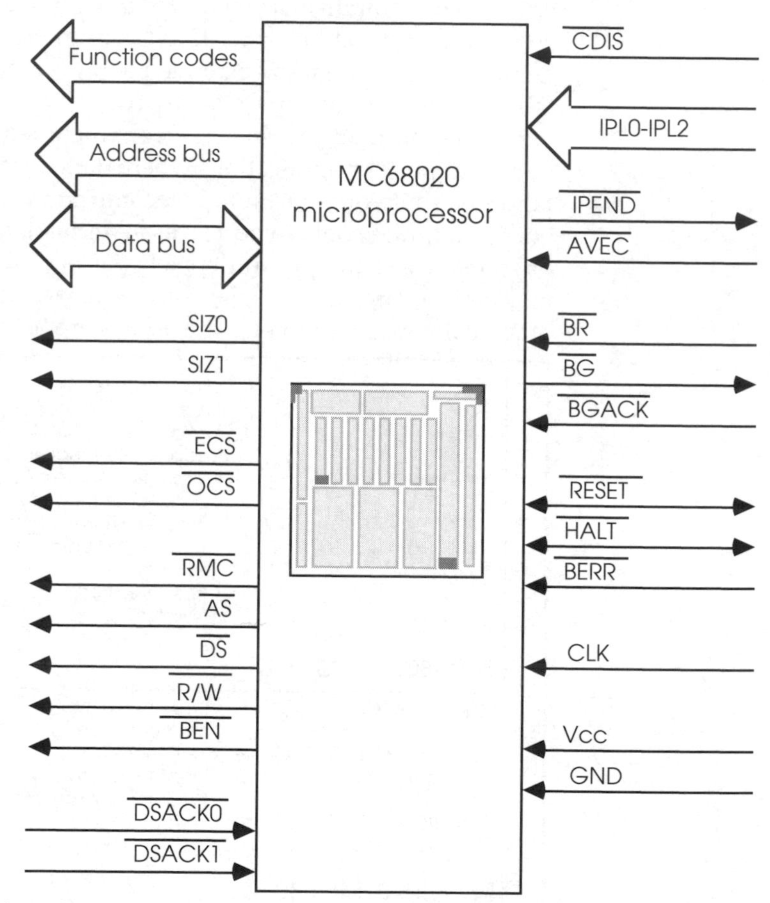

The MC68020 device pinout

In a multiprocessor system, this modification procedure must occur uninterrupted to prevent corruption. The CAS and CAS2 instruction meets this specification. The current pointer to the next list is read and stored in Dn. The new value is calculated and stored in Dm. The CAS instruction is then executed. The current pointer value is read and compared with Dn. If they are the same, no other updating by another processor has happened and Dm is written out to update the list. If they do not match, the value is copied into Dn, ready for a repeat run. This sequence is performed using an indivisible read-modify-write cycle. The condition codes are updated during each stage. The CAS2 instruction performs a similar function but with two sets of values. This instruction is also performed as a series of indivisible cycles but with different addresses appearing during the execution.

Bus interfaces

Many of the signals shown in the pin out diagram are the same as those of the MC68000 — the function codes FC0–2, interrupt pins IPL0–2 and the bus request pins, RESET*, HALT* and BERR* perform the same functions.

With the disappearance of the M6800 style interface, separate signals are used to indicate an auto-vectored interrupt. The AVEC* signal is used for this function and can be permanently asserted if only auto-vectored interrupts are required. The IPEND signal indicates when an interrupt has been internally recognised and awaits an acknowledgement cycle. RMC* indicates an indivisible read-modify-write cycle instead of simply leaving AS* asserted between the bus cycles. The address strobe is always released at the end of a cycle. ECS* and OCS* provide an early warning of an impending bus cycle, and indicate when valid address information is present on the bus prior to validation by the address strobe.

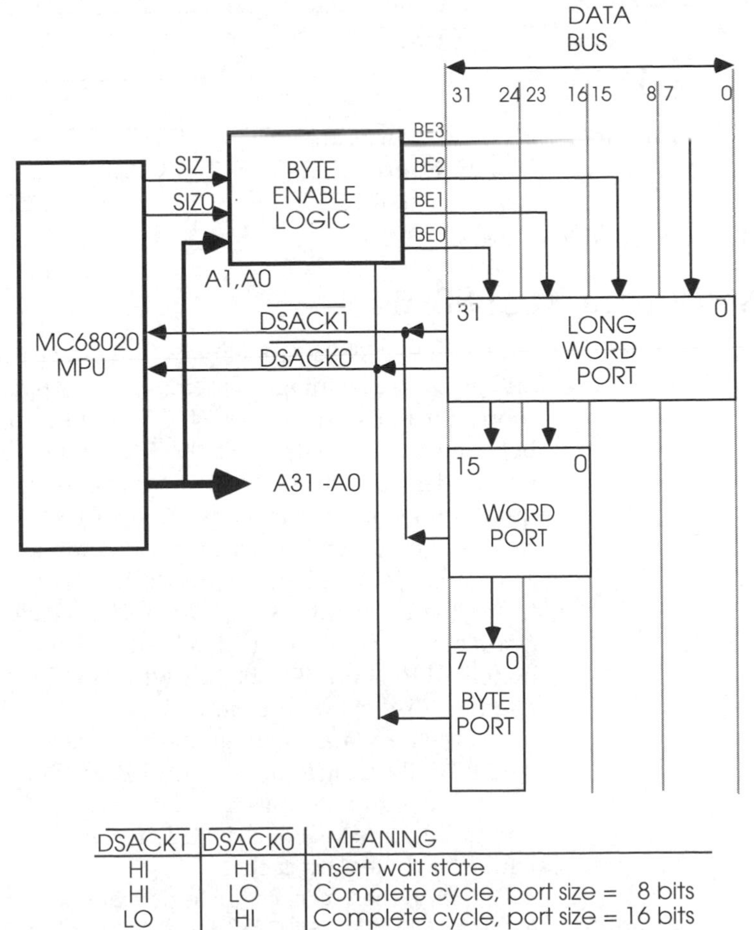

DSACK1	DSACK0	MEANING
HI	HI	Insert wait state
HI	LO	Complete cycle, port size = 8 bits
LO	HI	Complete cycle, port size = 16 bits
LO	LO	Complete cycle, port size = 32 bits

MC68020 dynamic bus sizing

The M68000 upper and lower data strobes have been replaced by A0 and the two size pins, SIZE0 and SIZE1. These indicate the amount of data left to transfer on the current bus cycle and, when used with address bits A0 and A1, can provide decode information so that the correct bytes within the 4 byte wide data bus can be enabled. The old DTACK* signal has been replaced by two new ones, DSACK0* and DSACK1*. They provide the old DTACK* function of indicating a successful bus cycle and are used in the dynamic bus sizing. The bus interface is asynchronous and similar to the M68000 cycle but with a shorter three–cycle sequence, as shown.

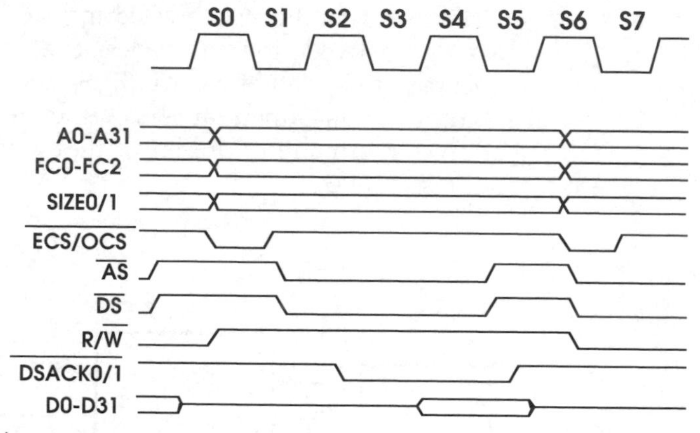

MC68020 bus cycle timings

Motorola MC68030

The MC68030 appeared some 2–3 years after the MC68020 and used the advantage of increased silicon real estate to integrate more functions on to a MC68020-based design. The differences between the MC68020 and the MC68030 are not radical — the newer design can be referred to as evolutionary rather than a quantum leap. The device is fully MC68020 compatible with its full instruction set, addressing modes and 32 bit wide register set. The initial clock frequency was designed to 20 MHz, some 4 MHz faster than the MC68020, and this has yielded commercially available parts running at 50 MHz. The transistor count has increased to about 300000 but with smaller geometries; die size and heat dissipation are similar.

Memory management has now been brought on-chip with the MC68030 using a subset of the MC68851 PMMU with a smaller 22 entry on-chip address translation cache. The 256 byte instruction cache of the MC68020 is still present and has been augmented with a 256 byte data cache.

Both these caches are logical and are organised differently from the 64 × 4 MC68020 scheme. A 16 × 16 organisation has been adopted to allow a new synchronous bus to burst fill cache lines. The cache lookup and address translations occur in parallel to improve performance.

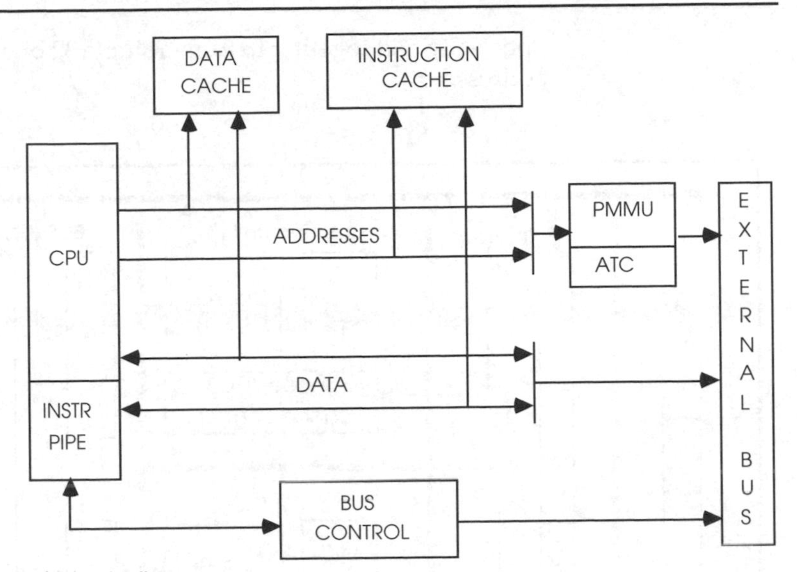

The MC68030 internal block diagram

The processor supports both the coprocessor interface and the MC68020 asynchronous bus with its dynamic bus sizing and misalignment support. However, it has an alternative synchronous bus interface which supports a two-clock access with optional single-cycle bursting. The bus interface choice can be made dynamically by external hardware.

The MC68040

The MC68040 incorporates separate integer and floating point units giving sustained performances of 20 integer MIPS and 3.5 double precision Linpack MFLOPS respectively, dual 4 kbyte instruction and data caches, dual memory management units and an extremely sophisticated bus interface unit. The block diagram shows how the processor is partitioned into several separate functional units which can all execute concurrently. It features a full Harvard architecture internally and is remarkably similar at the block level, to the PowerPC RISC processor.

The design is revolutionary rather than evolutionary: it takes the ideas of overlapping instruction execution and pipelining to a new level for CISC processors. The floating point and integer execution units work in parallel with the on-chip caches and memory management to increase the overlapping so that many instructions are executed in a single cycle, and thus give it its performance.

The pinout reveals a large number of new signals. One major difference about the MC68040 is its high drive capability. The processor can be configured on reset to drive either 55 or 5 mA per bus or control pin. This removes the need for external buffers, reducing chip count and the associated propagation delays, which often inflict a high speed design. The 32 bit address and 32 bit data buses are similar to its predecessors although the signals can be optionally tied together to form a single 32 bit multiplexed data/address bus.

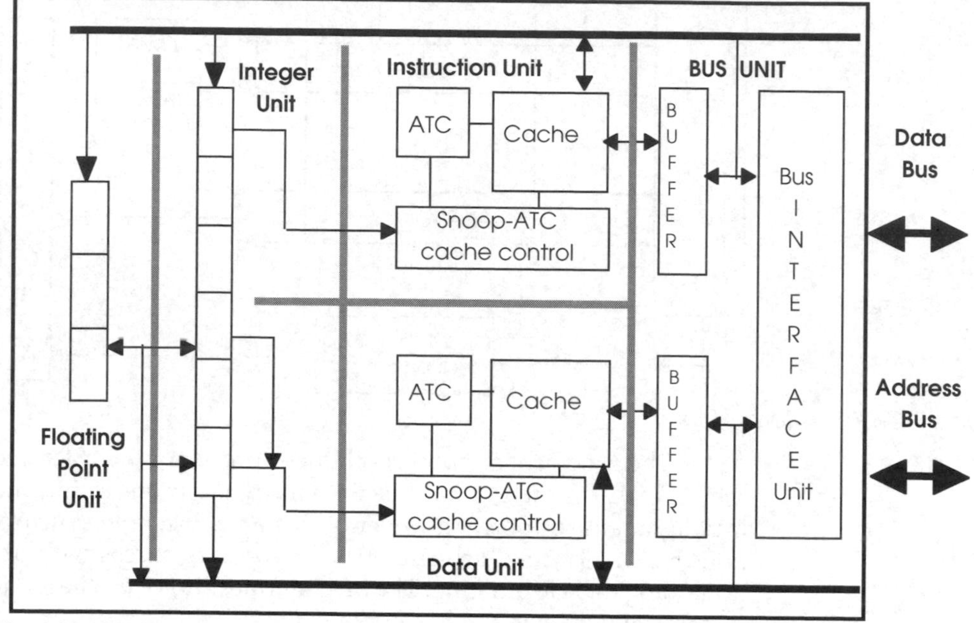

The MC68040 block diagram

The User Programmable Attributes (UPA0 and UPA1) are driven according to 2 bits within each page descriptor used by the onboard memory management units. They are primarily used to enable the MC68040 Bus Snooping protocols, but can also be used to give additional address bits, software control for external caches and other such functions. The two size pins, SIZ0 and SIZ1, no longer indicate the number of remaining bytes left to be transferred as they did on the MC68020 and MC68030, but are used to generate byte enables for memory ports. They now indicate the size of the current transfer. Dynamic bus sizing is supported via external hardware if required. Misaligned accesses are supported by splitting the transfer into a series of aligned accesses of differing sizes. The transfer type signals, TT1 and TT2, indicate the type of transfer that is taking place and the Transfer Modifier pins TM0-2 provide further information. These five pins effectively replace the three function code pins. The TLN0-1 pins indicate the current long word number within a burst fill access.

The synchronous bus is controlled by the Master and Slave transfer control signals: Transfer Start (TS*) indicates a valid address on the bus while the Transfer in Progress (TIP*) signal is asserted during all external bus cycles and can be used to power up/down external memory to conserve power in portable applications. These two Master signals are complemented by the slave signals: Transfer Acknowledge (TA*) successfully terminates the bus cycle, while Transfer Error Acknowledge (TEA*) terminates the cycle and the burst fill as a result of an error. If both these signals are asserted on the first access of the burst, the cycle is terminated and immediately rerun. On the second, third and fourth accesses, a retry attempt is not allowed and the processor simply assumes that an error has occurred and will terminate the burst as normal.

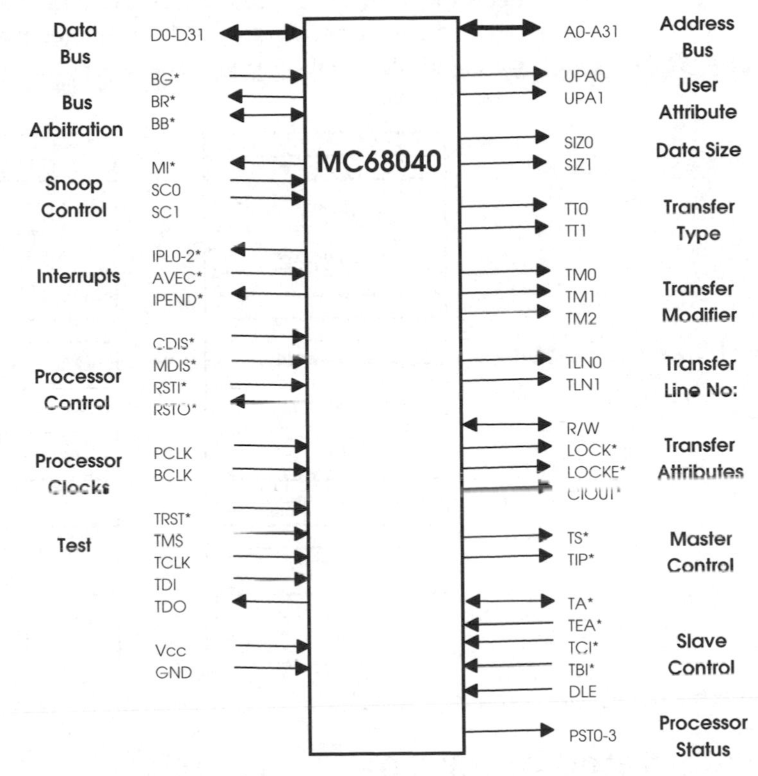

The MC68040 pinout

The processor can be configured to use a different signal, Data Latch Enable DLE to latch read data instead of the rising edge of the BCLK clock. The internal caches and memory management units can be disabled via the CDIS* and MDIS* pins respectively.

The programming model

To the programmer the programming model of the MC68040 is the same as its predecessors such as the MC68030. It has the same eight data and eight address registers, the vector same base register (VBR), the alternate function code registers although some codes are reserved, the same dual Supervisor stack pointer and the two cache control registers although only two bits are now used to enable or disable either of the two on-chip caches. Internally the implementation is different. Its instruction execution unit consists of a six–stage pipeline which sequentially fetches an instruction, decodes it, calculates the effective address, fetches an address operand, executes the instruction and finally writes back the results. To prevent pipeline stalling, an internal Harvard architecture is used to allow simultaneous instruction and operand fetches. It has been optimised for many instructions and addressing modes so that single-cycle execution can be achieved. The early pipeline stages are effectively duplicated to allow both paths of a branch instruction to be processed until the path decision is taken. This removes pipeline stalls and the subsequent performance degradation. While integer instructions are being executed, the floating point unit is free to execute floating point instructions.

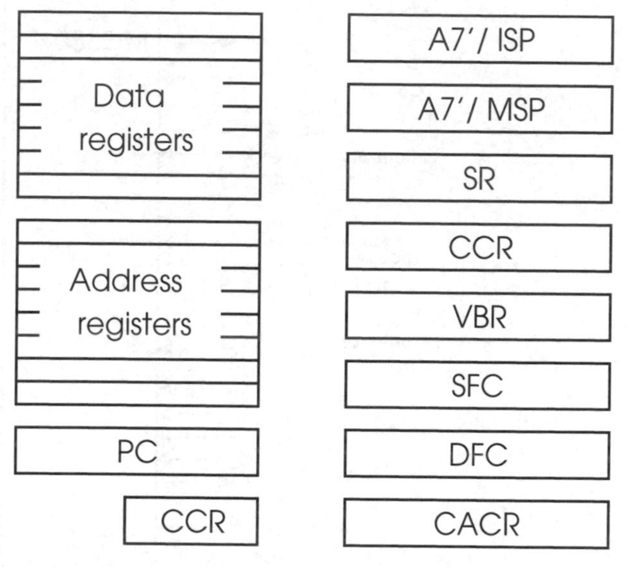

The MC68040 programming model

Integrated processors

With the ability of semiconductor manufacturers to be able to integrate several million transistors onto a single piece of silicon, it should come as no surprise that there are now processors available which take the idea of integration offered by a microcontroller, but use a high performance processor instead. The Intel 80186 started this process by combining DMA channels with an 8086 architecture. The most successful family so far has been the MC683xx family from Motorola. There are now several members of the family currently available.

They combine an M68000 or MC68020 (CPU32) family processor and its asynchronous memory interface, with all the standard interface logic of chip selects, wait state generators, bus and watchdog timers into a system interface module and use a second RISC type processor to handle specific I/O functions. This approach means that all the additional peripherals and logic needed to construct an MC68000-based system has gone. In many cases, the hardware design is almost at the 'join up the dots' level where the dots are the processor and memory pins.

This approach has been adopted by others and many different processor cores, such as SPARC and MIPs, are now available in similar integrated processors. PowerPC versions of the MC68360 are now in production where the MC68000-based CPU32 core is replaced with a 50 MHz PowerPC processor. For embedded systems, this is definitely the way of the future.

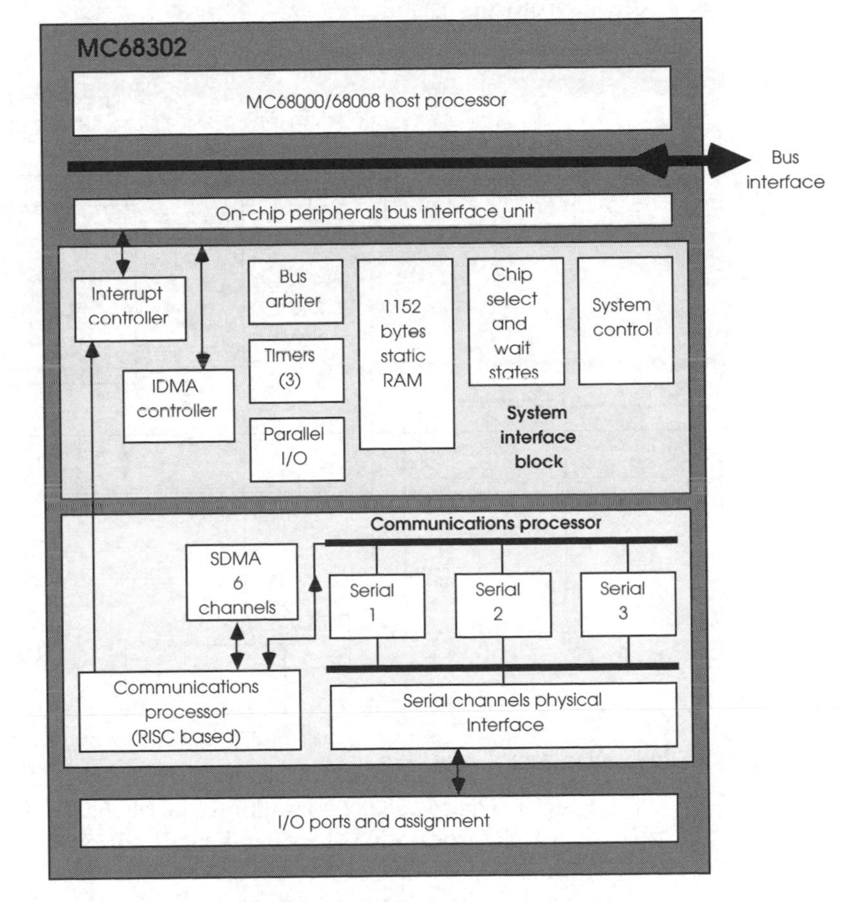

The MC68302 Integrated Multiprotocol Processor

The MC68302 uses a 16 MHz MC68000 processor core with power down modes and either an 8 or 16 bit external bus. The system interface block contains 1152 bytes of dual port RAM, 28 pins of parallel I/O, an interrupt controller and a DMA device, as well as the standard glue logic. The communications processor is a RISC machine that controls three multiprotocol channels, each with their own pair of DMA channels. The channels support BISYNC, DDCMP, V.110, HDLC synchronous modes and standard UART functions. This processor takes buffer structures from either the internal or external RAM and takes care of the day-to-day activities of the serial channels. It programs the DMA channel to transfer the data, performs the character and address comparisons and cyclic redundancy check (CRC) generation and checking. The processor has sufficient power to cope with a combined data rate of 2 Mbits per second across the three channels. Assuming an 8 bit character and a single interrupt to poll all three channels, the processor is handling the equivalent of an interrupt every 12 microseconds. In addition, it is performing all the data framing etc. While this is going on, the on-chip M68000 is free to perform other functions, like the higher layers of X.25 or other OSI protocols as shown.

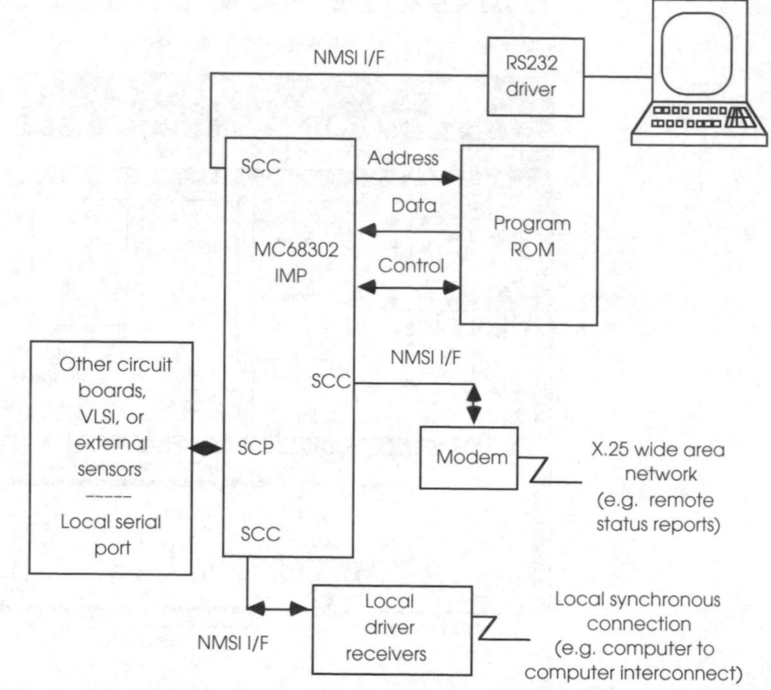

A typical X25-ISDN terminal interface

The MC68332 is similar to the MC68302, except that it has a CPU32 processor (MC68020-based) running at 16 MHz and a timer processor unit instead of a communications processor. This has 16 channels which are controlled by a RISC-like processor to perform virtually any timing function. The timing resolution is down to 250 nanoseconds with an external clock source or 500 nanoseconds with an internal one. The timer processor can perform the common timer algorithms on any of the 16 channels without placing any overhead on the CPU32.

A queued serial channel and 2 kbits of power down static RAM are also on-chip and for many applications, all that is required to complete a working system is an external program EPROM and a clock.

This is a trend that many other architectures are following especially with RISC processors. Apart from the high performance range of the processor market or where complete flexibility is needed, most processors today come with at least some basic peripherals such as serial and parallel ports and a simple or glueless interface to memory. In many cases, they dramatically reduce the amount of hardware design needed to add external memory and thus complete a simple design. This type of processor is gaining popularity with designers.

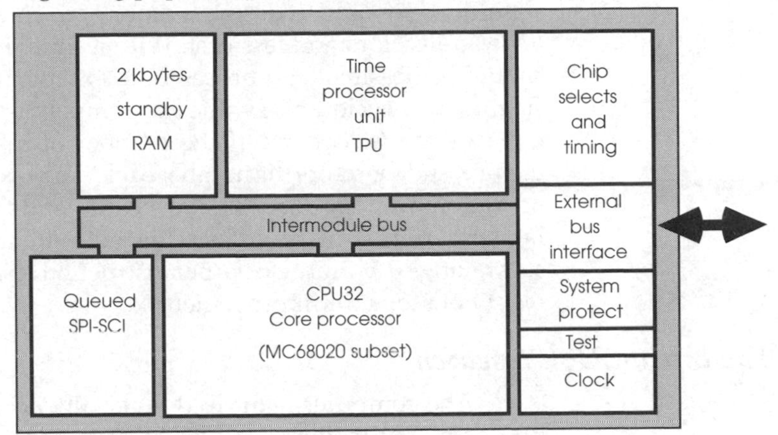

The MC68332 block diagram

RISC processors

Until 1986, the expected answer to the question 'which processor offers the most performance' would be the MC68020, the MC68030 or even the 386! Without exception, CISC processors such as these, had established the highest perceived performances. There were more esoteric processors, like the transputer, which offered large MIPS figures from parallel arrays but these were often considered only suitable for niche markets and applications. However, around this time, an interest in an alternative approach to microprocessor design started, which seemed to offer more processing power from a simpler design using less transistors. Performance increases of over five times the then current CISC machines were suggested. These machines, such as the Sun SPARC architecture and the MIPS R2000 processor, were the first of a modern generation of processors based on a reduced instruction set, generically called reduced instruction set processors (RISC).

The 80/20 rule

Analysis of the instruction mix generated by CISC compilers is extremely revealing. Such studies for CISC mainframes and mini computers shows that about 80% of the instructions generated and executed used only 20% of an instruction set. It was an obvious conclusion that if this 20% of instructions were speeded up, the performance benefits would be far greater. Further analysis shows that these instructions tend to perform the simpler operations and use only the simpler addressing modes. Essentially, all the effort invested in processor design to provide complex instructions and thereby reduce the compiler workload was being wasted. Instead of using them, their operation was synthesised from sequences of simpler instructions.

This has another implication. If only the simpler instructions are required, the processor hardware required to implement them could be reduced in complexity. It therefore follows that it should be possible to design a more performant processor with fewer transistors and less cost. With a simpler instruction set, it should be possible for a processor to execute its instructions in a single clock cycle and synthesise complex operations from sequences of instructions. If the number of instructions in a sequence, and therefore the number of clocks to execute the resultant operation, was less than the cycle count of its CISC counterpart, higher performance could be achieved. With many CISC processors taking 10 or more clocks per instruction on average, there was plenty of scope for improvement.

The initial RISC research

The computer giant IBM is usually acknowledged as the first company to define a RISC architecture in the 1970s. This research was further developed by the Universities of Berkeley and Stanford to give the basic architectural models. RISC can be described as a philosophy with three basic tenets:

1. All instructions will be executed in a single cycle

   This is a necessary part of the performance equation. Its implementation calls for several features — the instruction op code must be of a fixed width which is equal to or smaller than the size of the external data bus, additional operands cannot be supported and the instruction decode must be simple and orthogonal to prevent delays. If the op code is larger than the data width or additional operands must be fetched, multiple memory cycles are needed, increasing the execution time.

2. Memory will only be accessed via load and store instructions

   This naturally follows from the above. If an instruction manipulates memory directly, multiple cycles must be performed to execute it. The instruction must be fetched and memory manipulated. With a RISC processor, the memory resident data is loaded into a register, the register manipulated and, finally, its contents written out to main memory. This sequence takes a minimum of three instructions. With register-based manipulation, large numbers of general-purpose registers are needed to maintain performance.

3. All execution units will be hardwired with no microcoding

   Microcoding requires multiple cycles to load sequencers etc and therefore cannot be easily used to implement single-cycle execution units.

Two generic RISC architectures form the basis of nearly all the current commercial processors. The main differences between them concern register sets and usage. They both have a Harvard external bus architecture consisting of separate buses for instructions and data. This allows data accesses to be performed in parallel with instruction fetches and removes any instruction/data conflict. If these two streams compete for a single bus, any data fetches stall the instruction flow and prevent the processor from achieving its single cycle objective. Executing an instruction on every clock requires an instruction on every clock.

The Berkeley RISC model

The RISC 1 computer implemented in the late 1970s used a very large register set of 138 × 32 bit registers. These were arranged in eight overlapping windows of 24 registers each. Each window was split so that six registers could be used for parameter passing during subroutine calls. A pointer was simply changed to select the group of six registers. To perform a basic call or return simply needed a change of pointer. The large number of registers is needed to minimise the number of fetches to the outside world. With this simple window technique, procedure calls can be performed extremely quickly. This can be very beneficial for real-time applications where fast responses are necessary.

However, it is not without its disadvantages. If the procedure calls require more than six variables, one register must be used to point to an array stored in external memory. This data must be loaded prior to any processing and the register windowing loses much of its performance. If all the overlapping windows are used, the system resolves the situation by tracking the window usage so either a window or the complete register set can be saved out to external memory.

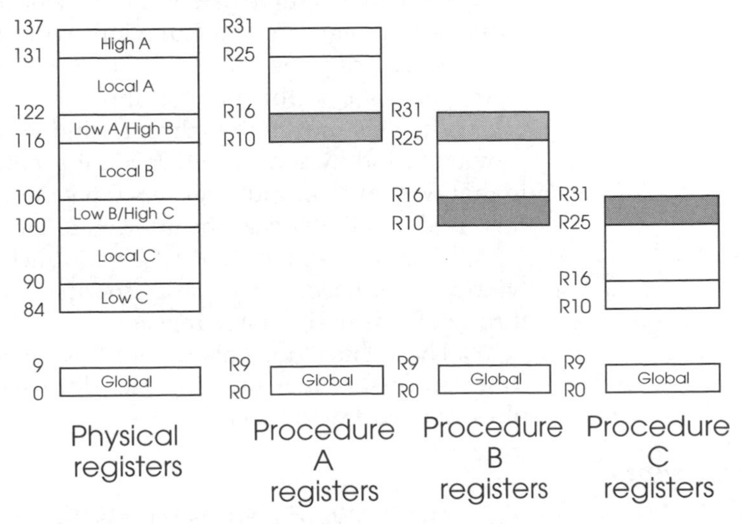

Register windowing

This overhead may negate any advantages that windowing gave in the first place. In real-time applications, the overhead of saving 138 registers to memory greatly increases the context switch and hence the response time. A good example of this approach is the Sun SPARC processor.

Sun SPARC RISC processor

The SPARC (scalable processor architecture) processor is a 32 bit RISC architecture developed by Sun Microsystems for their workstations but manufactured by a number of manufacturers such as LSI, Cypress, Fujitsu, Philips and Texas Instruments.

The basic architecture follows the Berkeley model and uses register windowing to improve context switching and parameter passing. The initial designs were based on a discrete solution with separate floating point units, memory management facilities and cache memory, but later designs have integrated these versions. The latest versions also support superscalar operation.

Architecture

The SPARC system is based on the Berkeley RISC architecture. A large 32 bit wide register bank containing 120 registers is divided into a set of seven register windows and a set of eight registers which are globally available. Each window set containing 24 registers is split into three sections to provide eight input, eight local and eight output registers. The registers in the output section provide the information to the eight input registers in the next window. If a new window is selected during a context switch or as a procedural call, data can be passed with no overhead by placing it in the output registers of the first window. This data is then available for the procedure or next context in its input registers. In this way, the windows are linked together to form a chain where the input registers for one window have the contents of the output registers of the previous window.

To return information back to the original or calling software, the data is put into the input registers and the return executed. This moves the current window pointer back to the previous window and the returned information is now available in that window's output registers. This method is the reverse of that used to initially pass the information in the first place.

The programmer and CPU can track and control which windows are used and what to do when all windows are full, through fields in the status register.

The architecture is also interesting in that it is one of the few RISC processors that uses logical addressed caches instead of physically addressed caches.

Interrupts

The SPARC processor supports 15 external interrupts which are generated using the four interrupt lines, IRL0 – IRL3. Level 15 is assigned as a non-maskable interrupt and the other 14 can be masked if required.

An external interrupt will generate an internal trap where the current and the next instructions are saved, the pipeline flushed and the processor switched into supervisor mode. The trap vector table which is located in the trap base register is then used to supply the address of the service routine. When the routine has completed, the REIT instruction is executed which restores the processor status and allows it to continue.

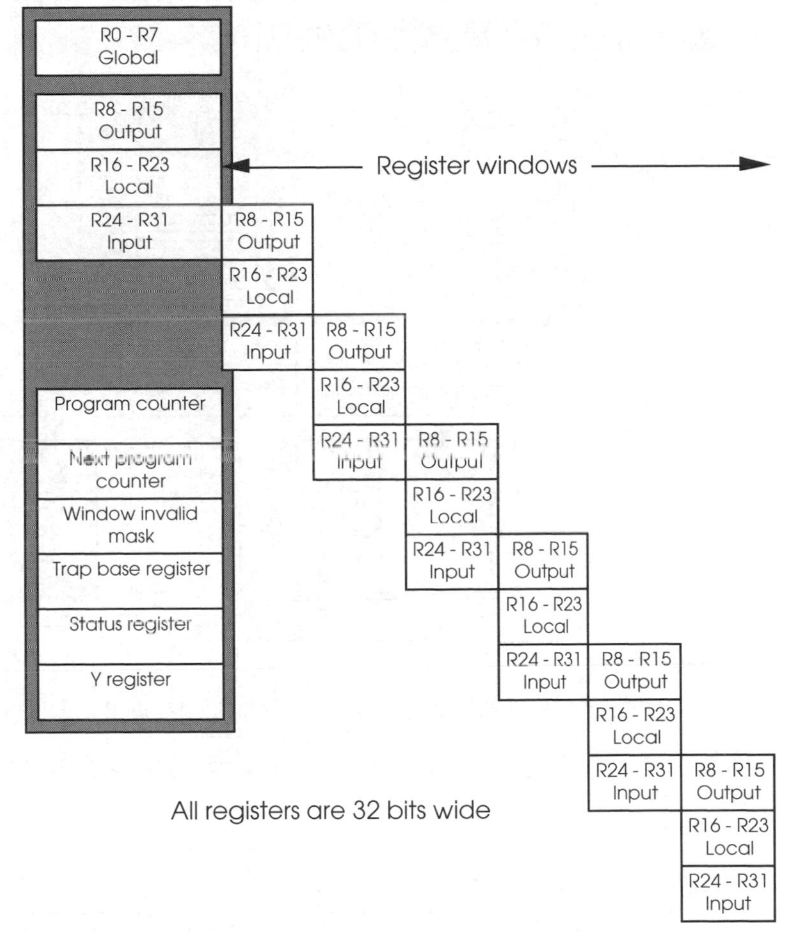

The SPARC register model

Instruction set

The instruction set comprises of 64 instructions. All access to memory is via load and store instructions as would be expected with a RISC architecture. All other instructions operate on the register set including the currently selected window. The instruction set is also interesting in that it has a multiply step command instead of the more normal multiply command. The multiply step command allows a multiply to be synthesised.

The Stanford RISC model

This model uses a smaller number of registers (typically 32) and relies on software techniques to allocate register usage during procedural calls. Instruction execution order is optimised by its compilers to provide the most efficient way of performing the software task. This allows pipelined execution units to be used within the processor design which, in turn, allow more powerful instructions to be used.

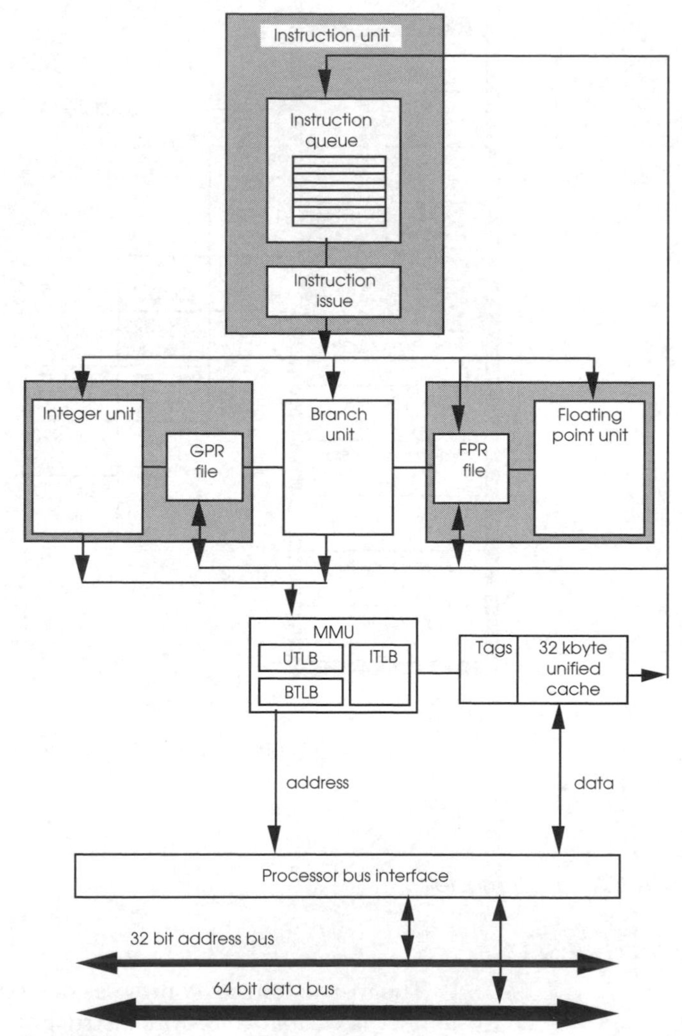

MPC601 internal block diagram

However, RISC is not the magic panacea for all performance problems within computer design. Its performance is extremely dependent on very good compiler technology to provide the correct optimisations and keep track of all the registers. Many of the early M68000 family compilers could not track all the 16 data and address registers and therefore would only use two or three. Some compilers even reduced register usage to one register and effectively based everything on stacks and queues. Secondly, the greater number of instructions it needed increased code size dramatically at a time when memory was both expensive and low in density. Without the compiler technology and cheap memory, a RISC system was not very practical and the ideas were effectively put back on the shelf.

The MPC601 was the first PowerPC processor available. It has three execution units: a branch unit to resolve branch instructions, an integer unit and a floating point unit.

The floating point unit supports IEEE format. The processor is superscalar. It can dispatch up to two instructions and process three every clock cycle. Running at 66 MHz, this gives a peak performance of 132 million instructions per second.

The branch unit supports both branch folding and speculative execution where the processor speculates which way the program flow will go when a branch instruction is encountered and starts executing down that route while the branch instruction is resolved.

The general-purpose register file consists of 32 separate registers, each 32 bits wide. The floating point register file also contains 32 registers, each 64 bits wide, to support double precision floating point. The external physical memory map is a 32 bit address linear organisation and is 4 Gbytes in size.

The MPC601's memory subsystem consists of a unified memory management unit and on-chip cache which communicates to external memory via a 32 bit address bus and a 64 bit data bus. At its peak, this bus can fetch two instructions per clock or 64 bits of data. It also supports split transactions, where the address bus can be used independently and simultaneously with the data bus to improve its utilisation. Bus snooping is also provided to ensure cache coherency with external memory.

The cache is 32 kbytes and supports both data and instruction accesses. It is accessed in parallel with any memory management translation. To speed up the translation process, the memory management unit keeps translation information in one of three translation lookaside buffers.

The MPC603 block diagram

The MPC603 was the second PowerPC processor to appear. Like the MPC601, it has the three execution units: a branch unit to resolve branch instructions, an integer unit and a floating point unit.

The floating point unit supports IEEE format. However, two additional execution units have been added to provide dedicated support for system registers and to move data between the register files and the two on-chip caches. The processor is superscalar and can dispatch up to three instructions and process five every clock cycle.

The branch unit supports both branch folding and speculative execution. It augments this with register renaming, which allows speculative execution to continue further than allowed on the MPC601 and thus increase the processing advantages of the processor.

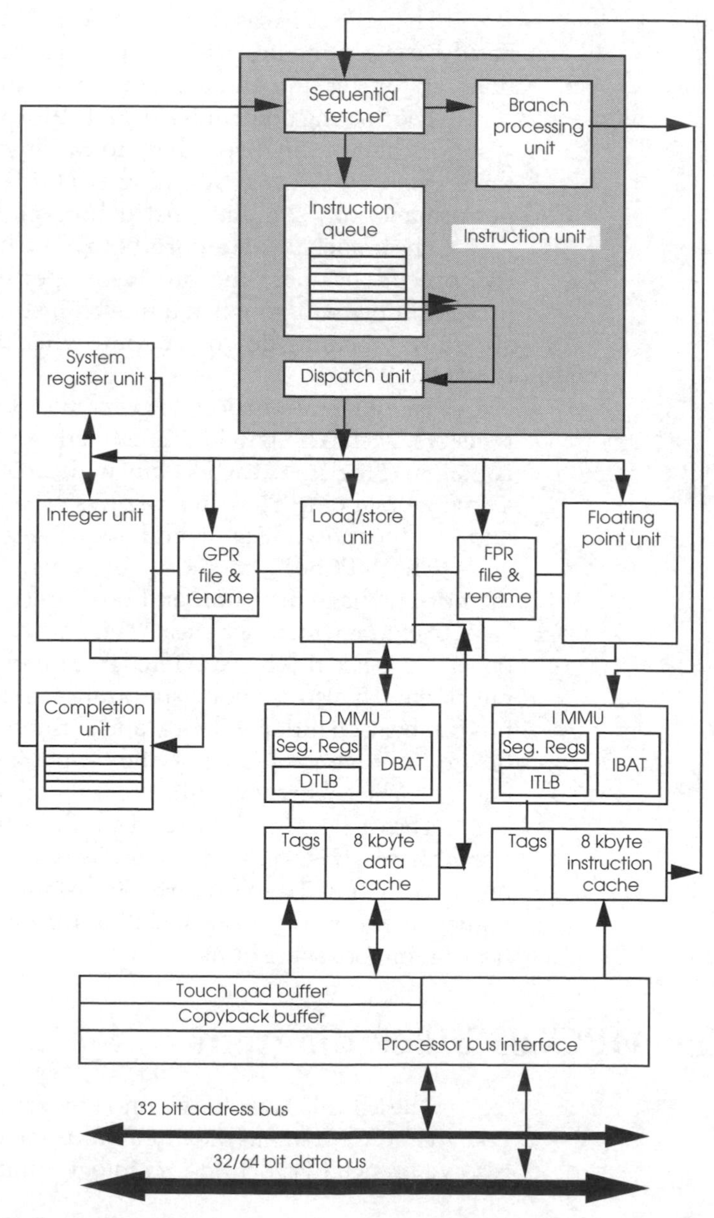

MPC603 internal block diagram

The general-purpose register file consists of 32 separate registers, each 32 bits wide. The floating point register file contains 32 registers, each 64 bits wide to support double precision floating point. The external physical memory map is a 32 bit address linear organisation and is 4 Gbytes in size.

The MPC603's memory subsystem consists of a separate memory management unit and on-chip cache for data and instructions which communicates to external memory via a 32 bit address bus and a 64 or 32 bit data bus. This bus can, at its peak, fetch two instructions per clock or 64 bits of data. Each cache is 8 kbytes in size, giving a combined on-chip cache size of 16 kbytes. The bus also supports split transactions, where the address bus can be used independently and simultaneously with the data bus to improve its utilisation. Bus snooping is also provided to ensure cache coherency with external memory.

As with the MPC601, the MPC603 speeds up the address translation process, by keeping translation information in one of four translation lookaside buffers, each of which is divided into two pairs, one for data accesses and the other for instruction fetches. It is different from the MPC601 in that translation tablewalks are performed by the software and not automatically by the processor.

The device also includes power management facilities and is eminently suitable for low power applications.

The ARM RISC architecture

It is probably fair to say that the ARM RISC architecture is really what RISC is all about. Small simple processors that provide adequate performance for their intended marketplace. For ARM, this was not the area of blinding performance but in the then embryonic mobile and handheld world where power consumption is as important as anything else. ARM also brought in the concept of the fabless semiconductor company where they licence their designs to others to build. As a result, if you want an ARM processor then you need to go to one of the 50+ licenced manufacturers. As a result, ARM processor architectures power the bulk of the digital mobile phones and organisers available today.

The ARM register set

The architecture uses standard RISC architecture techniques (load-store architecture, simple addressing modes based on register contents and instruction information only, fixed length instructions etc.,) and has a large 32 register file which is banked to provide a programming model of 16 registers with additional registers from the 32 used when the processor handles exceptions. This is called register banking where some of the spare registers are allocated as replacements for a selected set of the first 16 registers. This means that there is little need to save the registers during a context switch. This mechanism is very similar to register windowing in reality.

Two registers have special usage: register 14 is used as a link register and holds the address of the next instruction after a branch and link instruction. This permits the software flow to return using this link address after a subroutine has been executed. While it can be used as a general purpose register, care has to be taken that its contents are not destroyed so that when a return is executed, the program returns to the wrong address or even one that has no associated memory! Register 15 is used as the program counter. The ARM architecture uses a fixed 4 byte instruction word and supports the aligned instruction organisation only. This means that each instruction must start on a word boundary and also means that the lowest two bits in the program counter are always set to zero. In effect, this reduces the register to only 30 bits is size. Looking at the PC can be a little strange as it points not to the currently executing instruction but to two instructions after that. Useful to remember when debugging code.

While registers 14 and 15 are already allocated to special use, the remaining 14 registers may also be used for special functions. These definitions are not forced in hardware as is the case with the previous two examples, but are often enforced in software either through the use of a programming convention or by a compiler. Register 13 is frequently used as a stack pointer by convention but other registers could be used to fulfil this function or to provide additional stack pointers.

Exceptions

Exception processing with the ARM architecture is a little more complicated in that it supports several different exception processing modes and while it could be argued that these are nothing more than a user mode and several variants of a supervisor mode (like many other RISC architectures), they are sufficiently different to warrant their separate status.

The processor normally operates in the user mode where it will execute the bulk of any code. This gives access to the 16 register program file as previously described. To get into an exception mode from a user mode, there are only five methods to do so. The common methods such as an interrupt, a software interrupt, memory abort and the execution of an undefined instruction are all there. However a fifth is supported which is designed to reduce the latency time taken to process a fast interrupt. In all cases, the processor uses the register banking to preserve context before switching modes. Registers 13 and 14 are both automatically banked to preserve their contents so that they do not need to be saved. Once in the exception handler, register 14 is used to hold the return address ready for when the handler completes and returns and register 13 provides a unique stack pointer for the handler to use. Each exception will cause the current instruction to complete and then the execution flow will change to the address stored in the associated location in the vector table. This technique is similar to that used with both CISC and RISC processors.

If the handler needs to use any of the other registers, they must be saved before use and then restored before returning. To speed this process up, there is a fifth mode called the fast interrupt mode where registers 8 to 12 are also banked and these can be used by the handler without the overhead of saving and restoring. This is known as the fast interrupt mode.

Exception Modes

Priviledged Modes

Modes

USER	SYSTEM	Supervisor	Abort	Undefined	Interrupt	Fast Interrupt
R0	R0	R0	R0	R0	R0	R0
R1	R1	R1	R1	R1	R1	R1
R2	R2	R2	R2	R2	R2	R2
R3	R3	R3	R3	R3	R3	R3
R4	R4	R4	R4	R4	R4	R4
R5	R5	R5	R5	R5	R5	R5
R6	R6	R6	R6	R6	R6	R6
R7	R7	R7	R7	R7	R7	R7
R8	R8	R8	R8	R8	R8	R8_fiq
R9	R9	R9	R9	R9	R9	R9_fiq
R10	R10	R10	R10	R10	R10	R10_fiq
R11	R11	R11	R11	R11	R11	R11_fiq
R12	R12	R12	R12	R12	R12	R12_fiq
R13	R13	R13_svc	R13_abt	R13_und	R13_irq	R13_fiq
R14	R14	R14_svc	R14_abt	R14_und	R14_irq	R14_fiq
PC	PC	PC	PC	PC	PC	PC
CPSR	CPSR	CPSR	CPSR	CPSR	CPSR	CPSR
		SPSR_svc	SPSR_abt	SPSR_und	SPSR_irq	SPSR_fiq

White text on black indicates that the User/System register has been replaced by an alternative banked register.

The ARM processing modes and register banking

The exception modes do not stop there. There is also a sixth mode known as the system mode which is effectively an enhanced user mode in that it uses the user mode registers but is provided with privileged access to memory and any coprocessors that might be present.

The Thumb instructions

The ARM processor architecture typically uses 32 bit wide instructions. Now bearing in mind it is targeted at portable applications where power consumption is critical, the combination of RISC architectures coupled with 32 bit wide instructions leads to a characteristic known as code expansion. RISC works by simplifying operations into several simple instructions. Each instruction is 32 bits in size and a typical sequence may take three instructions. This means that 12 bytes of program storage are needed to store the instructions. Compare this to a CISC architecture that could do the same work with one instruction with 6 bytes. This means that the RISC solution requires twice as much program storage which means twice the memory cost and power consumption (this is a bit of a simplification but the increase is very significant). And yes, this argument is very similar to that put forward when the first microprocessors appeared where memory was very expensive and it was advantageous to uses as little of it as possible — hence the CISC architectures.

ARM's solution to this was to add a new set of instructions to the instructions set called the Thumb instructions. These are reduced in functionality but are only 16 bits in size and therefore take less space. As the processor always brings in data in 32 bit words, two Thumb instructions are brought in and executed in turn. Thumb instructions are always executed in a special Thumb mode which is controlled by a Thumb bit in the status register. This requires some software support so that the compilers can insure that the Thumb instruction sequences are only executed by the CPU when it is in its Thumb mode, but the benefit is a greatly reduced code size, approaching that offered by CISC processors.

Digital signal processors

Signal processors started out as special processors that were designed for implementing digital signal processing (DSP) algorithms. A good example of a DSP function is the finite impulse response (FIR) filter. This involves setting up two tables, one containing sampled data and the other filter coefficients that determine the filter response. The program then performs a series of repeated multiply and accumulates using values from the tables. The bandwidth of such filters depends on the speed of these simple operations. With a general-purpose architecture like the M68000 family the code structure would involve setting up two tables in external memory, with an address register allocated to each one to act as a pointer. The beginning and the end of the code would consist of the loop initialisation and control, leaving the multiply–accumulate operations for the central part. The M68000 instruction set does offer some facilities for efficient code: the incremental addressing allows pointers to progress down the tables automatically, and the decrement and branch instruction provides a good way of implementing the loop structures. However, the disadvantages are many: the multiply takes >40 clocks, the single bus is used for all the instruction fetches and table searches, thus consuming time and bandwidth. In addition the loop control timings vary depending on whether the branch is taken or not. This can make bandwidth predictions difficult to calculate. This results in very low bandwidths and is therefore of limited use within digital signal processing. This does not mean that an MC68000 cannot perform such functions: it can, providing performance is not of an issue.

RISC architectures like the PowerPC family can offer some immediate improvements. The capability to perform single cycle arithmetic is an obvious advantage. The Harvard architecture reduces the execution time further by allowing simultaneous data and instruction fetches. The PowerPC can, by virtue of its high performance, achieve performances suitable for many DSP applications. The system cost is high involving a multiple chip solution with very fast memory etc. In applications that need high speed general processing as well, it can also be a suitable solution. The ARM 9E processor with its DSP enhanced instructions (essentially speeded up multiply instructions) can also provide DSP levels of performance without the need of a DSP.

Another approach is to build a dedicated processor to perform certain algorithms. By using discrete building blocks, such as hardware multipliers, counters etc., a total hardware solution can be designed to perform such functions. Modulo counters can be used to form the loop structures and so on. The disadvantages are cost and a loss of flexibility. Such hardware solutions are difficult to alter or program. What is obviously required is a processor whose architecture is enhanced specifically for DSP applications.

DSP basic architecture

As an example of a powerful DSP processor, consider the Motorola DSP56000. It is used in many digital audio applications where it acts as a multi-band graphics equaliser or as a noise reduction system.

The processor is split into 10 functional blocks. It is a 24 bit data word processor to give increased resolution. The device has an enhanced Harvard architecture with three separate external buses: one for program and X and Y memories for data. The communication between these and the outside world is controlled by two external bus switches, one for data and the other for addresses. Internally, these two switches are functionally reproduced by the internal data bus switch and the address arithmetic unit (AAU). The AAU contains 24 address registers in three banks of 8. These are used to reference data so that it can be easily fetched to maintain the data flow into the data ALU.

The program address generator, decode controller and interrupt controller organise the instruction flow through the processor. There are six 24 bit registers for controlling loop counts, operating mode, stack manipulation and condition codes. The program counter is 24 bit although the upper 8 bits are only used for sign extension.

The main workhorse is the data ALU, which contains two 56 bit accumulators A and B which each consist of three smaller registers A0, A1, A2, B0, B1 and B2. The 56 bit value is stored with the most significant 24 bit word in A1 or B1, the least significant 24 bit word in A0 or B0 and the 8 bit extension word is stored in A2 or B2. The processor uses a 24 bit word which can provide a dynamic range of some 140 dB, while intermediate 56 bit results can extend this to 330 dB. In practice, the extension byte is used for over- and underflow. In addition there are four 24 bit registers X1, X0, Y1 and Y0. These can also be paired to form two 48 bit registers X and Y.

These registers can read or write data from their respective data buses and are the data sources for the multiply–accumulate (MAC) operation. When the MAC instruction is executed, two 24 bit values from X0, X1, Y1 or Y0 are multiplied together, and then added or subtracted from either accumulator A or B. This takes place in a single machine cycle of 75 ns at 27 MHz. While this is executing, two parallel data moves can take place to update the X and Y registers with the next values. In reality, four separate operations are taking place concurrently.

The data ALU also contains two data shifters for bit manipulation and to provide dynamic scaling of fixed point data without modifying the original program code by simply programming the scaling mode bits. The limiters are used to reduce any arithmetic errors due to overflow, for example. If overflow occurs, i.e. the resultant value requires more bits to describe it than are available, then it is more accurate to write the maximum valid number than the overflowed value. This maximum or limited value is substituted by the data limiter in such cases, and sets a flag in the condition code register to indicate what has happened.

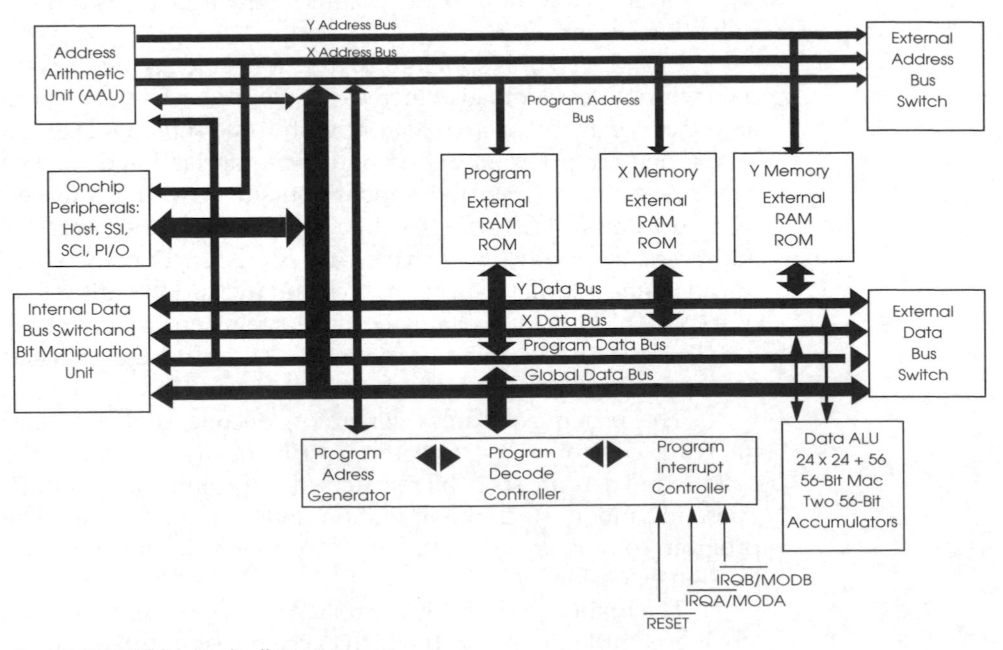

The DSP56000 block diagram

The external signals are split into various groups. There are three ports A, B and C and seven special bus control signals, two interrupt pins, reset, power and ground and, finally, clock signals. The device is very similar in design to an 8 bit microcontroller unit

(MCU), and it can be set into several different memory configurations.

The three independent memory spaces, X data, Y data and program are configured by the MB, MA and DE bits in the operating mode register. The MB and MA bits are set according to the status of the MB and MA pins during the processor´s reset sequence. These pins are subsequently used for external interrupts. Within the program space, the MA and MB bits determine where the program memory is and where the reset starting address is located. The DE bit either effectively enables or disables internal data ROMs which contain a set of μ and A Law expansion tables in the X data ROM and a four quadrant sine wave table in the Y data ROM. The on-chip peripherals are mapped into the X data space between $FFC0 and $FFFF. Each of the three spaces is 64 kbytes in size.

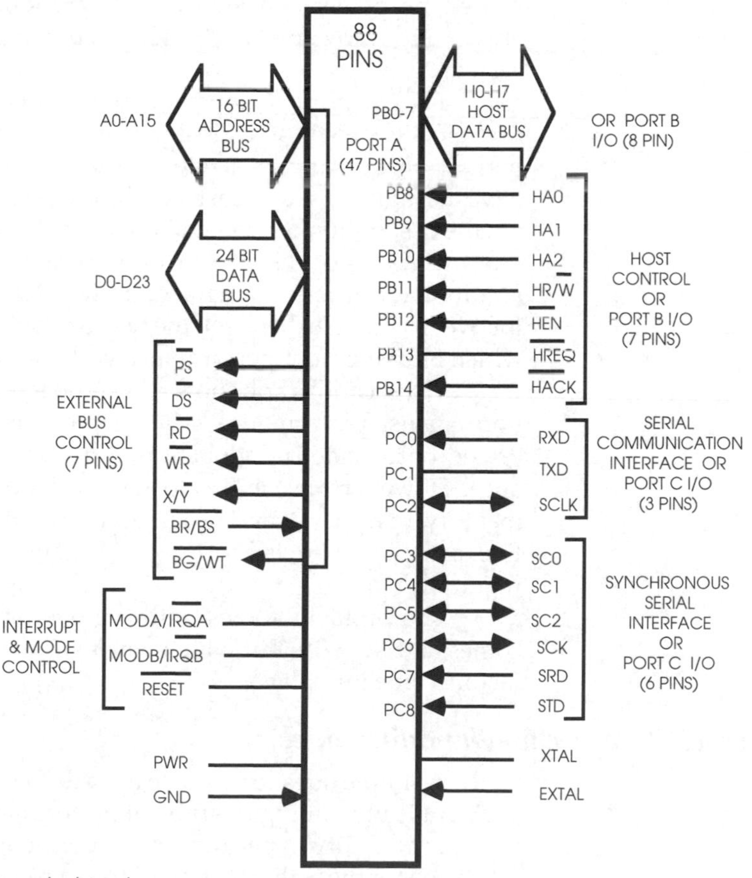

The DSP56000/1 external pinout

These memory spaces communicate to the outside world via a shared 16 bit address bus and a 24 bit data bus. Two additional signals, PS* and X/Y* identify which type of access is taking place. The DSP56000 can be programmed to insert a fixed number of wait states on external accesses for each memory space and I/O. Alternatively, an asynchronous handshake can be adopted by using the bus strobe and wait pins (BS* and WT*).

Using a DSP as a microcontroller is becoming another common trend. The processor has memory and peripherals which makes it look like a microcontroller — albeit one with a very fast processing capability and slightly different programming techniques. This, coupled with the increasing need for some form of DSP function such as filtering in many embedded systems, has meant that DSP controllers are a feasible choice for embedded designs.

Choosing a processor

So far in this chapter, the main processor types used in embedded systems along with various examples have been discussed. There are very many types available ranging in cost, processing power and levels of integration. The question then arises concerning how do you select a processor for an embedded system?

The two graphs show the major trends with processors. The first plots system cost against performance. It shows that for the highest performance discrete processors are needed and these have the highest system cost. For the lowest cost, microcontrollers are the best option but they do not offer the level of performance that integrated or discrete processors offer. Many use the 8 bit accumulator processor architecture which has now been around for over 20 years. In between the two are the integrated processors which offer medium performance with medium system cost.

The second graph shows the trend towards system integration against performance. Microcontrollers are the most integrated, but as stated previously, they do not offer the best performance. However, the ability to pack a whole system including memory, peripherals and processor into a single package is attractive, provided there is enough performance to perform the work required.

The problem comes with the overlap areas where it becomes hard to work out which way to move. This is where other factors come into play.

Does it have enough performance?

A simple question to pose but a difficult one to answer. The problem is in defining the type of performance that is needed. It may be the ability to perform integer or floating point arithmetic operations or the ability to move data from one location to another. Another option may be the interrupt response time to allow data to be collected or processed.

The problem is that unless the end system is understood it is difficult to know exactly how much performance is needed. Add to this the uncertainty in the software programming overhead, i.e. the performance loss in using a high level language compared to a low level assembler, and it is easy to see why the answer is not straightforward.

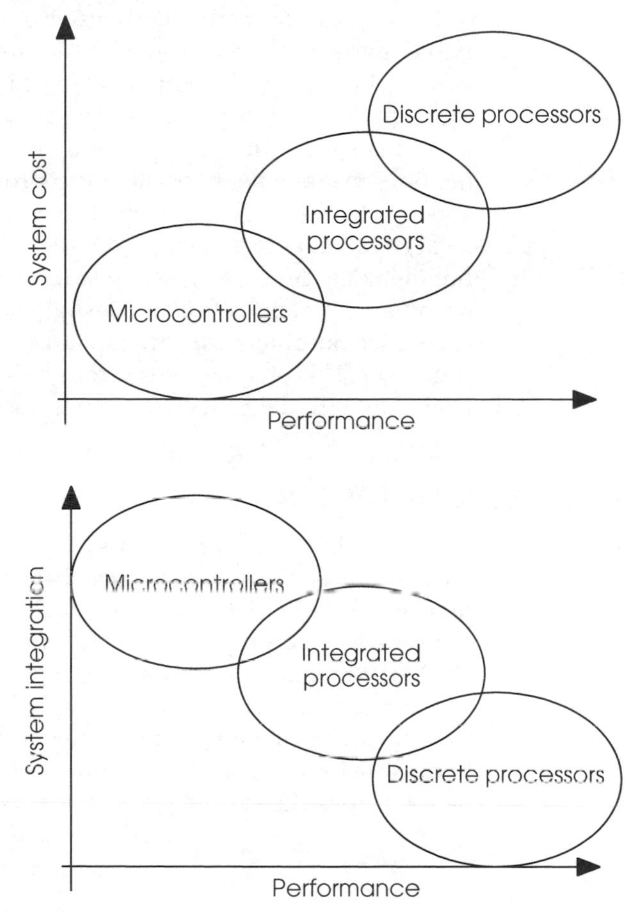

Processor selection graphs

In practice, most paper designs assume that about 20–40% of the processor performance will be lost to overheads in terms of MIPs and processing. Interrupt latencies can be calculated to give more accurate figures but as will be explained in Chapter 7, this has its own set of problems and issues to consider.

This topic of selecting and configuring a processor is discussed in many of the design notes and tutorials at the end of this book.

3 Memory systems

Within any embedded system, memory is an important part of the design, and faced with the vast variety of memory that is available today, choosing and selecting the right type for the right application is of paramount importance. Today's designs use more than just different types of memory and will include both memory management and memory protection units to partition and isolate the memory system. Memory caches are used to keep local copies of data and code so that it is accessed faster and does not delay the processor. As a result, the memory subsystem has become extremely complex. Designs only seen on mainframes and supercomputers are now appearing on the humble embedded processor. This chapter goes through the different types that are available and discusses the issues associated with them that influence the design.

Memory technologies

Within any embedded system design that uses external memory, it is almost a sure bet that the system will contain a mixture of non-volatile memory such as EPROM (erasable programmable read only memory) to store the system software and DRAM (dynamic random access memory) for use as data and additional program storage. With very fast systems, SRAM (static random access memory) is often used as a replacement for DRAM because of its faster speed or within cache memory subsystems to help improve the system speed offered by DRAM.

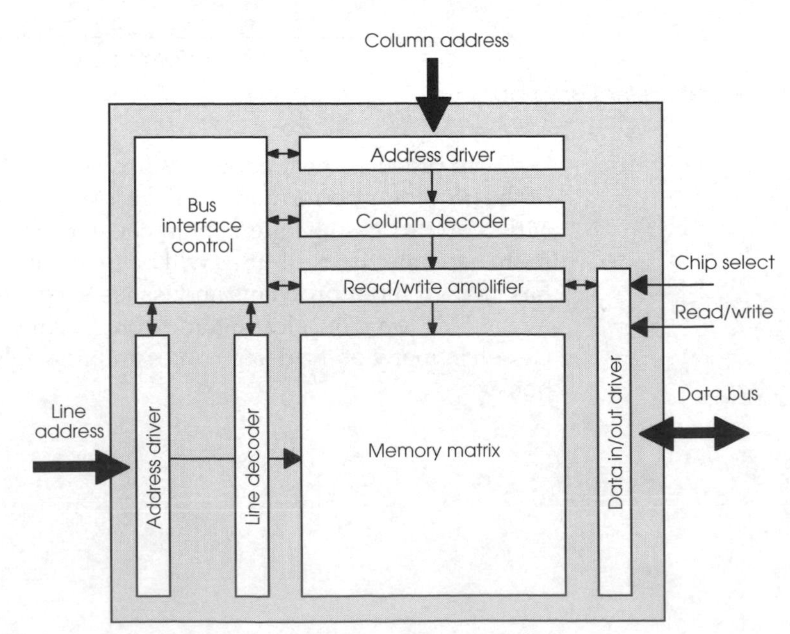

RAM block diagram

The main signals used with memory chips fall into several groups:

- Address bus

  The address bus is used to select the particular location within the memory chip. The signals may be multiplexed as in the case with DRAM or non-multiplexed as with SRAM.

- Data bus

  This bus provides the data to and from the chip. In some cases, the memory chip will use separate pins for incoming and outgoing data, but in others a single set of pins is used with the data direction controlled by the status of chip select signals, the read / write pin and output enable pins.

- Chip selects

  These can be considered as additional address pins that are used to select a specific chip within an array of memory devices. The address signals that are used for the chip selects are normally the higher order pins. In the example shown, the address decode logic has enabled the chip select for the second RAM chip — as shown by the black arrow — and it is therefore the only chip driving the data bus and supplying the data. As a result, each RAM chip is located in its own space within the memory map although it shares the same address bus signals with all the other RAM chips in the array.

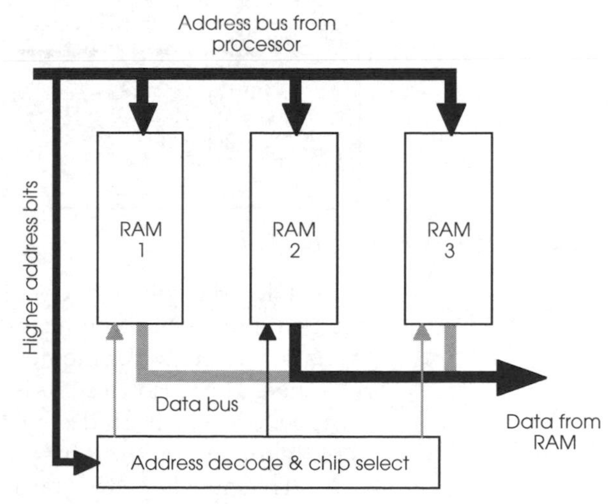

Address decode and chip select generation

- Control signals including read / write signals

  Depending on the functionality provided by the memory device, there are often additional control signals. Random access memory will have a read / write signal to indicate the type of access. This is missing from read only devices such as EPROM. For devices that have multiplexed address buses, as in the case with DRAM, there are control signals associated with this type of operation.

There are now several different types of semiconductor memory available which use different storage methods and have different interfaces.

DRAM technology

DRAM is the predominantly used memory technology for PCs and embedded systems where large amounts of low cost memory are needed. With most memory technologies, the cost per bit is dependent on two factors: the number of transistors that are used to store each bit of data and the type of package that is used. DRAM achieves its higher density and lower cost because it only uses a single transistor cell to store each bit of data. The data storage element is actually a small capacitor whose voltage represents a binary zero or one which is buffered by the transistor. In comparison, a SRAM cell contains at least four or five transistors to store a single bit of data and does not use a capacitor as the active storage element. Instead, the transistors are arranged to form a flip-flop logic gate which can be flipped from one binary state to the other to store a binary bit.

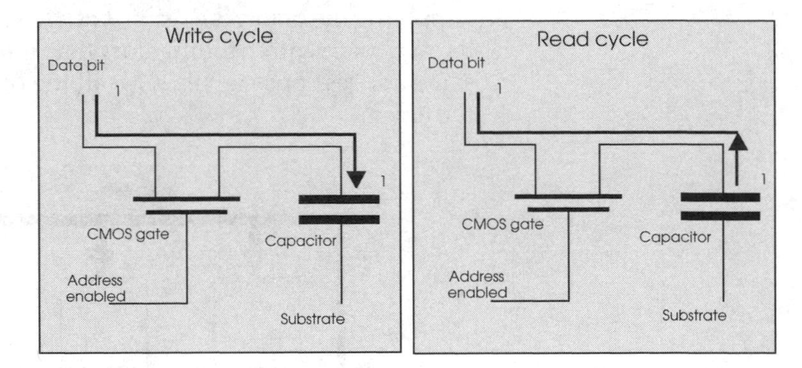

DRAM cell read and write cycles

DRAM technology does have its drawbacks with the major one being its need to be refreshed on a regular basis. The term 'dynamic' refers to the memory's constant need for its data to be refreshed. The reason for this is that each bit of data is stored using a capacitor, which gradually loses its charge. Unless it is frequently topped up (or refreshed), the data disappears.

This may appear to be a stupid type of memory — but the advantage it offers is simple — it takes only one transistor to store a bit of data whereas static memory takes four or five. The memory chip's capacity is dependent on the number of transistors that can be fabricated on the silicon and so DRAM offers about four times the storage capacity of SRAM (static RAM). The refresh overhead takes about 3–4% of the theoretical maximum processing available and is a small price to pay for the larger storage capacity. The refresh is performed automatically either by a hardware controller or through the use of software. These techniques will be described in more detail later on in this chapter.

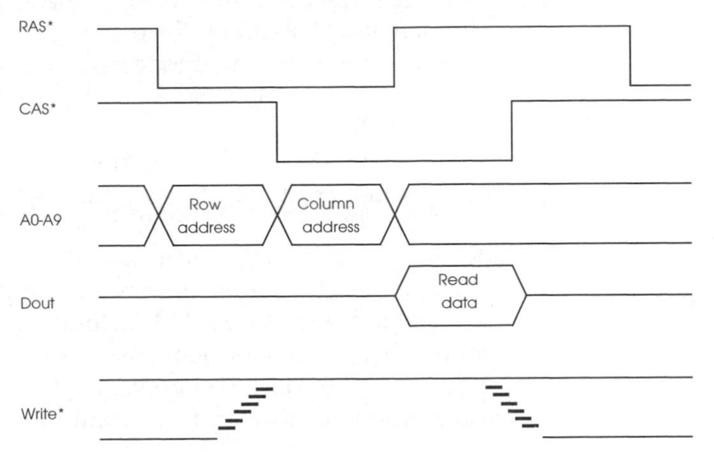

Basic DRAM interface

The basic DRAM interface takes the processor generated address, places half of the address (the high order bits) onto the memory address bus to form the row address and asserts the RAS* signal. This partial address is latched internally by the DRAM. The remaining half (the low order bits), forming the column address, are then driven onto the bus and the CAS* signal asserted. After the access time has expired, the data appears on the Dout pin and is latched by the processor. The RAS* and CAS* signals are then negated. This cycle is repeated for every access. The majority of DRAM specifications define minimum pulse widths for the RAS* and CAS* and these often form the major part in defining the memory access time. When access times are quoted, they usually refer to the time from the assertion of the RAS* signal to the appearance of the data. There are several variations on this type of interface, such as page mode and EDO. These will be explained later on in this chapter

Video RAM

A derivative of DRAM is the VRAM (video RAM), which is essentially a DRAM with the ability for the processor to update its contents at the same time as the video hardware uses the data to create the display. This is typically done by adding a large shift register to a normal DRAM. This register can be loaded with a row or larger amounts of data which can then be serially clocked out to the video display. This operation is in parallel with normal read/ write operations using a standard address/data interface.

SRAM

SRAM does not need to be refreshed and will retain data indefinitely — as long as it is powered up. In addition it can be designed to support low power operation and is often used in preference to DRAM for this reason. Although the SRAM cell contains more transistors, the cell only uses power when it is being switched. If the cell is not accessed then the quiescent current is extremely low. DRAM on the other hand has to be refreshed by external bus accesses and these consume a lot of power. As a result, the DRAM memory will have a far higher quiescent current than that of SRAM.

The SRAM memory interface is far simpler than that of DRAM and consists of a non-multiplexed address bus and data bus. There is normally a chip select pin which is driven from other address pins to select a particular SRAM when they are used in banks to provide a larger amount of storage.

Typical uses for SRAM include building cache memories for very fast processors, being used as main memory in portable equipment where its lower power consumption is important and as expansion memory for microcontrollers.

Pseudo-static RAM

Pseudo-static RAM is a memory chip that uses DRAM cells to provide a higher memory density but has the refresh control built into the chip and therefore acts like a static RAM. It has been used in portable PCs as an alternative to SRAM because of its low cost. It is not as common as it used to be due to the drop in cost of SRAM and the lower power modes that current synchronous DRAM technology offers.

Battery backed-up SRAM

The low power consumption of SRAM makes it suitable for conversion into non-volatile memories, i.e. memory that does not lose its data when the main power is removed by adding a small battery to provide power at all times. With the low quiescent current often being less than the battery's own leakage current, the SRAM can be treated as a non-volatile RAM for the duration of the battery's life. The CMOS (complementary metal oxide semiconductor) memory used by the MAC and IBM PC, which contains the configuration data, is SRAM. It is battery backed-up to ensure it is powered up while the computer is switched off.

Some microcontrollers with on-chip SRAM support the connection of an external battery to backup the SRAM contents when the main power is removed.

EPROM and OTP

EPROM is used to store information such as programs and data that must be retained when the system is switched off. It is used within PCs to store the Toolbox and BIOS routines and power on software in the MAC and IBM PC that is executed when the computer is switched on. These devices are read only and cannot be written to, although they can be erased by ultraviolet (UV) light and have a transparent window in their case for this purpose. This window is usually covered with a label to prevent accidental erasure, although it takes 15–30 minutes of intense exposure to do so.

There is a different packaged version of EPROM called OTP (one time programmable) which is an EPROM device packaged in a low cost plastic package. It can be programmed once only because there is no window to allow UV light to erase the EPROM inside. These are becoming very popular for small production runs.

Flash

Flash memory is a non-volatile memory which is electrically erasable and offers access times and densities similar to that of DRAM. It uses a single transistor as a storage cell and by placing a higher enough charge to punch through an oxide layer, the transistor cell can be programmed. This type of write operation can take several milliseconds compared to sub 100 ns for DRAM or faster for SRAM. Reads are of the order of 70–100 ns.

FLASH has been positioned and is gaining ground as a replacement for EPROMs but it has not succeeded in replacing hard disk drives as a general-purpose form of mass storage. This is due to the strides in disk drive technology and the relatively slow write access time and the wearout mechanism which limits the number of writes that can be performed. Having said this, it is frequently used in embedded systems that may need remote software updating. A good example of this is with modems where the embedded software is stored in FLASH and can be upgraded by downloading the new version via the Internet or bulletin board using the modem itself. Once downloaded, the new version can be transferred from the PC to the modem via the serial link.

EEPROM

Electrically erasable programmable read only memory is another non-volatile memory technology that is erased by applying a suitable electrical voltage to the device. These types of memory do not have a window to allow UV light in to erase them and thus offer the benefits of plastic packaging, i.e. low cost with the ability to erase and reprogram many times.

The erase/write cycle is slow and can typically only be performed on large blocks of memory instead of at the bit or byte level. The erase voltage is often generated internally by a charge pump but can be supplied externally. The write cycles do have a wearout mechanism and therefore the memory may only be guaranteed for a few hundred thousand erase/write cycles and this, coupled with the slow access time, means that they are not a direct replacement for DRAM.

Memory organisation

A memory's organisation refers to how the data is arranged within the memory chips and within the array of chips that is used to form the system memory. An individual memory's storage is measured in bits but can be organised in several different ways. A 1 Mbit memory can be available as a 1 Mbit × 1 device, where there is only a single data line and eight are needed in parallel to store one byte of data. Alternatives are the 256 kbits × 4, where there are four data lines and only two are needed to store a byte, and 128 kbit × 8, which has 8 data lines. The importance of these different organisations becomes apparent when upgrading memory and determining how many chips are needed.

The minimum number of chips that can be used is determined by the width of the data path from the processor and the number of data lines the memory chip has. For an MC68000 processor with a 16 bit wide data path, 16 × 1 devices, 4 × 4 or 2 × 8 devices would be needed. For a 32 bit processor, like the MC68020, MC68030, MC68040, 80386DX or 80486, this figure doubles. What is interesting is that the wider the individual memory chip's data storage, the smaller the number of chips that is required to upgrade. This does not mean that, for a given amount of memory, less × 4 and × 8 chips are needed when compared with × 1 devices, but that each minimum upgrade can be smaller, use fewer chips and be less expensive. With a 32 bit processor and using 1 Mbit × 1 devices, the minimum upgrade would need 32 chips and add 32 Mbytes. With a × 4 device, the minimum upgrade would only need 8 chips and add 8 Mbytes.

This is becoming a major problem as memories become denser and the smaller size chips are discontinued. This poses problems to designers that need to design some level of upgrade capability to cater for the possible — some would say inevitable — need for more memory to store the software. With the ***smallest*** DRAM chip that is still in production being a 16 Mbit device and the likelihood that this will be replaced by 64 and 128 Mbit devices in the not so distant future, the need for one additional byte could result in the addition of 8 or 16 Mbytes or memory. More importantly, if a × 1 organisation is used, then this means that an additional 8 chips are needed. By using a wider organisation, the number of chips is reduced. This is becoming a major issue and is placing a lot of pressure on designers to keep the memory budget under control. The cost of going over is becoming more and more expensive. With cheap memory, this could be argued as not being an issue but there is still the space and additional cost. Even a few cents multiplied by large production volumes can lead to large increases.

By 1 organisation

Today, single-bit memories are not as useful as they used to be and their use is in decline compared to wider data path devices. Their use is restricted to applications that need non-standard width memory arrays that these type of machines use, e.g. 12 bit, 17 bit etc. They are still used to provide a parity bit and can be found on SIMM memory modules but as systems move away from implementing parity memory — many PC motherboards no longer do so — the need for such devices will decline.

By 4 organisation

This configuration has effectively replaced the × 1 memory in microprocessor applications because of its reduced address bus loading and complexity — only 8 chips are needed to build a 32 bit wide data path instead of 32 and only two are needed for an 8 bit wide bus.

By 8 and by 9 organisations

Wider memories such as the × 8 and × 9 are beginning to replace the × 4 parts in many applications. Apart from higher integration, there are further reductions in address bus capacitance to build a 32 or 64 bit wide memory array. The reduction in bus loading can improve the overall access time by greatly reducing the address setup and stabilisation time, thus allowing more time within the memory cycle to access the data from the memories. This improvement can either reduce costs by using slower and cheaper memory, or allow a system to run faster given a specific memory part. The × 9 variant provides a ninth bit for parity protection. For microcontrollers, these parts allow memory to be increased in smaller increments.

By 16 and greater organisations

Wider memories with support for 16 bits or wider memory are already appearing but it is likely that they will integrate more of the interface logic so that the time consumed by latches and buffers during the memory access will be removed, thus allowing slower parts to be used in wait state-free designs.

Parity

The term parity has been mentioned in the previous paragraphs along with statements that certainly within the PC industry it is no longer mandatory and the trend is moving away from its implementations. Parity protection is an additional bit of memory which is used to detect single-bit errors with a block of memory. Typically, one parity bit is used per byte of data. The bit is set to a one or a zero depending on the number of bits that are set to one within the data byte. If this number is odd, the parity bit is set to a one and if the number is even, it is set to zero. This can be reversed to provide two parity schemes known as odd and even parity.

If a bit is changed within the word through an error or fault, then the parity bit will no longer be correct and a comparison of the parity bit and the calculated parity value from the newly read data will disagree. This can then be used to flag a signal back to the processor, such as an error. Note that parity does not allow the error to be corrected nor does it protect from all multiple bit failures such as two set or cleared bits failing together. In addition it requires a parity controller to calculate the value of the parity bit on write cycles and calculate and compare on read cycles. This additional work can slow down memory access and thus the processor performance.

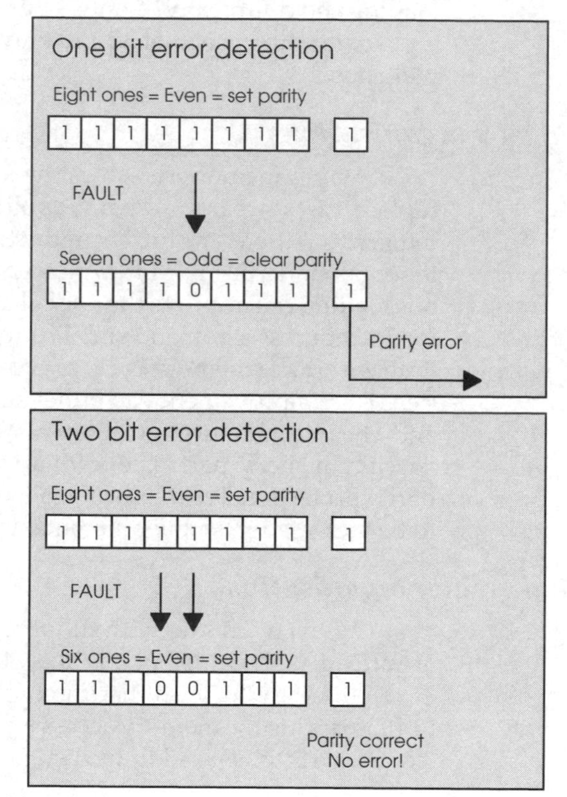

Parity detection for one and two bit errors

However, for critical embedded systems it is important to know if there has been a memory fault and parity protection may be a requirement.

Parity initialisation

If parity is used, then it may be necessary for software routines to write to each memory location to clear and/or set up the parity hardware. If this is not done, then it is possible to generate false parity errors.

Error detecting and correcting memory

With systems that need very high reliability, it is possible through increasing the number of additional bits per byte and by using special coding techniques to increase the protection offered by parity protection. There are two types of memory design that do this:

- Error detecting memory

  With this type of memory, additional bits are added to the data word to provide protection from multiple bit failures. Depending on the number of bits that are used and the coding techniques that use the additional bits, protection can be provided for a larger number of error conditions. The disadvantages are the additional memory bits needed along with the complex controllers required to create and compare the codes.

- Error detecting and correction

  This takes the previous protection one step further and uses the codes not only to detect the error but correct it as well. This means that the system will carry on despite the error whereas the previous scheme would require the system to be shut down as it could not rely on the data. EDC systems, as they are known, are expensive but offer the best protection against memory errors.

Access times

As well as different sizes and organisations, memory chips have different access times. The access time is the maximum time taken by the chip to read or write data and it is important to match the access time to the design. (It usually forms part of the part number: MCM51000AP10 would be a 100 ns access time memory and MCM51000AP80 would be an 80 ns version.) If the chip is too slow, the data that the processor sees will be invalid and corrupt, resulting in software problems and crashes. Some designs allow memories of different speed to be used by inserting wait states between the processor and memory so that sufficient time is given to allow the correct data to be obtained. These often require jumper settings to be changed or special setup software to be run, and depend on the manufacture and design of the board.

If a processor clock speed is increased, the maximum memory access time must be reduced — so changing to a faster processor may require these settings to be modified. This is becoming less of a problem with the advent of decoupled processors where the CPU speed can be set as a ratio of the bus speed and by changing an initialisation routine, a faster CPU can be used with the same external bus as a slower one. This is similar to the overdrive processors that clock the internal CPU either 22 or 4 times faster. They are using the same trick except that there is no additional software change needed.

Packages

The major option with memories is packaging. Some come encapsulated in plastic, some in a ceramic shell and so on. There are many different types of package options available and, obviously, the package must match the sockets on the board. Of the many different types, four are commonly used: the dual in line package, zig–zag package, SIMM and SIP. The most common package encountered with the MAC is the SIMM, although all the others are used, especially with third party products.

Dual in line package

This package style, as its name implies, consists of two lines of legs either side of a plastic or ceramic body. It is the most commonly used package for the BIOS EPROMs, DRAM and SRAM. It is available in a variety of sizes with 24, 26 and 28 pin packages used for EPROMs and SRAMs and 18 and 20 pin packages for 1 Mbit × 1 and 256 kbit × 4 DRAMs. However, it has virtually been replaced by the use of SIMM modules and is now only used for DRAM on the original MAC 128K and 512K models and for DRAM and for EPROM on models up to the MAC IIx.

Zig–zag package

This is a plastic package used primarily for DRAM. Instead of coming out of the sides of the package, the leads protrude from the pattern and are arranged in a zig-zag — hence the name. This type of package can be more difficult to obtain, compared with the dual in line devices, and can therefore be a little more expensive. This format is often used on third party boards.

SIMM and DIMM

SIMM is not strictly a package but a subassembly. It is a small board with finger connection on the bottom and sufficient memory chips on board to make up the required configuration, such as 256 Kbit × 8 or × 9, 1 Mbit × 8 or × 9, 4 Mbit, and so on. SIMMs have rapidly gained favour and many new designs use these boards instead of individual memory chips. They require special sockets, which can be a little fragile and need to be handled correctly. There are currently two types used for PCs: the older 30 pin SIMM which uses an 8 or 9 bit (8 bits plus a parity bit) data bus and a more recent 72 pin SIMM which has a 32 or 36 bit wide data bus. The 36 bit version is 32 bits data plus four parity bits. Apple has used both types and a third which has 64 pins but like the IBM PC world standardised on the 72 pin variety which suited the 32 bit processors at the time. With the advent of wider bus CPUs, yet another variation appeared called the DIMM. This typically has 168 bits but looks like a larger version of the SIMM. With the wider buses came an increase in memory speeds and a change in the supply voltages. One method of getting faster speeds and reduced power consumption is to reduce the supply voltage. Instead of the signal levels going from 0 to 5 volts, today's CPUs and corresponding memories use a 3.3 volt supply (or even lower). As a result, DIMMs are now described by the speed, memory type and voltage supply e.g. 3.3 volt 133 MHz SDRAM DIMM.

30 and 72 pin SIMMs

The older 30 pin SIMMs are normally used in pairs for a 16 bit processor bus (80386SX, MC68000) and in fours for 32 bit buses (80386DX, 80486, MC68030, MC68040) while the 72 pin SIMMs are normally added singly although some higher performance boards need a pair of 72 pin SIMMs to support bank switching.

SIP

This is the same idea as SIMM, except that the finger connections are replaced by a single row of pins. SIP has been overtaken by SIMM in terms of popularity and is now rarely seen.

DRAM interfaces

The basic DRAM interface

The basic DRAM interface takes the processor generated address, places the high order bits onto the memory address bus to form the row address and asserts the RAS* signal. This partial address is latched internally by the DRAM. The remaining low order bits, forming the column address, are then driven onto the bus and the CAS* signal asserted. After the access time has expired, the data appears on the Dout pin and is latched by the processor. The RAS* and CAS* signals are then negated. This cycle is repeated for every access. The majority of DRAM specifications define minimum pulse widths for the RAS* and CAS* and these often form the major part in defining the memory access time. To remain compatible with the PC–AT standard, memory refresh is performed every 15 microseconds.

This direct access method limits wait state-free operation to the lower processor speeds. DRAM with 100 ns access time would only allow a 12.5 MHz processor to run with zero wait states. To achieve 20 MHz operation needs 40 ns DRAM, which is unavailable today, or fast static RAM which is at a price. Fortunately, the embedded system designer has more tricks up his sleeve to improve DRAM performance for systems, with or without cache.

Page mode operation

One way of reducing the effective access time is to remove the RAS* pulse width every time the DRAM was accessed. It needs to be pulsed on the first access, but subsequent accesses to the same page (i.e. with the same row address) would not require it and so are accessed faster. This is how the 'page mode' versions of most 256 kb, 1 Mb and 4 Mb memory work. In page mode, the row address is supplied as normal but the RAS* signal is left asserted. This selects an internal page of memory within the DRAM where any bit of data can be accessed by placing the column address and asserting CAS*. With 256 kb size memory, this gives a page of 1 kbyte (512 column bits per DRAM row with 16 DRAMs in the array). A 2 kbyte page is available from 1 Mb DRAM and a 4 kbyte page with 4 Mb DRAM.

This allows fast processors to work with slower memory and yet achieve almost wait state-free operation. The first access is slower and causes wait states but subsequent accesses within the selected page are quicker with no wait states.

However, there is one restriction. The maximum time that the RAS* signal can be asserted during page mode operation is often specified at about 10 microseconds. In non-PC designs, the refresh interval is frequently adjusted to match this time, so a refresh cycle will always occur and prevents a specification violation. With the PC standard of 15 microseconds, this is not possible. Many chip sets neatly resolve the situation by using an internal counter which times out page mode access after 10 microseconds.

Page interleaving

Using a page mode design only provides greater performance when the memory cycles exhibit some form of locality, i.e. stay within the page boundary. Every access outside the boundary causes a page miss and two or three wait states. The secret, as with caches, is to increase the hits and reduce the misses. Fortunately, most accesses are sequential or localised, as in program subroutines and some data structures. However, if a program is frequently accessing data, the memory activity often follows a code–data–code–data access pattern. If the code areas and data areas are in different pages, any benefit that page mode could offer is lost. Each access changes the page selection, incurring wait states. The solution is to increase the number of pages available. If the memory is divided into several banks, each bank can offer a selected page, increasing the number of pages and, ultimately, the number of hits and performance. Again, extensive hardware support is needed and is frequently provided by the PC chip set.

Page interleaving is usually implemented as a one, two or four way system, depending on how much memory is installed. With a four way system, there are four memory banks, each with their own RAS* and CAS* lines. With 4 Mbyte DRAM, this would offer 16 Mbytes of system RAM. The four way system allows four pages to be selected within page mode at any one time. Page 0 is in bank 1, page 1 in bank 2, and so on, with the sequence restarting after four banks.

With interleaving and Fast Page Mode devices, inexpensive 85 ns DRAM can be used with a 16 MHz processor to achieve a 0.4 wait state system. With no page mode interleaving, this system would insert two wait states on every access. With the promise of faster DRAM, future systems will be able to offer 33–50 MHz with very good performance — without the need for cache memory and its associated costs and complexity.

Burst mode operation

Some versions of the DRAM chip, such as page mode, static column or nibble mode devices, do not need to have the RAS / CAS cycle repeated and can provide data much faster if only the new column address is given. This has allowed the use of a burst fill memory interface, where the processor fetches more data than it needs and keeps the extra data in an internal cache ready for future use. The main advantage of this system is in reducing the need for fast static RAMs to realise the processor's performance. With 60 ns page mode DRAM, a 4-1-1-1 (four clocks for the first access, single cycle for the remaining burst) memory system can easily be built. Each 128 bits of data fetched in such a way takes only seven clock cycles, compared with five in the fastest possible system. If bursting was not supported, the same access would take 16 clocks. This translates to a very effective price performance — a 4-1-1-1 DRAM system gives about 90% of the performance of a more expensive 2-1-1-1 static RAM design. This interface is used on the higher performance processors where it is used in conjunction with on-chip caches. The burst fill is used to load a complete line of data within the cache.

This allows fast processors to work with slower memory and yet achieve almost wait state-free operation. The first access is slower and causes wait states but subsequent accesses within the selected page are quicker with no wait states.

EDO memory

EDO stands for extended data out memory and is a form of fast page mode RAM that has a quicker cycling process and thus faster page mode access. This removes wait states and thus improves the overall performance of the system. The improvement is achieved by fine tuning the CAS* operation.

With fast page mode when the RAS* signal is still asserted, each time the CAS* signal goes high the data outputs stop asserting the data bus and go into a high impedance mode. This is used to simplify the design by using this transistion to meet the timing requirements. It is common with this type of design to permanently ground the output enable pin. The problem is that this requires the CAS* signal to be asserted until the data from the

DRAM is latched by the processor or bus master. This means that the next access cannot be started until this has been completed, causing delays.

EDO memory does not cause the outputs to go to high impedance and it will continue to drive data even if the CAS* signal is removed. By doing this, the CAS* precharge can be started for the next access while the data from the previous access is being latched. This saves valuable nanoseconds and can mean the removal of a wait state. With very high performance processors, this is a big advantage and EDO type DRAM is becoming the *de facto* standard for PCs and workstations or any other application that needs high performance memory.

DRAM refresh techniques

DRAM needs to be periodically refreshed and to do this there are several methods that can be used. The basic technique involves accessing the DRAM using a special refresh cycle. During these refresh cycles, no other access is permitted. The whole chip must be refreshed within a certain time period or its data will be lost. This time period is known as the refresh time. The number of accesses needed to complete the refresh is known as the number of cycles and this number divided into the refresh time gives the refresh rate. There are two refresh rates in common use: standard, which is 15.6 μs and extended, which is 125 μs. Each refresh cycle is approximately twice the length of a normal access — a 70 ns DRAM typically has a refresh cycle time of 130 ns — and this times the number of cycles gives the total amount of time lost in the refresh time to refresh. This figure is typically 3–4% of the refresh time. During this period, the memory is not accessible and thus any processor will have to wait for its data. This raises some interesting potential timing problems.

Distributed versus burst refresh

With a real-time embedded system, the time lost to refresh must be accounted for. However, its effect is dependent on the method chosen to perform all the refresh cycles within the refresh time. A 4 M by 1 DRAM requires 1024 refresh cycles. Are these cycles executed in a burst all at once or should they be distributed across the whole time? Bursting means that the worst case delay is 1024 times larger than that of a single refresh cycle that would be encountered in a distributed system. This delay is of the order of 0.2 ms, a not inconsiderable time for many embedded systems! The distributed worst case delay due to refresh is about 170 ns.

Most systems use the distributed method and depending on the size of time critical code, calculate the number of refresh cycles that are likely to be encountered and use that to estimate the delay caused by refresh cycles. It should be remembered that in both cases, the time and access overhead for refresh is the same.

Software refresh

It is possible to use software to perform the refresh by using a special routine to periodically circle through the memory and thus cause its refresh. Typically a timer is programmed to generate an interrupt. The interrupt handler would then perform the refresh. The problem with this arrangement is that any delay in performing the refresh potentially places the whole memory and its contents at risk. If the processor is stopped or single stepped, its interrupts disabled or similar, the refresh is halted and the memory contents lost. The disadvantage in this is that it makes debugging such a system extremely difficult. Many of the debugging techniques cannot be used because they stop the refresh. If the processor crashes, the refresh is stopped and the contents are lost.

There have been some neat applications where software refresh is used. The Apple II personal computer used a special memory configuration so that every time the DRAM blocks that were used for video memory were accessed to update the screen, they effectively refreshed the DRAM.

RAS only refresh

With this method, the row address is placed on the address bus, RAS* is asserted but CAS* is held off. This generates the recycle address. The address generation is normally done by an external hardware controller, although many early controllers required some software assistance. The addressing order is not important but what is essential is that all the rows are refreshed within the refresh time.

CAS before RAS (CBR) refresh

This is a later refresh technique that is now commonly used. It has lower power consumption because it does not use the address bus and the buffers can be switched off. It works by using an internal address counter stored on the memory chip itself which is periodically incremented. Each incrementation starts a refresh cycle internally. The mechanism works as its name suggests by asserting CAS* before RAS*. Each time that RAS* is asserted, a refresh cycle is performed and the internal counter incremented.

Hidden refresh

This is a technique where a refresh cycle is added to the end of a normal read cycle. The term hidden refers to the fact that the refresh cycle is hidden in a normal read and not to any hiding of the refresh timing. It does not matter which technique you use, refresh will still cost time and performance! What happens is that the RAS* signal goes high and is then asserted low. This happens at the end of the read cycle when the CAS* signal is still asserted. This is a similar situation to the CBR method. Like it, this toggling of the RAS* signal at the end of the read cycle starts a CBR refresh cycle internally.

Memory management

Memory management used to be the preserve of workstations and PCs where it is used to help control and manage the resources within the system. It inevitably caused memory access delays and extra cost and because of this, was rarely used in embedded systems. Another reason was that many of the real-time operating systems did not support it and without the software support, there seemed little need to have it within the system. While some form of memory management can be done in software, memory management is usually implemented with additional hardware called a MMU (memory management unit) to meet at least one of four system requirements:

1. **The need to extend the current addressing range.**

The often perceived need for memory management is usually the result of prior experience or background, and centres on extending the current linear addressing range. The Intel 80x86 architecture is based around a 64 kbyte linear addressing segment which, while providing 8 bit compatibility, does require memory management to provide the higher order address bits necessary to extend the processor's address space. Software must track accesses that go beyond this segment, and change the address accordingly. The M68000 family has at least a 16 Mbyte addressing range and does not have this restriction. The PowerPC family has an even larger 4 Gbyte range. The DSP56000 has a 128 kword (1 word = 24 bits) address space, which is sufficient for most present day applications, however, the intermittent delays that occur in servicing an MMU can easily destroy the accuracy of the algorithms. For this reason, the linear addressing range may increase, but it is unlikely that paged or segmented addressing will appear in DSP applications.

2. **To remove the need to write relocatable or position-independent software.**

Many systems have multitasking operating systems where the software environment consists of modular blocks of code running under the control of an operating system. There are three ways of allocating memory to these blocks. The first simply distributes blocks in a pre-defined way, i.e. task A is given the memory block from $A0000 to $A8000, task B is given from $C0000 to $ D8000, etc. With these addresses, the programmer can write the code to use this memory. This is fine, providing the distribution does not change and there is sufficient design discipline to adhere to the plan. However, it does make all the code hardware and position dependent. If another system has a slightly different memory configuration, the code will not run correctly.

To overcome this problem, software can be written in such a way that it is either relocatable or position independent. These two terms are often interchanged but there is a difference: both can execute anywhere in the memory map, but relocatable code must maintain the same address offsets between its data and code segments. The main technique is to avoid the use of absolute addressing modes, replacing them with relative addressing modes.

If this support is missing or the compiler technology cannot use it, memory management must be used to translate the logical program addresses and map them into physical memory. This effectively realigns the memory so that the processor and software think that the memory is organized specially for them, but in reality is totally different.

3. To partition the system to protect it from other tasks, users, etc.

To provide stability within a multitasking or multiuser system, it is advisable to partition the memory so that errors within one task do not corrupt others. On a more general level, operating system resources may need separating from applications. The M68000 processor family can provide this partitioning through the use of the function codes or by the combination of the user / supervisor signals and Harvard architecture. This partitioning is very coarse, but is often all that is necessary in many cases. For finer grain protection, memory management can be used to add extra description bits to an address to declare its status. If a task attempts to access memory that has not been allocated to it, or its status does not match (e.g. writing to a read only declared memory location), the MMU can detect it and raise an error to the supervisor level. This aspect is becoming more important and has even sprurred manufacturers to define stripped down MMUs to provide this type of protection.

4. To allow programs to access more memory than is physically present in the system.

With the large linear addressing offered by today´s 32 bit microprocessors, it is relatively easy to create large software applications which consume vast quantities of memory. While it may be feasible to install 64 Mbytes of RAM in a workstation, the costs are expensive compared with a 64 Mbyte winchester disk. As the memory needs go up, this differential increases. A solution is to use the disk storage as the main storage medium, divide the stored program into small blocks and keep only the blocks in the processor system memory that are needed.

As the program executes, the MMU can track how the program uses the blocks, and swap them to and from the disk as needed. If a block is not present in memory, this causes a page fault and forces some exception processing which performs the swapping operation. In this way, the system appears to have large amounts of system RAM when, in reality, it does not. This virtual memory technique is frequently used in workstations and in the UNIX operating system.

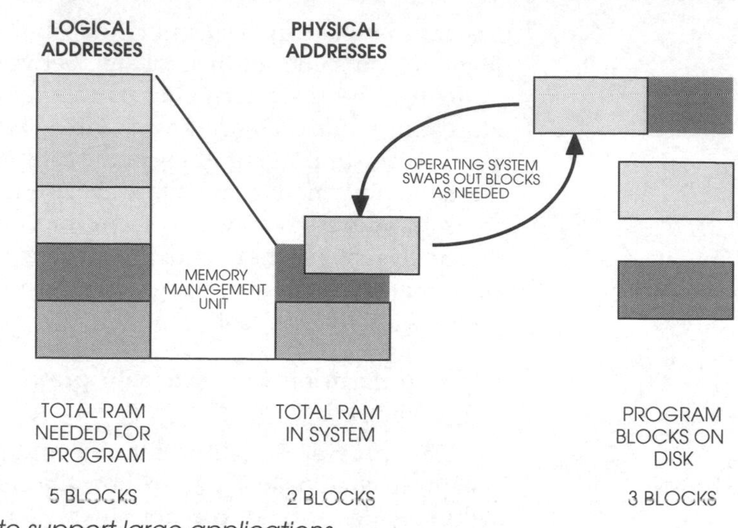

Using virtual memory to support large applications

Disadvantages of memory management

Given that memory management is necessary and beneficial, what are the trade-offs? The most obvious is the delay it inserts into the memory access cycle. Before a translation can take place, the logical address from the processor must appear. The translation usually involves some form of table look up, where the contents of a segment register or the higher order address bits are used to locate a descriptor within a memory block. This descriptor provides the physical address bits and any partitioning information such as read only etc. These signals are combined with the original lower order address bits to form the physical memory address. This look up takes time, which must be inserted into the memory cycle, and usually causes at least one wait state. This slows the processor and system performance down.

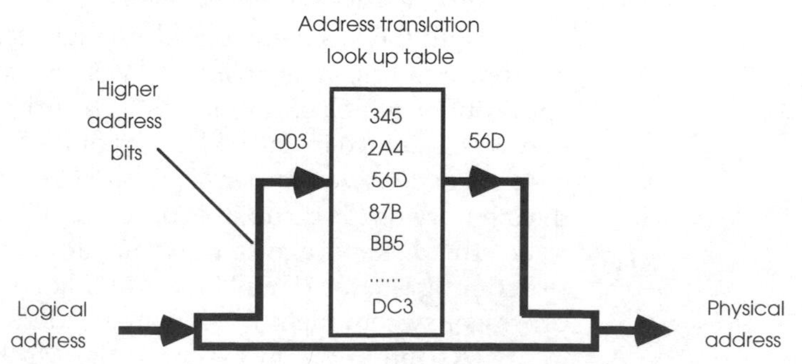

Address translation mechanism

In addition, there can be considerable overheads in managing all the look up tables and checking access rights etc. These overheads appear on loading a task, during any memory allocation and when any virtual memory system needs to swap memory blocks out to disk. The required software support is usually performed by an operating system. In the latter case, if the system memory is very small compared with the virtual memory size and application, the memory management driver will consume a lot of processing and time in simply moving data to and from the disk. In extreme cases, this overhead starts to dominate the system which is working hard but achieving very little. The addition of more memory relieves the need to swap and returns more of the system throughput to executing the application.

Segmentation and paging

There are two methods of splitting the system memory into smaller blocks for memory management. The size of these blocks is quite critical within the design. Each block requires a translation descriptor and therefore the size of the block is important. If the granularity is too small (i.e. the blocks are 1–2 kbytes), the number of descriptors needed for a 4 Gbyte system is extremely large. If the blocks are too big, the number of descriptors reduces but granularity increases. If a program just needs a few bytes, a complete block will have to be allocated to it and this wastes the unused memory. Between these two extremes lies the ideal trade-off.

A segmented memory management scheme has a small number of descriptors but solves the granularity problem by allowing the segments to be of a variable size in a block of contiguous memory. Each segment descriptor is fairly complex and the hardware has to be able to cope with different address translation widths. The memory usage is greatly improved, although the task of assigning memory segments in the most efficient way is difficult.

This problem occurs when a system has been operating for some time and the segment distribution is now right across the memory map. The free memory has been fragmented into small chunks albeit in large numbers. In such cases, the total system memory may be more than sufficient to allocate another segment, but the memory is non-contiguous and therefore not available. There is nothing more frustrating, when using such systems, as the combination of '2 Mbytes RAM free' and 'Insufficient memory to load' messages when trying to execute a simple utility. In such cases, the current tasks must be stopped, saved and restarted to repack them and free up the memory. This problem can also be found with file storage systems which need contiguous disk sectors and tracks.

A paged memory system splits memory needs into multiple, same sized blocks called pages. These are usually 1–2 kbytes in size, which allows them to take easy advantage of fragmented memory. However, each page needs a descriptor, which greatly increases the size of the look up tables. With a 4 Gbyte logical address space and 1 kbyte page size, the number of descriptors needed is over 4 million. Each descriptor would be 32 bits (22 bits translation address, 10 bits for protection and status) in size and the corresponding table would occupy 16 Mbytes! This is a little impractical, to say the least. To decrease the amount of storage needed for the page tables, multi-level tree structures are used. Such mechanisms have been implemented in the MC68851 paged memory management unit (PMMU), the MC68030 processor, PowerPC and ARM 920 processors, to name but a few.

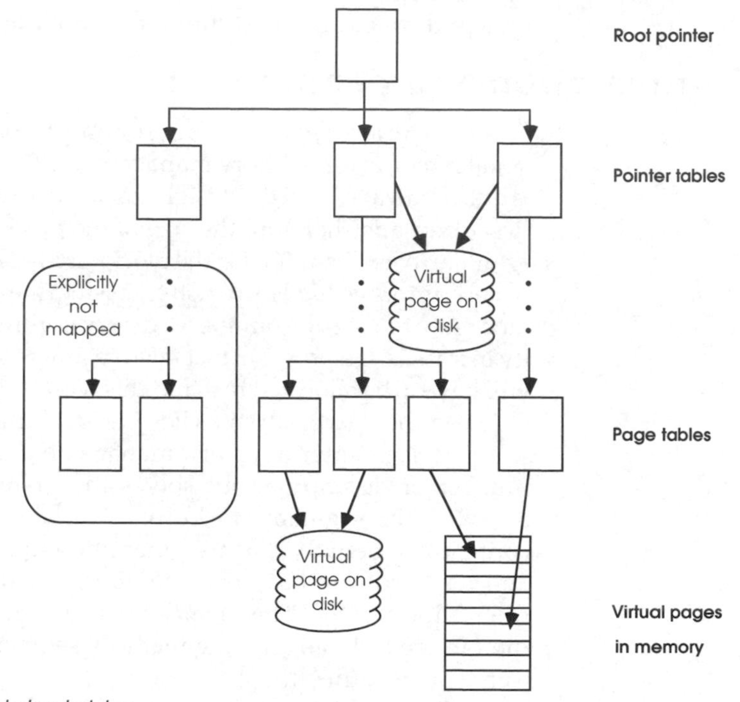

Using trees for descriptor tables

Trees work by dividing the logical address into fields and using each of the fields to successively reference into tables until the translation address is located. This is then concatenated with the lower order page address bits to complete a full physical address. The root pointer forms the start of the tree and there may be separate pointers for user and supervisor use.

The root pointer points to separate pointer tables, which in turn point to other tables and so on, until the descriptor is finally reached. Each pointer table contains the address of the next location. Most systems differ in the number of levels and the page sizes that can be implemented. Bits can often be set to terminate the table walk when large memory areas do not need to be uniquely defined at a lower or page level. The less levels, the more efficient the table walking.

The next diagram shows the three-level tree used by the MC88200 CMMU. The logical 32 bit address is extended by a user/supervisor bit which is used to select one of two possible segment table bases from the user and supervisor area registers. The segment number is derived from bits 31 to 22 and is concatenated with a 20 bit value from the segment table base and a binary '00' to create a 32 bit address for the page number table. The page table base is derived similarly until the complete translation is obtained. The remaining 12 bits of the descriptors are used to define the page, segment or area status in terms of access and more recently, cache coherency mechanisms. If the attempted access does not match with these bits (e.g. write to a write protected page), an error will be sent back to the processor.

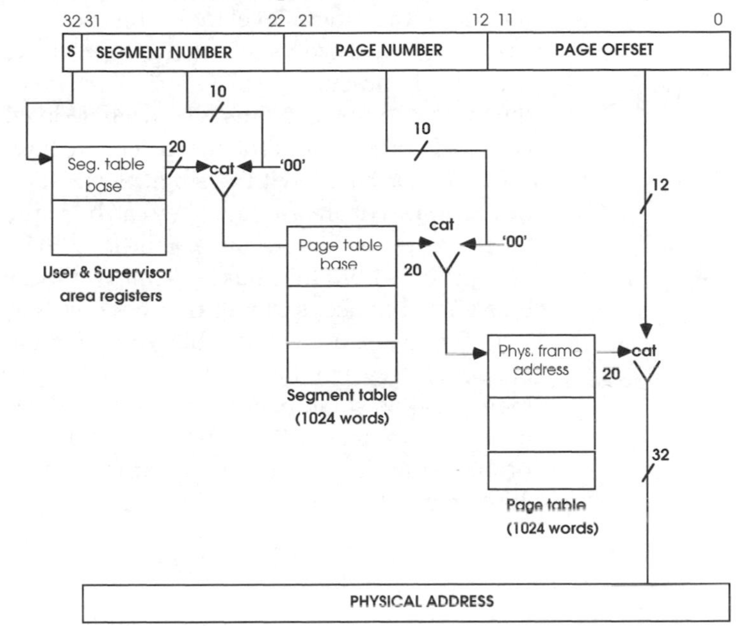

The MC88200 table walking mechanism

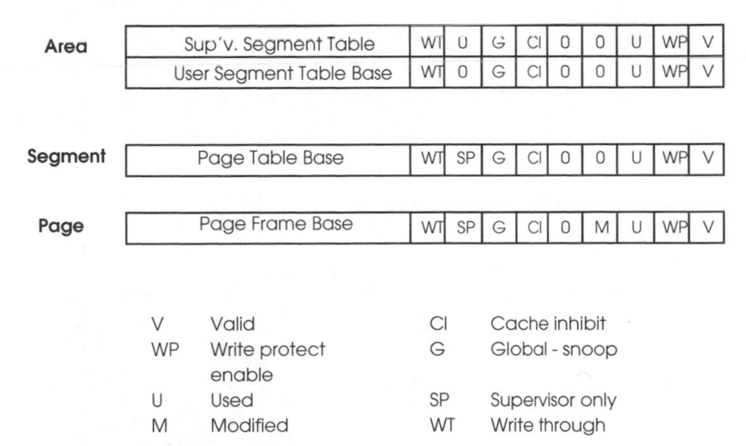

MC88200 memory management descriptors

The next two diagrams show a practical implementation using a two-level tree from an MC68851 PMMU. This is probably the most sophisticated of any of the MMU schemes to have appeared in the last 20 years. It could be argued that it was over-engineered as subsequent MMUs have effectively all been subsets of its features. One being the number of table levels it supported. However as an example, it is a good one to use as its basic principles are used even today.

The table walking mechanism compresses the amount of information needed to store the translation information. Consider a task that occupies a 1 Mbyte logical address map consisting of a code, data and stack memory blocks. Not all the memory is used or allocated and so the page tables need to identify which of the 2 kbyte pages are not valid and the address translation values for the others. With a single-level table, this would need 512 descriptors, assuming a 2 kbyte page size. Each descriptor would need to be available in practice. Each 2 kbyte block has its own descriptor in the table and this explains why a single level table will have 512 entries. Even if a page is not used, it has to have an entry to indicate this. The problem with a single level table is the number of descriptors that are needed. They can be reduced by increasing the page size — a 4 kbyte page will have the number of required descriptors — but this makes memory allocation extravagant. If a data structure needs 5 bytes then a whole page has to be allocated to it. If it is just a few bytes bigger than a page, a second page is needed. If a system has a lot of small data structures and limited memory then small pages are probably the best choice to get the best memory use. If the structures are large or memory in not a problem then larger page sizes are more efficient from the memory management side of the design.

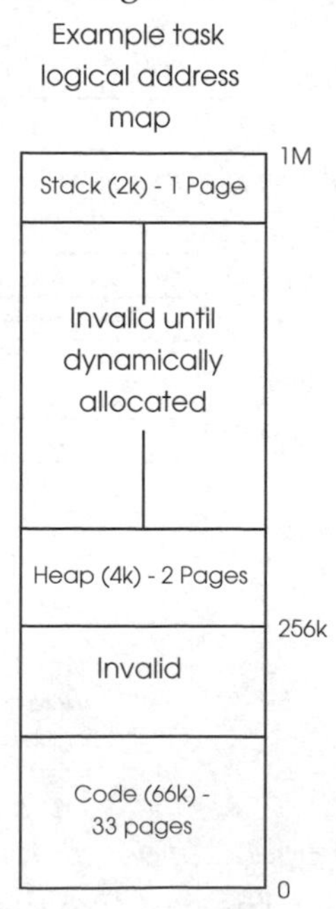

The required memory map

With the two-level scheme shown, only the pages needed are mapped. All other accesses cause an exception to allocate extra pages as necessary. This type of scheme is often called demand paged memory management. It has the advantage of immediately removing the need for descriptors for invalid or unused pages which reduces the amount of data needed to store the descriptors.

This means that with the two-level table mechanism shown in the diagram, only 44 entries are needed, occupying only 208 bytes, with additional pages increasing this by 4 bytes. The example shows a relatively simple two-level table but up to five levels are often used. This leads to a fairly complex and time consuming table walk to perform the address translation.

To improve performance, a cache or buffer is used to contain the most recently used translations, so that table walks only occur if the entry is not in the address translation cache (ATC) or translation look-aside buffer (TLB). This makes a lot of sense — due to the location of code and data, there will be frequent accesses to the same page and by caching the descriptor, the penalty of a table walk is only paid on the first access. However, there are still some further trade-offs to consider.

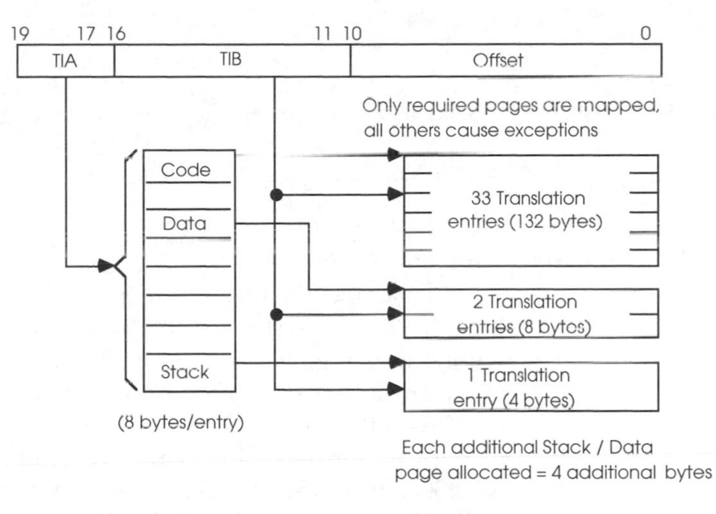

Example multilevel tree

Memory protection units

There has been a trend in recent processor designs to include a tripped down memory management unit that allows the memory to be partitioned and protected without any address translation. This removes the time consuming address translation mechanism which reduces the memory access time and the amount of hardware needed when compared with a full MMU implementation. In addition with system on a chip designs, this can reduce the chip size, cost and power consumption although it is fair to say that the size of these units are small compared to that of the whole chip and especially any on-chip memory. It is also possible to use the MMU as a memory protection unit by disabling the address translation or by arranging for the translation to be non-existent i.e. the physical and logical addresses are the same.

The basic idea behind a memory protection unit is to police the memory subsystem so that only approved memory accesses can take place. If a memory access is made to a protected area by software that does not have the correct access rights, an error signal is generated which can be used to start supervisor level software to decide what to do.

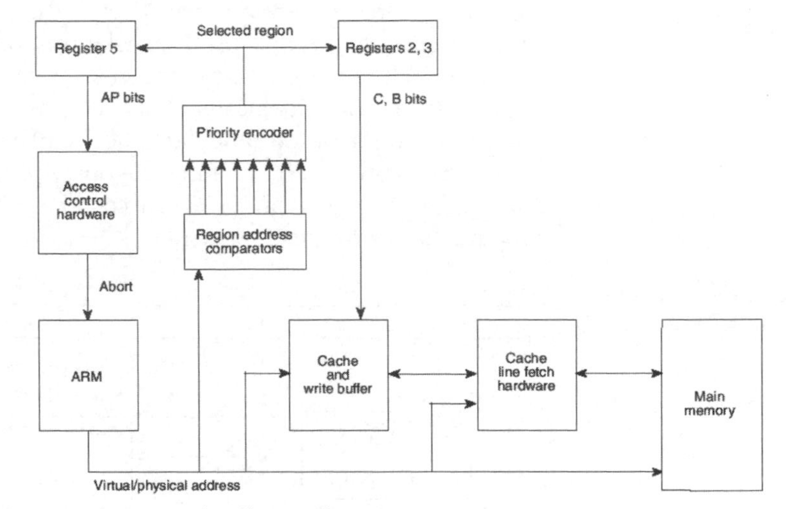

The ARM architecture memory protection unit

The ARM architecture memory protection unit performs this function. It can divide the memory range into eight separate regions. Each region can be as small as 4 kbytes up to 4 Gbyte and its starting address must be on a region boundary. If region is set to 4 Kbytes then it can start on an address like 0x45431000 but an 8 kbyte region cannot. Its nearest valid address would be 0x45430000 or 0x45432000. Each region has an associated cacheable bit, a bufferable bit and access permission bits. These control whether the data stored in the region is cacheable (C bit), can be buffered in the processor's write buffer (B bit) and the type of access permitted (AP bits). These are in fact very similar to the permission bits used in the corresponding ARM MMU architecture and are stored in control registers. The regions are numbered and this defines a priority level for resolving which permission bits take precedence if regions overlap. For example region 2 may not permit data caching while region 6 does. If region 6 overlaps region 2, then the memory accesses in the overlapped area will be cached. This provides an additional level of control.

The sequence for a memory access using the protection unit is shown in the diagram and is as follows:

- The CPU issues an address which is compared to the addresses that define the regions.
- If the address is not in any of these regions, the memory access is aborted.
- If the address is inside of one or more of the regions then the highest number region will supply the permission bits and

these will be evaluated. If the access permission bits do not match, the access is aborted. If they do match, the sequence will continue. The C and B bits are then used to control the behaviour of the cache and write buffer as appropriate and eventually the memory access will complete successfully, depending on how the C and B bits are set.

In practice, MMUs and memory protection units are becoming quite common in embedded systems. Their use can provide a greater level of security by trapping invalid memory accesses before they corrupt other data structures. This means that an erroneous task can be detected without bringing down the rest of the system. With a multitasking system, this means that a task may crash but the rest of the system will not. It can also be used to bring down a system gracefully as well.

Cache memory

With the faster processors available today, the wait states incurred in external memory accesses start to dramatically reduce performance. To recover this, many designs implement a cache memory to buffer the processor from such delays. Once predominantly used with high end systems, they are now appearing both on chip and as external designs.

Cache memory systems work because of the cyclical structures within software. Most software structures are loops where pieces of code are repeatedly executed, albeit with different data. Cache memory systems store these loops so that after the loop has been fetched from main memory, it can be obtained from the cache for subsequent executions. The accesses from cache are faster than from main memory and thus increase the system's throughput.

There are several criteria associated with cache design which affect its performance. The most obvious is cache size — the larger the cache, the more entries are stored and the higher the hit rate. For the 80x86 processor architecture, the best price performance is obtained with a 64 kbyte cache. Beyond this size, the cost of getting extra performance is extremely high.

The set associativity is another criterion. It describes the number of cache entries that could possibly contain the required data. With a direct map cache, there is only a single possibility, with a two way system, there are two possibilities, and so on. Direct mapped caches can get involved in a bus thrashing situation, where two memory locations are separated by a multiple of the cache size. Here, every time word A is accessed, word B is discarded from the cache. Every time word B is accessed, word A is lost, and so on. The cache starts thrashing and overall performance is degraded. With a two way design, there are two possibilities and this prevents bus thrashing. The cache line refers to the number of consecutive bytes that are associated with each cache entry. Due to the sequential nature of instruction flow, if a cache hit occurs at the beginning of the line, it is highly probable that the rest of the line will be accessed as well. It is therefore prudent to burst fill cache lines whenever a miss forces a main memory access. The differences between set associativity and line length are not as clear as cache size. It is difficult to say what the best values are for a particular system. Cache performances are extremely system and software dependent and, in practice, system performance increases of 20–30% are typical.

Cache size and organization

There are several criteria associated with cache design which affect its performance. The most obvious is cache size — the larger the cache, the more entries that are stored and the higher the hit rate. However, as the cache size increases, the return gets smaller and smaller. In practice, the cache costs and complexity place an economic limit on most designs. As the size of programs increase, larger caches are needed to maintain the same hit rate and hence the 'ideal cache size is always twice that available' comment. In reality, it is the combination of size, organization and cost that really determines the size and its efficiency.

Consider a basic cache operation. The processor generates an address which is fed into the cache memory system. The cache stores its data in an array with an address tag. Each tag is compared in turn with the incoming address. If they do not match, the next tag is compared. If they do match, a cache hit occurs, the corresponding data within the array is passed on the data bus to the processor and no further comparisons are made. If no match is found (a cache miss), the data is fetched from external memory and a new entry is created in the array. This is simple to implement, needing only a memory array and a single comparator and counter. Unfortunately, the efficiency is not very good due to the serial interrogation of the tags.

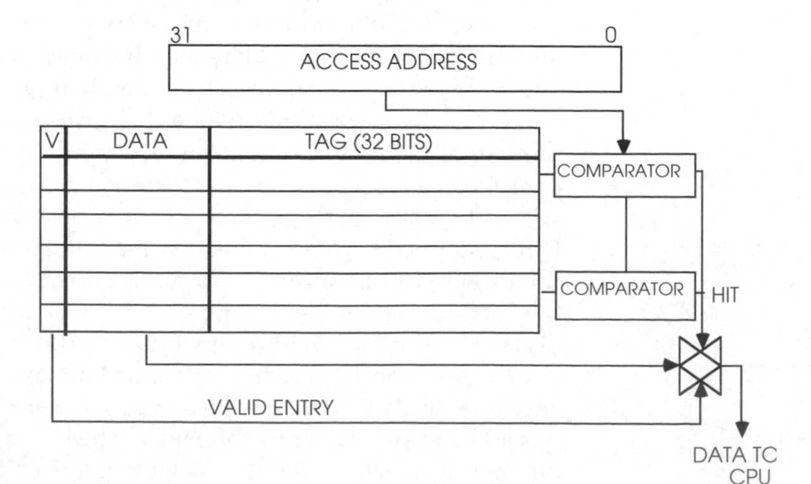

A fully associative cache design

A better solution is to have a comparator for each entry, so all entries can be tested simultaneously. This is the organization used in a fully associative cache. In the example, a valid bit is added to each entry in the cache array, so that invalid or unused entries can be easily identified. The system is very efficient from a software perspective — any entry can be used to store data from any address. A software loop using only 20 bytes (10 off 16 bit instructions) but scattered over a 1,024 byte range would run as efficiently as another loop of the same size but occupying consecutive locations.

The disadvantage of this approach is the amount of hardware needed to perform the comparisons. This increases in proportion to the cache size and therefore limits these fully associative caches to about 10 entries. The fully associative cache locates its data by effectively asking n questions where n is the number of entries within it. An alternative organization is to assume certain facts derived from the data address so that only one location can possibly have the data and only one comparator is needed, irrespective of the cache size. This is the idea behind the direct map cache.

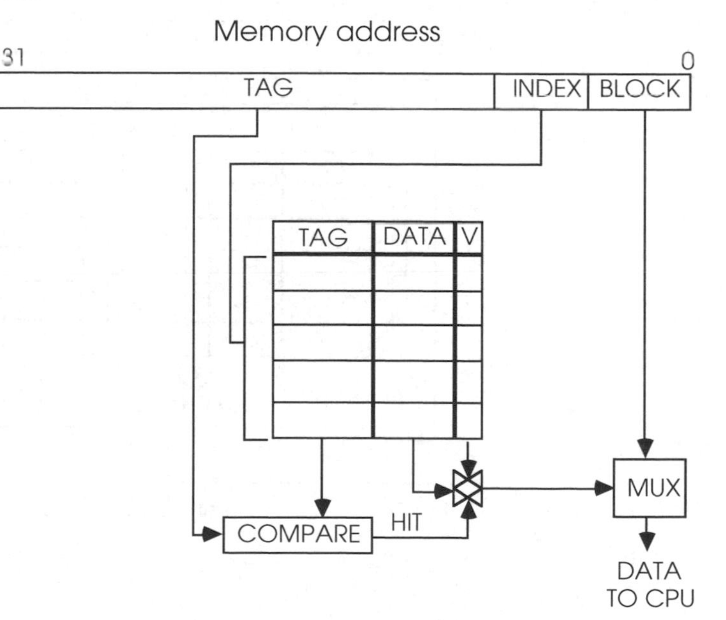

Direct mapped cache

The address is presented as normal but part of it is used to index into the tag array. The corresponding tag is compared and, if there is a match, the data is supplied. If there is a cache miss, the data is fetched from external memory as normal and the cache updated as necessary. The example shows how the lower address bits can be used to locate a byte, word or long word within the memory block stored within the array. This organization is simple from the hardware design perspective but can be inefficient from a software viewpoint.

The index mechanism effectively splits external memory space into a series of consecutive memory pages, with each page the same size as the cache. Each page is mapped to resemble the cache and therefore each location in the external memory page can only correspond with its own location in the cache. Data that is offset by the cache size thus occupies the same location within the cache, albeit with different tag values. This can cause bus thrashing. Consider a case where words A and B are offset by the cache size. Here, every time word A is accessed, word B is discarded from the cache. Every time word B is accessed, word A is lost. The cache starts thrashing and the overall performance is degraded. The MC68020 is a typical direct mapped cache.

A way to solve this is to split the cache so there are two or four possible entries available for use. This increases the comparator count but provides alternative locations and prevents bus thrashing. Such designs are described as '*n* way set associative', where *n* is the number of possible locations. Values of 2, 4, 8 are quite typical of such designs.

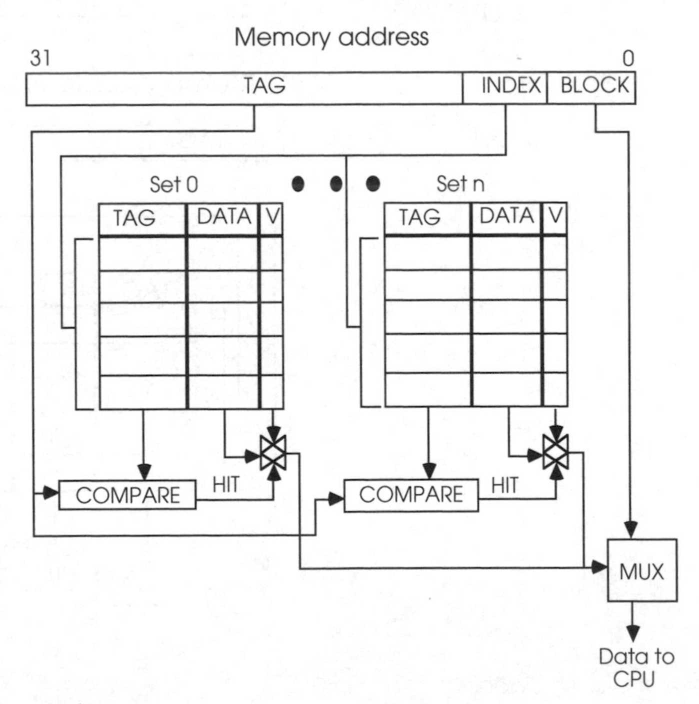

A set associative cache design

Many RISC-based caches are organized as a four way set associative cache where a particular address will have four possible locations in the cache memory. This has advantages for the software environment in that context switching code with the same address will not necessarily overwrite each other and keep destroying the contents of the cache memories.

The advantage of a set associative cache is its ability to prevent thrashing at the expense of extra hardware. However, all the caches so far described can be improved by further reorganizing so that each tag is associated with a line of long words which can be burst filled using a page memory interface.

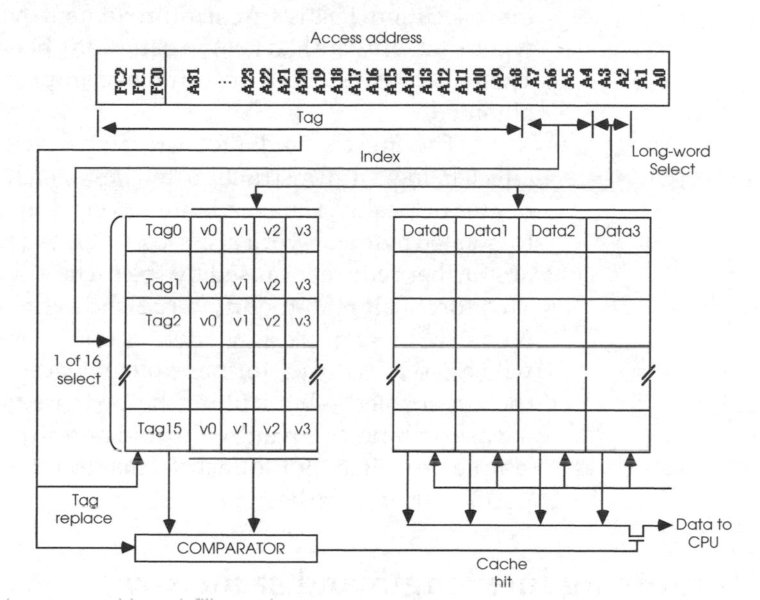

The MC68030 direct mapped burst fill cache

The logic behind this idea is based on the sequential nature of instruction execution and data access. Instruction fetches and execution simply involve accesses to sequential memory locations until a program flow change happens due to the execution of a branch or jump instruction. Data accesses often follow this pattern during stack and data structure manipulation.

It follows that if a cache is organized with, say, four long words within each tag line, a hit in the first long word would usually result in hits in the rest of the line, unless a flow change took place. If a cache miss was experienced, it would be beneficial to bring in the whole line, providing this could be achieved in less time than to bring in the four long words individually. This is exactly what happens in a page mode interface. By combining these, a more efficient cache can be designed which even benefits in line code. This is exactly how the MC68030 cache works.

Address bits 2 and 3 select which long word is required from the four stored in the 16 byte wide line. The remaining higher address bits and function codes are the tag which can differentiate between supervisor or user accesses etc. If there is a cache miss, the processor uses its synchronous bus with burst fill to load up the complete line.

	W/O Burst	W/ Burst
Instruction Cache	46%	82%
Data Cache - Reads	60%	72%
Data Cache - R & W	40%	48%

Estimated hit rates for the MC68030 caches

With a complete line updated in the cache, the next three instructions result in a hit, providing there is no preceding flow change. These benefit from being cached, even though it is their first execution. This is a great improvement over previous designs, where the software had to loop before any benefit could be gained. The table above shows the estimated improvements that can be obtained.

The effect is similar to increasing the cache size. The largest effect being with instruction fetches which show the greatest degree of locality. Data reads are second, but with less impact due to isolated byte and word accesses. Data read and write operations are further reduced, caused by the cache's write-through policy. This forces all read-modify-write and write operations to main memory. In some designs, data accesses are not sequential, in which case, system performance actually degrades when the data cache is enabled — burst filling the next three long words is simply a waste of time, bus bandwidth and performance. The solution is simple — switch off the cache! This design is used in most high performance cache designs.

Optimising line length and cache size

This performance degradation is symptomatic of external bus thrashing due to the cache line length and / or burst fill length being wrong and leading to system inefficiencies. It is therefore important to get these values correct. If the burst fill length is greater than the number of sequential instructions executed before a flow change, data is fetched which will not be used. This consumes valuable external bus bandwidth. If the burst length is greater than the line length, multiple cache lines have to be updated, which might destroy a cache entry for another piece of code that will be executed later. This destroys the efficiency of the cache mechanism and increases the cache flushing, again consuming external bus bandwidth. Both of these contribute to the notorious 'bus thrashing' syndrome where the processor spends vast amounts of time fetching data that it never uses. Some cache schemes allow line lengths of 1, 4, 8, 16 or 32 to be selected, however, most systems use a line and burst fill length of 4. Where there are large blocks of data to be moved, higher values can improve performance within these moves, but this must be offset by any affect on other activities.

Cache size is another variable which can affect performance. Unfortunately, it always seems to be the case that the ideal cache is twice the size of that currently available! The biggest difficulty is that cache size and efficiency are totally software dependant — a configuration that works for one application is not necessarily the optimum for another.

The table shows some efficiency figures quoted by Intel in their 80386 Hardware Reference Manual and from this data, it is apparent that there is no clear cut advantage of one configuration over another. It is very easy to get into religious wars of cache organisation where one faction will argue that their particular organisation is right and that everything else is wrong. In practice, it is incredibly difficult to make such claims without measuring and benchmarking a real system. In addition, the advantages can be small compared to other performance techniques such as software optimisation. In the end, the bigger the cache the better, irrespective of its set-associativity or not is probably the best maxim to remember.

Size (k)	Associativity	Line size (bytes)	Hit rate (%)	Performance ratio versus DRAM
1	direct	4	41	0.91
8	direct	4	73	1.25
16	direct	4	81	1.35
32	direct	4	86	1.38
32	2-way	4	87	1.39
32	direct	8	91	1.41
64	direct	4	88	1.39
64	2-way	4	89	1.40
64	4-way	4	89	1.40
64	direct	8	92	1.42
64	2-way	8	93	1.42
128	direct	4	89	1.39
128	2-way	4	89	1.40
128	direct	8	93	1.42

(source: 80386 Hardware Reference Manual)

Cache performance

Logical versus physical caches

Cache memory can be located either side of a memory management unit and use either physical or logical addresses as its tag data. In terms of performance, the location of the cache can dramatically affect system performance. With a logical cache, the tag information refers to the logical addresses currently in use by the executing task. If the task is switched out during a context switch, the cache tags are no longer valid and the cache, together with its often hard-won data must be flushed and cleared. The processor must now go to main memory to fetch the first instructions and wait until the second iteration before any benefit is obtained from the cache. However, cache accesses do not need to go through the MMU and do not suffer from any associated delay.

Physical caches use physical addresses, do not need flushing on a context switch and therefore data is preserved within the cache. The disadvantage is that all accesses must go through the memory management unit, thus incurring delays. Particular care must also be exercised when pages are swapped to and from disk.

If the processor does not invalidate any associated cache entries, the cache contents will be different from the main memory contents by virtue of the new page that has been swapped in.

Of the two systems, physical caches are more efficient, providing the cache coherency problem is solved and MMU delays are kept to a minimum. RISC architectures like the PowerPC solve the MMU delay issue by coupling the MMU with the cache system. An MMU translation is performed in conjunction with the cache look up so that the translation delay overlaps the memory access and is reduced to zero. This system combines the speed advantages of a logical cache with the data efficiency of a physical cache.

Most internal caches are now designed to use the physical address (notable exceptions are some implementations of the SPARC architecture which use logical internal caches).

Unified versus Harvard caches

There is another aspect of cache design that causes great debate among designers and this concerns whether the cache is unified or separate. A unified cache, as used on the Intel 80486DX processors and the Motorola MPC601 PowerPC chip, uses the same cache mechanism to store both data and instructions. The separate or Harvard cache architecture has separate caches for data and instructions. The argument for the unified cache is that its single set of tags and comparators reduces the amount of silicon needed to implement it and thus for a given die area, a larger cache can be provided compared to separate caches. The argument against is that a unified cache usually has only a single port and therefore simultaneous access to both instructions and data will result in one or the other being delayed while the first access is completed. This delay can halt or slow down the processor's ability to execute instructions.

Conversely, the Harvard approach uses more silicon area for the second set of tags and comparators but does allow simultaneous access. In reality, the overall merits of each approach depend on several factors, and depending where the cross-over points lie, the factors will be in favour of one or other. If software needs to exploit superscalar operation then the Harvard architecture is less likely to impede superscalar execution. If the application has large data and code structures, then a larger unified cache may be better. As with most cache organisation decisions, the only clear way to make a decision is to evaluate using the end application and the test software.

Cache coherency

The biggest challenge with cache design is how to solve the problem of data coherency, while remaining hardware and software compatible. The issue arises when data is cached which can then be modified by more than one source. An everyday analogy is that of a businessman with two diaries — one kept by his secretary in the office and the other kept by him. If he is out of the office and makes an appointment, the diary in the office is no longer valid and his secretary can double book him assuming, incorrectly, that the office diary is correct.

This problem is normally only associated with data but can occur with instructions within an embedded application. The stale data arises when a copy is held both in cache and in main memory. If either copy is modified, the other becomes stale and system coherency is destroyed. Any changes made by the processor can be forced through to the main memory by a 'write-through' policy, where all writes automatically update cache and main memory. This is simple to implement but does couple the processor unnecessarily to the slow memory. More sophisticated techniques, like 'copy-back' and 'modified write-back' can give more performance (typically 15%, although this is system and software dependent) but require bus snooping support to detect accesses to the main memory when the valid data is in the cache.

The 'write-through' mechanism solves the problem from the processor perspective but does not solve it from the other direction. DMA (Direct Memory Access) can modify memory directly without any processor intervention. Consider a task swapping system. Task A is in physical memory and is cached. A swap occurs and task A is swapped out to disk and replaced by task B at the same location. The cached data is now stale. A software solution to this involves flushing the cache when the page fault happens so the previous contents are removed. This can destroy useful cached data and needs operating system support, which can make it non–compatible. The only hardware solution is to force any access to the main memory via the cache, so that the cache can update any modifications.

This provides a transparent solution — but it does force the processor to compete with the DMA channels and restricts caching to the main memory only, with a resultant impact on performance.

While many system designs use cache memory to buffer the fast processor from the slower system memory, it should be remembered that access to system memory is needed on the first execution of an instruction or software loop and whenever a cache miss occurs. If this access is too slow, these overheads greatly diminish the efficiency of the cache and, ultimately, the processor's performance. In addition, switching on caches can cause software that works perfectly to crash and, in many configurations, the caches remain switched off to allow older software to execute correctly.

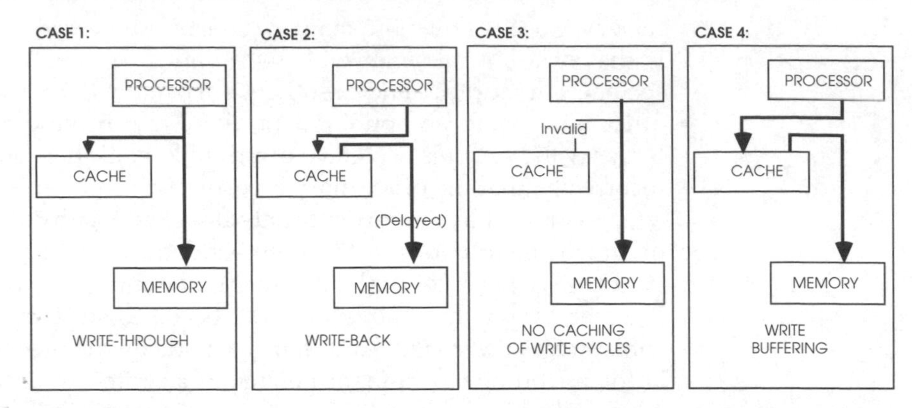

Different write schemes

Other problems can occur when data that is not intended to be cached is cached by the system. Shared memory or I/O ports are two areas that come immediately to mind. Shared memory relies on the single memory structure to contain the recent data. If this is cached then any updates may not be made to the shared memory. Any other CPU or DMA that accesses the shared memory will not get the latest data and the stale data may cause the system to crash. The same problem can happen with I/O ports. If accesses are cached then reading an I/O port to get status information will return with the cached data which may not be consistent with the data at the I/O port. It is important to be able to control which memory regions are cached and which are not. It should be no surprise that MMUs and memory protection units are used to perform this function and allow the control of the caches to be performed automatically based on memory addresses and associated access bits.

A lot is made of cache implementations — but unless the main system memory is fast and software reliable, system and software performance will degrade. Caches help to regain performance lost through system memory wait states but they are never 100% efficient. A system with no wait states always provides the best performance. Add to that the need for control and the selection of the right cache coherency policy for the system and designing for any system that has caches requires a detailed understanding of what is going on to get the best out of the system.

Case 1: write-through

In this case, all data writes go through to main memory and update the system as well as the cache. This is simple to implement but couples the processor unnecessarily to slow memory. If data is modified several times before another master needs it, the write-through policy consumes external bus bandwidth supplying data that is not needed. This is not terribly efficient. In its favour, the scheme is very simple to implement, providing there is only a single cache within the system.

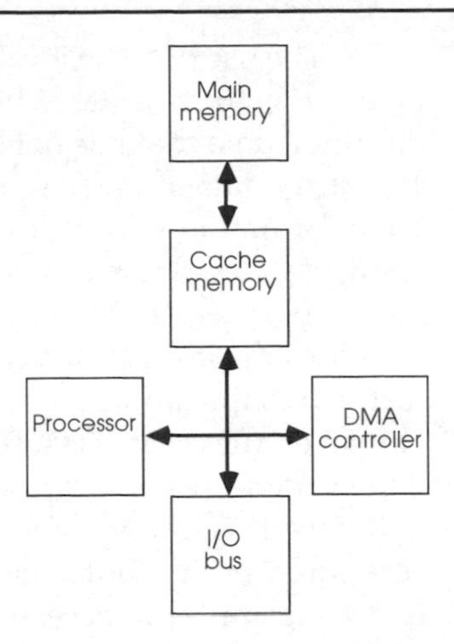

A coherent cache architecture

If there are more than two caches, the stale data problem reappears in another guise. Consider such a system where two processors with caches have a copy of a global variable. Neither processor accesses main memory when reading the variable, as the data is supplied by the respective caches. Processor A now modifies the variable — its cache is updated, along with the system memory. Unfortunately, processor B's cache is left with the old stale data, creating a coherency problem. A similar problem can occur within paging systems.

It also does not address the problem with I/O devices either although the problem will occur when the I/O port is read for a second and subsequent times as the cache will supply the data on these accesses instead of the I/O port itself.

DMA (direct memory access) can modify memory directly without any processor intervention. Consider a UNIX paging system. Page A is in physical memory and is cached. A page fault occurs and page A is swapped out to disk and replaced by page B at the same location. The cached data is now stale. A software solution to this involves flushing the cache when the page fault happens so the previous contents are removed. This can destroy useful cached data and needs operating system support, which can make it non-compatible. The only hardware solution is to force any access to the main memory via the cache, so that the cache can update any modifications. This provides a transparent solution, but it does force the processor to compete with the DMA channels, and restricts caching to the main memory only, with the subsequent reduced performance.

Case 2: write-back

In this case, the cache is updated first but the main memory is not updated until later. This is probably the most efficient method of caching, giving 15–20% improvement over a straight write-through cache. This scheme needs a bus snooping mechanism for coherency and this will be described later.

The usual cache implementation involves adding dirty bits to the tag to indicate which cache lines or partial lines hold modified data that has not been written out to the main memory. This dirty data must be written out if there is any possibility that the information will be lost. If a cache line is to be replaced as a result of a cache miss and the line contains dirty data, the dirty data must be written out before the new cache line can be accepted. This increases the impact of a cache miss on the system. There can be further complications if memory management page faults occur. However, these aspects must be put into perspective — yes, there will be some system impact if lines must be written out, but this will have less impact on a wider scale. It can double the time to access a cache line, but it has probably saved more performance by removing multiple accesses through to the main memory. The trick is to get the balance in your favour.

Case 3: no caching of write cycles

In this method, the data is written through but the cache is not updated. If the previous data had been cached, that entry is marked invalid and is not used. This forces the processor to access the data from the main memory. In isolation, this scheme does seem to be extremely wasteful, however, it often forms the backbone of a bus snooping mechanism.

Case 4: write buffer

This is a variation on the write-through policy. Writes are written out via a buffer to the main memory. This enables the processor to update the 'main memory' very quickly, allowing it to carry on processing data supplied by the cache. While this is going on, the buffer transfers the data to the main memory. The main advantage is the removal of memory delays during the writes. The system still suffers from coherency problems caused through multiple caches.

Another term associated with these techniques is write allocation. A write-allocate cache allocates entries in the cache for any data that is written out. The idea behind this is simple — if data is being transferred to external memory, why not cache it, so that when it is accessed again, it is already waiting in the cache. This is a good idea if the cache is large but it does run the risk of overwriting other entries that may be more useful. This problem is particularly relevant if the processor performs block transfers or memory initialisation. Its main use is within bus snooping mechanisms where a first write-allocate policy can be used to tell other caches that their data is now invalid. The most important need with these methods and ideas is bus snooping.

Bus snooping

With bus snooping, a memory cache monitors the external bus for any access to data within the main memory that it already has. If the cache data is more recent, the cache can either supply it direct or force the other master off the bus, update main memory and start a retry, thus allowing the original master access to valid data. As an alternative to forcing a retry, the cache containing the valid data can act as memory and supply the data directly. As previously discussed, bus snooping is essential for any multimaster system to ensure cache coherency.

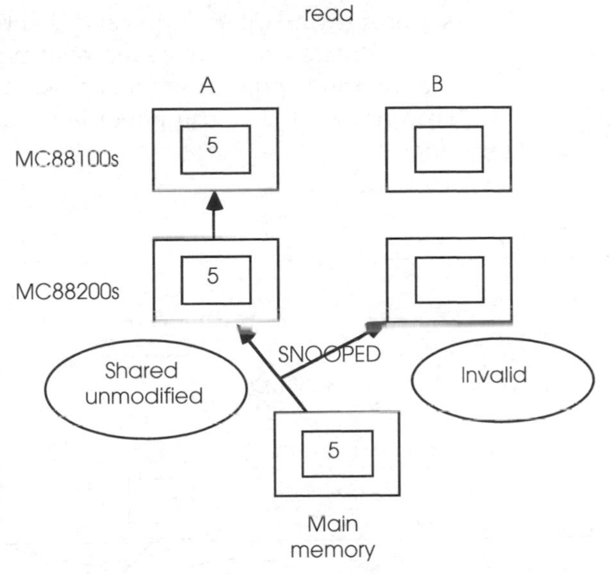

M88000 cache coherency - i

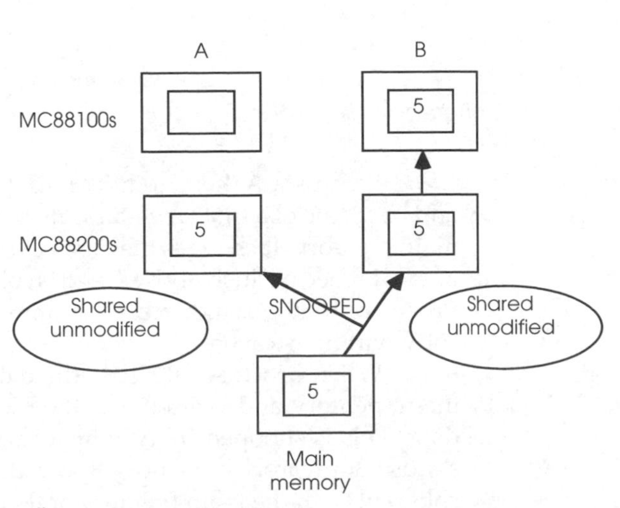

M88000 cache coherency - ii

The bus snooping mechanism used by the MC88100/ MC88200 uses a combination of write policies, cache tag status and bus monitoring to ensure coherency. Nine diagrams show a typical sequence. In the first figure on the previous page, processor A reads data from the main memory and this data is cached. The main memory is declared global and is shared by processors A and B. Both these caches have bus snooping enabled for this global memory. This causes the cached data to be tagged as shared unmodified; i.e. another master may need it and the data is identical to that of main memory. A´s access is snooped by processor B, which does nothing as its cache entry is invalid. It should be noted that snooping does not require any direct processor of software intervention and is entirely automatic.

Processor B accesses the main memory, as shown in the next diagram and updates its cache as well. This is snooped by A but the current tag of shared unmodified is still correct and nothing is done.

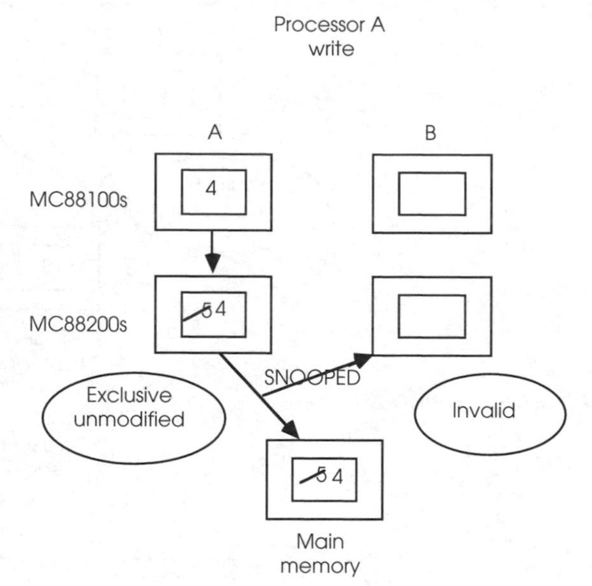

M88000 cache coherency - iii

Processor A then modifies its data as shown in diagram (iii) and by virtue of a first write-allocate policy, writes through to the main memory. It changes the tag to exclusive unmodified; i.e. the data is cached exclusively by A and is coherent with main memory. Processor B snoops the access and immediately invalidates its old copy within its cache.

When processor B needs the data, it is accessed from the main memory and written into the cache as shared unmodified data. This is snooped by A, which changes its data to the same status. Both processors now know that the data they have is coherent with the main memory and is shared.

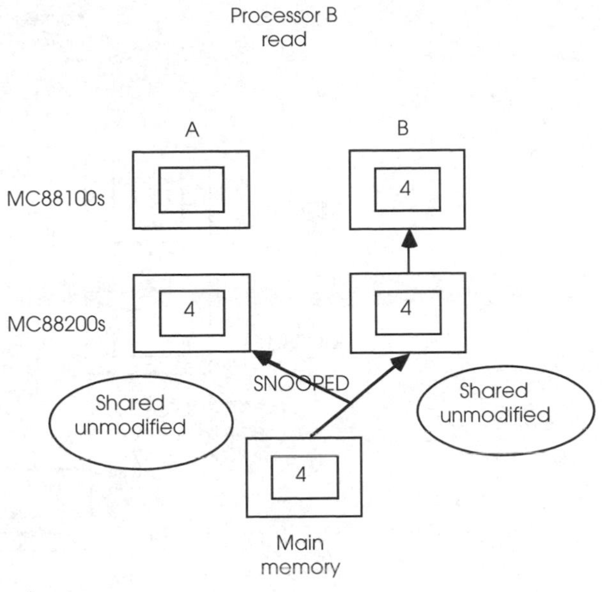

M88000 cache coherency - iv

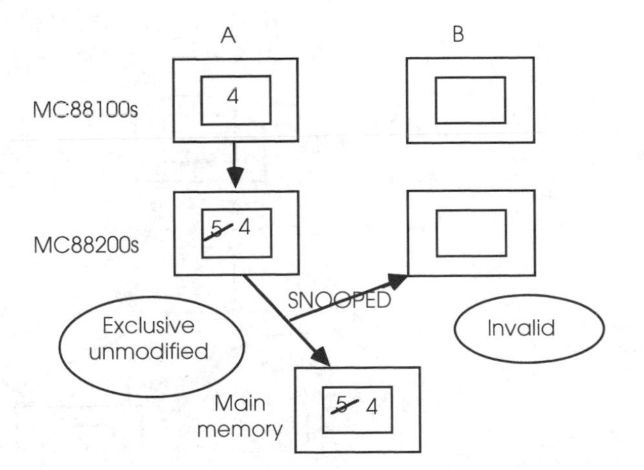

M88000 cache coherency - v

Processor A now modifies the data which is written out to the main memory and snooped by B which marks its cache entry as invalid. Again, this is a first write-allocate policy in effect.

Processor A modifies the data again but, by virtue of the copyback selection, the data is not written out to the main memory. Its cache entry is now tagged as exclusive modified; i.e. this may be the only valid copy within the system.

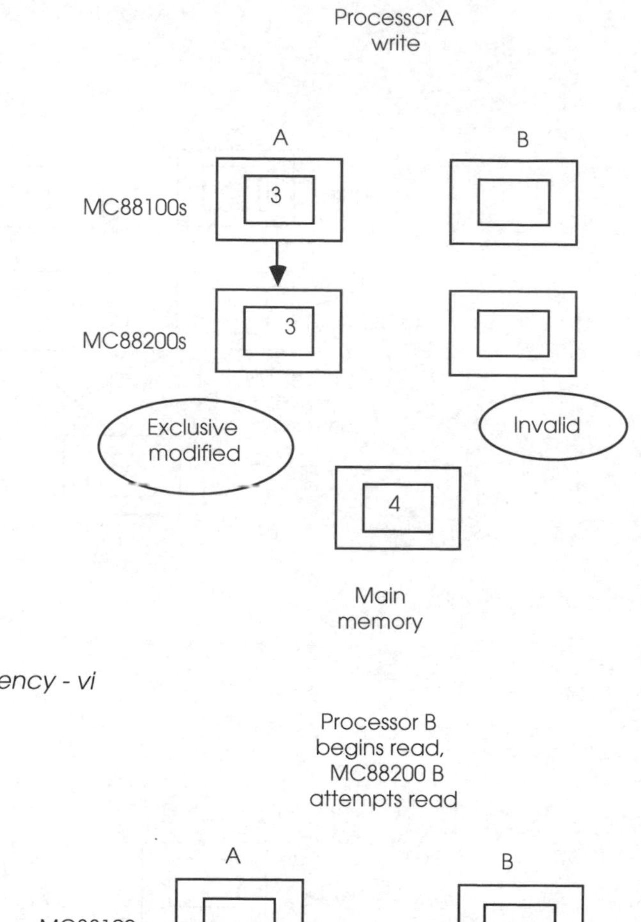

M88000 cache coherency - vi

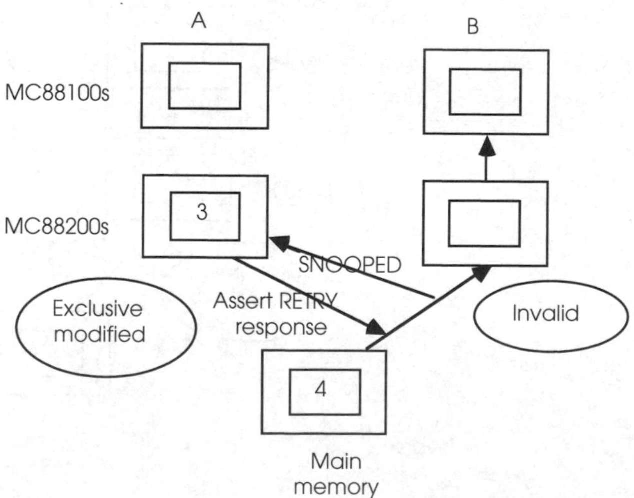

M88000 cache coherency - vii

Processor B tries to get the data and starts an external memory access, as shown. Processor A snoops this access, recognises that it has the valid copy and so asserts a retry response to processor B, which comes off the bus and allows processor A to update the main memory and change its cache tag status to shared unmodified.

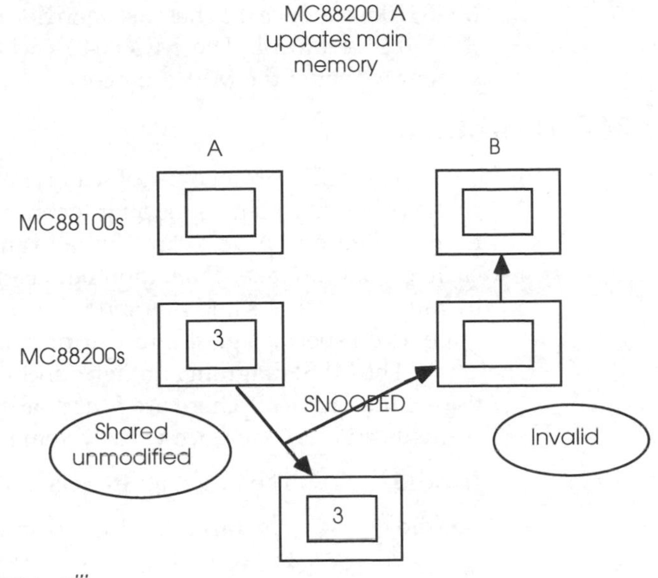

M88000 cache coherency - viii

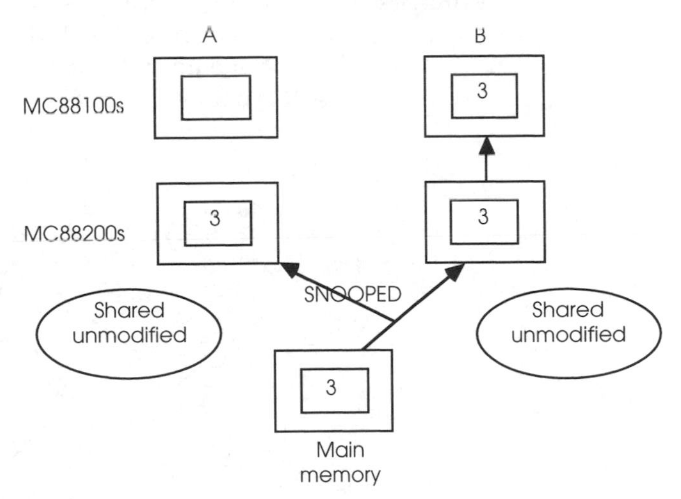

M88000 cache coherency - ix

Once completed, processor B is allowed back onto the bus to complete its original access, this time with the main memory containing the correct data.

This sequence is relatively simple, compared with those encountered in real life where page faults, cache flushing, etc., further complicate the state diagrams. The control logic for the CMMU is far more complex than that of the MC88100 processor itself and this demonstrates the complexity involved in ensuring cache coherency within multiprocessor systems.

The problem of maintaining cache coherency has led to the development of two standard mechanisms — MESI and MEI. The

MC88100 sequence that has just been discussed is similar to that of the MESI protocol. The MC68040 cache coherency scheme is similar to that of the MEI protocol.

The MESI protocol

The MESI protocol is a formal mechanism for controlling cache coherency using snooping techniques. Its acronym stands for modified, exclusive, shared, invalid and refers to the states that cached data can take. Transition between the states is controlled by memory accesses and bus snooping activity. This information appears on special signal pins during bus transactions.

The MESI diagram is generic and shows the general operation of the protocol. There are four states that describe the cache contents and its coherence with system memory:

Invalid The target address is not cached.

Shared The target address is in the cache and also in at least one other. It is coherent with system memory.

Exclusive The target address is in the cache but the data is coherent with system memory.

Modified The target address is in the cache, but the contents has been modified and is not coherent with system memory. No other cache in the system has this data.

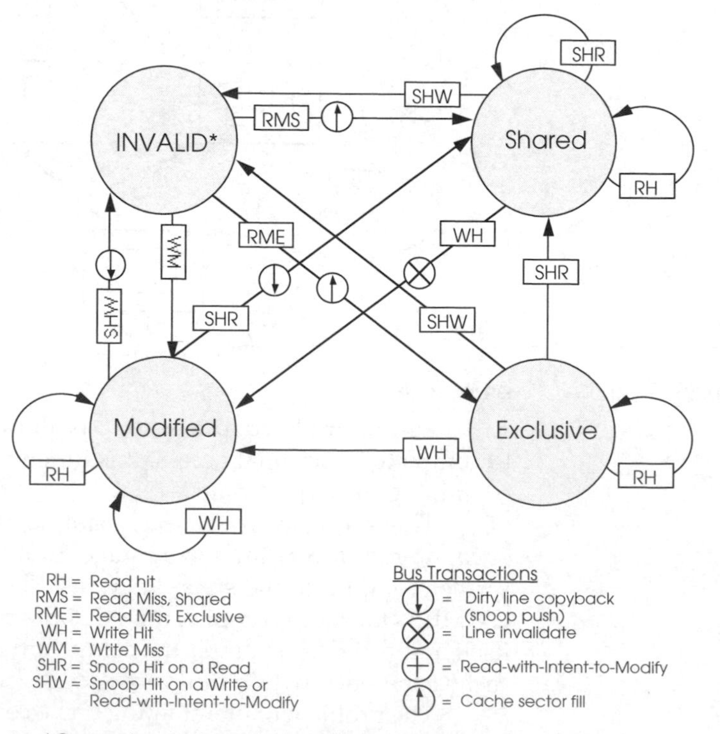

MESI cache coherency protocol

The movement from one state is governed by memory actions, cache hits and misses and snooping activity. For example, if a processor needs to write data to a memory address that has a write-back policy and cache coherency enabled as part of its page descriptors — controlled by the WIM bits — and causes a cache miss, the processor will move from an invalid state to a modified state by performing a 'read with intent to modify' bus cycle.

The MESI protocol is widely used in multiprocessor designs, for example, in the Futurebus+ interconnection bus. The MPC601 uses this protocol.

The MEI protocol

The MEI protocol — modified, exclusive, invalid — does not implement the shared state and so does not support the MESI shared state where multiple processors can cache shared data.

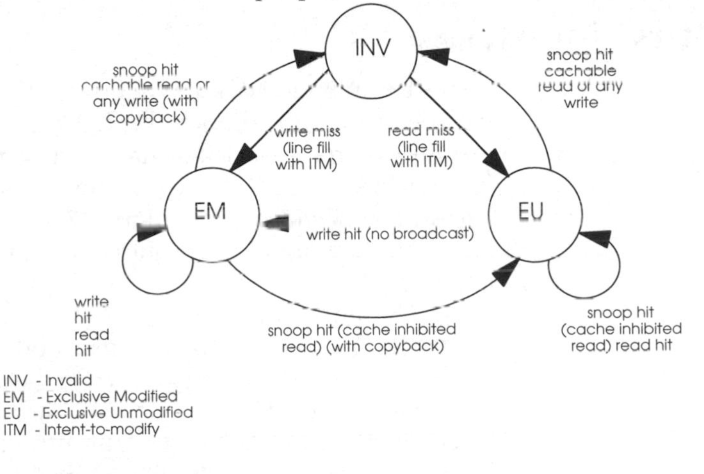

MPC603 MEI coherency diagram

The MPC603 uses this simplified form of protocol to support other intelligent bus masters such as DMA controllers. It is not as good as the MESI bus for true multiprocessor support. On the other hand, it is less complex and easier to implement. The three states are defined as follows:

Invalid The target address is not cached.

Exclusive unmodified The target address is in the cache but the data is coherent with system memory.

Exclusive modified The target address is in the cache, but the contents have been modified and are not coherent with system memory. No other cache in the system has this data.

Note that the cache coherency implementation is processor specific and may change. The two mechanisms described here are the two most commonly used methods for processors and are likely to form the basis of future designs.

Two final points: these schemes require information to be passed by the external buses to allow other bus masters to identify the transitions. This requires the hardware design to implement them. If this is not done, these schemes will not work and the software environment may require extensive change and the imposition of constraints. Cache coherency may need to be restricted to cache inhibited or write-through. DMA accesses could only be made to cache inhibited memory regions. The supervisor must take responsibility for these decisions and implementations to ensure correct operation. In other words, do not assume that cache coherency software for one hardware design will work on another. It will, if the bus interface design is the same. It will not if they are different.

Finally, cache coherency also means identifying the areas of memory which are not to be cached.

Burst interfaces

The adoption of burst interfaces by virtually all of today's high performance processors has led to the development of special memory interfaces which include special address generation and data latches to help the designer. Burst interfaces take advantage of page and nibble mode memories which supply data on the first access in the normal time, but can supply subsequent data far quicker.

The burst interface, which is used on processors from Motorola, Intel, AMD, MIPs and many other manufacturers gains its performance by fetching data from the memory in bursts from a line of sequential locations. It makes use of a burst fill technique where the processor will access typically four words in succession, enabling a complete cache line to be fetched or written out to memory. The improved speed is obtained by taking advantage of page mode or static column memory. These type of memories offer faster access times — single cycle in many cases — after the initial access is made.

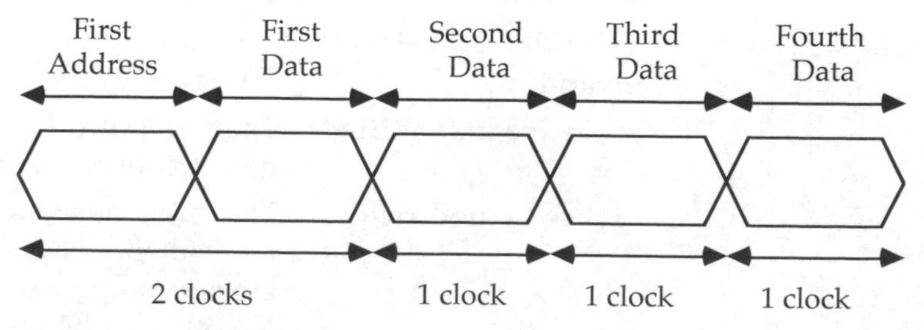

A burst fill interface

The advantage is a reduction in clock cycles needed to fetch the same amount of data. To fetch four words with a three clock memory cycle takes 12 clocks. Fetching the same amount of data using a 2-1-1-1 burst (two clocks for the first access, single cycles for the remainder) takes only five clocks. This type of interface gives a good fit with the page mode DRAM where the first access is used to set up the page access and the remainder of the burst accesses addresses within the page, thus taking advantage of the faster access.

Burst fill offers advantages of faster and more efficient memory accesses, but there are some fundamental changes in its operation when compared with single access buses. This is particularly so with SRAM when it is used as part of a cache:

- The address is only supplied for the first access in a burst and not for the remaining accesses. External logic is required to generate the additional addresses for the memory interface.
- The timing for each data access in the burst sequence is unequal: typical clock timings are 2-1-1-1 where two clocks are taken for the first access, but subsequent accesses are single cycle.
- The subsequent single cycle accesses compress address generation, set-up and hold and data access into a single cycle, which can cause complications generating write pulses to write data into the SRAM, for example.

These characteristics lead to conflicting criteria within the interface: during a read cycle, the address generation logic needs to change the address to meet set-up and hold times for the next access, while the current cycle requires the address to remain constant during its read access. With a write cycle, the need to change the address for the next cycle conflicts with the write pulse and constant address required for a successful write.

Meeting the interface needs

For a designer implementing such a system there are four methods of improving the SRAM interface and specification to meet the timing criteria:

- Use faster memory.
- Use synchronous memory with on-chip latches to reduce gate delays.
- Choose parts with short write pulse requirements and data set-up times.
- Integrate address logic on-chip to remove the delays and give more time.

While faster and faster memories are becoming available, they are more expensive, and memory speeds are now becoming limited by on- and off-chip buffer delays rather than the cell access times. The latter three methods depend on semiconductor manufacturers recognising the designer's difficulties and providing static RAMs which interface better with today's high performance processors.

This approach is beneficially for many high speed processors, but it is not a complete solution for the burst interfaces. They still need external logic to generate the cyclical addresses from the presented address at the beginning of the burst memory access. This increases the design complexity and forces the use of faster memories than is normally necessary simply to cope with the propagation delays. The obvious step is to add this logic to the latches and registers of a synchronous memory to create a protocol specific memory that supports certain bus protocols. The first two members of Motorola's protocol specific products are the MCM62940 and MCM62486 32k × 9 fast static RAMs. They are, as their part numbering suggests, designed to support the MC68040 and the Intel 80486 bus burst protocols. These parts offer access times of 15 and 20 ns.

The first access may take two processor clocks but remaining accesses can be made in a single cycle. There are some restrictions to this: the subsequent accesses must be in the same memory page and the processor must have somewhere to store the extra data that can be collected. The obvious solution is to use this burst interface to fill a cache line. The addresses will be in the same page and by storing the extra data in a cache allows a processor to use it at a later date without consuming additional bus bandwidth. The main problem faced by designers with these interfaces is the generation of the new addresses. In most designs the processor will only issue the first address and will hold this constant during the extra accesses. It is up to the interface logic to take this address and increment it with every access. With processors like the MC68030, this function is a straight incremental count. With the MC68040, a wrap-around burst is used where the required data is fetched first and the rest of the line fetched, wrapping around to the line beginning if necessary. Although more efficient for the processor, the wrap-around logic is more complicated.

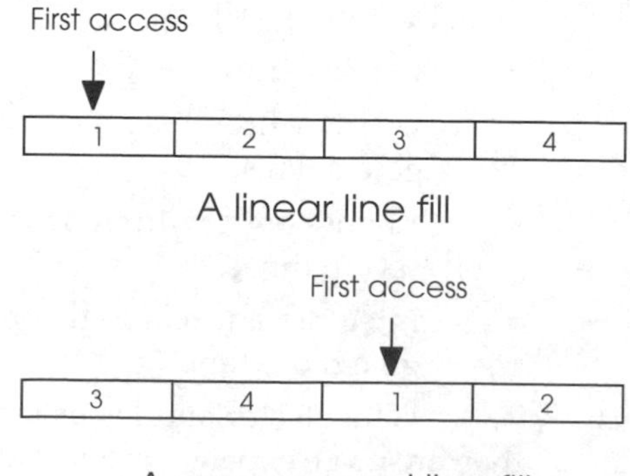

Linear and wrap-around line fills

The solution is to add this logic along with latches and registers to a memory to create a specific part that supports certain bus protocols. The first two members of Motorola's protocol specific products are the MCM62940 and MCM62486 32k × 9 fast static RAMs. They are, as their part numbering suggests, designed to support the MC68040 and the Intel 80486 bus burst protocols. These parts offer access times of 15 and 20 ns.

The MCM62940 has an on-chip burst counter that exactly matches the MC68040 wrap-around burst sequence. The address and other control data can be stored either by using the asynchronous or synchronous signals from the MC68040 depending on the design and its needs. A late write abort is supported which is useful in cache designs where cache writes can be aborted later in the cycle than normally expected, thus giving more time to decide whether the access should result in a cache hit or be delayed while stale data is copied back to the main system memory.

The MCM62486 has an on-chip burst counter that exactly matches the Intel 80486 burst sequence, again removing external logic and time delays and allowing the memory to respond to the processor without the need for the wait state normally inserted at the cycle start. In addition, it can switch from read to write mode while maintaining the address and count if a cache read miss occurs, allowing cache updating without restarting the whole cycle.

Big and little endian

There are two methods of organising data within memory depending on where the most significant bit is located. The Intel 80x86 and Motorola 680x0 and PowerPC processors use different organisations and this can cause problems.

The PowerPC architecture uses primarily a big endian byte order, i.e. an address points to the most significant byte of a value in memory. This can cause problems with other processors that use the alternative little endian organisation, where an address points to the least significant byte.

The PowerPC architecture solves this problem by providing a mode switch which causes all data memory references to be performed in a little-endian fashion. This is done by swapping address bit lines instead of using data multiplexers. As a result, the byte swapping is not quite what may be expected and varies depending on the size of the data. It is important to remember that swapping the address bits only reorders the bytes and not the individual bits within the bytes. The bit order remains constant.

The diagram shows the different storage formats for big and little endian double words, words, half words and bytes. The most significant byte in each pair is shaded to highlight its position. Note that there is no difference when storing individual bytes.

An alternative solution for processors that do not implement the mode swapping is to use the load and store instructions that byte reverse the data as it moves from the processor to the memory and vice versa.

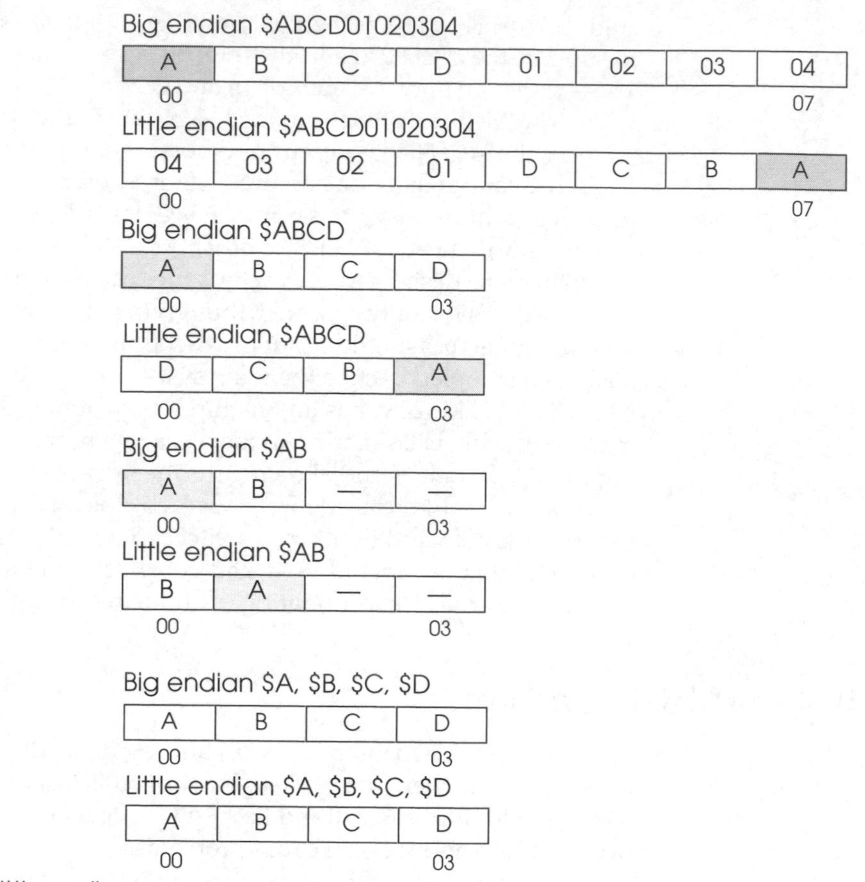

Big versus little endian memory organisation

Dual port and shared memory

Dual port and shared memory are two types of memory that offer similar facilities, i.e. the ability of two processors to access the same memory and thus share data and/or programs. It is often used as a communication mechanism between processors. The difference between them concerns how they cope with two simultaneous accesses.

With dual port memory, such bus contention is resolved within the additional circuitry that is contained with the memory chip or interface circuitry. This usually consists of buffers that are used as temporary storage for one processor while the other accesses the memory. Both the memory accesses are completed as if there were only a single access.

The buffered information is transferred when the memory is available. If both accesses are writes to the same memory address, the first one to access the memory is normally given priority but this should not be assumed. Many systems consider this a programming error and use semaphores in conjunction with special test and set instructions to prevent this happening.

Shared memory resolves the bus contention by holding one of the processors off by inserting wait states into its memory access. This results in lost performance because the held off processor cannot do anything and has to wait for the other to complete. As a result, both processors lose performance because they are effectively sharing the same bus.

Shared memory is easier to design and is often used when large memory blocks are needed. Dual port memory is normally implemented with special hardware and is limited to relatively small memory blocks of a few kbytes.

Bank switching

Bank switching simply involves having several banks of memory with the same address locations. At any one time, only one bank of memory is enabled and accessible by the microprocessor. Bank selection is made by driving the required bank select line. These lines come from either an external parallel port or latch whose register(s) appear as a bank switching control register within the processors's normal address map.

In the example, the binary value 1000 has been loaded into the bank selection register. This asserts the select line for bank 'a' which is subsequently accessed by the processor. Special VLSI (very large scale integration) parts were even developed which provided large number of banks and additional control facilities: the Motorola MC6883 SAM is an example used in the Dragon MC6809-based home computer from the early 1980s.

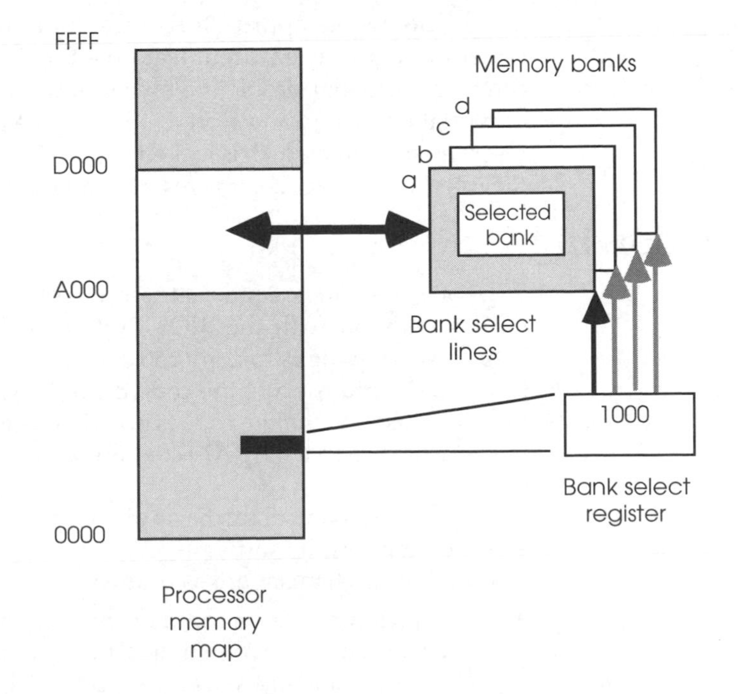

Bank switching

Memory overlays

Program overlays are similar to bank switching except that some form of mass storage is used to contain the different overlays. If a particular subroutine is not available, the software stores part of its memory as a file on disk and loads a new program section from disk into its place. Several hundred kilobytes of program, divided into smaller sections, can be made to overlay a single block of memory.

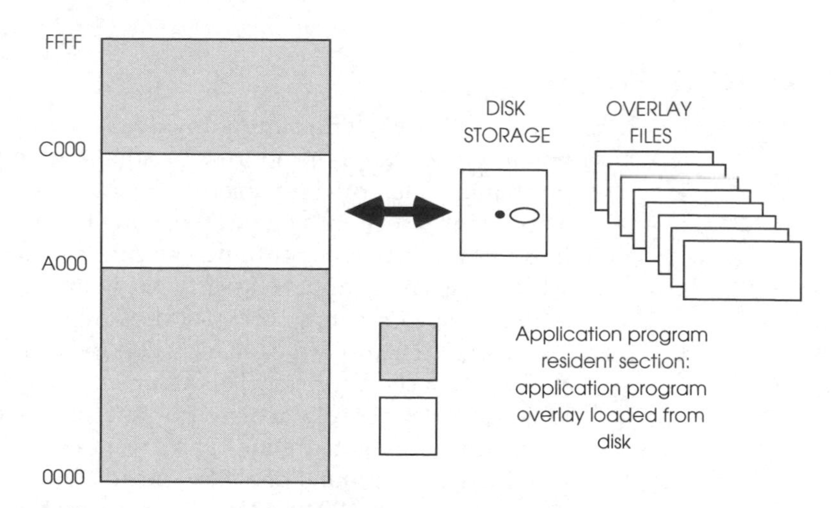

Program overlays, making a large program fit into the 64 kilobyte memory

This whole approach requires careful design so that the system integrity is ensured. Data passing between the overlays can be a particular problem which requires careful design. Typically, data is passed and stored either on the processor stack or in reserved memory which is locked in and does not play any part in the overlay process, i.e. it is resident all the time.

Shadowing

This is a technique that is probably best known from its implementation with the BIOS ROMs used in a PC. The idea behind shadowing is to copy the contents of the slow ROM into faster RAM and execute the code from the RAM. As a result the time taken to execute the code is greatly reduced. The shadowing refers to the fact that the RAM contains a copy of the original ROM contents.

This mechanism can be implemented either with hardware assist or entirely in software. The basic principles behind the shadowing mechanism are as follows:

- Typically the ROM contains the start up code as well as the system software. When the CPU is reset it will start executing this start-up code. As part of the initialisation, the contents of the ROM is copied into the RAM area where it can be executed. This part is common to both implementations.

- With a hardware assisted implementation, the address decode logic is used to switch the address decode to select the RAM instead of the ROM. As a result, any access to the ROM will be automatically switched to the RAM and will execute faster and without any change in the software as the addressing has not changed. This also provides an option to execute the code out of ROM or RAM and this can be used to isolate problems when executing out of RAM. If there are software-based timing routines in the ROM code, then this will execute faster when they are executed out of RAM and can cause problems. Virtually all IBM PCs implement their shadowing for the BIOS ROMs using this technique. It is also possible to use a MMU to perform the address translation if needed.
- In the pure software-based system, the software that is copied is now in a different memory location and providing the software was compiled and linked to execute in this location, there is no need to use any memory address translation. The code can simply be executed. In this case, the ROM is simply used to contain the code and in practice, running the software from this location is not intended. It is possible with position independent code and by changing the entry points into the code to execute it from the ROM but this requires some careful software design and management to ensure that this can be done. These techniques are covered in Chapter 7.

Example interfaces

MC68000 asynchronous bus

The MC68000 bus is fundamentally different to the buses used on the MC6800 and MC6809 processors. Their buses were synchronous in nature and assumed that both memory and peripherals could respond within a cycle of the bus. The biggest drawback with this arrangement concerned system upgrading and compatibility. If one component was uprated, the rest of the system needed uprating as well. It was for this reason that all M6800 parts had a system rating built into their part number. If a design specified an MC6809B, then it needed 2 MHz parts and subsequently, could not use an 'A' version which ran at 1 MHz. If a design based around the 1 MHz processor and peripherals was upgraded to 2 MHz, all the parts would need replacing. If a peripheral was not available at the higher speed, the system could not be upgraded. With the increasing processor and memory speeds, this restriction was unacceptable.

The MC68000 bus is truly asynchronous: it reads and writes data in response to inputs from memory or peripherals which may appear at any stage within the bus cycle. Provided certain signals meet certain set-up times and minimum pulse widths, the processor can talk to anything. As the bus is truly asynchronous it will wait indefinitely if no reply is received. This can cause similar symptoms to a hung processor; however, most system designs use a watchdog timer and the processor bus error signal to resolve this problem.

A typical bus cycle starts with the address, function codes and the read/write line appearing on the bus. Officially, this data is not valid until the address strobe signal AS* appears but many designs start decoding prior to its appearance and use the AS* to validate the output. The upper and lower data strobes, together with the address strobe signal (both shown as DS*), are asserted to indicate which bytes are being accessed on the bus. If the upper strobe is asserted, the upper byte is selected. If the lower strobe is asserted, the lower byte is chosen. If both are asserted together, a word is being accessed.

Once complete, the processor waits until a response appears from the memory or peripheral being accessed. If the rest of the system can respond without wait states (i.e. the decoding and access times will be ready on time) a DTACK* (Data Transfer ACKnowledge) signal is returned. This occurs slightly before clock edge S4. The data is driven onto the bus, latched and the address and data strobes removed to acknowledge the receipt of the DTACK* signal by the processor. The system responds by removing DTACK* and the cycle is complete. If the DTACK* signal is delayed for any reason, the processor will simply insert wait states into the cycle. This allows extra time for slow memory or peripherals to prepare data.

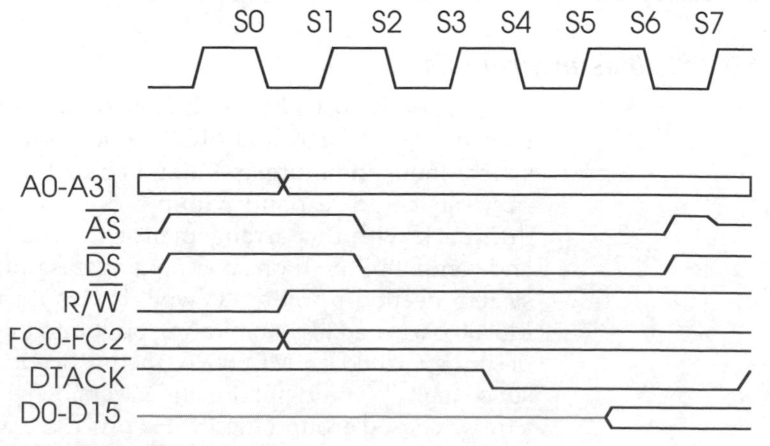

An MC68000 asynchronous bus cycle

The advantages that this offers are many fold. First, the processor can use a mixture of peripherals with different access speeds without any particular concern. A simple logic circuit can generate DTACK* signals with the appropriate delays as shown. If any part of the system is upgraded, it is a simple matter to adjust the DTACK* generation accordingly. Many M68000 boards provide jumper fields for this purpose and a single board and design can support processors running at 8, 10, 12 or 16 MHz. Secondly, this type of interface is very easy to interface to other buses and peripherals. Additional time can be provided to allow signal translation and conversion.

M6800 synchronous bus

Support for the M6800 synchronous bus initially offered early M68000 system designers access to the M6800 peripherals and allowed them to build designs as soon as the processor was available. With today's range of peripherals with specific M68000 interfaces, this interface is less used. However, the M6800 parts are now extremely inexpensive and are often used in cost-sensitive applications.

The additional signals involved are the E clock, valid memory address (VMA*) and valid peripheral address (VPA*). The cycle starts in a similar way to the M68000 asynchronous interface except that DTACK* is not returned. The address decoding generates a peripheral chip select which asserts VPA*. This tells the M68000 that a synchronous cycle is being performed.

The address decoding monitors the E clock signal, which is derived from the main system clock, but is divided down by 10 with a 6:4 mark/space ratio. It is not referenced from any other signal and is free running. At the appropriate time (i.e. when E goes low) VMA* is asserted. The peripheral waits for E to go high and transfers the data. When E goes low, the processor negates VMA* and the address and data strobes to end the cycle.

For systems running at 10 MHz or lower, standard 1 MHz M6800 parts can be used. For higher speeds, 1.5 or 2 MHz versions must be employed. However, higher speed parts running at a lower clock frequency will not perform the peripheral functions at full performance.

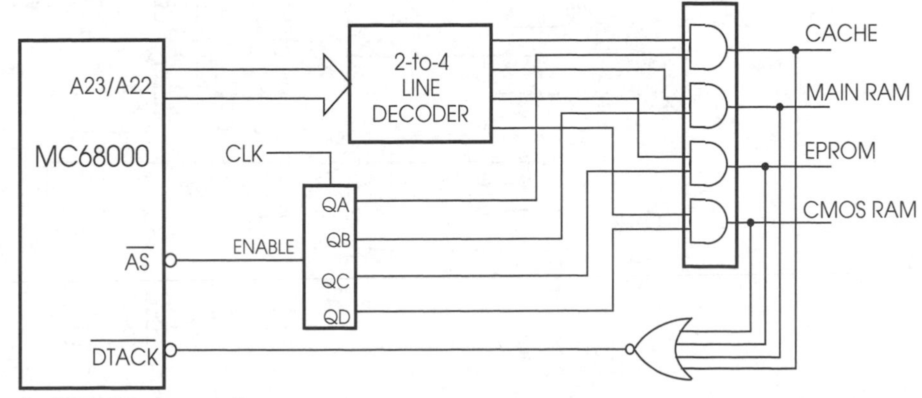

Example DTACK generation*

The MC68040 burst interface

Earlier in this chapter, some of the problems faced by a designer using SRAM with a burst interface were discussed. The MC68040 burst interface shows how these conflicts arise and their solution. It operates on a 2-1-1-1 basis, where two clock periods are allocated for the first access, and the remaining accesses are performed each in a single cycle. The first function the interface must perform is to generate the toggled A2 and A3 addresses from the first address put out by the MC68040. This involves creating a modulo 4 counter where the addresses will increment and wrap around. The MC68040 uses the burst access to fetch four long words for the internal cache line. It will start anywhere in the line so that the first data that is accessed can be passed to the processor while the rest of the data is fetched in parallel. This improves performance by fetching the immediate data first, but it does complicate the address generation logic — a standard 2 bit counter is not applicable. A typical circuit is shown.

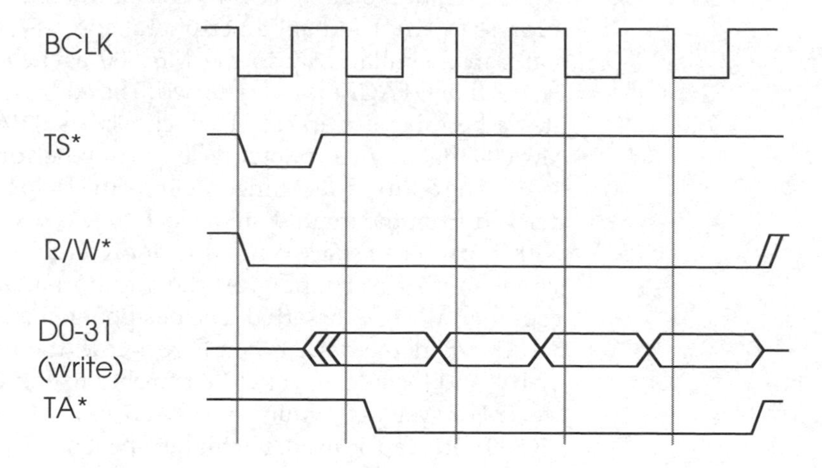

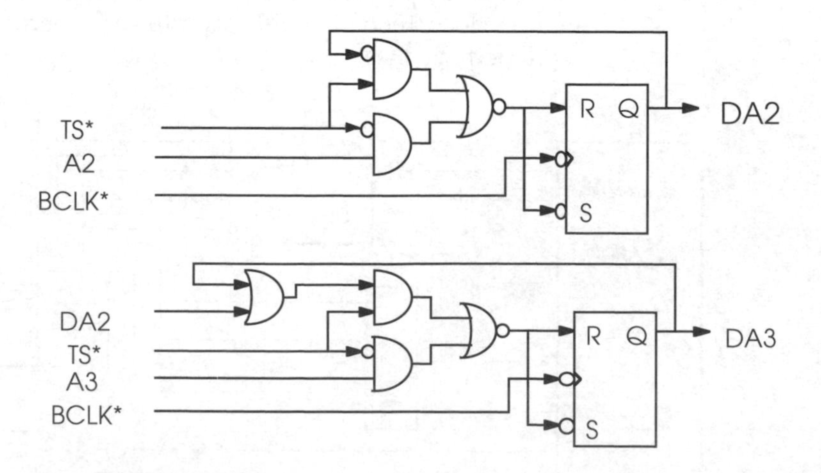

Modulo 4 counter (based on a design by John Hansen, Motorola Austin)

Given the generated addresses, the hardest task for the interface is to create the write pulse needed to write data to the FSRAMs. The first hurdle is to ensure that the write pulse commences after the addresses have been generated. The easiest way of doing this is to use the two phases of the BCLK* to divide the timing into two halves. During the first part, the address is latched by the rising edge of BCLK*.

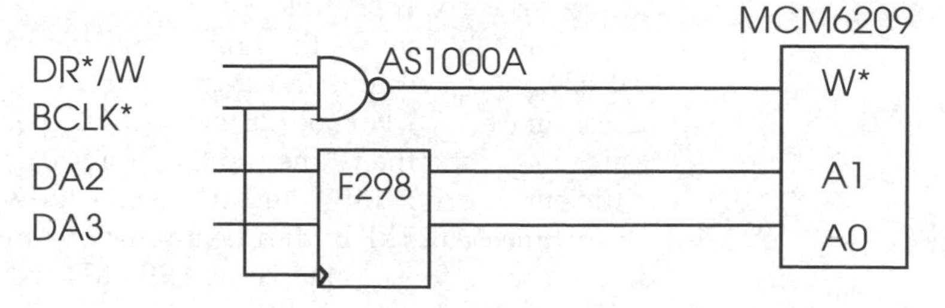

*Latching the address and gating W**

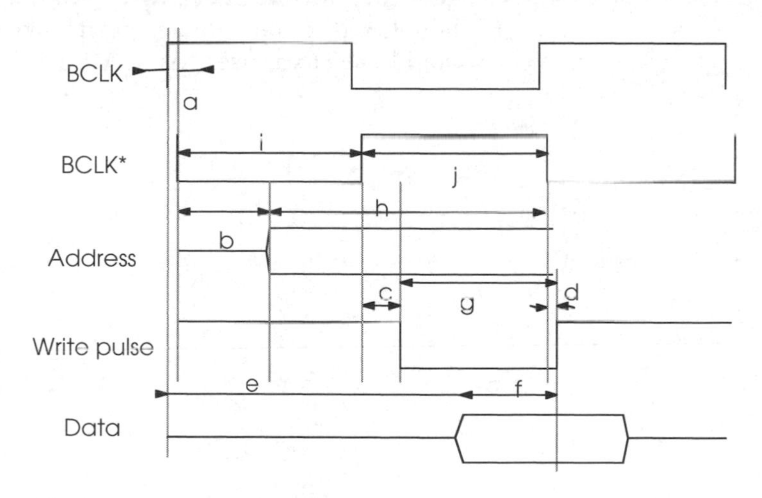

Timing	Description
a	Clock skew between BCLK and its inverted signal BCLK*.
b	Delay between BCLK* and valid address — determined by latch delay.
c	Gate delay in generating Write pulse from rising BCLK* edge.
d	Gate delay in terminating Write pulse from falling BCLK* edge.
e	Time from rising edge of BCLK to valid data from MC68040.
f	Data set-up time for write referenced from = i+j-e+a.
g	Write pulse width = j-c+d.
h	Valid address, i.e. memory access time.
i,j	Cycle times for BCLK and BCLK*.

Write pulse timings

Latching DA2 and DA3 holds the address valid while allowing the modulo 4 counter to propagate the next value through. The falling edge of BCLK* is then used to gate the read/write signal to create a write pulse. The write pulse is removed before the next address is latched. This guarantees that the write pulse will be generated after the address has become valid. This circuit neatly solves the competing criteria of bringing the write pulse high before the address can be changed and the need to change the address as early as possible

The table shows the timing and the values for the write pulse, t_{WLWH}, write data set-up time, t_{DVWH} and the overall access time t_{AVAV}. For both 25 and 33 MHz speeds, the access time is always greater than 20 ns and therefore 20 ns FSRAM would be sufficient. The difficulty comes in meeting the write pulse and data set-up times. At 25 MHz, the maximum write pulse is 17 ns and the data set-up is 9 ns. Many 20 ns FSRAMs specify the minimum write pulse width with the same value as the overall access time. As a result 20 ns access time parts would not meet this specification. The data set-up is also longer and it is likely that 15 ns or faster parts would have to be used. At 33 MHz, the problem is worse.

4 Basic peripherals

This chapter describes the basic peripherals that most microcontrollers provide. It covers parallel ports which are the simplest I/O devices, timer counters for generating and measuring time- and count-based events, serial interfaces and DMA controllers.

Parallel ports

Parallel ports provide the ability to input or output binary data with a single bit allocated to each pin within the port. They are called parallel ports because the initial chips that provided this support grouped several pins together to create a controllable data port similar to that used for data and address buses. It transfers multiple bits of information simultaneously, hence the name parallel port. Although the name implies that the pins are grouped together, the individual bits and pins within the port can usually be used independently of each other.

These ports are used to provide parallel interfaces such as the Centronics printer interface, output signals to LEDs and alphanumeric displays and so on. As inputs, they can be used with switches and keyboards to support control panels.

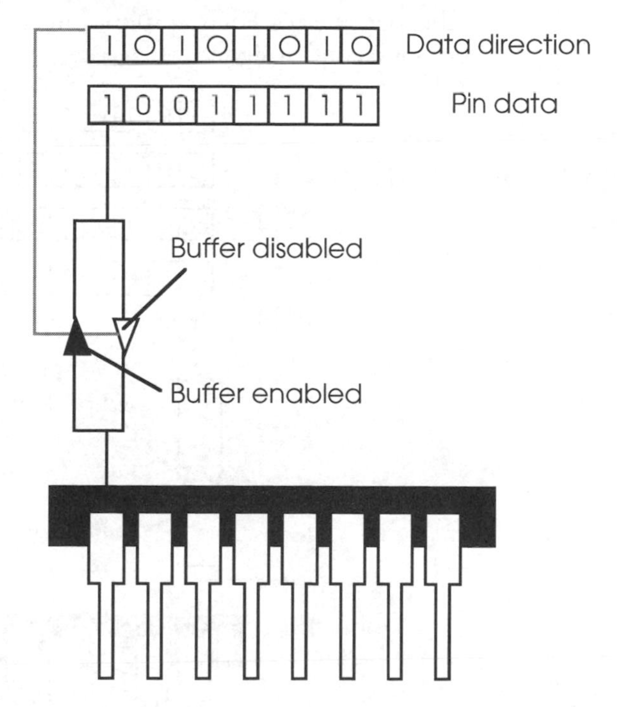

A simple parallel I/O port

The basic operation is shown in the diagram which depicts an 8 pin port. The port is controlled by two registers: a data direction register which defines whether each pin is an output or an input and a data register which is used to set an output value by writing to it and to obtain an input value by reading from it. The actual implementation typically uses a couple of buffers which are enabled depending on the setting of the corresponding bit in the data direction register.

This simple model is the basis of many early parallel interface chips such as the Intel 8255 and the Motorola 6821 devices. The model has progressed with the incorporation of a third register or an individual control bit that provides a third option of making the pin become high impedance and thus neither an input or output. This can be implemented by switching off both buffers and putting their connections to the pin in a high impedance state. Output ports that can do this are often referred to as tri-state because they can either be logic high, logic low or high impedance. In practice, this is implemented on-chip as a single buffer with several control signals from the appropriate bits within the control registers. This ability has led to the development of general-purpose ports which can have additional functionality to that of a simple binary input/output pin.

Multi-function I/O ports

With many parallel I/O devices that are available today, either as part of the on-chip peripheral set or as an external device, the pins are described as general-purpose and can be shared with other peripherals. For example, a pin may be used as part of a serial port as a control signal.

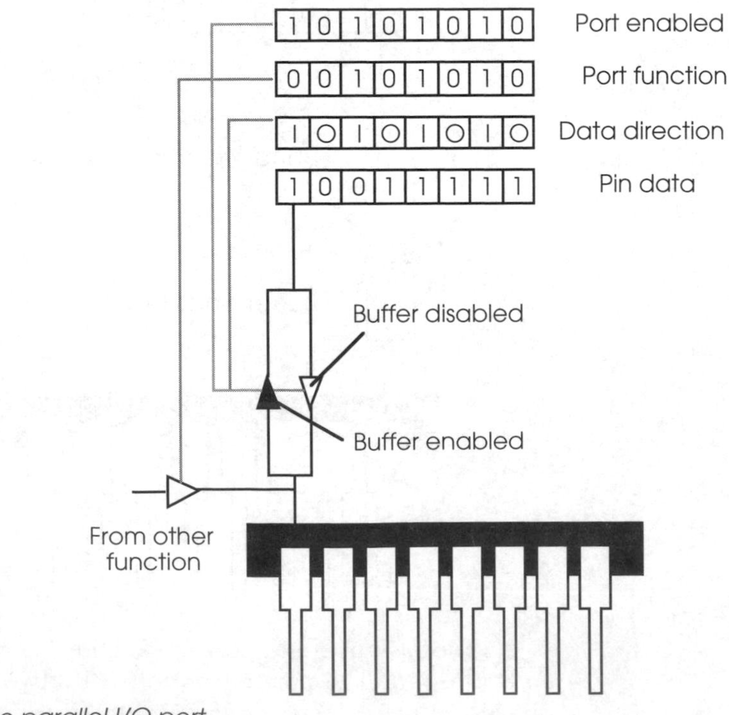

A general-purpose parallel I/O port

It may be used as a chip select for the memory design or simply as an I/O pin. The function that the pin performs is set up internally through the use of a function register which internally configures how the external pin is connected internally. If this is not set up correctly, then despite the correct programming of the other registers, the pin will not function as expected.

Note: This shared use does pose a problem for designers in that many manufacturer data sheets will specify the total number of I/O pins that are available. In practice, this is often reduced because pins need to be assigned as chip selects and to other essential functions. As a result, the number that is available for use as I/O pins is greatly reduced.

Pull-up resistors

It is important to check if a parallel I/O port or pin expects an external pull-up resistor. Some devices incorporate it internally and therefore do not need it. If it is needed and not supplied, it can cause incorrect data on reading the port and prevent the port from turning off an external device.

Timer/counters

Digital timer/counters are used throughout embedded designs to provide a series of time or count related events within the system with the minimum of processor and software overhead. Most embedded systems have a time component within them such as timing references for control sequences, to provide system ticks for operating systems and even the generation of waveforms for serial port baud rate generation and audible tones.

They are available in several different types but are essentially based around a simple structure as shown.

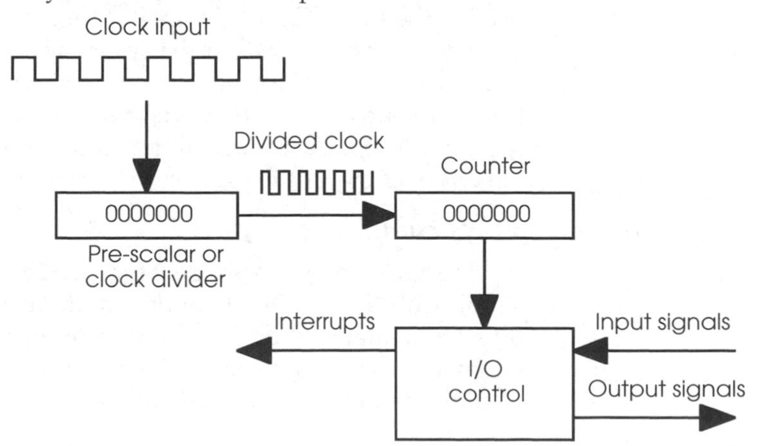

Generic timer/counter

The central timing is derived from a clock input. This clock may be internal to the timer/counter or be external and thus connected via a separate pin. The clock may be divided using a simple divider which can provide limited division normally based on a power of two or through a pre-scalar which effectively scales down or divides the clock by the value that is written into the pre-scalar register. The divided clock is then passed to a counter which is normally configured in a count-down operation, i.e. it is loaded with a preset value which is clocked down towards zero. When a zero count is reached, this causes an event to occur such as an interrupt of an external line changing state. The final block is loosely described as an I/O control block but can be more sophisticated than that. It generates interrupts and can control the counter based on external signals which can gate the count-down and provide additional control. This expands the functionality that the timer can provide as will be explained later.

Types

Timer/counters are normally defined in terms of the counter size that they can provide. They typically come in 8, 16 and 24 bit variants. The bit size determines two fundamental properties:

- The pre-scalar value and hence the frequency of the slowest clock that can be generated from a given clock input.
- The counter size determines the maximum value of the counter-derived period and when used with an external clock, the maximum resolution or measurement of a time-based event.

These two properties often determine the suitability of a device for an application.

8253 timer modes

A good example of a simple timer is the Intel 8253 which is used in the IBM PC. The device has three timer/counters which provide a periodic 'tick' for the system clock, a regular interrupt every 15 μs to perform a dynamic memory refresh cycle and, finally, a source of square waveforms for use as audio tones with the built-in speaker. Each timer/counter supports six modes which cover most of the simple applications for timer/counters.

Interrupt on terminal count

This is known as mode 0 for the 8253 and is probably the simplest of its operations to understand. An initial value is loaded into the counter register and this then immediately starts to count down at the frequency determined by the clock input. When the counter reaches zero, an interrupt is generated.

Programmable one-shot

With mode 1, it is possible to create a single pulse with a programmable duration. The pulse length is first loaded into the counter. Nothing further happens until the external gate signal is pulled high. This rising edge starts the counter to count down towards zero and the counter output signal goes high to start the external pulse. When the counter reaches zero, the counter output goes low thus ending the pulse.

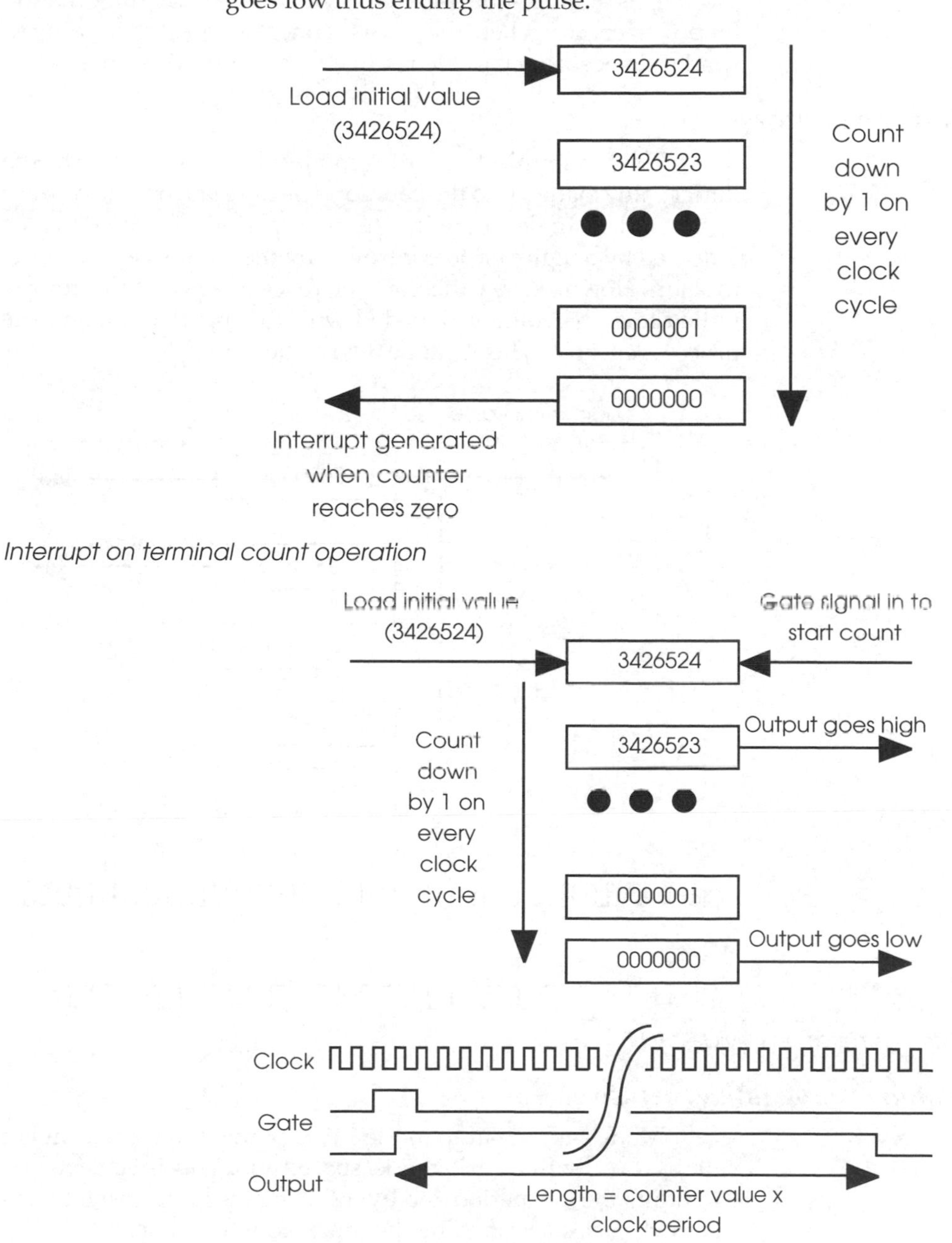

Interrupt on terminal count operation

Programmable one-shot timer counter mode

The pulse duration is determined by the initial value loaded into the counter times the clock period. While this is a common timer/counter mode, many devices such as the 8253 incorporate a reset. If the gate signal is pulled low and then high again to create a new rising edge while the counter is counting down, the current count value is ignored and replaced by the initial value and the count continued. This means that the output pulse length will be extended by the time between the two gate rising edges.

This mode can be used to provide pulse width modulation for power control where the gate is connected to a zero crossing or similar detector or clock source to create a periodic signal.

Rate generator

This is a simple divide by *N* mode where *N* is defined by the initial value loaded into the counter. The output from the counter consists of a single low with the time period of a single clock followed by a high period controlled by the counter which starts to count down. When the counter reaches zero, the output is pulled low, the counter reloaded with the initial value and the process repeated. This is mode 3 with the 8253.

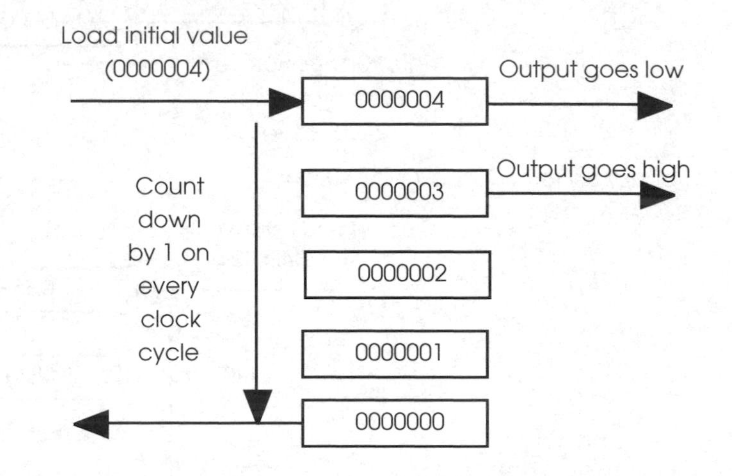

Rate generation (divide by N)

Square wave rate generator

Mode 4 is similar to mode 3 except that the waveform is a square wave with a 50:50 mark/space ratio. This is achieved by extending the low period and by reducing the high period to half the clock cycles specified by the initial counter value.

Software triggered strobe

When mode 4 is enabled, the counter will start to count as soon as it is loaded with its initial value. When it reaches zero, the output is pulsed low for a single clock period and then goes high again. If the counter is reloaded by software before it reaches zero, the output does not go low. This can be used as a software-based watchdog timer where the output is connected to a non-maskable interrupt line or a system reset.

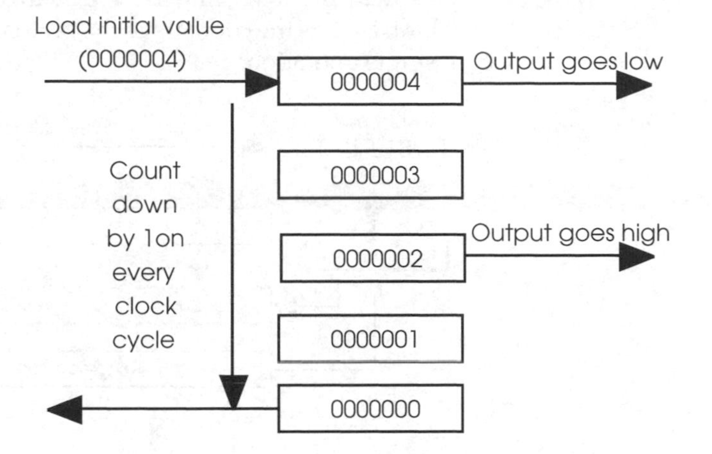

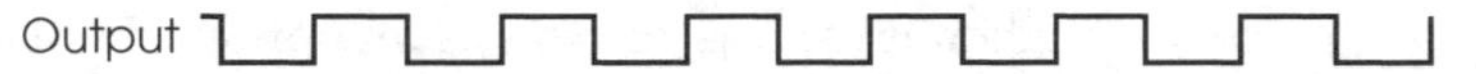

Square wave generation

Hardware triggered strobe

Mode 5 is similar to mode 4 except that the retriggering is done by the external gate pin acting as a trigger signal.

Generating interrupts

The 8253 has no specific interrupt pins and therefore the timer OUT pin is often used to generate an external interrupt signal. With the IBM PC, this is done by connecting the OUT signal from timer/counter 0 to the IRQ 0 signal and setting the timer/counter to run in mode 3 to generate a square wave. The input clock is 1.19318 MHz and by using a full 16 bit count value, is divided by 65536 to provide a 18.3 Hz timer tick. This is counted by the software to provide a time of day reference and to provide a system tick.

MC68230 modes

The Motorola MC68230 is a good example of a more powerful timer architecture that can provide a far higher resolution than the Intel 8253. The timer is based around a 24 bit architecture which is split into three 8 bit components. The reason for this is that the device uses an 8 bit bus to communicate with the host processor such as a MC68000 CPU. This means that the counter cannot be loaded directly from the processor in a single bus cycle. As a result, three preload registers have been added to the basic architecture previously described. These are preloaded using three separate accesses prior to writing to the Z control bit in the control register. This transfers the contents of the preload register to the counter as a single operation.

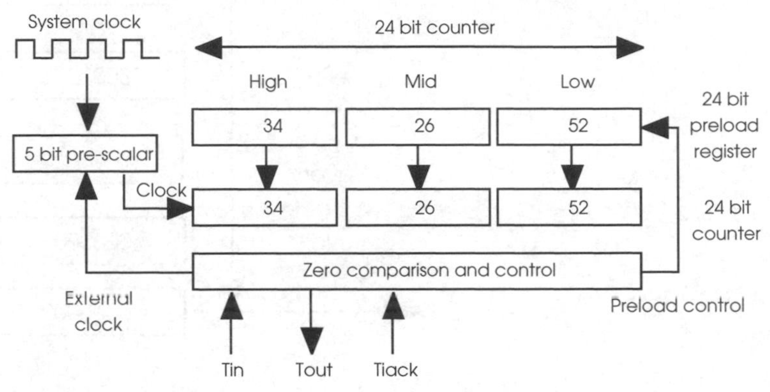

The MC68230 timer/counter architecture

Instead of writing to the counter to either reset it or initialise it, the host processor uses a combination of preload registers and the Z bit to control the timer. The timer can be made to preload when it reaches zero or, as an option, simply carry on counting. This gives a bit more flexibility in that timing can be performed after the zero count as well as before it.

This architecture also has a 5 bit pre-scalar which is used to divide the incoming clock which can be sourced from the system clock or externally via the Tin signal. The pre-scalar can be loaded with any 5 bit value to divide the clock before it drives the counter.

Timer processors

An alternative to using a timer / counter is the development of timer computers where a processor is used exclusively to manage and implement complex timing functions over multiple timer channels. The MC68332 is a good example of such a processor. It has a CPU32 processor (MC68020 based) running at 16 MHz and a timer processor unit instead of a communications processor. This has 16 channels which are controlled by a RISC-like processor to perform virtually any timing function. The timing resolution is down to 250 nanoseconds with an external clock source or 500 nanoseconds with an internal one. The timer processor can perform the common timer algorithms on any of the 16 channels without placing any overhead on the CPU32.

A queued serial channel and 2 kbits of power-down static RAM are also on-chip and for many applications all that is required to complete a working system is an external program EPROM and a clock. The timer processor has several high level functions which can easily be accessed by the main processor by programming a parameter block. For example, the missing tooth calculation for generating ignition timing can be easily performed through a combination of the timer processor and the CPU32 core. A set of parameters is calculated by the CPU32 and loaded into a parameter block which commands the timer processor to perform the algorithm. Again, no interrupt routines or periodic peripheral bit manipulation is needed by the CPU32.

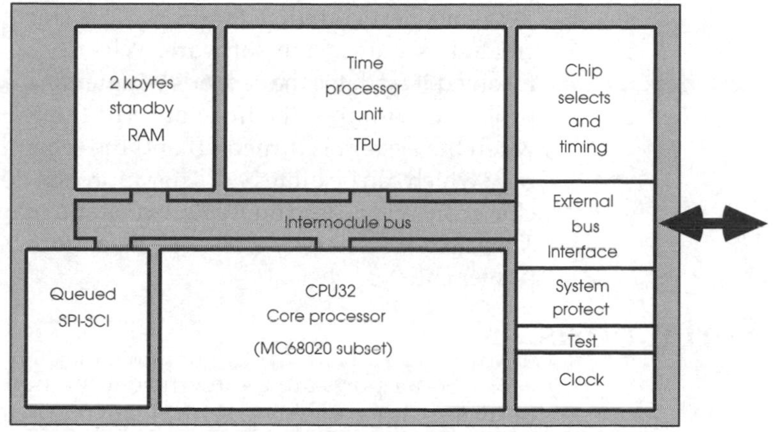

The MC68332 block diagram

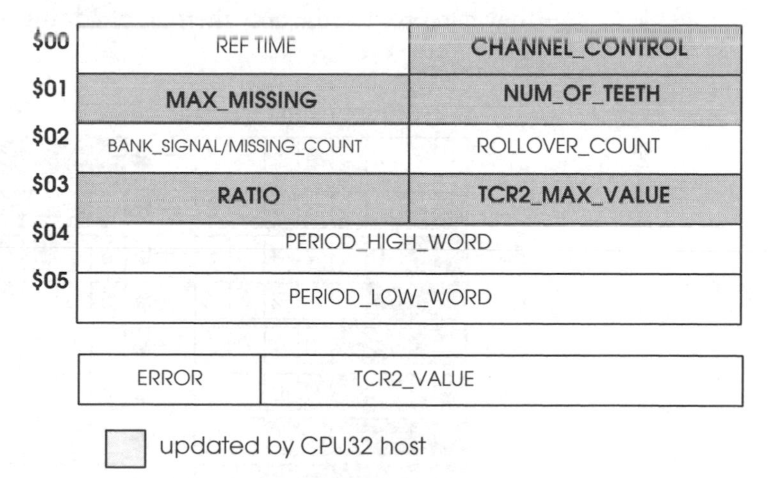

The parameter block for a period measurement with missing transition detection

Real-time clocks

There is a special category of timer known as a real-time clock whose function is to provide an independent time keeper that can provide time measurements in terms of the current time and date as opposed to a counter value. The most popular device is probably the MC146818 and its derivatives and clones that were used in the first IBM PC. These devices are normally driven off a 32 kHz watch crystal and are battery backed-up to maintain the data and time. The battery back-up was done externally with a battery or large capacitor but has also been incorporated into the chip in the case of the versions supplied by Dallas Semiconductor.

These devices can also provide a system tick signal for use by the operating system.

Simulating a real-time clock in software

These can be simulated in software by programming a timer to generate a periodic event and simply using this as a reference and counting the ticks. The clock functions are then created as part of the software. When enough ticks have been received it updates the seconds counter and so on. There are two problems with this: the first concerns the need to reset the clock when the system is turned off and the second concerns the accuracy which can be quite bad. This approach does save on a special clock chip and is used on VCRs, ovens and many other appliances. This also explains why they need resetting when there has been a power cut!

Serial ports

Serial ports are a pin efficient method of communicating between other devices within an embedded system. With microcontrollers which do not have an external extension bus, they can provide the only method of adding additional functionality.

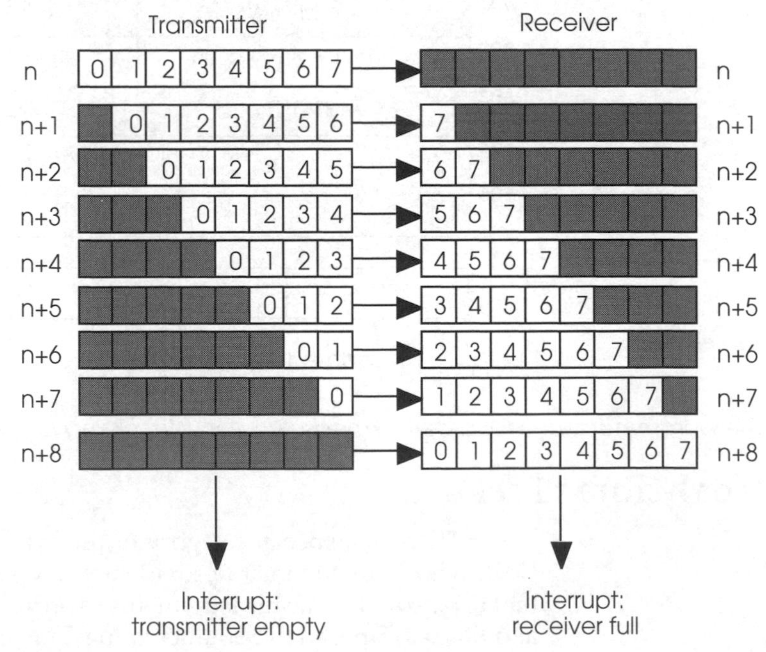

Basic serial port operation

The simplest serial ports are essentially a pair of shift registers that are connected together with one input (receiver) connected to the output of the other to create a transmitter. They are clocked together by a common clock and thus data is transmitted from one register to the other. The time taken is dependent on the clock frequency and the number of bits that are transferred. The shift registers are normally 8 bits wide. When the transmitter is emptied, it can be made to generate a local interrupt which can indicate to the processor that the byte has been transferred and/ or that the next byte should be loaded into the register. The receiver can also generate an interrupt when the complete byte is received to indicate that it is ready for reading.

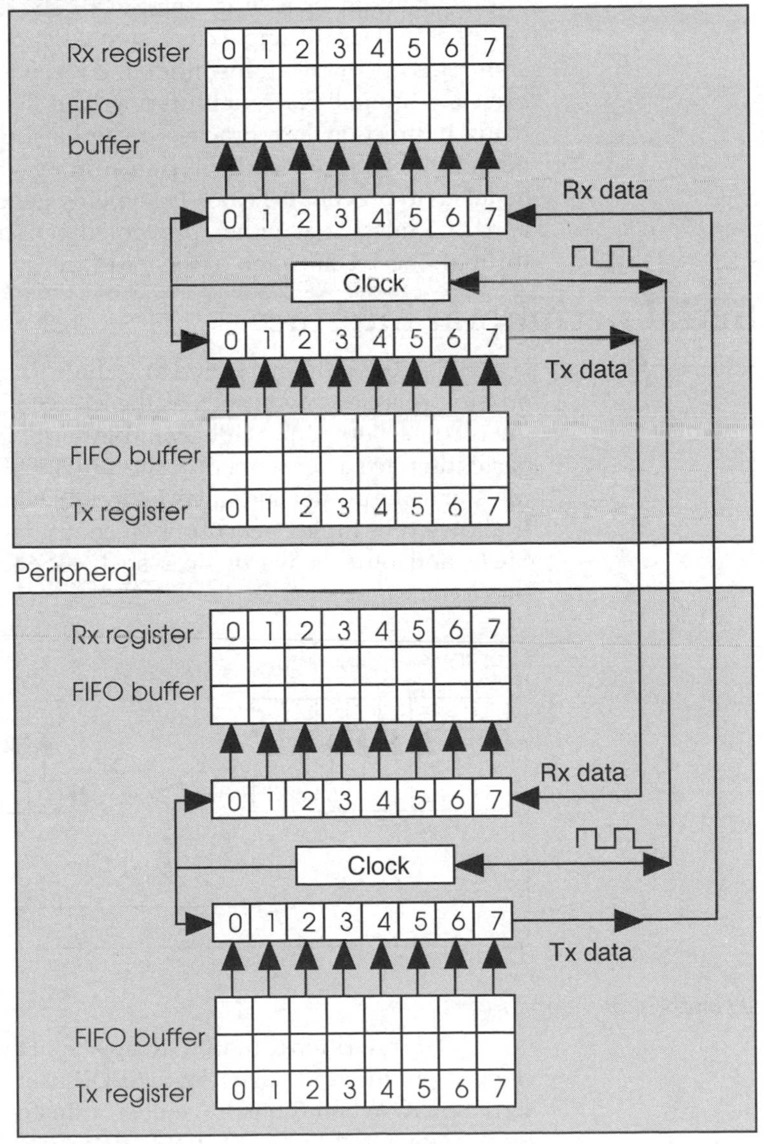

Serial interface with FIFO buffering

Most serial ports use a FIFO as a buffer so that data is not lost. This can happen if data is transmitted before the preceding byte has been read. With the FIFO buffer, the received byte is transferred to it when the byte is received. This frees up the shift register to receive more bits without losing the data. The FIFO buffer is read to receive the data hence the acronym's derivation — first in, first out. The reverse can be done for the transmitter so that data can be sent to the transmitter before the previous value has been sent.

The size of the FIFO buffer is important in reducing processor overhead and increasing the serial port's throughput as will be explained in more detail later on. The diagram shows a generic implementation of a serial interface between a processor and peripheral. It uses a single clock signal which is used to clock the shift registers in each transmitter and receiver. The shift registers each have a small FIFO for buffering. The clock signal is shown as being bidirectional: in practice it can be supplied by one of the devices or by the device that is transmitting. Obviously care has to be taken to prevent the clock from being generated by both sides and this mistake is either prevented by software protocol or through the specification of the interface.

Serial peripheral interface

This bus is often referred to as the SPI and is frequently used on Motorola processors such as the MC68HC05 and MC68HC11 microcontrollers to provide a simple serial interface. It uses the basic interface as described in the previous section with a shift register in the master and slave devices driven by a common clock. It allows full-duplex synchronous communication between the MCU and other slave devices such as peripherals and other processors.

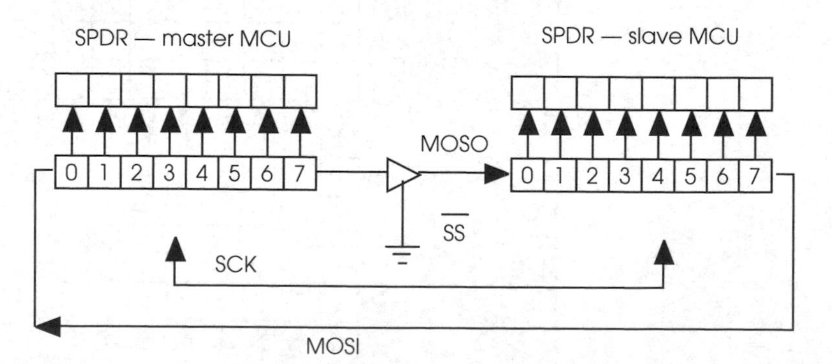

SPI internal architecture

Data is written to the SPDR register in the master device and clocked out into the slave device SPDR using the common clock signal SCK. When 8 bits have been transferred, an interrupt is locally generated so that the data can be read before the next byte is clocked through. The SS or slave select signal is used to select which slave is to receive the data. In the first example, shown with only one slave, this is permanently asserted by grounding the signal. With multiple slaves, spare parallel I/O pins are used to select the slave prior to data transmission. The diagram below shows such a configuration. If pin 1 on the master MCU is driven low, slave 1 is selected and so on. The unselected slaves tri-state the SPI connections and do not receive the clocked data and take no part in the transfer.

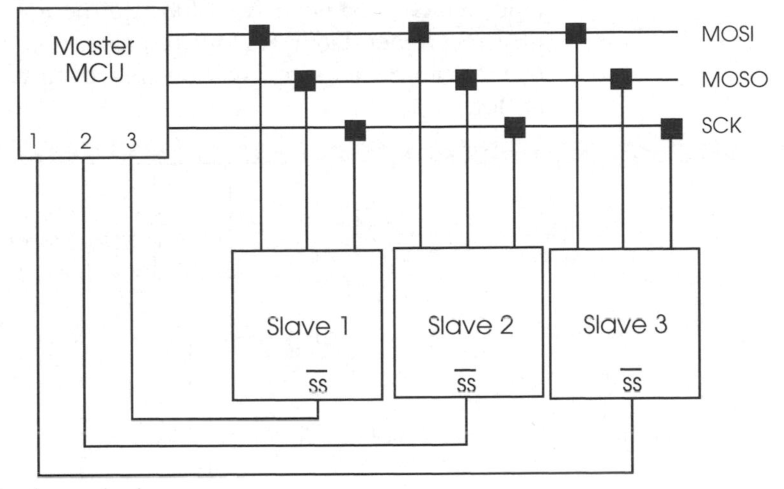

Supporting multiple slave devices

It should not be assumed that an implementation buffers data. As soon as the master writes the data into the SPDR it is transmitted as there is no buffering. As soon as the byte has been clocked out, an interrupt is generated indicating that the byte has been transferred. In addition, the SPIF flag in the status register (SPSR) is set. This flag must be cleared by the ISR before transmitting the next byte.

The slave device does have some buffering and the data is transferred to the SPDR when a complete byte is transferred. Again, an interrupt is generated when a byte is received. It is essential that the interrupt that is generated by the full shift register is serviced quickly to transfer the data before the next byte is transmitted and transferred to the SPDR. This means that there is an eight clock time period for the slave to receive the interrupt and transfer the data. This effectively determines the maximum data rate.

I²C bus

The inter-IC, or I²C bus as it is more readily known, was developed by Philips originally for use within television sets in the mid-1980s. It is probably the most known simple serial interface currently used. It combines both hardware and software protocols to provide a bus interface that can talk to many peripheral devices and can even support multiple bus masters. The serial bus itself only uses two pins for its implementation.

The bus consists of two lines called SDA and SCL. Both bus masters and slave peripheral devices simply attach to these two lines as shown in the diagram. For small numbers of devices and where the distance between them is small, this connection can be direct. For larger numbers of devices and/or where the track length is large, Philips can provide a special buffer chip (P82B715) to increase the current drive. The number of devices is effectively determined by the line length, clock frequency and load capacitance which must not exceed 400 pF although derating this to 200 pF is recommended. With low frequencies, connections of several metres can be achieved without resorting to special drivers or buffers.

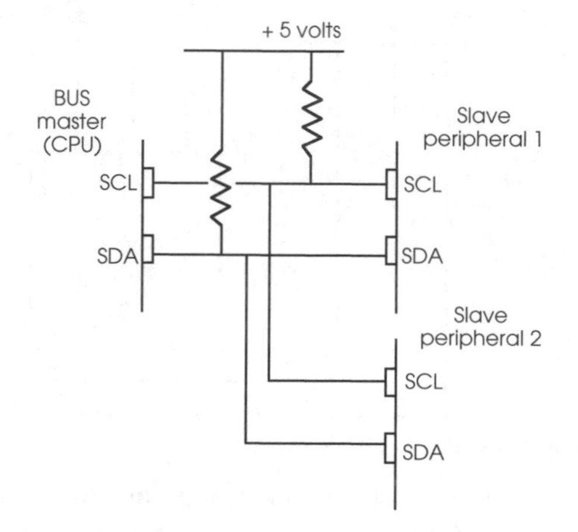

I^2C electrical connections

The drivers for the signals are bidirectional and require pull-up resistors. When driven they connect the line to ground to create a low state. When the drive is removed, the output will be pulled up to a high voltage to create a high signal. Without the pull-up resistor, the line would float and can cause indeterminate values and thus cause errors.

The SCL pin provides the reference clock for the transfer of data but it is not a free running clock as used by many other serial ports. Instead it is clocked by a combination of the master and slave device and thus the line provides not only the clock but also a hardware handshake line.

The SDA pin ensures the serial data is clocked out using the SCL line status. Data is assumed to be stable on the SDA line if SCL is high and therefore any changes occur when the SCL is low. The sequence and logic changes define the three messages used.

Message	1st event	2nd event
START	SDA H\L	SCL H\L
STOP	SCL L\H	SDA L\H
ACK	SDA H\L	SCL H\L

The table shows the hardware signalling that is used for the three signals, START, STOP and ACKNOWLEDGE. The START and ACKNOWLEDGE signals are similar but there is a slight difference in that the START signal is performed entirely by the master whereas the ACKNOWLEDGE signal is a handshake between the slave and master.

Data is transferred in packets with a packet containing one or more bytes. Within each byte, the most significant bit is transmitted first. A packet, or telegram as it is sometimes referred to, is defined as the data transmitted between START and STOP signals sent from the master. Within the packet transmission, the slave will acknowledge each byte by using the ACKNOWLEDGE signal. The basic protocol is shown in the diagram.

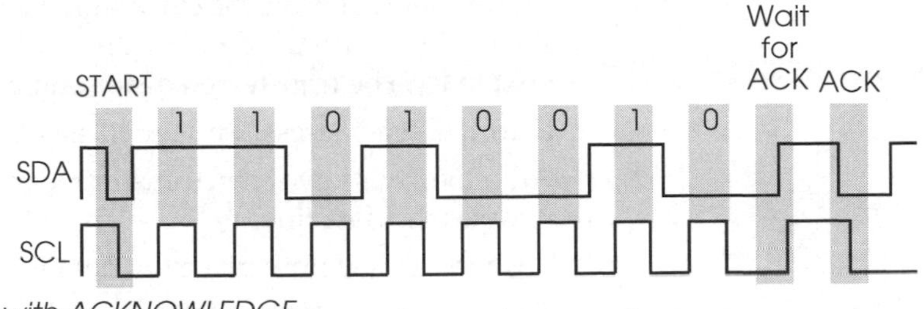

Write byte transfer with ACKNOWLEDGE

The 'wait for ACK' stage looks like another data bit except that it is located as the ninth bit. With all data being transmitted as bytes, this extra one bit is interpreted by the peripheral as an indication that the slave should acknowledge the byte transfer. This is done by the slave pulling the SDA line low after the master has released the data and clock line. The ACK signal is physically the same as the START signal and is ignored by the other peripherals because the data packet has not been terminated by the STOP command. When the ACKNOWLEDGE command is issued, it indicates that the transfer has been completed. The next byte of data can start transmission by pulling the SCL signal down low.

Read and write access

While the previous paragraphs described the general method of transferring data, there are some specific differences for read and write accesses. The main differences are concerned with who controls which line during the handshake.

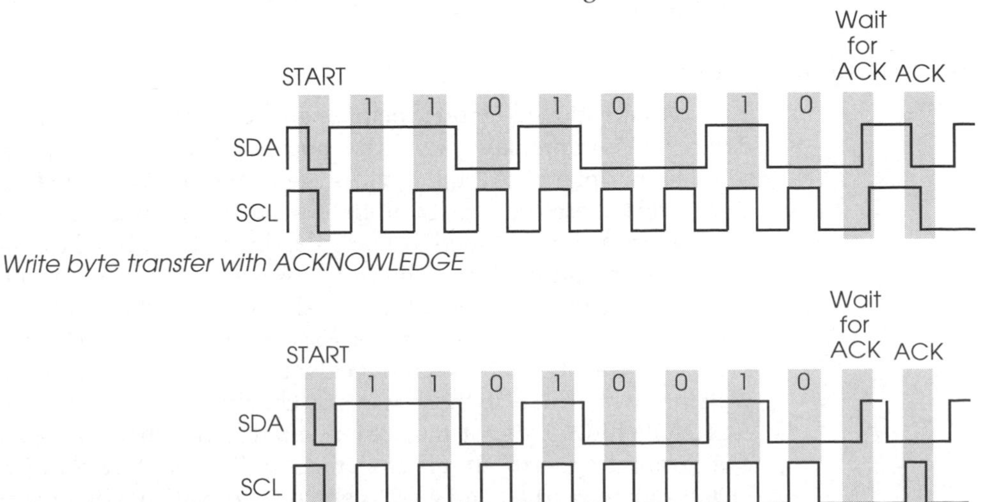

Write byte transfer with ACKNOWLEDGE

Read byte transfer with ACKNOWLEDGE

During a write access the sequence is as follows:

- After the START and 8 bits have been transmitted, the master releases the data line followed by the clock line. At this point it is waiting for an acknowledgement.
- The addressed slave will pull the data line down to indicate the ACKNOWLEDGE signal.
- The master will drive the clock signal low and in return, the slave will release the data line, ready for the first bit of the next byte to be transferred or to send a STOP signal.

During a read access the sequence is as follows:

- After the 8 bits have been transmitted by the slave, the slave releases the data line.
- The master will now drive the data line low.
- The master will then drive the clock line high and low to create a clock pulse.
- The master will then release the data line ready for the first bit of the next byte or a STOP signal.

It is also possible to terminate a transfer using a STOP instead of waiting for an ACKNOWLEDGE. This is sometimes needed by some peripherals which do not issue an ACKNOWLEDGE on the last transfer. The STOP signal can even be used in mid transmission of the byte if necessary.

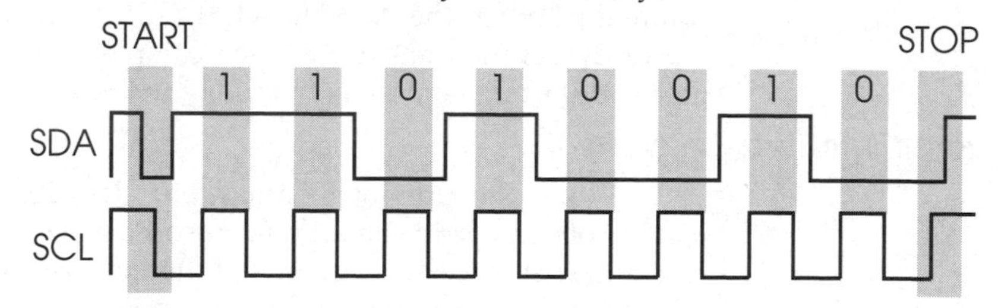

A Write byte transfer with STOP

Addressing peripherals

As mentioned before, the bus will support multiple slave devices. This immediately raises the question of how the protocol selects a peripheral. All the devices are connected onto the two signals and therefore can see all the transactions that occur. The slave selection is performed by using the first byte within the data packet as an address byte. The protocol works as shown in the diagram. The master puts out the START signal and this tells all the connected slave devices to start accepting the data. The address byte is sent out and each slave device compares the address with its own value. If there is a match, then it will send the ACKNOWLEDGE signal. If there is no match, then there has been a programming error. In this case, there will be no ACKNOWLEDGE signal returned and effectively the SDA signal will remain high.

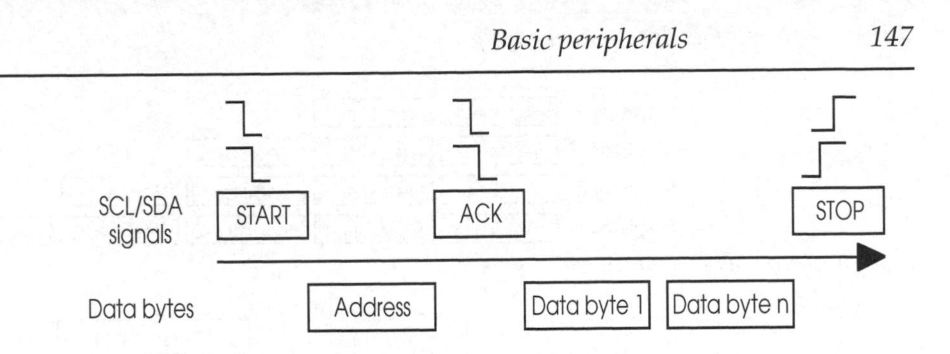

A complete data packet including addressing and signalling

The address value is used to select the device and to indicate the type of operation that the master requests: read if the eighth bit is set to one or write if set to zero. This means that the address byte allows 128 devices with read/write capability to be connected. The address value for a device is either pre-programmed, i.e. assigned to that device number, or can be programmed through the use of external pins. Care must be made to ensure that no two devices have the same address.

In practice, the number of available addresses is less because some are reserved and others are used as special commands. To provide more addressing, an extended address has been developed that uses two bytes: the first byte uses a special code (five ones) to distinguish it from a single byte address. In this way both single byte and double byte address slaves can be used on the same bus.

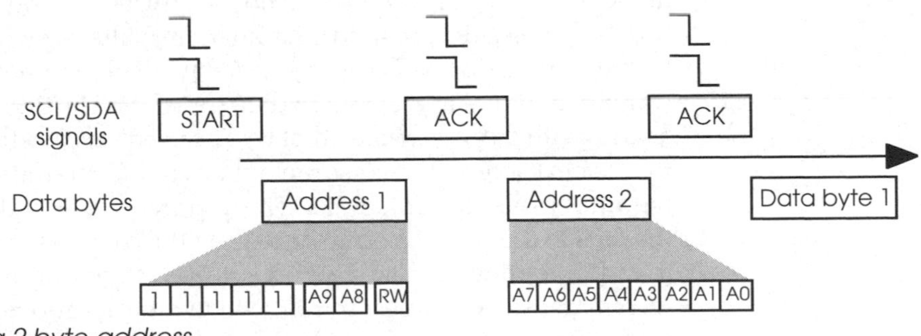

Sending a 2 byte address

Sending an address index

So far the transfers that have been described have assumed that the peripheral only has one register or memory location. This is rarely the case and thus several addressing schemes have been developed to address individual locations within the peripheral itself.

For peripherals with a small number of locations, a simple technique is simply to incorporate an auto-incrementing counter within the peripheral so that each access selects the next register. As the diagram shows, the contents of register 4 can be accessed by performing four successive transfers after the initial address has been sent.

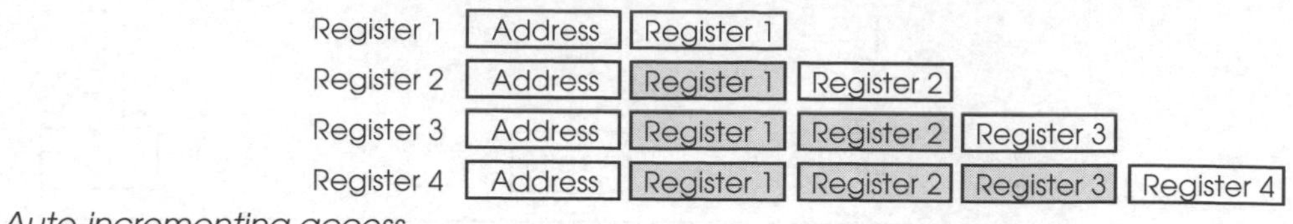

Auto-incrementing access

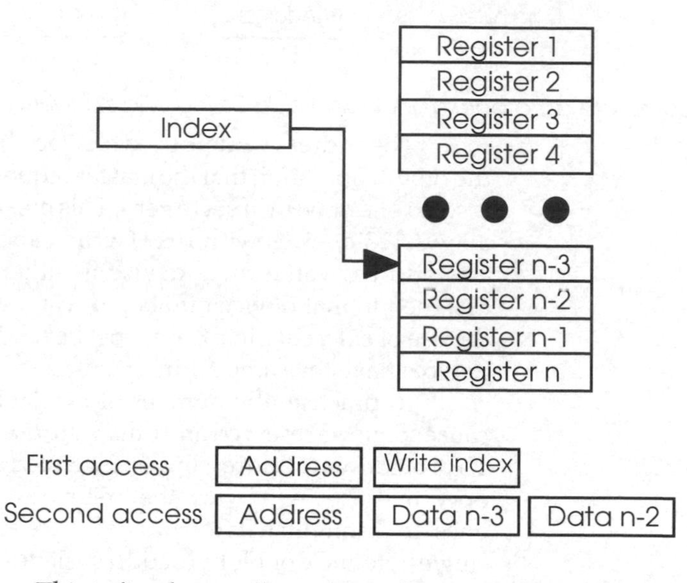

Combined format

This is fine for small numbers of registers but with memory devices such as EEPROM this is not an efficient method of operation. An alternative method uses an index value which is written to the chip, prior to accessing the data. This is known as the combined format and uses two data transfers. The first transfer is a write with the index value that is to be used to select the location for the next access. To access memory byte 237, the data byte after the address would contain 237. The next transfer would then send the data in the case of a write or request the contents in the case of a read. Some devices also support auto-incrementing and thus the second transfer can access multiple sequential locations starting at the previously transmitted index value.

Timing

One of the more confusing points about the bus is the timing or lack of it. The clock is not very specific about its timings and does not need a specified frequency or even mark to space ratios. It can even be stopped and restarted at a later point in time if needed. The START, STOP and ACKNOWLEDGE signals have a minimum delay time between the clock and data edges and pulse widths but apart from this, everything is very free and easy.

This is great in one respect but can cause problems in another. Typically the problem is concerned with waiting for the ACKNOWLEDGE signal before proceeding. If this signal is not returned then the bus will be locked up until the master terminates the transfers with a STOP signal. It is important therefore not to miss the transition. Unfortunately, the time taken for a slave to respond is dependent on the peripheral and with devices like EEPROM, especially during a write cycle, this time can be extremely long.

As a result, the master should use a timer/counter to determine when sufficient time has been given to detect the ACKNOWLEDGE before issuing a STOP.

This can be done in several ways: polling can be used with a counter to determine the timeout value. The polling loop is completed when either the ACKNOWLEDGE is detected to give a success or if the polling count is exceeded. An alternative method is to use a timer to tell the master when to check for the acknowledgement. There are refinements that can be added where the timeout values are changed depending on the peripheral that is being accessed. A very sophisticated system can use a combination of timer and polling to check for the signal *n* times with an interval determined by the timer. Whichever method is chosen, it is important that at least one is implemented to ensure that the bus is not locked up.

Multi-master support

The bus supports the use of multiple masters but does not have any in-built mechanism for controlling access. Instead, it uses a technique called collision detect to determine if two masters start to use the bus at the same time. A master waits until the bus is clear, i.e. there is no current transfer, and then issues a START signal. If another master has done the same then the signals that appear on the line will be corrupted. If a master wants a line to be high and the other wants to drive it low, then the line will go low. With the bidirectional ports that are used, each master can monitor the line and confirm that it is in the expected state. If it is not, then a collision has occurred and the master should discontinue transmission and allow the other master to continue.

It is important that timeouts for acknowledgement are incorporated to ensure that the bus cannot be locked up. In addition, care must be taken with combined format accesses to prevent a second master from resetting the index on the peripheral. If master A sets the index into an EEPROM peripheral to 53 and before it starts the next START-address-data transfer, a second master gets the bus and sets the index to its value of 97, the first master will access incorrect data. The problem can be even worse as the diagram shows. When master B overwrites the index value prior to master A's second access, it causes data corruption for both parties. Master A will access location 97 and due to auto-incrementing, master B will access location 98 — neither of which is correct! The bus does not provide a method of solving this dilemma and the only real solutions are not to share the peripheral between the devices or use a semaphore to protect access. The protection of resources is a perennial problem with embedded systems and will be covered in more detail later on.

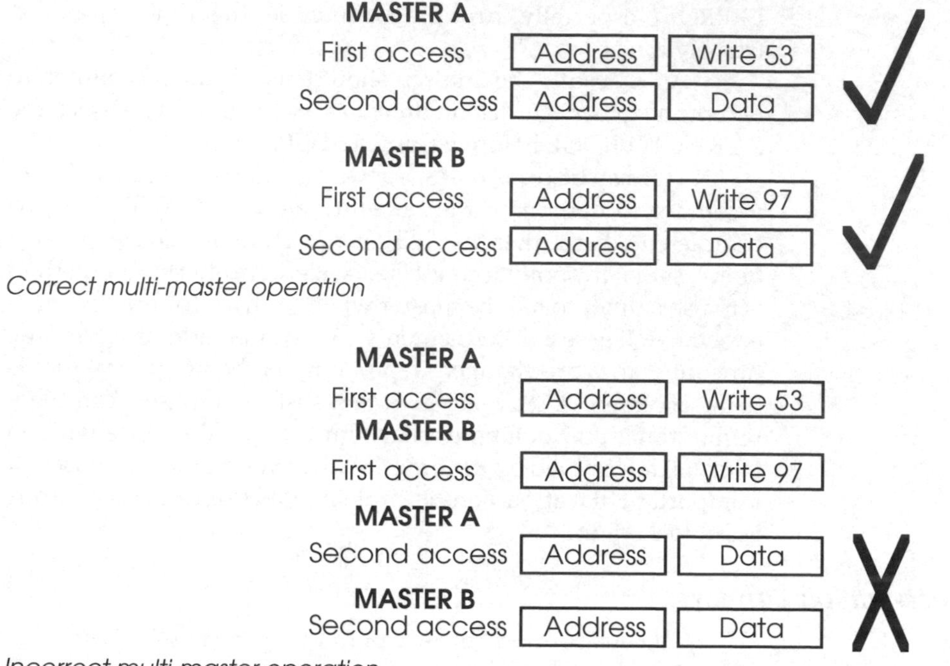

Correct multi-master operation

Incorrect multi-master operation

M-Bus (Motorola)

M-Bus is an ideal interface for EEPROMs, LCD controllers, A/D converters and other components that could benefit from fast serial transfers. This two-wire bidirectional serial bus allows a master and a slave to rapidly exchange data. It allows for fast communication with no address translation. It is very similar in operation to I^2C and thus M-Bus devices can be used with these type of serial ports. The maximum transfer rate is 100 kb/s.

What is an RS232 serial port?

Up until now, the serial interfaces that have been described have used a clock signal as a reference and therefore the data transfers are synchronous to that clock. For the small distances between chips, this is fine and the TTL or CMOS logic voltages are sufficient to ensure operation over the small connection distances. However, this is not the case if the serial data is being transmitted over many metres. The low voltage logic levels can be affected by the cable capacitance and thus a logic one at the transmitter may be seen as an indeterminate voltage at the receiver end. Clock edges can become skewed and out of sync with the data causing the wrong data to be accepted. As a result, a slightly different serial port is used for connecting over longer distances, generically referred to an RS232.

For most people, the mention of RS232 immediately brings up the image and experiences of connecting peripherals to the ubiquitous IBM PC. The IBM PC typically has one or two serial ports, *COM1* and *COM2*, which are used to transfer data between the PC and printers, modems and even other computers. The term 'serial' comes from the fact that only one data line is used to transmit and receive data and thus the information must be sent and received a bit at a time. Instead of transmitting the 8 bits that make up a byte using eight data lines at once, one data line is used to send 8 bits, one at a time. In practice, several lines are used to provide separate lines for data transmit and receive, and to provide a control line for hardware handshaking. One important difference is that the data is transmitted asynchronously i.e. there is no separate reference clock. Instead the data itself provides its own reference clock in terms of its format.

The serial interface can be divided into two areas. The first is the physical interface, commonly referred to as RS232 or EIA232, which is used to transfer data between the terminal and the computer. The electrical interface uses a combination of +5, +12 and –12 volts for the electrical interface. This used to require the provision of additional power connections but there are now available interface chips that take a 5 volt supply (MC1489) and generate internally the other voltages that are needed to meet the interface specification. Typically, a logic one is signalled by a +3 to +15 volts level and a logic zero by –3 to –15 volts. Many systems use +12 and –12 volts.

Note: The term RS232 strictly specifies the physical interface and not the serial protocol. Partly because RS232 is easier to say than universal asynchronous communication using an RS232 interface, the term has become a general reference to almost any asynchronous serial communication.

The second area controls the flow of information between the terminal and computer so that neither is swamped with data it cannot handle. Again, failure to get this right can cause data corruption and other problems.

When a user presses a key, quite a lengthy procedure is carried out before the character is transmitted. The pressed key generates a specific code which represents the letter or other character. This is converted to a bit pattern for transmission down the serial line via the serial port on the computer system. The converted bit pattern may contain a number of start bits, a number of bits (5, 6, 7 or 8) representing the data, a parity bit for error checking and a number of stop bits. These are all sent down the serial line by a UART (universal asynchronous receiver transmitter) in the terminal at a predetermined speed or baud rate.

The start bits are used to indicate that the data being transmitted is the start of a character. The stop bits indicate that character has ended and thus define the data sequence that contains the data. The parity bit can either be disabled, i.e. set to zero or configured to support odd or even parity. The bit is set to indicate that the total number of bits that have been sent is either an odd or even number. This allows the receiving UART to detect a single bit error during transmission or reception. The bit sequencing and resultant waveform is asynchronous in that there is not a reference clock transmitted. The data is detected by using a local clock reference, i.e. from the baud rate generator and the start/stop bit edges. This is why it is so important not only to configure the data settings but to set the correct baud rate settings so that the individual bits are correctly interpreted. As a result, both the processor and the peripheral it is communicating with must use the same baud rate and the same combination of start, stop, data and parity bits to ensure correct communication. If different combinations are used, data will be wrongly interpreted.

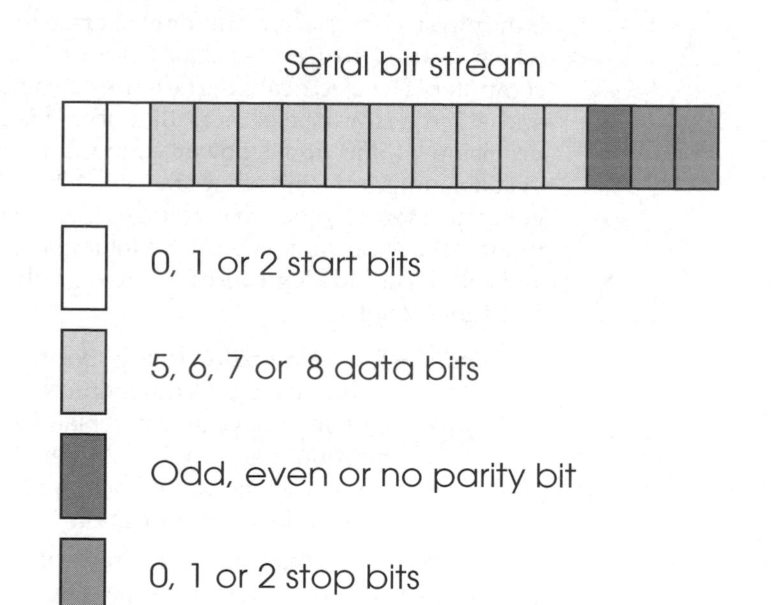

Serial bit streams

If the terminal UART is configured in half duplex mode, it echoes the transmitted character so it can be seen on the screen. Once the data is received at the other end, it is read in by another UART and, if this UART is set up to echo the character, it sends it back to the terminal. (If both UARTs are set up to echo, multiple characters are transmitted!) The character is then passed to the application software or operating system for further processing.

If the other peripheral or processor is remote, the serial line may include a modem link where the terminal is connected to a modem and a telephone line, and a second modem is linked to the computer at the other end. The modem is frequently controlled by the serial line, so if the terminal is switched off, the modem effectively hangs up and disconnects the telephone line. Modems can also echo characters and it is possible to get four characters on the terminal screen in response to a single key stroke.

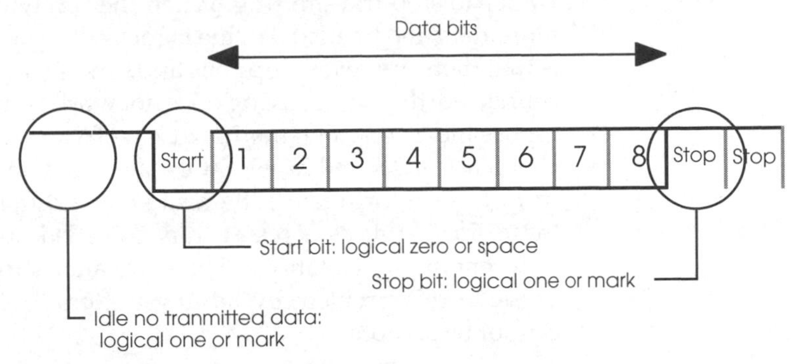

Asynchronous data format

The actual data format for the sequence is shown in the diagram. When no data is transmitted, the TXD signal is set to a logical one. When data is transmitted, a start bit is sent by setting the line to a logical zero. Data is then sent by setting the data to a zero or one accordingly and finally the stop bits are sent by forcing the line to a logical one. The stop bits essentially look the same as the idle bits when no data is being transmitted. The timing is defined by the baud rate that both the receiver and transmitter are using. The baud rate used to be supplied by an external timer/ counter called a baud rate generator that generates a clock signal at the right frequency. This function is now performed on-chip with modern controller chips and usually can work with the system clock or with a simple watch crystal instead of one with a specific frequency.

Note: If the settings are slightly incorrect, i.e. the number of stop and data bits is wrong, then it is possible for the data to appear to be received correctly. For example, if data is transmitted at 7 data bits with 2 stop bits and received as 8 data bits with 1 stop bit, the receiver would get the 7 data bits and set the eighth data bit to a one. If this character was then displayed on the screen, it could appear in the correct format due to the fact that many character sets ignore the eighth bit. In this case, the software and system would appear to work. If the data was used in some other protocol where the eighth bit was either used or assumed to be set to zero, the program and system would fail!

Asynchronous flow control

Flow control is necessary to prevent either the terminal or the computer from sending more data than the other can cope with. If too much is sent, it either results in missing characters or in a data overrun error message. The first flow control method is hardware handshaking, where hardware in the UART detects a potential overrun and asserts a handshake line to tell the other UART to stop transmitting. When the receiving device can take more data, the handshake line is released. The problem with this is that there are several options and, unless the lines are correctly connected, the handshaking does not work correctly and data loss is possible. The second method uses software to send flow control characters XON and XOFF. XOFF stops a data transfer and XON restarts it. Unfortunately, there are many different ways of using these lines and, as a result, this so-called standard has many different implementations. There are alternative methods of addressing this problem by adding buffers to store data when it cannot be accepted.

The two most common connectors are the 25 pin D type and the 9 pin D type. These are often referred to as DB-25 and DB-9 respectively. Their pin assignments are as follows:

DB-25	***Signal***	***DB-9***
1	Chassis ground	Not used
2	Transmit data — *TXD*	3
3	Receive data — *RXD*	2
4	Request to send — *RTS*	7
5	Clear to send — *CTS*	8
6	Data set ready — *DSR*	6
7	Signal ground — *GND*	5
8	Data carrier detect — *DCD*	1
20	Data terminal ready — *DTR*	4
22	Ring indicator — *RI* or *RING*	9

There are many different methods of connecting these pins and this has caused many problems especially for those faced with the task of implementing the software for a UART in such a configuration. To implement hardware handshaking, individual I/O pins are used to act as inputs or outputs for the required signals. The functionality of the various signals is as follows:

TXD Transmit data. This transmits data and would normally be connected to the RXD signal on the other side of the connection.

RXD Receive data. This transmits data and would normally be connected to the TXD signal on the other side of the connection. In this way, there is a cross-over connection.

RTS Request to send. This is used in conjunction with CTS to indicate that this side is ready to send and needs confirmation that the other side is ready.

CTS Clear to send. This is the corresponding signal to RTS and is sent by the other side on receipt of the RTS to indicate that it is ready to receive data.

DSR Data set ready. This is used in conjunction with DTR to indicate that each side is powered on and ready.

DCD Data carrier detect. This is normally used to determine which side is in control of the hardware handshake protocol.

DTR Data terminal ready. This is used in conjunction with DSR to indicate that each side is powered on and ready.

RI Ring indicator.This is asserted when a connected modem has detected an incoming call.

Much of the functionality of these signals has been determined by the need to connect to modems initially to allow remote communication across telephone lines. While modem links are still important, many serial lines are used in modemless links to peripherals such as printers. In these cases, the interchange of signals which the modem performs must be simulated within the cabling and this is done using a null modem cable. The differences are best shown by looking at some example serial port cables.

Modem cables

These are known as modem or straight through cables because the connections are simply one to one with no crossing over or other more complex wiring. They are used to link PCs with modems, printers, plotters and other peripherals. However, do not use them when linking a PC to another PC or computer — they won't work! For those links, a null modem cable is needed.

Null modem cables

Null modem cables are used to link PCs together. They work by switching over the transmit and receive signals and the handshaking connections so that each PC 'sees' a modem at the other end. There are many configurations depending on the number of wires that are needed within the cable.

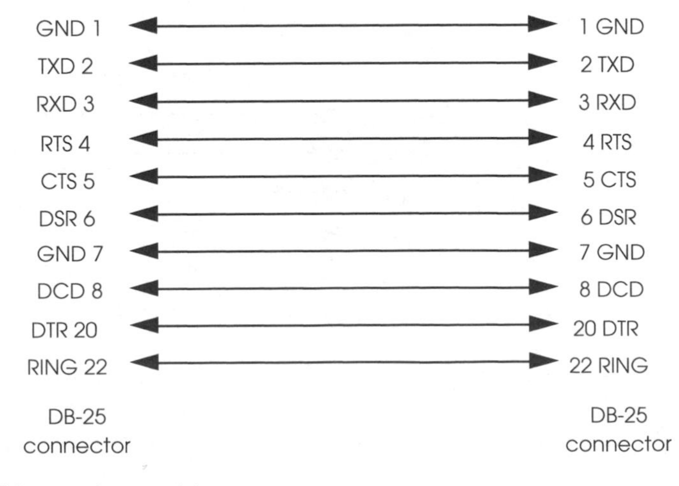

An IBM PS/2 and PC XT to modem cable

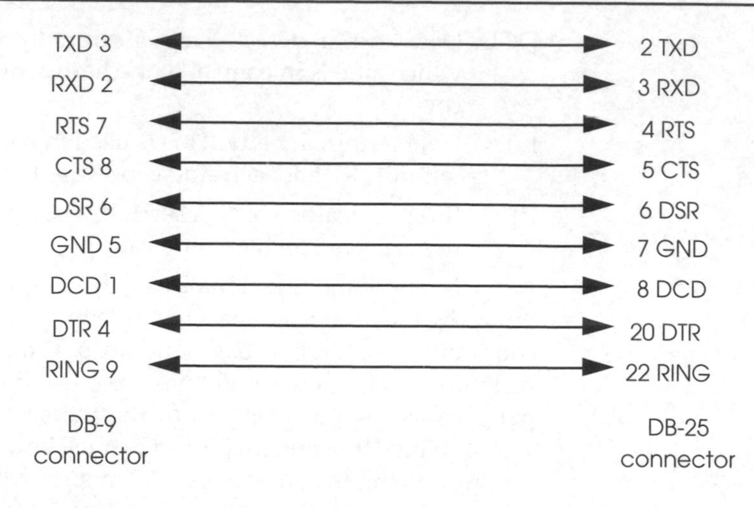

An IBM PC AT to modem cable

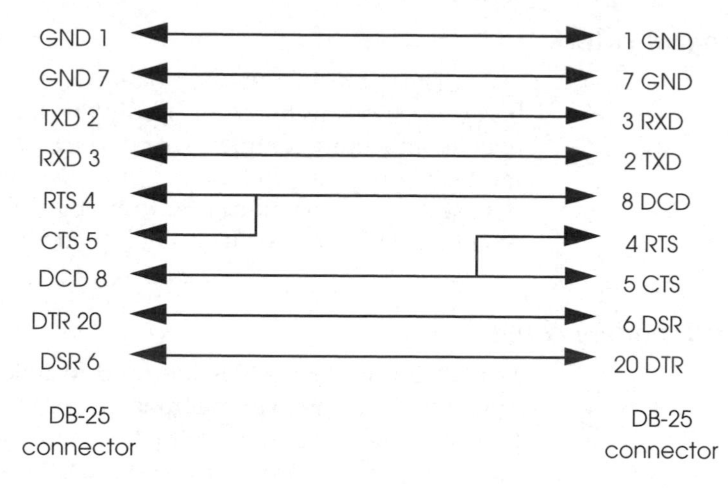

An IBM DB-25 to DB-25 standard null modem cable

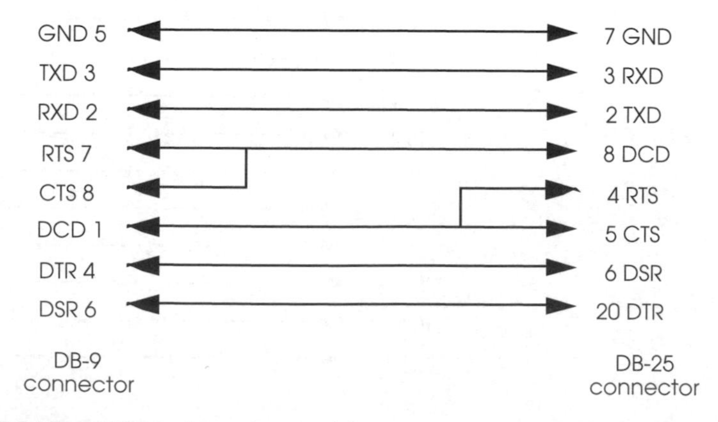

An IBM PC AT to IBM PC XT or PS/2 null modem cable

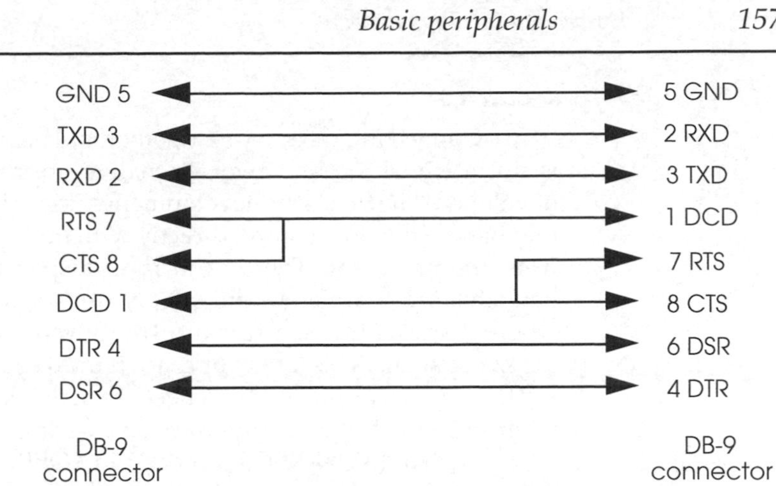

An IBM PC AT to PC AT null modem cable

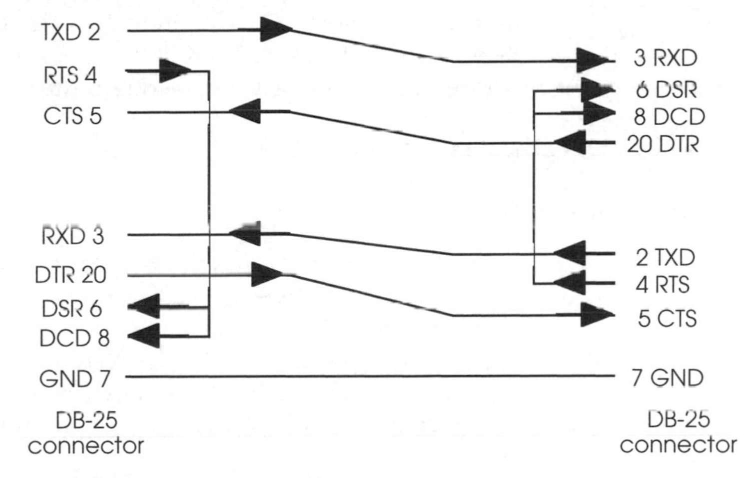

An IBM six core DB-25 to DB-25 null modem cable

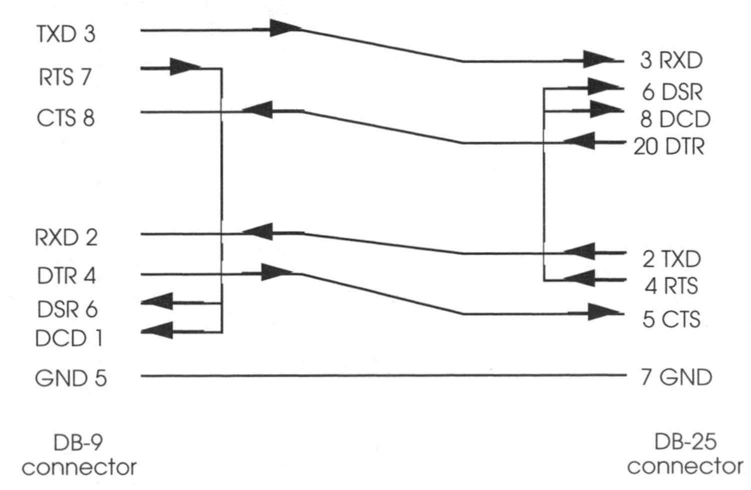

An IBM six core DB-9 to DB-25 null modem cable

XON-XOFF flow control

Connecting these wires together and using the correct pins is not a trivial job and two alternative approaches have been developed. The first is the development of more intelligent UARTs that handle the flow control directly with little or no intervention from the processor. The second is to dispense with hardware handshaking completely and simply use software handshaking where characters are sent to control the flow of characters between the two systems. This latter approach is used by Apple Macintosh, UNIX and many other systems because of its reduced complexity in terms of the hardware interface and wiring.

With the XON-XOFF protocol, an XOFF character (control-S or ASCII code 13) is sent to the other side to tell it to stop sending any more data. The halted side will wait until it receives an XON character (control-Q or ASCII code 11) before recommencing the data transmission. This technique does assume that each side can receive the XOFF character and that transmission can be stopped before overflowing the other side's buffer.

UART implementations

8250/16450/16550

Probably the most commonly known and used UART is the 8250 and its derivatives as used to provide serial ports *COM1* and *COM2* on an IBM PC. The original design used an Intel 8250 which has been largely replaced by the National Semiconductor 16450 and 16550 devices or by cloned devices within a super I/O chip which combines all the PC's I/O devices into a single piece of silicon.

Signal	Pin	Pin	Signal
D0	1	40	VCC
D1	2	39	*RI
D2	3	38	*DCD
D3	4	37	*DSR
D4	5	36	*CTS
D5	6	35	MR
D6	7	34	*OUT1
D7	8	33	*DTR
RCLK	9	32	*RTS
SIN	10	31	*OUT2
SOUT	11	30	INTR
CS0	12	29	NC (-RXRDY)
CS1	13	28	A0
*CS2	14	27	A1
*BAUDOUT	15	26	A2
XIN	16	25	*ADS
XOUT	17	24	CSOUT (-TXRDY)
*WR	18	23	DDIS
WR	19	22	RD
VSS	20	21	*RD

* indicates an active low signal

UART pinout

The original devices used voltage level shifters to provide the + and –12 volt RS232 signalling voltage levels but this function is sometimes included within the UART as well.

The pinout shows the hardware signals that are used and these fall into two groups: those that are used to provide the UART interface to the processor and those that are the UART signals. Some of the signals are active low, i.e. when they are at a zero voltage level, they represent a logical one. These signals are indicated by an asterisk.

The interface signals

The UART interface signals are for the 8250 UART and its derivatives are as follows:

*ADS — This is the address strobe signal and is used to latch the address and chip select signals during a processor access. The latching takes place on the positive edge of the and assumes that the other signals are stable at this point. This signal can be ignored by permanently asserting it. In this case, the address and chip selects must be set up and stable for the whole cycle with the processor and peripheral clock signals providing the timing references. The IBM PC uses the chip in this way.

*BAUDOUT — This is the 16x clock signal from the transmitter section of the UART. The clock frequency is the main clock frequency divided by the values stored in the baud generator divisor latches. It is normally used — as in the IBM PC, for example — to route the transmit clock back into the receive section by connecting this pin to the RCLK pin. By doing this, both the transmit and receive baud rates are the same and use the same clock frequency. To create an asynchronous system such as 1200/75 which is used for teletext links, an external transmit clock is used to feed RCLK instead.

CS0,1 and 2 — These signals are used to select the UART and are derived from the rest of the processor's address signals. The lower 3 bits of the CPU address bus are connected to the A0–A2 pins to select the internal registers. The rest of the address bus is decoded to generate a chip select signal. This can be a single entity, in which case two of the chip selects are tied to the appropriate logic level. If the signal is low, then CS0 and CS1 would be tied high. The provision of these three chip selects provides a large amount of flexibility. The truth table is shown below.

CS0	CS1	CS2	Action
High	High	Low	Selected
Low	Low	Low	Dormant
Low	Low	High	Dormant
Low	Low	Low	Dormant
Low	High	High	Dormant
Low	High	Low	Dormant
High	Low	High	Dormant
High	Low	Low	Dormant

D0–D7 These signals form the 8 bit bus that is connected between the peripheral and the processor. All transfers between the UART and processor are byte based.

DDIS This goes low whenever the CPU is reading data from the UART. It can be used to control bus arbitration logic.

INTR This pin is normally connected to an interrupt pin on the processor or in the case of the IBM PC, the interrupt controller. It is asserted when the UART needs data to be transferred to or from the internal buffers, or if an error condition has occurred such as a data overrun. The ISR has to investigate the UART's status registers to determine the actual service(s) requested by the peripheral.

MR This is the master reset pin and is used to reset the device and restore the internal registers to their power-on default values. This is normally connected to the system/processor reset signal to ensure that the UART is reset when the system is.

*OUT1 This is a general-purpose I/O pin whose state can be set by programming bit 2 of the MCR to a '1'.

*OUT2 This is another general-purpose I/O pin whose state can be set by programming bit 3 of the MCR '1'. In the IBM PC it is used to gate the interrupt signal from the UART to the interrupt controller. In this way, interrupts from the UART can be externally disabled.

RCLK This is the input for the clock for the receiver section of the chip. See *BAUDOUT on the previous page for more details.

RD, *RD These are read strobes that are used to indicate the type of access that the CPU needs to perform. If RD is high or *RD is low, the CPU access is a read cycle.

SIN — This is the serial data input pin for the receiver.

SOUT — This is the serial data output pin for the transmitter.

*RXRDY,*TXRDY
These pins are used for additional DMA control and can be used to initiate DMA transfers to and from the read and write buffers. They are not used within the IBM PC design where the CPU is responsible for moving data to and from the UART.

WR, *WR — These are read strobes that are used to indicate the type of access that the CPU needs to perform. If WR is high or *WR is low, the CPU access is a write cycle.

XIN, XOUT — These pins are used to either connect an external crystal or connect to an external clock. The frequency is typically 8 MHz.

A0–2 — These are the three address signals which are used in conjunction with DLAB to select the internal registers. They are normally connected to the lower bits of the processor address bus. The upper bits are normally decoded to create a set of chip select signals to select the UART and locate it at a specific address location.

DLAB	A2	A1	A0	Register
0	0	0	0	READ: receive buffer WRITE: transmitter holding
0	0	0	1	Interrupt enable
x	0	1	0	READ: Interrupt identification WRITE: FIFO control *
x	0	1	1	Line control
x	1	0	0	Modem control
x	1	0	1	Line status
x	1	1	0	Modem status
x	1	1	1	Scratch
1	0	0	0	Divisor latch (LSB)
1	0	0	1	Divisor latch (MSB)

*undefined with the 16450.

Register descriptions

The main difference between the various devices concerns the buffer size that they support and, in particular, the effect that it has on the effective throughput of the UART.

The UART relies on the CPU to transfer data and therefore the limit on the serial data throughput that can be sustained is determined by the time it takes to interrupt the CPU and for the appropriate interrupt service routine to identify the reason for the interrupt — it may have been raised as a result of an error — and then transfer the data if the interrupt corresponds to a data ready for transfer request. Finally, the processor returns from the interrupt.

The time to perform this task is determined by the processor type and memory speed. The time then defines the maximum rate that data can be received. If the interrupt service routine takes longer than the time to receive the next data, there is a large risk that a data overrun will occur where data is received before the previous byte is read by the processor. To address this issue, a buffer is often used. With the later versions of the UART such as the 16450 and 16550, the FIFO buffer size has been increased. The largest buffer (16 bytes) is available on the 16550 and this device is frequently used for high speed data communications.

The 16 byte buffer means that if the processor is late for the first byte, any incoming data will simply be buffered and not cause a data overrun. As a result, the interrupt service routine need only be executed 1/16 of the times for a single buffer UART. This dramatically reduces the CPU processing needed for high speed data transfer.

There is a downside: the data now arrives in a packet with up to 16 bytes and must be processed slightly differently. With a byte at a time, the decoding of the data (i.e. is it a command or is it data that a higher level protocol may impose?) is easy to decode. With a packet of up to 16 bytes, the bytes have to be parsed to separate them out. This means that the decoding software is slightly more complex to handle both the parsing and the mechanisms to store and track the incoming data packets. An example of this in included in the chapter on buffers.

The Motorola MC68681

Within the Motorola product offering, the MC68681 has become a fairly standard UART that has been used in many MC680x0 designs. It has a quadruple buffered receiver and a double buffered transmitter. The maximum transfer rates that can be achieved are high: 9.8 Mbps with a 25 MHz clock with no clock division (×1 mode) and 612 kbps with the same clock with a divide by 16 setting (×16 mode). Each transmitter and receiver is independently programmable using one of 19 fixed rates.

It has a sophisticated interrupt structure that supports seven maskable interrupt conditions:

- Change of state on CTSx*

  This is used to support hardware handshaking. If the CTS signal changes, an interrupt can be generated to instruct the processor to stop or start sending data. This fast response coupled with the buffering ensures that data is not lost.

- Break condition (either channel)

  The break condition is either used to request connection, i.e. send a break from a terminal to start a remote login or is symptomatic of a lost or dropped connection.

- Ready receive / FIFO full (either channel)

  As previously discussed, interrupts are ideal for the efficient handling and control of receive buffers. This interrupt indicates that there is data ready.
- Transmitter ready (either channel)

  This is similar to the previous interrupt and is used to indicate that the transmitter is ready to take data for transmission.

DMA controllers

Direct memory access (DMA) controllers are frequently an elegant hardware solution to a recurring software / system problem of providing an efficient method of transferring data from a peripheral to memory.

In systems without DMA, the solution is to use the processor to either regularly poll the peripheral to see if it needs servicing or to wait for an interrupt to do so. The problem with these two methods is that they are not very efficient. Polling, by its very nature, is going to check the status and find that no action is required more times than it will find that servicing is needed. If this is not the case, then data can be lost through data over- and under-run. This means that it spends a lot of time in non-constructive work. In many embedded systems, this is not a problem but in low power systems, for example, this unnecessary work processing and power consumption cannot be tolerated.

Interrupts are a far better solution. An interrupt is sent from the peripheral to the processor to request servicing. In many cases, all that is needed is to simply empty or load a buffer. This solution starts becoming an issue as the servicing rate increases. With high speed ports, the cost of interrupting the processor can be higher than the couple of instructions that it executes to empty a buffer. In these cases, the limiting factor for the data transfer is the time to recognise, process and return from the interrupt. If the data needs to be processed on a byte by byte basis in real-time, this may have to be tolerated but with high speed transfers this is often not the case as the data is treated in packets.

This is where the DMA controller comes into its own. It is a device that can initiate and control bus accesses between I/O devices and memory, and between two memory areas. With this type of facility, the DMA controller acts as a hardware implementation of the low-level buffer filling or emptying interrupt routine.

There are essentially three types of DMA controller which offer different levels of sophistication concerning memory address generation. They are often classified in terms of their addressing capability into 1D, 2D and 3D types. A 1D controller would only have a single address register, a 2D device two and a 3D device three or more.

A generic DMA controller

A generic controller consists of several components which control the operation:

- Address generator
  This is probably the most important part of a DMA controller and typically consists of a base address register and an auto-incrementing counter which increments the address after every transfer. The generated addresses are used within the actual bus transfers to access memory and/or peripherals. When a predefined number of bytes have been transferred, the base address is reloaded and the count cleared to zero ready to repeat the operation.
- Address bus
  This is where the address created by the address generator is used to access a specific memory location or peripheral.
- Data bus
  This is the data bus that is used to transfer data from the DMA controller to the destination location. In some cases, the data transfer may be made direct from the peripheral to the memory with the DMA controller directly selecting the peripheral.
- Bus requester
  This is used to request the bus from the main CPU. In older designs, the processor bus was not designed to support multiple masters and there were no bus request signals. In these cases, the processor clock was extended or delayed to steal memory cycles from the processor for the DMA controller to use.
- Local peripheral control
  This allows the DMA controller to select the peripheral and get it to accept or provide data directly or for a peripheral to request a data transfer, depending on the DMA controller's design. This is necessary to support the single or implied address mode which is explained in more detail later on.
- Interrupt signals
  Most DMA controllers can interrupt the processor when the data transfers are complete or if an error has occurred. This prompts the processor to either reprogram the DMA controller for a different transfer or acts as a signal that a new batch of data has been transferred and is ready for processing.

Operation

Using a DMA controller is reasonably simple provided the programming defines exactly the data transfer operations that the processor expects. Most errors lie in correct programming and in failing to understand how the device operates. The key phases of its operation are:

- Program the controller

  Prior to using the DMA controller, it must be configured with parameters that define the addressing such as base address and byte count that will be used to transfer the data. In addition, the device will be configured in terms of its communication with the processor and peripheral. Processor communication will normally include defining the conditions that will generate an interrupt. The peripheral communication may include defining which request pin is used by the peripheral and any arbitration mechanism that is used to reconcile simultaneous requests for DMA from two or more peripherals. The final part of this process is to define how the controller will transfer blocks of data: all at once or individually or some other combination.

Source address	FF FF 01 04
Base address	00 00 23 00
Count	00 00 00 10
Bytes transferred	00 00 00 00
Status	OK

DMA controller registers

- Start a transfer

  A DMA transfer is normally initiated in response to a peripheral request to start a transfer. It usually assumes that the controller has been correctly configured to support this request. With a peripheral and processor, the processor will normally request a service by asserting an interrupt pin which is connected to the processor's interrupt input(s). With a DMA controller, this peripheral interrupt signal can be used to directly initiate a transfer or if it is left attached to the processor, the interrupt service routine can start the DMA transfers by writing to the controller.

- Request the bus

  The next stage is to request the bus from the processor. With most modern processors supporting bus arbitration directly, the DMA controller issues a bus request signal to the processor which will release the bus when convenient and allow the DMA controller to proceed. Without this support, the DMA controller has to cycle steal from the processor so that it is held off the bus while the DMA controller uses it. As will be described later on in this chapter, most DMA controllers provide some flexibility concerning how they use and compete with bus bandwidth with the processor and other bus masters.

- Issue the address

  Assuming the controller has the bus, it will then issue the bus to activate the target memory location. A variety of interfaces are used — usually dependent on the number of pins that are available and include both non-multiplexed and multiplexed buses. In addition, the controller provides other signals such as read / write and strobe signals that can be used to work with the bus. DMA controllers tend to be designed for a specific processor family bus but most recent devices are also generic enough to be used with nearly any bus.
- Transfer the data

  The data is transferred either from a holding buffer within the DMA controller or directly from a peripheral.
- Update address generator

  Once the data transfer has been completed, the address generator uses the completion to calculate the address for the next transfer and update the byte / transfer counters.
- Update processor

  Depending on how the DMA controller has been programmed it can notify the processor using interrupts of events within the transfer process such as an address error or the completion of a data or block transfer.

DMA controller models

There are various modes or models that DMA controllers can support ranging from simple to complex addressing modes and single and double data transfers.

Single address model

With the single address model, the DMA controller uses its address bus to address the memory location that will participate in the bus memory cycle. The controller uses a peripheral bus — in some cases a single select and a read / write pin — to select the peripheral device so its data bus becomes active. The select signal from the processor often has to generate an address to access the specific register within the peripheral such as the buffer register. If the peripheral is prompting the transfer, the peripheral would pull down a request line — typically its interrupt line is used for this purpose.

In this way, data can be transferred between the memory and peripheral as needed, without the data being transferred through the DMA controller and thus taking two cycles. This model is also known as the implicit address because the second address is implied and not directly given, i.e. there is no source address supplied.

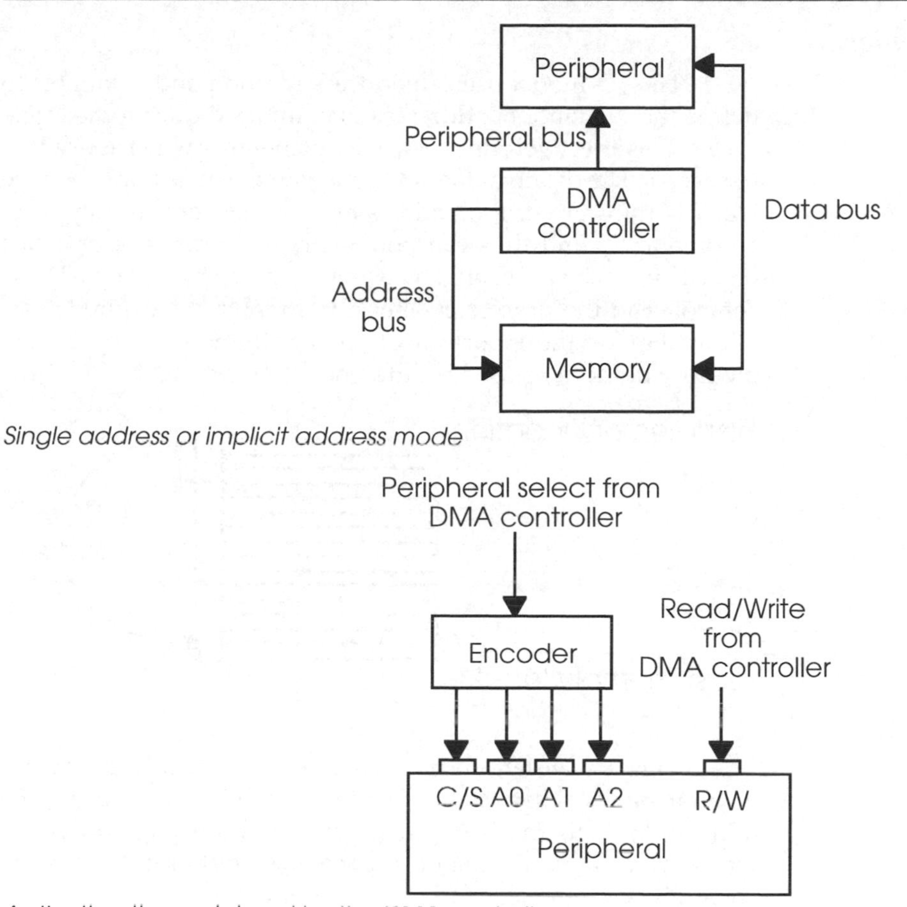

Single address or implicit address mode

Activating the peripheral by the DMA controller

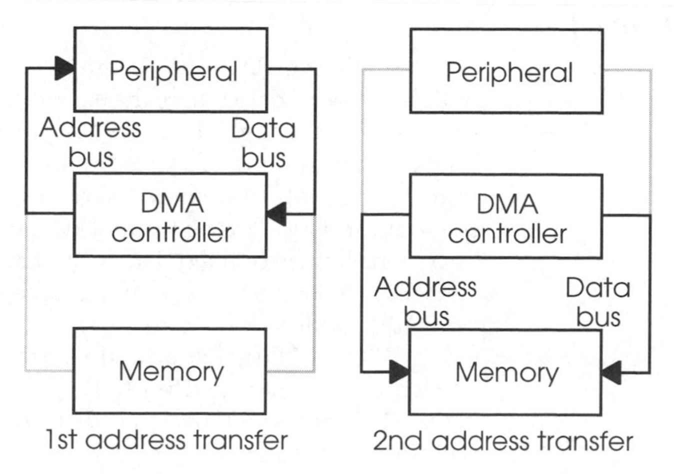

Dual address transfer

Dual address model

The dual address mode uses two addresses and two accesses to transfer data between a peripheral or memory and another memory location. This consumes two bus cycles and uses a buffer within the DMA controller to temporarily hold data.

1D model

The 1D model uses an address location and a counter to define the sequence of addresses that are used during the DMA cycles. This effectively defines a block of memory which is used for the access. The disadvantage of this arrangement is that when the block is transferred, the address and counter are usually reset automatically and thus can potentially overwrite the previous data. This can be prevented by using an interrupt from the DMA controller to the processor when the counter has expired. This allows the CPU the opportunity to change the address so that next memory block to be used is different.

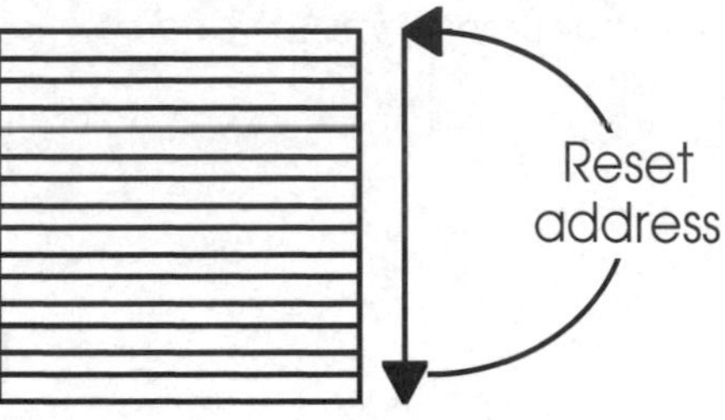

A circular buffer

This model left on its own can be used to implement a circular buffer where the automatic reset is used to bring the address back to the beginning. Circular buffering can be an efficient technique in terms of both the size of buffering and timing constraints.

2D model

While the 1D model is simple, there are times especially with high speed data where the addressing mode is not powerful enough even though it can be augmented through processor intervention. A good example of this is with packet-based communication protocols where the data is wrapped up with additional information in the form of headers. The packets typically have a maximum or fixed data format and thus large amounts of consecutive data have to be split and header and trailer information either added or removed.

With the 2D model, an address stride can be specified which is used to calculate an offset to the base address at the end of a count. This allows DMA to occur in non-consecutive blocks of memory. Instead of the base address being reset to the original address, it has the stride added to it. In addition the count register is normally split into two: one register to specify the count for the block and a second register to specify the total number of blocks or bytes to be transferred. Using these new features, it is easy to set up a DMA controller to transfer data and split into blocks ready for the insertion of header information. The diagram shows how this can be done.

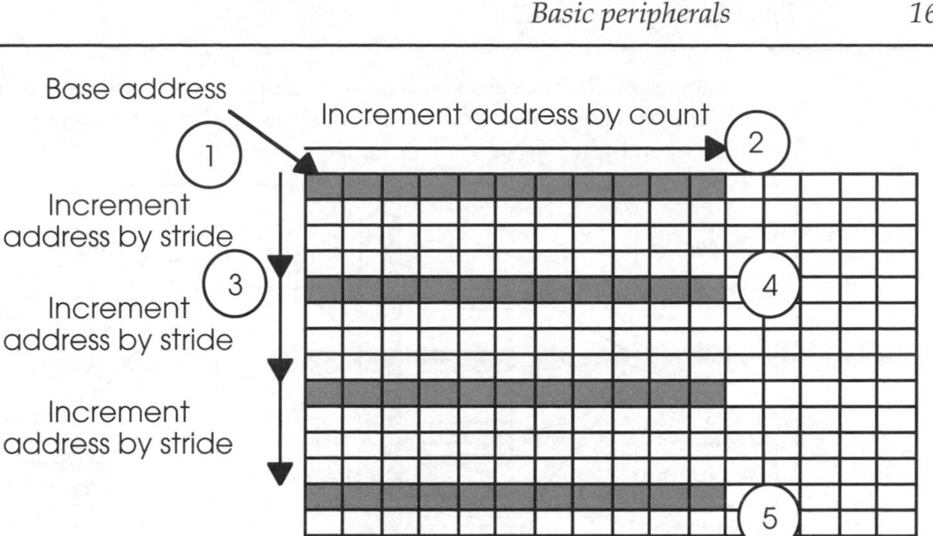

1. Use base address to start transfer. Increment until counter expires.
2. Reset counter and change base address using stride.
3. Use base address and increment until counter expires.
4. Reset counter and change base address using stride.
5. Repeat until total number of requested bytes transferred. (Total count = 11 x 4 = 44 bytes)

2D addressing structure

3D model

The third type of controller takes the idea of address strides a step further by defining the ability to change the stride automatically so that blocks of different sizes and strides can be created. It is possible to simulate this with a 2D controller and software so that the processor reprograms the device to simulate the automatic change of stride.

Channels and control blocks

By now, it should be reasonably clear that DMA controllers need to be pre-programmed with a block of parameters to allow them to operate. The hardware interface that they use is common to almost every different set of parameters — the only real difference is when a single or dual address mode is used with the need to directly access a peripheral as well as the memory address bus.

It is also common for a peripheral to continually use a single set of parameters. As a result, the processor has to continually reprogram the DMA controller prior to use if it is being shared between several peripherals. Each peripheral would have to interrupt the processor prior to use of the DMA to ensure that it was programmed. Instead of removing the interrupt burden from the processor, the processor still has it — albeit it is now programming the DMA controller and not moving data. Moving data could be an even lighter load!

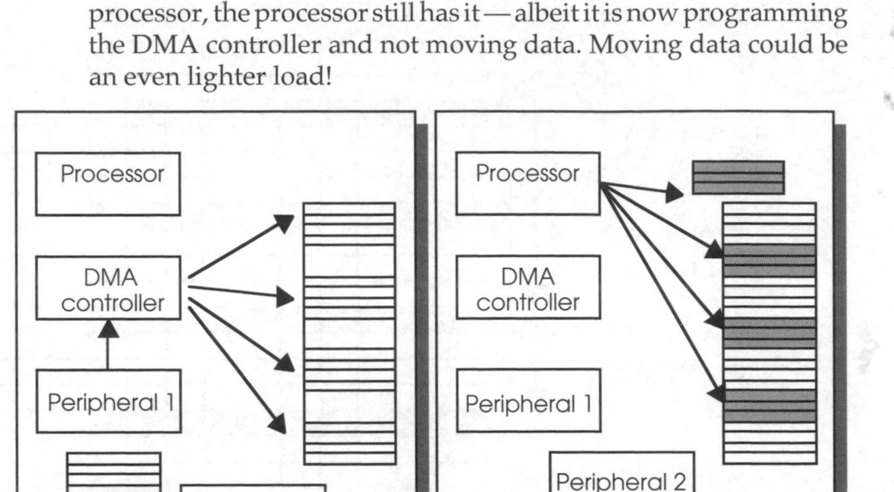

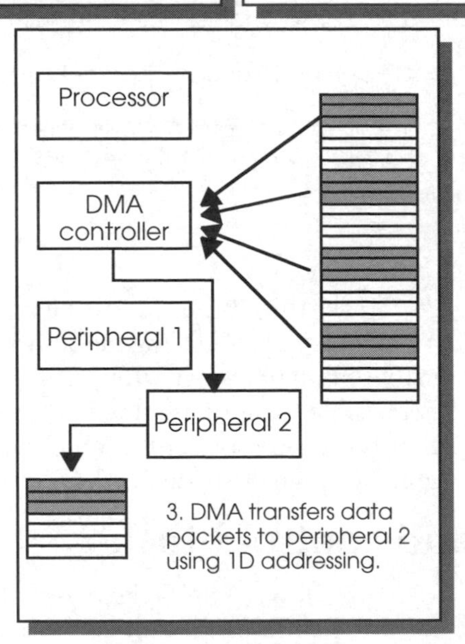

Using 2D addressing to create space for headers

To overcome this the idea of channels of control blocks was developed. Here the registers that contain the parameters are duplicated with a set for each channel. Each peripheral is assigned an external request line which when asserted will cause the DMA controller to start a DMA transfer in accordance with the parameters that have been assigned with the request line. In this way, a single DMA controller can be shared with multiple peripherals, with each peripheral having its own channel. This is how the DMA controller in the IBM PC works. It supports four channels (0 to 3).

An extension to this idea of channels, or control blocks as they are also known, is the idea of chaining. With chaining, channels are linked together so that more complex patterns can be created. The first channel controls the DMA transfers until it has completed its allotted transfers and then control is passed to the next control block that has been chained to it. This simple technique allows very complex addressing patterns to be created such as described in the paragraphs on 3D models.

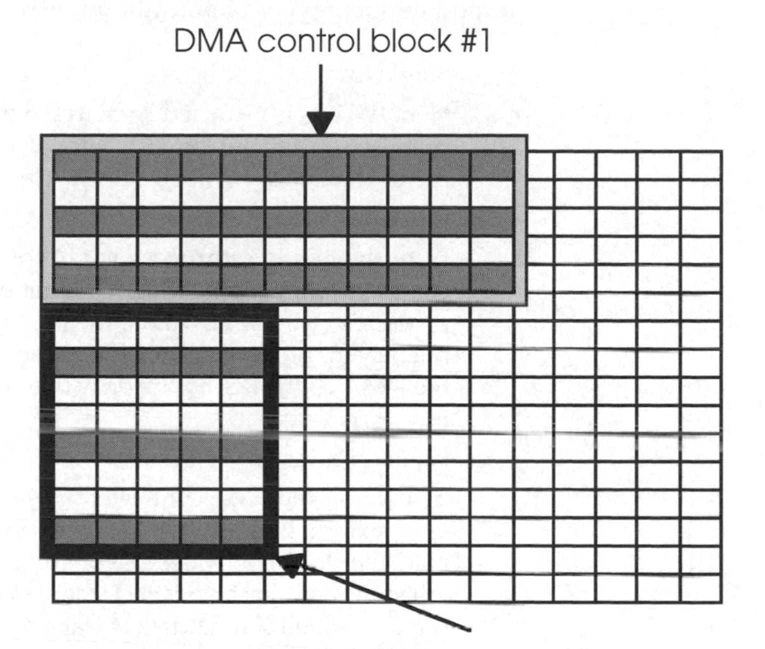

Using control blocks

There is one problem with the idea of channels and external pins: what happens if multiple requests are received by the DMA controller at the same time? To resolve this situation, arbitration is used to prioritise multiple requests. This may be a strict priority scheme where one channel has the highest priority or can be a fairer system such as a round-robin where the priority is equally distributed to give a fairer allocation of priority.

Sharing bus bandwidth

The DMA controller has to compete with the processor for external bus bandwidth to transfer data and as such can affect the processor's performance directly. With processors that do not have any cache or internal memory, such as the 80286 and the MC68000, their bus utilisation is about 80–95% of the bandwidth and therefore any delay in accessing external memory will result in a decreased processor performance budget and potentially longer interrupt latency — more about this in the chapter on interrupts.

For devices with caches and/or internal memory, their external bus bandwidth requirements are a lot lower and thus the DMA controller can use bus cycles without impeding the processor's performance. This last statement depends on the chances of the DMA controller using the bus at the same time as the processor. This in turn depends on the frequency and size of the DMA transfers. To provide some form of flexibility for the designer so that a suitable trade-off can be made, most DMA controllers support different types of bus utilisation.

- Single transfer

  Here the bus is returned back to the processor after every transfer so that the longest delay it will suffer in getting access to memory will be a bus cycle.

- Block transfer

  Here the bus is returned back to the processor after the complete block has been sent so that the longest delay the processor will suffer will be the time of a bus cycle multiplied by the number of transfers to move the block. In effect, the DMA controller has priority over the CPU in using the bus.

- Demand transfer

  In this case, the DMA controller will hold the bus for as long as an external device requests it to do so. While the bus is held, the DMA controller is at liberty to transfer data as and when needed. In this respect, there may be gaps when the bus is retained but no data is transferred.

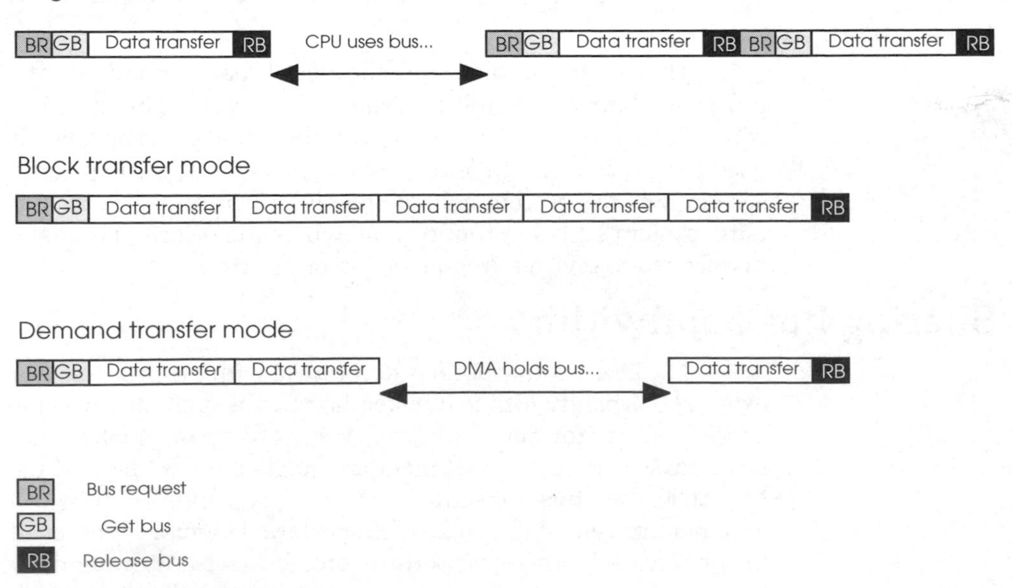

DMA transfer modes

DMA implementations

Intel 8237

This device is used in the IBM PC and is therefore probably the most used DMA controller in current use. Like most peripherals today, it has moved from being a separate entity to be part of the PC chip set that has replaced the 100 or so devices in the original design with a single chip.

It can support four main transfer modes including single and block transfers, the demand mode and a special cascade mode where additional 8237 DMA controllers can be cascaded to expand the four channels that a single device can support. It can transfer data between a peripheral and memory and by combining two channels together, perform memory to memory transfers although this is not used or supported within the IBM PC environment. In addition, there is a special verify transfer mode which is used within the PC to generate dummy addresses to refresh the DRAM memory. This is done in conjunction with a 15 μs interrupt derived from a timer channel on the PC motherboard.

To resolve simultaneous DMA requests, there is an internal arbitration scheme which supports either a fixed or rotating priority scheme.

Motorola MC68300 series

Whereas five or 10 years ago, DMA controllers were freely available as separate devices, the increasing ability to integrate functionality has led to their demise as separate entities and most DMA controllers are either integrated onto the peripheral or as in this case onto the processor chip. The MC68300 series combine an MC68000 / MC68020 type of processor with peripherals and DMA controllers.

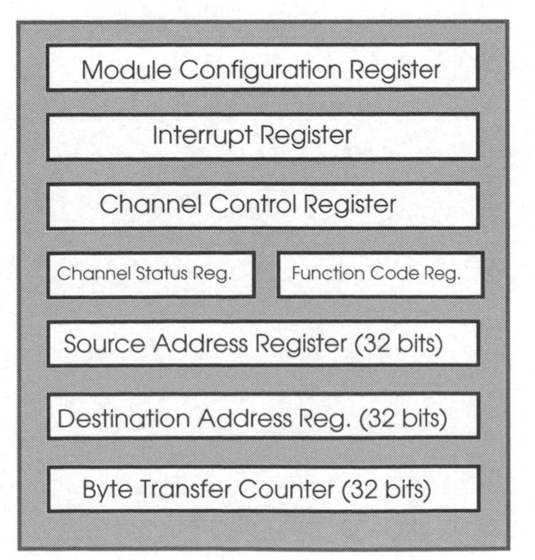

MC683xx generic DMA controller

It consists of a two channel fully programmable DMA controller that can support high speed data transfer rates of 12.5 Mbytes/s in dual address transfer mode or 50 Mbytes/s in single address mode at a 25 MHz clock rate. The dual address mode is considerably slower because two cycles have to be performed as previously described. By virtue of its integration onto the processor chip with the peripherals and internal memory, it can DMA data between internal and external resources. Internal cycles can be programmed to occupy 25, 50, 75, or 100% of the available internal bus bandwidth while external cycles support burst and single transfer mode.

The source and destination registers can be independently programmed to remain constant or incremented as required.

Using another CPU with firmware

This is a technique that is sometimes used where a DMA controller is not available or is simply not fast or sophisticated enough. The DMA CPU requires its own local memory and program so that it can run in isolation and not burden the main memory bus. The DMA CPU is sent messages which instruct it on how to perform its DMA operations. The one advantage that this offers is that the CPU can be programmed with higher level software which can be used to process the data as well as transfer it. Many of the processors used in embedded systems fall into this category of device.

5 Interfacing to the analogue world

This chapter discusses the techniques used to interface to the outside world which unfortunately is largely analogue in nature. It discusses the process of analogue to digital conversion and basic power control techniques to drive motors and other similar devices from a microcontroller.

Analogue to digital conversion techniques

The basic principle behind analogue to digital conversion is simple and straightforward: the analogue signal is sampled at a regular interval and each sample is divided or quantised by a given value to determine the number of given units of value that approximate to the analogue value. This number is the digital equivalent of the analogue signal.

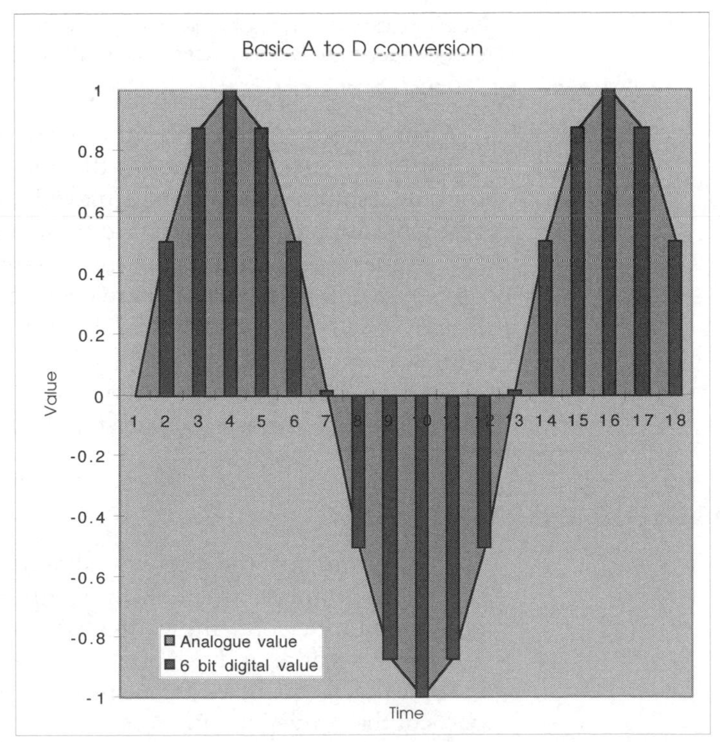

Basic A to D conversion

The combination graph shows the general principle. The grey curve represents an analogue signal which, in this case, is a sine wave. For each cycle of the sine wave, 13 digital samples are taken which encode the digital representation of the signal.

Quantisation errors

Careful examination of the combination chart reveals that all is not well. Note the samples at time points 7 and 13. These should be zero — however, the conversion process does not convert them to zero but to a slightly higher value. The other points show similar errors; this is the first type of error that the conversion process can cause. These errors are known as quantisation errors and are caused by the fact that the digital representation is step based and consists of a selection from one of a fixed number of values. The analogue signal has an infinite range of values and the difference between the digital value the conversion process has selected and the analogue value is the quantisation error.

Size	Resolution	Storage (1s)	Storage (60s)	Storage (300s)
4 bit	0.0625000000	22050	1323000	6615000
6 bit	0.0156250000	33075	1984500	9922500
8 bit	0.0039062500	44100	2646000	13230000
10 bit	0.0009765625	55125	3307500	16537500
12 bit	0.0002441406	66150	3969000	19845000
16 bit	0.0000152588	88200	5292000	26460000
32 bit	0.0000000002	176400	10584000	52920000

Resolution assumes an analogue value range of 0 to 1
Storage requirements are in bytes and a 44.1 kHz sample rate

Digital bit size, resolution and storage

The size of the quantisation error is dependent on the number of bits used to represent the analogue value. The table shows the resolution that can be achieved for various digital sizes. As the table depicts, the larger the digital representation, the finer the analogue resolution and therefore the smaller the quantisation error and resultant distortion. However, the table also shows the increase which occurs with the amount of storage needed. It assumes a sample rate of 44.1 kHz, which is the same rate as used with an audio CD. To store five minutes of 16 bit audio would take about 26 Mbytes of storage. For a stereo signal, it would be twice this value.

Sample rates and size

So far, much of the discussion has been on the sample size. Another important parameter is the sampling rate. The sampling rate is the number of samples that are taken in a time period, usually one second, and is normally measured in hertz, in the same way that frequencies are measured. This determines several aspects of the conversion process:

- It determines the speed of the conversion device itself. All converters require a certain amount of time to perform the conversion and this conversion time determines the maximum rate at which samples can be taken. Needless to say, the fast converters tend to be the more expensive ones.

- The sample rate determines the maximum frequency that can be converted. This is explained later in the section on Nyquist's theorem.
- Sampling must be performed on a regular basis with exactly the same time period between samples. This is important to remove conversion errors due to irregular sampling.

Irregular sampling errors

The line chart shows the effect of irregular sampling. It effectively alters the amplitude or magnitude of the analogue signal being measured. With reference to the curve in the chart, the following errors can occur:

- If the sample is taken early, the value converted will be less than it should be. Quantisation errors will then be added to compound the error.
- If the sample is taken late, the value will be higher than expected. If all or the majority of the samples are taken early, the curve is reproduced with a similar general shape but with a lower amplitude. One important fact is that the sampled curve will not reflect the peak amplitudes that the original curve will have.
- If there is a random timing error — often called jitter — then the resulting curve is badly distorted, again as shown in the chart.

Other sample rate errors can be introduced if there is a delay in getting the samples. If the delay is constant, the correct characteristics for the curve are obtained but out of phase. It is interesting that there will always be a phase error due to the conversion time taken by the converter. The conversion time will delay the digital output and therefore introduces the phase error — but this is usually very small and can typically be ignored.

The phase error shown assumes that all delays are consistent. If this is not the case, different curves can be obtained as shown in the next chart. Here the samples have been taken at random and at slightly delayed intervals. Both return a similar curve to that of the original value — but still with significant errors.

In summary, it is important that samples are taken on a regular basis with consistent intervals between them. Failure to observe these design conditions will introduce errors. For this reason, many microprocessor-based implementations use a timer and interrupt service routine mechanism to gather samples. The timer is set-up to generate an interrupt to the processor at the sampling rate frequency. Every time the interrupt occurs, the interrupt service routine reads the last value for the converter and instructs it to start a new conversion before returning to normal execution. The instructions always take the same amount of time and therefore sampling integrity is maintained.

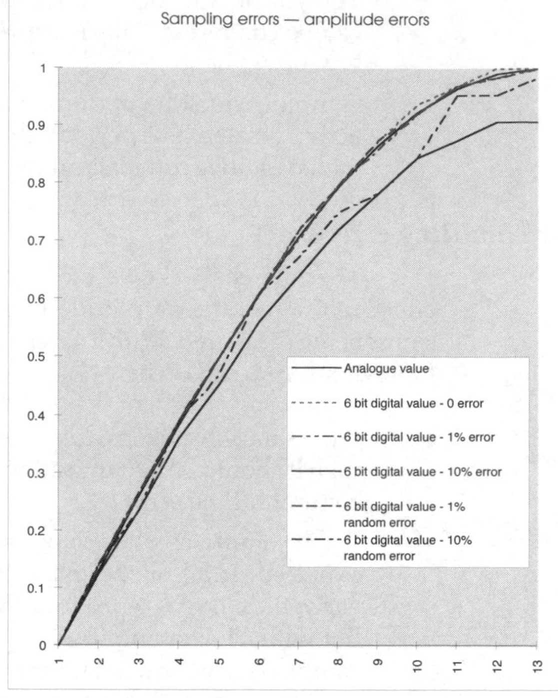

Sampling errors — amplitude errors

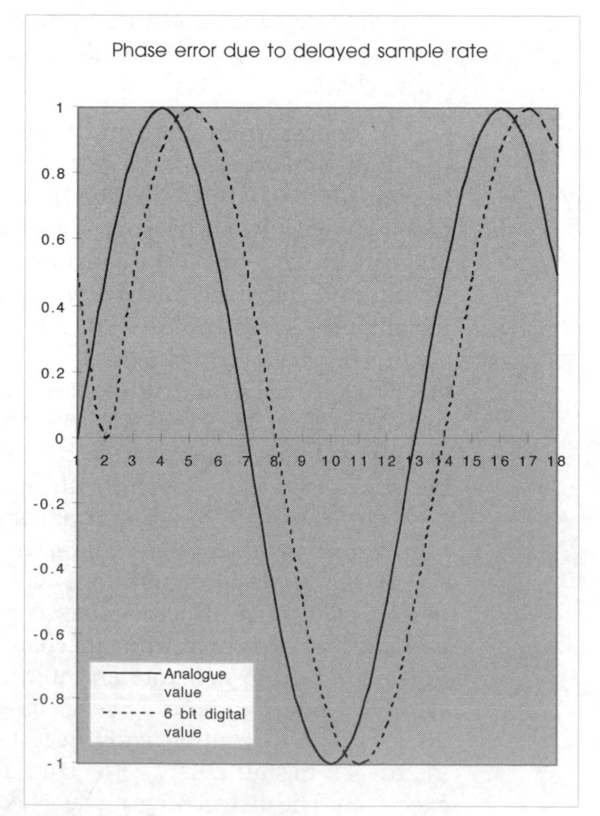

Phase errors due to delayed sample rate

Nyquist's theorem

The sample rate also must be chosen carefully when considering the maximum frequency of the analogue signal being converted. Nyquist's theorem states that the *minimum* sampling rate frequency should be twice the maximum frequency of the analogue signal. A 4 kHz analogue signal would need to be sampled at twice that frequency to convert it digitally. For example, a hi-fi audio signal with a frequency range of 20 to 20 kHz would need a minimum sampling rate of 40 kHz.

Higher frequency sampling introduces a frequency component which is normally filtered out using an analogue filter.

Codecs

So far the discussion has been based on analogue to digital (A to D) and digital to analogue (D to A) converters. These are the names used for generic converters. Where both A to D and D to A conversion is supported, they can also be called codecs. This name is derived from **co**der-**dec**oder and is usually coupled with the algorithm that is used to perform the coding. Generic A to D conversion is only one form of coding; many others are used within the industry where the analogue signal is converted to the digital domain and then encoded using a different technique. Such codecs are often prefixed by the algorithm used for the encoding.

Linear

A linear codec is one that is the same as the standard A to D and D to A converters so far described, i.e. the relationship between the analogue input signal and the digital representation is linear. The quantisation step is the same throughout the range and thus the increase in the analogue value necessary to increment the digital value by one is the same, irrespective of the analogue or digital values. Linear codecs are frequently used for digital audio.

A-law and μ-law

For telecommunications applications with a limited bandwidth of 300 to 3100 Hz, logarithmic codecs are used to help improve quality. These codecs, which provide an 8 bit sample at 8 kHz, are used in telephones and related equipment. Two types are in common use: the a-law codec in the UK and the μ-law codec in the US. By using a logarithmic curve for the quantisation, where the analogue increase to increment the digital value varies depending on the size of the analogue signal, more digital bits can be allocated to the more important parts of the analogue signal and thus improve their resolution. The less important areas are given less bits and, despite having coarser resolution, the quality reduction is not really noticeable because of the small part they contribute to the signal. Conversion between a linear digital signal and a-law/μ-law or between an a-law and μ-law signal is easily performed using a look-up table.

PCM

The linear codecs that have been so far described are also known as PCM — pulse code modulation codecs. This comes from the technique used to reconstitute the analogue signal by supplying a series of pulses whose amplitude is determined by the digital value. This term is frequently used within the telecommunications industry.

There are alternative ways of encoding using PCM which can reduce the amount of data needed or improve the resolution and accuracy.

DPCM

Differential pulse coded modulation (DPCM) is similar to PCM, except that the value encoded is the difference between the current sample and the previous sample. This can improve the accuracy and resolution by having a 16 bit digital dynamic range without having to encode 16 bit samples. It works by increasing the dynamic range and defining the differential dynamic range as a partial value of it. By encoding the difference, the smaller digital value is not exceeded but the overall value can be far greater. There is one proviso: the change in the analogue value from one sample to another must be less than the differential range and this determines the maximum slope of any waveform that is encoded. If the range is exceeded, errors are introduced.

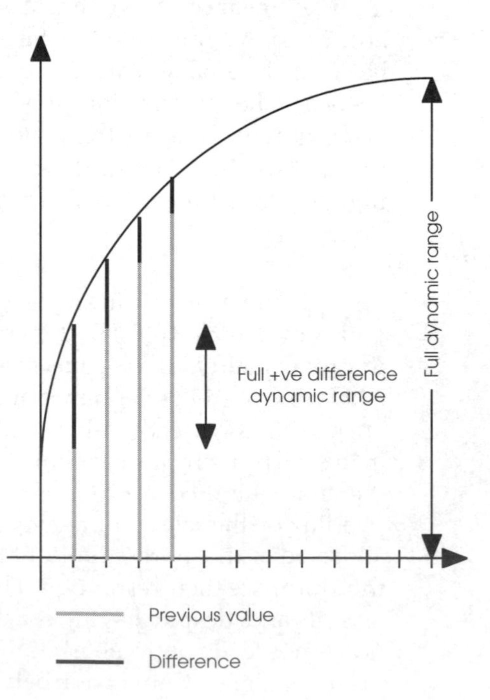

DPCM encoding

The diagram shows how this encoding works. The analogue value is sampled and the previous value subtracted. The result is then encoded using the required sample size and allowing for a plus and minus value. With an 8 bit sample size, 1 bit is used as a sign bit and the remaining 7 bits are used to encode data. This allows the previous value to be used as a reference, even if the next value is smaller. The 8 bits are then stored or incorporated into a bitstream as with a PCM conversion.

To decode the data, the reverse operation is performed. The signed sample is added to the previous value, giving the correct digital value for decoding. In both the decode and encode process, values which are far larger than the 8 bit sample are stored. This type of encoding is easily performed with a microprocessor with 8 bit data and a 16 bit or larger accumulator.

A to D and D to A converters do not have to cope with the full resolution and can simply be 8 bit decoders. These can be used provided analogue subtractors and adders are used in conjunction with them. The subtractor is used to reduce the analogue input value before inputting to the small A to D converter. The adder is used to create the final analogue output from the previous analogue value and the output from the D to A converter.

ADPCM

Adaptive differential pulse code modulation (ADPCM) is a variation on the previous technique and frequently used in telecommunications. The difference is encoded as before but instead of using all the bits to encode the difference, some bits are used to encode the quantisation value that was used to encode the data. This means that the resolution of the difference can be adjusted — adapted — as needed and, by using non-linear quantisation values, better resolution can be achieved and a larger dynamic range supported.

Power control

Most embedded designs need to be able to switch power in some way or another, if only to drive an LED or similar indicator. This, on first appearances, appears to be quite simple to do but there are some traps that can catch designers out. This section goes through the basic principles and techniques.

Matching the drive

The first problem that faces any design is matching the logic level voltages with that of a power transistor or similar device. It is often forgotten or assumed that with logic devices, a logical high is always 5 volts and that a logical low is zero. A logical high of 5 volts is more than enough to saturate a bipolar transistor and turn it on. Similarly, 0 volts is enough to turn off such a transistor.

Unfortunately, the specifications for TTL compatible logic levels are not the same as indicated by these assumptions. The voltage levels are define a logic low output as any voltage below a maximum which is usually 0.4 volts and a logic high output as a voltage above 2.4 volts assuming certain bus capacitance and load currents and a supply voltage of 4.5 to 5.5 volts. These figures are typical and can vary.

If the output high is used to drive a bipolar transistor, then the 2.4 volt value is high enough to turn on the transistor. The only concern is the current drive that the output can provide. This value times the gain of the transistor determines the current load that the transistor can provide. With an output low voltage of 0.4 volts, the situation is less clear and is dependent on the biasing used on the transistor. It is possible that instead of turning the transistor off completely, it partially turns the device off and some current is still provided.

With CMOS logic levels, similar problems can occur. Here the logic high is typically two thirds of the supply voltage or higher and a logic low is one third of the supply voltage or lower. With a 5 volt supply, this works out at 3.35 volts and 1.65 volts for the high and low states. In this case, the low voltage is above the 0.7 volts needed to turn on a transistor and thus the transistor is likely to be switched on all the time irrespective of the logic state. These voltage mismatches can also cause problems when combining CMOS and TTL devices using a single supply. With bipolar transistors there are several techniques that can be used to help avoid these problems:

- Use a high gain transistor

  The higher the gain of the transistor, the lower the drive needed from the output pin and the harder the logic level will be. If the required current is high, then the voltage on the output is more likely to reach its limits. With an output high, it will fall to the minimum value. With an output low, it will rise to the maximum value.

  Darlington transistor pairs are often used because they have a far higher gain compared to a single transistor.

- Use a buffer pack

  Buffer packs are logic devices that have a high drive capability and can provide higher drive currents than normal logic outputs. This increased drive capability can be used to drive an indicator directly or can be further amplified.

- Use a field effect transistor (FET)

  These transistors are voltage controlled and have a very high effective gain and thus can be used to switch heavy loads easily from a logic device. There are some problems, however, in that the gate voltages are often proportions of the supply voltages and these do not match with the logic voltage levels that are available. As a result, the FET does not switch correctly. This problem has been solved by the introduction of logic level switching FETs that will switch using standard logic voltages. The advantage that these offer is that they can simply have their gate directly connected to the logic output. The power supply and load are connected through the FET which acts as a switch.

Using H bridges

Using logic level FETs is a very simple and effective way of providing DC power control. With the FET acting like a power switch whose state reflects the logic level output from the digital controller, it is possible to combine several switches to create H bridges which allow a DC motor to be switched on and reversed in direction. This is done by using two outputs and four FETs acting as switches.

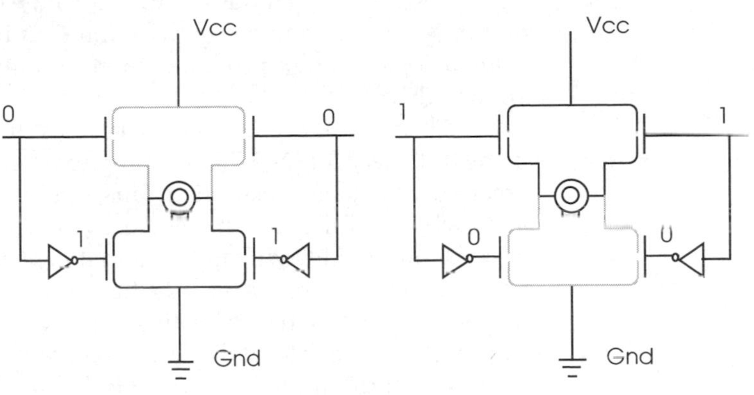

Switching the motor off using an H bridge

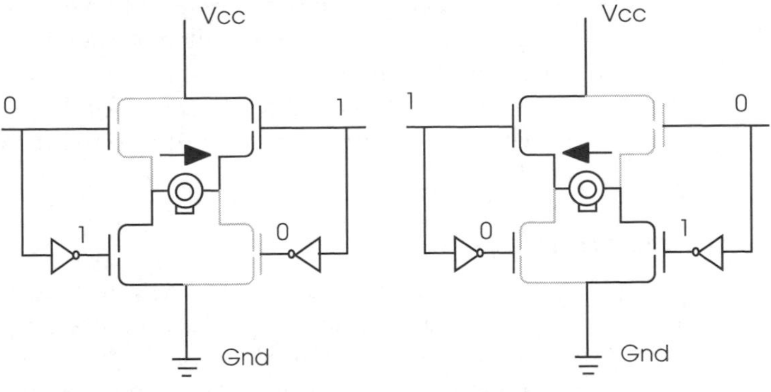

Switching the motor on with an H bridge

The FETs are arranged in two pairs so that by switching one on and the other off, one end of the motor can be connected to ground (0 volts) or to the voltage supply Vcc. Each FET in the pair is driven from a common input signal which is inverted on its way to one of the FETs. This ensures that only one of the pairs switches on in response to the input signal. With the two pairs, two input signals are needed. When these signals are the same, i.e. 00 or 11, either the top or bottom pairs of FETS are switched on and no voltage differential is applied across the motor, so nothing happens. This is shown in the first diagram where the switched-on paths are shown in black and the switched-off paths are in grey.

If the input signals are different then a top and a bottom FET is switched on and the voltage is applied across the motor and it revolves. With a 01 signal it moves in one direction and with a 10 signal it moves in the reverse direction.

This type of bridge arrangement is frequently used for controlling DC motors or any load where the voltage may need reversing.

Driving LEDs

Light emitting diodes (LEDs) are often used as indicators in digital systems and in many cases can simply be directly driven from a logic output provided there is sufficient current and voltage drive.

The voltage drive is necessary to get the LED to illuminate in the first place. LEDs will only light up when their diode reverse breakdown voltage is exceeded. This is usually about 2 to 2.2 volts and less than the logic high voltage. The current drive determines how bright the LED will appear and it is usual to have a current limiting resistor in series with the LED to prevent it from drawing too much current and overheating. For a logic device with a 5 volt supply a 300 Ω resistor will limit the current to about 10 mA. The problem comes if the logic output is only 2.4 or 2.5 volts and not the expected 5 volts. This means that the resistor is sufficient to drop enough voltage so that the LED does not light up. The solution is to use a buffer so that there is sufficient current drive or alternatively use a transistor to switch on the LED. There are special LED driver circuits packs available that are designed to connect directly to an LED without the need for the current limiting resistor. The resistor or current limiting circuit is included inside the device.

Interfacing to relays

Another method of switching power is to use a mechanical relay where the logic signal is used to energise the relay. The relay contacts make or break accordingly and switch the current. The advantage of a relay is that it can be used to switch either AC or DC power and there is no electrical connections between the low power relay coil connected to the digital circuits and the power load that is being switched. As a result, they are frequently used to switch high loads.

Relays do suffer from a couple of problems. The first is that the relay can generate a back voltage across its terminals when the energising current is switched off, i.e. when the logic output switches from a high to a low. This back EMF as it is known can be a high voltage and cause damage to the logic circuits. A logic output does not expect to see an input voltage differential of several tens of volts! The solution is to put a diode across the relay circuits so that in normal operation, the diode is reverse biased and does nothing. When the back EMF is generated, the diode starts to conduct and the voltage is shorted out and does no damage. This problem is experienced with any coil, including those in DC motors. It is advisable to fit a diode when driving these components as well.

The other problem is that the switch contacts can get sticky where they are damaged with the repeated current switching. This can erode the contacts and cause bad contacts or in some cases can cause local overheating so that the contacts weld themselves together. The relay is now sticky in that the contacts will not change when the coil is de-energised.

Interfacing to DC motors

So far with controlling DC motors, the emphasis has been simple on-off type switching. It is possible with a digital system to actually provide speed control using a technique called pulse width modulation.

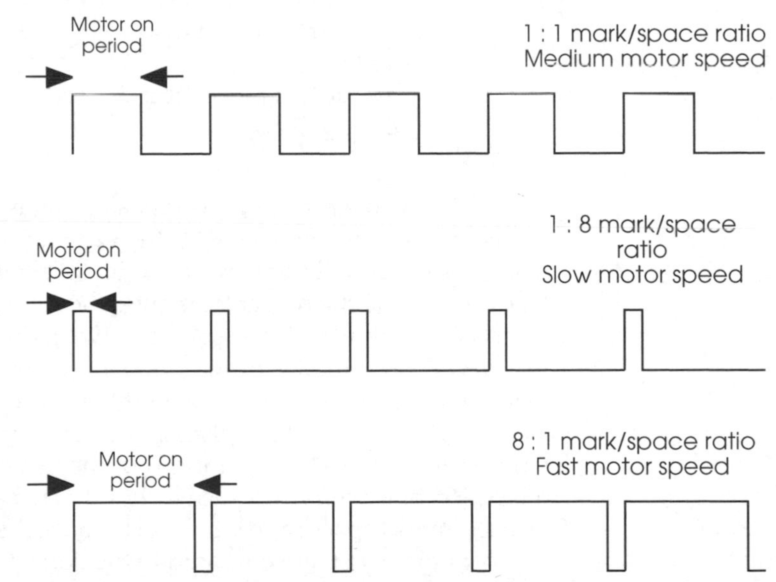

Using different PWM waveforms to control a DC motor speed

With a DC motor, there are two techniques for controlling the motor speed: the first is to reduce the DC voltage to the motor. The higher the voltage, the faster it will turn. At low voltages, the control can be a bit hit and miss and the power control is inefficient. The alternative technique called pulse width modulation (PWM) will control a motor speed not by reducing the voltage to the motor but by reducing the time that the motor is switched on.

This is done by generating a square wave at a frequency of several hundred hertz and changing the mark/space ratio of the wave form. With a large mark and a low space, the voltage is applied to the motor for almost all of the cycle time, and thus the motor will rotate very quickly. With a small mark and a large space, the opposite is true. The diagram shows the waveforms for medium, slow and fast motor control.

The only difference between this method of control and that for a simple on-off switch is the timing of the pulses from the digital output to switch the motor on and off. There are several methods that can be used to generate these waveforms.

Software only

With a software-only system, the waveform timing is done by creating some loops that provide the timing functions. The program pseudo code shows a simple structure for this. The first action is to switch the motor on and then to start counting through a delay loop. The length of time to count through the delay loop determines the motor-on period. When the count is finished, the motor is switched off. The next stage is to count through a second delay loop to determine the motor-off period.

```
repeat (forever)
{
     switch on motor
     delay loop1
     switch off motor
     delay loop2
}
```

This whole procedure is repeated for as long as the motor needs to be driven. By changing the value of the two delays, the mark/space ratio of the waveform can be altered. The total time taken to execute the repeat loop gives the frequency of the waveform. This method is processor intensive in that the program has to run while the motor is running. On first evaluation, it may seem that while the motor is running, nothing else can be done. This is not the case. Instead of simply using delay loops, other work can be inserted in here whose duration now becomes part of the timing for the PWM waveform. If the work is short, then the fine control over the mark/space ratio is not lost because the contribution that the work delay makes compared to the delay loop is small. If the work is long, then the minimum motor-on time and thus motor speed is determined by this period.

```
repeat (forever)
{
     switch on motor
     perform task a
     delay loop1
     switch off motor
     delay loop2
}
```

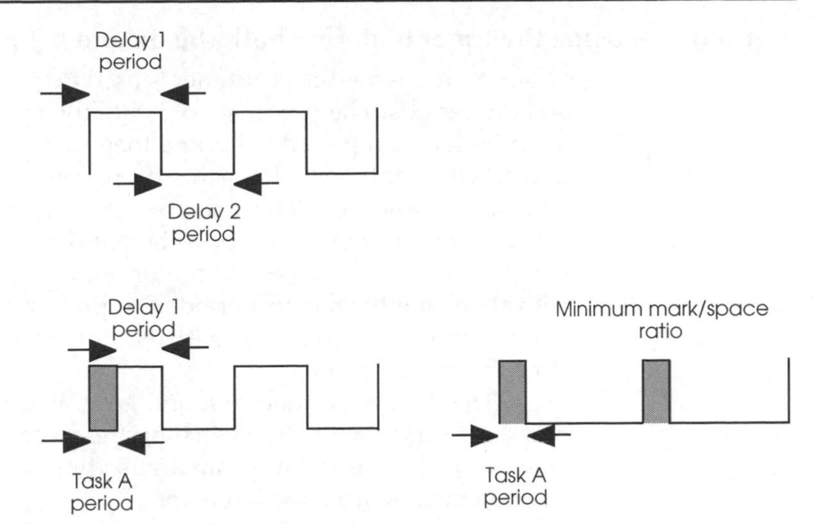

The timing diagrams for the software PWM implementation

The timing diagrams for the software loop PWM waveforms are shown in the diagrams above. In general, software only timing loops are not efficient methods of generating PWM waveforms for motor control. The addition of a single timer greatly improves the mechanism.

Using a single timer

By using a single timer, PWM waveforms can be created far easier and free up the processor to do other things without impacting the timing. There are several methods that can be used to do this. The key principle is that the timer can be programmed to create a periodic interrupt.

Method 1 — using the timer to define the on period

With this method, the timer is used to generate the on period. The processor switches the motor on and then starts the timer to count down. While the timer is doing this, the processor is free to do what ever work is needed. The timer will eventually time out and generate a processor interrupt. The processor services the interrupt and switches the motor off. It then goes into a delay loop still within the service routine until the time period arrives to switch the motor on again. The processor switches the motor on, resets the timer and starts it counting and continues with its work by returning from the interrupt service routine.

Method — using the timer to define frequency period

With this method, the timer is used to generate a periodic interrupt whose frequency is set by the timer period. When the processor services the interrupt, it uses a software loop to determine the on period. The processor switches on the motor and uses the software delay to calculate the on period. When the delay loop is completed, it switches off the motor and can continue with other work until the timer generates the next interrupt.

Method 3 — using the timer to define both the on and off periods

With this method, the timer is used to generate both the on and off periods. The processor switches the motor on, loads the timer with the on-period value and then starts the timer to count down. While the timer is doing this, the processor is free to do what ever work is needed. The timer will eventually time out and generate a processor interrupt, as before. The processor services the interrupt and switches the motor off. It then loads the timer with the value for the off period. The processor then starts the timer counting and continues with its work by returning from the interrupt service routine.

The timer now times out and generates an interrupt. The processor services this by switching the motor on, loading the timer with the one delay value and setting the timer counting before returning from the interrupt.

As a result, the processor is only involved when interrupted by the timer to switch the motor on or off and load the timer with the appropriate delay value and start it counting. Of all these three methods, this last method is the most processor efficient. With methods 1 and 2, the processor is only free to do other work when the mark/space ratio is such that there is time to do it. With a long motor-off period, the processor performs the timing in software and there is little time to do anything else. With a short motor-off period, there is more processing time and far more work can be done. The problem is that the work load that can be achieved is dependent on the mark/space ratio of the PWM waveform and engine speed. This can be a major restriction and this is why the third method is most commonly used.

Using multiple timers

With two timers, it is possible to generate PWM waveforms with virtually no software intervention. One timer is setup to generate a periodic output at the frequency of the required PWM waveform. This output is used to trigger a second timer which is configured as a monostable. The second timer output is used to provide the motor-on period. If these timers are set to automatically reload, the first timer will continually trigger the second and thus generate a PWM waveform. By changing the delay value in the second timer, the PWM mark/space ratio can be altered as needed.

6 Interrupts and exceptions

Interrupts are probably the most important aspect of any embedded system design and potentially can be responsible for many problems when debugging a system. Although they are simple in concept, there are many pitfalls that the unwary can fall into. This chapter goes through the principles behind interrupts, the different mechanisms that are used with various processor architectures and provides a set of do's and don'ts to help guide the designer.

What is an interrupt?

We all experience interrupts at some point during our lives and find that they either pose no problem at all or they can very quickly cause stress and our performance decreases. For example, take a car mechanic working in a garage who not only has to work on the cars but also answer the phone. The normal work of servicing a car continues throughout the day and the only other task is answering the phone. Not a problem, you might think — but each incoming phone call is an interrupt and requires the mechanic to stop the current work, answer the call and then resume the current work. The time it takes to answer the call depends on what the current activity is. If the call requires the machanic to simply put down a tool and pick up the phone, the overhead is short. If the work is more involved, and the mechanic needs to support a component's weight so it can be let go and then need to clean up a little before picking up the phone, the overhead can be large. It can be so long that the caller rings off and the phone call is missed. The mechanic then has to restart the work. If the mechanic receives a lot of phone calls, it is possible that more time is spent in getting ready to answer the call and restarting the work than is actually spent performing the work. In this case, the current work will not be completed on time and the overall performance will be greatly reduced.

With an embedded design, the mechanic is the processor and the current work is the foreground or current task that it is executing. The phone call is the interrupt and the time taken to respond to it is the interrupt latency. If the system is not designed correctly, coping with the interrupts can prevent the system from completing its work or miss an interrupt. In either case, this usually causes problems with the system and it will start to misbehave. In the same way that humans get irrational and start to go away from normal behaviour patterns when continually interrupted while trying to complete some other task, embedded systems can also start misbehaving! It is therefore essential to understand how to use interrupts and perhaps when not to, so that the embedded system can work correctly.

The impact of interrupts and their processing does not stop there either. It can also affect the overall design and structure of the system, particularly of the software that will be running on it. In a well designed embedded system, it is important to actively design it with interrupts in mind and to define how they are going to be used. The first step is to define what an interrupt is.

An interrupt is an event from either an internal or external source where a processor will stop its current processing and switch to a different instruction sequence in response to an event that has occurred either internally or externally. The processor may or may not return to its original processing. So what does this offer the embedded system designer? The key advantage of the interrupt is that it allows the designer to split software into two types: background work where tasks are performed while waiting for an interrupt and foreground work where tasks are performed in response to interrupts. The interrupt mechanism is normally transparent to the background software and it is not aware of the existence of the foreground software. As a result, it allows software and systems to be developed in a modular fashion without having to create a spaghetti bolognese blob of software where all the functions are thrown together. The best way of explaining this is to consider several alternative methods of writing software for a simple system.

The system consists of a processor that has to periodically read in data from a port, process it and write it out. While waiting for the data, it is designed to perform some form of statistical analysis.

The spaghetti method

In this case, the code is written in a straight sequence where occasionally the analysis software goes and polls the port to see if there is data. If there is data present, this is processed before returning to the analysis. To write such code, there is extensive use of branching to effectively change the flow of execution from the background analysis work to the foreground data transfer operations. The periodicity is controlled by two factors:

- The number of times the port is polled while executing the analysis task. This is determined by the data transfer rate.
- The time taken between each polling operation to execute the section of the background analysis software.

With a simple system, this is not too difficult to control but as the complexity increases or the data rates go up requiring a higher polling rate, this software structure rapidly starts to fall about and become inefficient. The timing is software based and therefore will change if any of the analysis code is changed or extended. If additional analysis is done, then more polling checks need to be inserted. As a result, the code often quickly becomes a hard to understand mess.

The situation can be improved through the use of subroutines so that instead of reproducing the code to poll and service the ports, subroutines are called and while this does improve the structure and quality of the code, it does not remove the fundamental problem of a software timed design. There are several difficulties with this type of approach:

- The system timing and synchronisation is completely software dependent which means that it now assumes certain processor speeds and instruction timing to provide a required level of performance.
- If the external data transfers are in bursts and they are asynchronous, then the polling operations are usually inefficient. A large number of checks will be needed to ensure that data is not lost. This is the old polling vs. interrupt argument reappearing.
- It can be very difficult to debug because there are multiple element/entry points within the code that perform the same operation. As a result, there are two asynchronous operations going on in the system. The software execution and asynchronous incoming data will mean that the routes from the analysis software to the polling and data transfer code will be used almost at random. The polling/data transfer software that is used will depend on when the data arrived and what the background software was doing. In this way, it makes reproducing errors extremely difficult to achieve and frequently can be responsible for intermittent problems that are very difficult to solve because they are difficult to reproduce.
- The software/system design is now time referenced as opposed to being event driven. For the system to work, there are time constraints imposed on it such as the frequency of polling which cannot be broken. As a result, the system can become very inefficient. To use an office analogy, it is not very efficient to have to send a nine page fax if you have to be present to insert each page separately. You either stay and do nothing while you wait for the right moment to insert the next page or you have to check the progress repeatedly so that you do not miss the next slot.

Using interrupts

An interrupt is, as its name suggests, a way of stopping the current software thread that the processor is executing, changing to a different software routine and executing it before restoring the processor's status to that prior to the interrupt so that it can continue processing.

Interrupts can happen asynchronously to the operation and can thus be used very efficiently with systems that are event as opposed to time driven. However, they can be used to create time driven systems without having to resort to software-based timers.

To convert the previous example to one using interrupts, all the polling and data port code is removed from the background analysis software. The data transfer code is written as part of the interrupt service routine (ISR) associated with the interrupt generated by the data port hardware. When the port receives a byte of data, it generates an interrupt. This activates the ISR which processes the data before handing execution back to the background task. The beauty of this type of operation is that the background task can be written independently of the data port code and that the whole timing of the system is now moved from being dependent on the polling intervals to one of how quickly the data can be accessed and processed.

Interrupt sources

There are many sources for interrupts varying from simply asserting an external pin to error conditions within the processor that require immediate attention.

Internal interrupts

Internal interrupts are those that are generated by on-chip peripherals such as serial and parallel ports. With an external peripheral, the device will normally assert an external pin which is connected to an interrupt pin on the processor. With internal peripherals, this connection is already made. Some integrated processors allow some flexibility concerning these hardwired connections and allow the priority level to be adjusted or even masked out or disabled altogether.

External interrupts

External interrupts are the common method of connecting external peripherals to the processor. They are usually provided through external pins that are connected to peripherals and are asserted by the peripheral. For example, a serial port may have a pin that is asserted when there is data present within its buffers. The pin could be connected to the processor interrupt pin so that when the processor sees the data ready signal as an interrupt. The corresponding interrupt service routine would then fetch the data from the peripheral before restoring the previous processing.

Exceptions

Many processor architectures use the term exception as a more generic term for an interrupt. While the basic definition is the same (an event that changes the software flow to process the event) an exception is extended to cover any event, including internal and external interrupts, that causes the processor to change to a service routine. Typically, exception processing is normally coupled with a change in the processor's mode. This will be described in more detail for some example processors later in this chapter.

The range of exceptions can be large and varied. A MC68000 has a 256 entry vector table which describes about 90 exception conditions with the rest reserved for future expansion. An 8 bit micro may have only a few.

Software interrupts

The advantage of an interrupt is that it includes a mechanism to change the program flow and in some processor architectures, to change into a more protected state. This means that an interrupt could be used to provide an interface to other software such as an operating system. This is the function that is provided by the software interrupt. It is typically an instruction or set of instructions that allows a currently executing software sequence to change flow and return using the more normal interrupt mechanism. With devices like the Z80 this function is provided by the SWI (software interrupt instruction). With the MC68000 and PowerPC architectures, the TRAP instruction is used.

To use software interrupts efficiently, additional data to specify the type of request and/or data parameters has to be passed to the specific ISR that will service the software interrupt. This is normally done by using one or more of the processor's registers. The registers are accessible by the ISR and can be used to pass status information back to the calling software.

It could be argued that there is no need to use software interrupts because branching to different software routines can be achieved by branches and jumps. The advantage that a software interrupt offers is in providing a bridge and routine between software running in the normal user mode and other software running in a supervisor mode. The different modes allow the resources such as memory and associated code and data to be protected from each other. This means that if the user causes a problem or makes an incorrect call, then the supervisor code and data are not at risk and can therefore survive and thus have a chance to restore the system or at least shut it down in an orderly manner.

Non-maskable interrupts

A non-maskable interrupt (NMI) is as its name suggests an external interrupt that cannot be masked out. It is by default at the highest priority of any interrupt and will always be recognised and processed. In terms of a strict definition, it is masked out when the ISR starts to process the interrupt so that it is not repeatedly recognised as a separate interrupt and therefore the non-maskable part refers to the ability to mask the interrupt prior to its assertion.

The NMI is normally used as a last resort to generate an interrupt to try and recover control. This can be presented as either a reset button or connected to a fault detection circuit such as a memory parity or watchdog timer. The 80x86 NMI as used on the IBM PC is probably the most known implementation of this function. If the PC memory subsystem detects a parity error, the parity circuitry asserts the NMI. The associated ISR does very little except stop the processing and flash up a window on the PC saying that a parity error has occurred and please restart the machine.

Recognising an interrupt

The start of the whole process is the recognition of an interrupt. Internal interrupts are normally defined by the manufacturer and are already hardwired. External interrupts, however, are not and can use a variety of mechanisms.

Edge triggered

With the edge triggered interrupt, it is the clock edge that is used to generate the interrupt. The transition can either be from a logical high to low or vice versa. With these systems, the recognition process is usually in two stages. The first stage is the external transition that typically latches an interrupt signal. This signal is then checked on an instruction boundary and, if set, starts the interrupt process. At this point, the interrupt has been successfully recognised and the source removed.

Level triggered

With a level triggered interrupt, the trigger is dependent on the logic level. Typically, the interrupt pin is sampled on a regular basis, e.g. after every instruction or on every clock edge. If it is set to the appropriate logic level, the interrupt is recognised and acted upon. Some processors require the level to be held for a minimum number of clocks or for a certain pulse width so that extraneous pulses that are shorter in duration than the minimum pulse width are ignored.

Maintaining the interrupt

So far, the recognition of an interrupt has concentrated on simply asserting the interrupt pin. This implies that provided the minimum conditions have been met, the interrupt source can be removed. Many microprocessor manufacturers recommend that this is not done and that the interrupt should be maintained until it has been explicitly serviced and the source told to remove it.

Internal queuing

This last point also raises a further potential complication. If an interrupt is asserted so that it conforms with the recognition conditions, removed and reasserted, the expectation would be that the interrupt service routine would be executed twice to service each interrupt. This assumes that there is an internal counter within the processor that can count the number of interrupts and thus effectively queue them. While this might be expected, this is not the case with most processors. The first interrupt would be recognised and, until it is serviced, all other interrupts generated using the pin are ignored. This is one reason why many processors insist on the maintain until serviced approach with interrupts. Any subsequent interrupts that have the same level will be maintained after the first one has been serviced and its signal removed. When the exception processing is completed, the remaining interrupts will be recognised and processed one by one until they are all serviced.

The interrupt mechanism

Once an interrupt or exception has been recognised, then the processor goes through an internal sequence to switch the processing thread and activate the ISR or exception handler that will service the interrupt or exception. The actual process and, more importantly, the implied work that the service routine must perform varies from processor architecture to architecture. The general processing for an MC68000 or 80x86 which uses a stack frame to hold essential data is different from a RISC processor that uses special internal registers.

Before describing in detail some of the most used mechanisms, let's start with a generic explanation of what is involved. The first part of the sequence is the recognition of the interrupt or exception. This in itself does not necessarily immediately trigger any processor reaction. If the interrupt is not an error condition or the error condition is not connected with the currently executing instruction, the interrupt will not be internally processed until the currently executing instruction has completed. At this point, known as an instruction boundary, the processor will start to internally process the interrupt. If, on the other hand, the interrupt is due to an error with the currently executing instruction, the instruction will be aborted to reach the instruction boundary.

At the instruction boundary, the processor must now save certain state information to allow it to continue its previous execution path prior to the interrupt. This will typically include a copy of the condition code register, the program counter and the return address. This information may be extended to include internal state information as well. The register set is not normally included.

The next phase is to get the location of the ISR to service the interrupt. This is normally kept in a vector table somewhere in memory and the appropriate vector can be supplied by the peripheral or preassigned, or a combination of both approaches. Once the vector has been identified, the processor starts to execute the code within the ISR until it reaches a return from interrupt type of instruction. At this point, the processor, reloads the status information and processing continues the previous instruction stream.

Stack-based processors

With stack-based processors, such as the Intel 80x86, Motorola M68000 family and most 8 bit microcontrollers based on the original microprocessor architectures such as the 8080 and

MC6800, the context information that the processor needs to preserve is saved on the external stack.

When the interrupt occurs, the processor context information such as the return address, copies of the internal status registers and so on are stored out on the stack in a stack frame. These stack frames can vary in size and content depending on the source of the interrupt or exception.

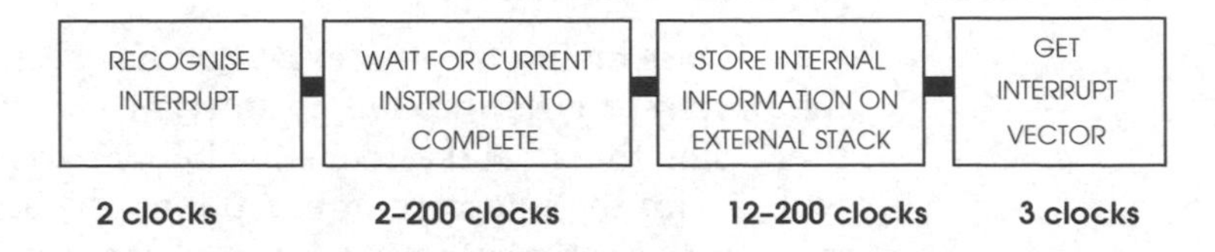

A typical processor interrupt sequence

When the interrupt processing is completed, the information is extracted back from the stack and used to restore the processing prior to the interrupt. It is possible to nest interrupts so that several stack frames and interrupt routines must be executed prior to the program flow being restored. The number of routines that can be nested in this way depends on the storage space available. With external stacks, this depends in turn on the amount of available memory.

Other processors use an internal hardware stack to reduce the external memory cycles necessary to store the stack frame. These hardware stacks are limited in the number of interrupts or exceptions that can be nested. It then falls to the software designer to ensure that this limit is not exceeded. To show these different interrupt techniques, let's look at some processor examples.

MC68000 interrupts

The MC68000 interrupt and exception processing is based on using an external stack to store the processor's context information. This is very common and similar methods are provided on the 80x86 family and many of the small 8 bit microcontrollers.

Seven interrupt levels are supported and are encoded onto three interrupt pins IP0–IP2. With all three signals high, no external interrupt is requested. With all three asserted, a non-maskable level 7 interrupt is generated. Levels 1–6, generated by other combinations, can be internally masked by writing to the appropriate bits within the status register.

The interrupt cycle is started by a peripheral generating an interrupt. This is usually encoded using a LS148 seven to three priority encoder. This converts seven external pins into a 3 bit binary code. The appropriate code sequence is generated and drives the interrupt pins. The processor samples the levels and requires the levels to remain constant to be recognised. It is recommended that the interrupt level remains asserted until its interrupt acknowledgement cycle commences to ensure recognition.

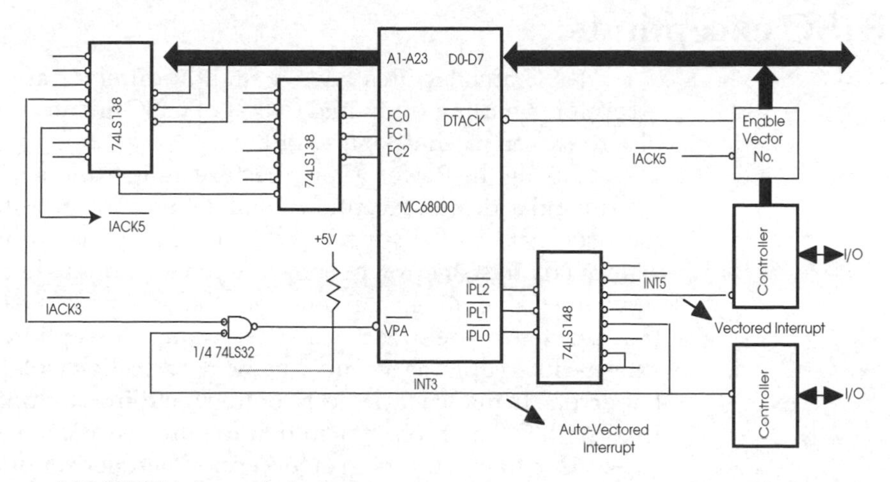

An example MC68000 interrupt design

Once the processor has recognised the interrupt, it waits until the current instruction has been completed and starts an interrupt acknowledgement cycle. This starts an external bus cycle with all three function code pins driven high to indicate an interrupt acknowledgement cycle.

The interrupt level being acknowledged is placed on address bus bits A1–A3 to allow external circuitry to identify which level is being acknowledged. This is essential when one or more interrupt requests are pending. The system now has a choice over which way it will respond:

- If the peripheral can generate an 8 bit vector number, this is placed on the lower byte of the address bus and DTACK* asserted. The vector number is read and the cycle completed. This vector number then selects the address and subsequent software handler from the vector table.
- If the peripheral cannot generate a vector, it can assert VPA* and the processor will terminate the cycle using the M6800 interface. It will select the specific interrupt vector allocated to the specific interrupt level. This method is called auto-vectoring.

To prevent an interrupt request generating multiple acknowledgements, the internal interrupt mask is raised to the interrupt level, effectively masking any further requests. Only if a higher level interrupt occurs will the processor nest its interrupt service routines. The interrupt service routine must clear the interrupt source and thus remove the request before returning to normal execution. If another interrupt is pending from a different source, it can be recognised and cause another acknowledgement to occur.

A typical circuit is shown. Here, level 5 has been allocated as a vectored interrupt and level 3 auto-vectored. The VPA* signal is gated with the level 3 interrupt to allow level 3 to be used with vectored or auto-vectored sources in future designs.

RISC exceptions

RISC architectures have a slightly different approach to exception handling compared to that of CISC architectures. This difference can catch designers out.

Taking the PowerPC architecture as an example, there are many similarities: an exception is still defined as a transition from the user state to the supervisor state in response to either an external request or error, or some internal condition that requires servicing. Generating an exception is the only way to move from the user state to the supervisor state. Example exceptions include external interrupts, page faults, memory protection violations and bus errors. In many ways the exception handling is similar to that used with CISC processors, in that the processor changes to the supervisor state, vectors to an exception handler routine, which investigates the exception and services it before returning control to the original program. This general principle still holds but there are fundamental differences which require careful consideration.

When an exception is recognised, the address of the instruction to be used by the original program when it restarts and the machine state register (MSR) are stored in the supervisor registers, SRR0 and SRR1. The processor moves into the supervisor state and starts to execute the handler, which resides at the associated vector location in the vector table. The handler can, by examining the DSISR and FPSCR registers, determine the exact cause and rectify the problem or carry out the required function. Once completed, the rfi instruction is executed. This restores the MSR and the instruction address from the SRR0 and SRR1 registers and the interrupted program continues.

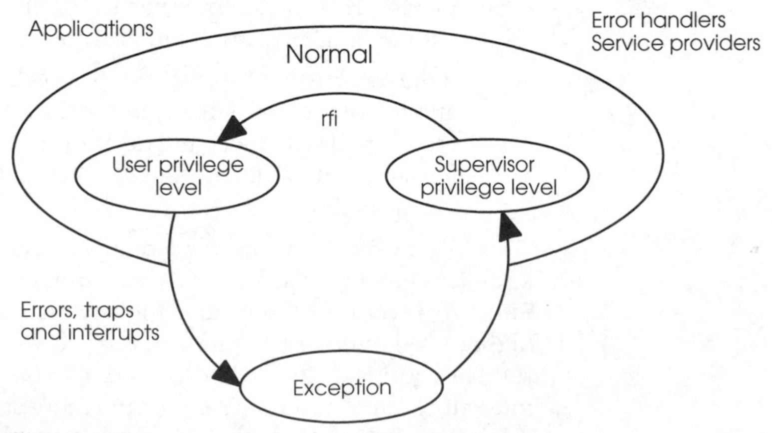

The exception transition model

There are four general types of exception: asynchronous precise or imprecise and synchronous precise and imprecise. Asynchronous and synchronous refer to when the exception is caused: a synchronous exception is one that is synchronised, i.e. caused by the instruction flow. An asynchronous exception is one where an external event causes the exception; this can effectively occur at any time and is not dependent on the instruction flow. A precise exception is where the cause is precisely defined and is usually recoverable. A memory page fault is a good example of this. An imprecise exception is usually a catastrophic failure, where the processor cannot continue processing or allow a particular program or task to continue. A system reset or memory fault while accessing the vector table falls into this category.

Synchronous precise

All instruction caused exceptions are handled as synchronous precise exceptions. When such an exception is encountered during program execution, the address of either the faulting instruction or the one after it is stored in SRR0. The processor will have completed all the preceding instructions; however, this does not guarantee that all memory accesses caused by these instructions are complete. The faulting instruction will be in an indeterminate state, i.e. it may have started and be partially or completely completed. It is up to the exception handler to determine the instruction type and its completion status using the information bits in the DSISR and FPSCR registers.

Synchronous imprecise

This is generally not supported within the PowerPC architecture and is not present on the MPC601, MPC603 or MCP604 implementations. However, the PowerPC architecture does specify the use of synchronous imprecise handling for certain floating point exceptions and so this category may be implemented in future processor designs.

Asynchronous precise

This exception type is used to handle external interrupts and decrementer-caused exceptions. Both can occur at any time within the instruction processing flow. All instructions being processed before the exceptions are completed, although there is no guarantee that all the memory accesses have completed. SRR0 stores the address of the instruction that would have been executed if no interrupt had occurred.

These exceptions can be masked by clearing the EE bit to zero in the MSR. This forces the exceptions to be latched but not acted on. This bit is automatically cleared to prevent this type of interrupt causing an exception while other exceptions are being processed.

The number of events that can be latched while the EE bit is zero is not stated. This potentially means that interrupts or decrementer exceptions could be missed. If the latch is already full, any subsequent events are ignored. It is therefore recommended that the exception handler performs some form of handshaking to ensure that all interrupts are recognised.

Asynchronous imprecise

Only two types of exception are associated with this: system resets and machine checks. With a system reset all current processing is stopped, all internal registers and memories are reset; the processor executes the reset vector code and effectively restarts processing. The machine check exception is only taken if the ME bit of the MSR is set. If it is cleared, the processor enters the checkstop state.

Recognising RISC exceptions

Recognising an exception in a superscalar processor, especially one where the instructions are executed out of program order, can be a little tricky — to say the least. The PowerPC architecture handles synchronous exceptions (i.e. those caused by the instruction stream) in strict program order, even though instructions further on in the program flow may have already generated an exception. In such cases, the first exception is handled as if the following instructions have never been executed and the preceding ones have all completed.

There are occasions when several exceptions can occur at the same time. Here, the exceptions are handled on a priority basis using the priority scheme shown in the table below. There is additional priority for synchronous precise exceptions because it is possible for an instruction to generate more than one exception. In these cases, the exceptions would be handled in their own priority order as shown below.

Class	Priority	Description
Async imprecise	1	System reset
	2	Machine check
Sync precise	3	Instruction dependent
Async precise	4	External interrupt
	5	Decrementer interrupt

Exception class priority

If, for example, with the single-step trace mode enabled, an integer instruction executed and encountered an alignment error, this exception would be handled before the trace exception. These synchronous precise priorities all have a higher priority than the level 4 and 5 asynchronous precise exceptions, i.e. the external interrupt and decrementer exceptions.

When an exception is recognised, the continuation instruction address is stored in SRR0 and the MSR is stored in SRR1. This saves the machine context and provides the interrupted program with the ability to continue. The continuation instruction may not have started, or be partially or fully complete, depending on the nature of the exception. The FPSCR and DSISR registers contain further diagnostic information for the handler. When in this state, external interrupts and decrementer exceptions are disabled. The EE bit is cleared automatically to prevent such asynchronous events from unexpectedly causing an exception while the handler is coping with the current one.

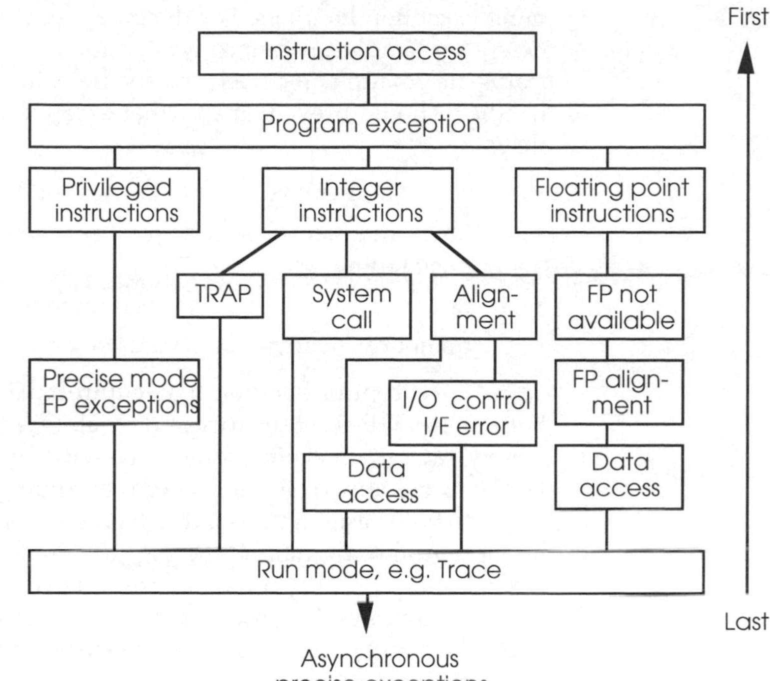

Precise exceptions priority

It is important to note that the machine status or context which is necessary to allow normal execution to continue is automatically stored in SRR0 and SRR1 — which overwrites the previous contents. As a result, if another exception occurs during an exception handler execution, the result can be catastrophic: the exception handler's machine status information in SRR0 and SRR1 would be overwritten and lost. In addition, the status information in FPSCR and DSISR is also overwritten. Without this information, the handler cannot return to the original program. The new exception handler takes control, processes its exception and, when the rfi instruction is executed, control is passed back to the first exception handler. At this point, this handler does not have its own machine context information to enable it to return control to the original program. As a result the system will, at best, have lost track of that program; at worst, it will probably crash.

This is not the case with the stack-based exception handlers used on CISC processors. With these architectures, the machine status is stored on the stack and, provided there is sufficient stack available, exceptions can safely be nested, with each exception context safely and automatically stored on the stack.

It is for this reason that the EE bit is automatically cleared to disable the external and decrementer interrupts. Their asynchronous nature means that they could occur at any time and if this happened at the beginning of an exception routine, that routine's ability to return control to the original program would be lost. However, this does impose several constraints when programming exception handlers. For the maximum performance in the exception handler, it cannot waste time by saving the machine status information on a stack or elsewhere. In this case, exception handlers should prevent any further exceptions by ensuring that they:

- reside in memory and not be swapped out;
- have adequate stack and memory resources and not cause page faults;
- do not enable external or decrementer interrupts;
- do not cause any memory bus errors.

For exception handlers that require maximum performance but also need the best security and reliability, they should immediately save the machine context, i.e. SRR registers FPSCR and DSISR, preferably on a stack before continuing execution.

In both cases, if the handler has to use or modify any of the user programming model, the registers must be saved prior to modification and they must be restored prior to passing control back. To minimise this process, the supervisor model has access to four additional general-purpose registers which it can use independently of the general-purpose register file in the user programming model.

Enabling RISC exceptions

Some exceptions can be enabled and disabled by the supervisor by programming bits in the MSR. The EE bit controls external interrupts and decrementer exceptions. The FE0 and FE1 bits control which floating point exceptions are taken. Machine check exceptions are controlled via the ME bit.

Returning from RISC exceptions

As mentioned previously, the rfi instruction is used to return from the exception handler to the original program. This instruction synchronises the processor, restores the instruction address and machine state register and the program restarts.

The vector table

Once an exception has been recognised, the program flow changes to the associated exception handler contained in the vector table.

The vector table is a 16 kbyte block (0 to $3FFF) that is split into 256 byte divisions. Each division is allocated to a particular exception or group of exceptions and contains the exception handler routine associated with that exception. Unlike many other architectures, the vector table does not contain pointers to the routines but the actual instruction sequences themselves. If the handler is too large to fit in the division, a branch must be used to jump to its continuation elsewhere in memory.

The table can be relocated by changing the EP bit in the machine state register (MSR). If cleared, the table is located at $0000000. If the bit is set to one (its state after reset) the vector table is relocated to $FFF00000. Obviously, changing this bit before moving the vector table can cause immense problems!

Vector Offset (hex)	Exception	
0 0000	Reserved	
0 0100	System Reset	Power-on, Hard & Soft Resets
0 0200	Machine Check	Eabled through MSR (ME)
0 0300	Data Access	Data Page Fault/Memory Protection
0 0400	Instruction Access	Instr. Page Fault/Memory Protection
0 0500	External Interrupt	$\overline{INT}$
0 0600	Alignment	Access crosses Segment or Page
0 0700	Program	Instr. Traps, Errors, Illegal, Privileged
0 0800	Floating-Point Unavailiable	MSR(FP)=0 & F.P. Instruction encountered
0 0900	Decrementer	Decrementer Register passes through 0
0 0A00	Reserved	
0 0B00	Reserved	
0 0C00	System Call	'sc' instruction
0 0D00	Trace	Single-step instruction trace
0 0E00	Floating-Point Assist	A floating-point exception

The basic PowerPC vector table

Identifying the cause

Most programmers will experience exception processing when a program has crashed or a cryptic message is returned from a system call. The exception handler can provide a lot of information about what has gone wrong and the likely cause. In this section, each exception vector is described and an indication of the possible causes and remedies given.

The first level investigation is the selection of the appropriate exception handler from the vector table. However, the exception handler must investigate further to find out the exact cause before trying to survive the exception. This is done by checking the information in the FPSCR, DSISR, DAR and MSR registers, which contain different information for each particular vector.

Fast interrupts

There are other interrupt techniques which greatly simplify the whole process but in doing so provide very fast servicing at the expense of several restrictions. These so-called fast interrupts are often used on DSP processors or microcontrollers where a small software routine is executed without saving the processor context.

This type of support is available on the DSP56000 signal processors, for example. External interrupts normally generate a fast interrupt routine exception. The external interrupt is synchronised with the processor clock for two successive clocks, at which point the processor fetches the two instructions from the appropriate vector location and executes them.

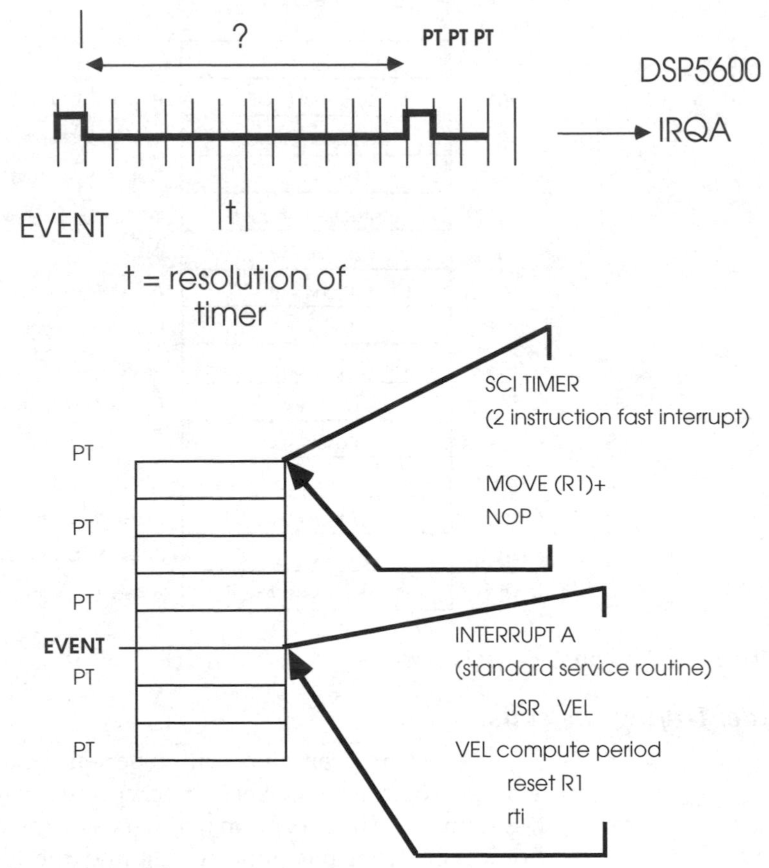

Using interrupts to time an event

Once completed, the program counter simply carries on as if nothing has happened. The advantage is that there is no stack frame building or any other such delays. The disadvantage concerns the size of the routine that can be executed and the resources allocated. When using such a technique, it is usual to allocate a couple of address registers for the fast interrupt routine to use. This allows coefficient tables to be built, almost in parallel with normal execution.

The SCI timer is programmed to generate a two instruction fast interrupt which simply auto-increments register R1. This acts as a simple counter which times the period between events. The event itself generates an IRQA interrupt, which forces a standard service routine. The exception handler jumps to routine VEL which processes the data (i.e. takes the value of R and uses it to compute the period), resets R1 and returns from the interrupt.

Interrupt controllers

In many embedded systems there are more external sources for interrupts than interrupt pins on the processor. In this case, it is necessary to use an interrupt controller to provide a larger number of interrupt signals. An interrupt controller performs several functions:

- It provides a large number of interrupt pins that can be allocated to many external devices. Typically this is at least eight and higher numbers can be supported by cascading two or more controllers together. This is done on the IBM PC AT where two 8 port controllers are cascaded to give 15 interrupt levels. One level is lost to provide the cascade link.
- It orders the interrupt pins in a priority level so that a high level interrupt will inhibit a lower level interrupt.
- It may provide registers for each interrupt pin which contain the vector number to be used during an acknowledge cycle. This allows peripherals that do not have the ability to provide a vector to do so.
- They can provide interrupt masking. This allows the system software to decide when and if an interrupt is allowed to be serviced. The controller, through the use of masking bits within a controller, can prevent an interrupt request from being passed through to the processor. In this way, the system has a multi-level approach to screening interrupts. It uses the screening provided by the processor to provide coarse grain granularity while the interrupt controller provides a finer level.

Instruction restart and continuation

The method of continuing the normal execution after exception processing due to a mid-instruction fault, such as that caused by a bus error or a page fault, can be done in one of two ways. Instruction restart effectively backs up the machine to the point in the instruction flow where the error occurred. The processor re-executes the instruction and carries on. The instruction continuation stores all the internal data and allows the errant bus cycle to be restarted, even if it is in the middle of an instruction.

The continuation mechanism is undoubtedly easier for software to handle, yet pays the penalty of having extremely large stack frames or the need to store large amounts of context information to allow the processor to continue mid-instruction. The restart mechanism is easier from a hardware perspective, yet can pose increased software overheads. The handler has to determine how far back to restart the machine and must ensure that resources are in the correct state before commencing.

The term 'restart' is important and has some implications. Unlike many CISC processors (for example, the MC68000, MC68020 and MC68030) the instruction does not continue; it is restarted from the beginning. If the exception occurred in the middle of the instruction, the restart repeats the initial action. For many instructions this may not be a problem — but it can lead to some interesting situations concerning memory and I/O accesses.

If the instruction is accessing multiple memory locations and fails after the second access, the first access will be repeated. The store multiple type of instruction is a good example of this, where the contents of several registers are written out to memory. If the target address is an I/O peripheral, an unexpected repeat access may confuse it.

While the majority of the M68000 and 80x86 families are of the continuation type. The MC68040 and PowerPC families along with most microcontrollers — especially those using RISC architectures — are of the restart type. As processors increase in speed and complexity, the penalty of large stack frames shifts the balance in favour of the restart model.

Interrupt latency

One of the most important aspects of using interrupts is in the latency. This is usually defined as the time taken by the processor from recognition of the interrupt to the start of the ISR. It consists of several stages and is dependent on both hardware and software factors. Its importance is that it defines several aspects of an embedded system with reference to its ability to respond to real-time events. The stages involved in calculating a latentcy are:

- The time taken to recognise the interrupt

  Do not asssume that this is instantaneous as it will depend on the processor design and its own interrupt recognition mechanism. As previously mentioned, some processors will repeatedly sample an interrupt signal to ensure that it is a real one and not a false one.

- The time taken by the CPU to complete the current instruction

  This will also vary depending on what the CPU is doing and its complexity. For a simple CISC processor, this time will vary as its instructions all take a different number of clocks to complete. Usually the most time-consuming instructions are those that perform multiplication or division or some complex data manipulation such as bit field operations. For RISC processors with single cycle execution, the time is usually that to clear the execution pipeline and is 1 or 2 clocks. For complex processors that execute multiple instructions per clocks, this calculation can get quite difficult. The other problem is identifying which instruction will be executing when the interrupt is recognised. In practice, this is impossible to do and the worst case execution time is used. This can often be one or more orders of magnitude greater than a typical instruction. A 32 bit division could take several hundred clocks while a simple add or subtract could be 1 cycle. Some compilers will restrict the use of multiple cycle instructions so that these worst case figures can be improved by only using fast executing instructions. As part of this, division and multiplication are often performed by software routines using a sequence of faster add, subtract and bit manipulation instructions. Of course, if the code is being written in assembly, then this can be finely tuned by examining the instruction mix.

- The time for the CPU to perform a context switch

  This is the time taken by the processor to save its internal context information such as its program counter, internal data registers and anything else it needs. For CISC processors, this can involve creating blocks of data on the stack by writing the information externally. For RISC processors this may mean simply switching registers internally without explicitly saving any information. Register windowing or shadowing is normally used.

- The time taken to fetch the interrupt vector

  This is normally the time to fetch a single value from memory but even this time can be longer than you think! We will come back to this topic.

- The time taken to start the interrupt service routine execution

  Typically very short. However remember that because the pipeline is cleared, the instruction will need to be clocked through to execute it and this can take a few extra clocks, even with a RISC architecture.

In practice, processor architectures will have different precedures for all these stages depending on their design philosphy and sophistication.

With a simple microcontroller, this calculation is simple: take the longest execution time for any instruction, add to it the number of memory accesses that the processor needs times the number of clocks per access, add any other delays and you arrive with a worst case interrupt latency. This becomes more difficult when other factors are added such as memory management and caches. While the basic calculations are the same, the number of clocks involved in a memory access can vary dramatically — often by an order of magnitude! To correctly calculate the value, the interupt latency calculations also have to take into account the cost of external memory access and this can sometimes be the overwhelmingly dominant factor.

While all RISC systems should be designed with single cycle memory access for optimum performance, the practicalities are that memory cycles often incur wait states or bus delays. Unfortunately for designers, RISC architectures cannot tolerate such delays — one wait state halves the performance, two reduces performance to a third. This can have a dramatic effect on real-time performance. All the advantages gained with the new architecture may be lost.

The solution is to use caches to speed up the memory access and remove delays. This is often used in conjunction with memory management to help control the caches and ensure data coherency, as well as any address translation. However, there are some potential penalties for any system that uses caches and memory management which must be considered.

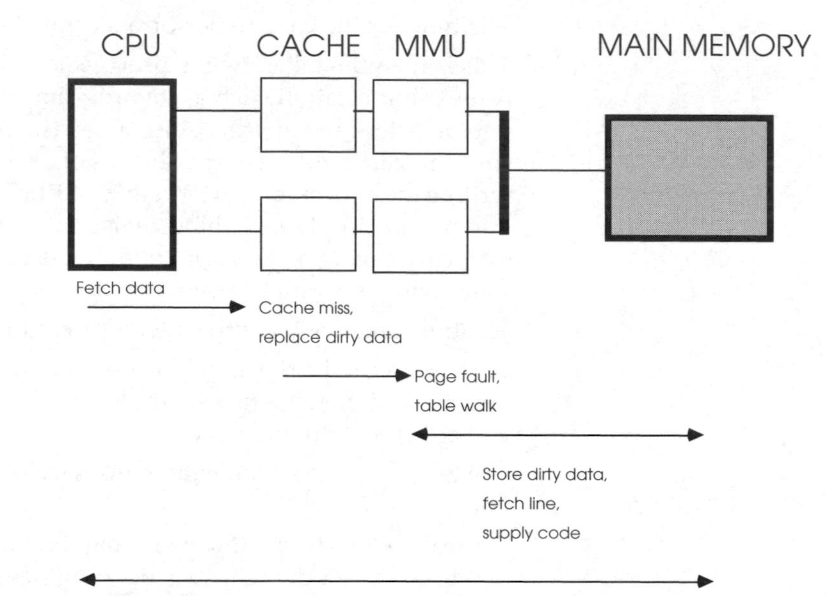

The impact of a cache miss

Consider the system in the diagram. The processor is using a Harvard architecture with combined caches and memory management units to buffer it from the slower main memory. The caches are operating in copyback mode to provide further speed improvements. The processor receives an interrupt and immediately starts exception processing. Although the internal context is preserved in shadow registers, the vector table, exception routines and data exist in external memory.

In this example, the first data fetch causes a cache miss. All the cache lines are full and contain some dirty data, therefore the cache must update main memory with a cache line before fetching the instruction. This involves an address translation, which causes a page fault. The MMU now has to perform an external table walk before the data can be stored. This has to complete before the cache line can be written out which, in turn, must complete before the first instruction of the exception routine can be executed. The effect is staggering — the quick six cycle interrupt latency is totally overshadowed by the 12 or so memory accesses that must be completed simply to get the first instruction. This may be a worst case scenario, but it has to be considered in any real-time design.

This problem can be contained by declaring exception routines and data as non-cachable, or through the use of a BATC or transparent window to remove the page fault and table walk. These techniques couple the CPU directly to external memory which, if slow, can be extremely detrimental to performance. Small areas of very fast memory can be reserved for exception handling to solve this problem; locking routines in cache can also be solutions, at the expense of performance in other system functions. It should be of no surprise that many of today's RISC controllers are using local memory (tightly coupled memory or TCM) to store interrupt routines and thus allow the system designer to address the latency problems of caches and memory management. This is added to the processor and resides on-chip. Access does not require any off-chip bus cycles and so it will not slow down the processor whenever it is used. By locating the time critical routines and data in this tightly coupled memory, it is possible to not only reduce the latency and ISR execution time but also make the execution more consistent in its timing.

Do's and Don'ts

This last section describes the major problems that are encountered with interrupt and exceptions, and, more importantly, how to avoid them.

Always expect the unexpected interrupt

Always include a generic handler for all unused/unexpected exceptions. This should record as much information about the processor context such as the exception/vector table number, the return address and so on. This allows unexpected exceptions to be detected and recognised instead of causing the processor and system to crash with little or no hope of finding what has happened.

Don't expect too much from an interrupt

Bear in mind that an interrupt is not for free and the simple act of generating one and returning even if the interrupt service routine does nothing, will consume performance. There will be a point where the number of interrupts are so high that the system spends more time with the overhead than actually processing or doing something with the data. It is important to balance the cost of an interrupt's overhead against the processing it will do. A good way of thinking about this is using a truck to carry bricks from A to B. If the truck carries one brick at a time, the time taken to load the truck, push it and unload it will mean it will be slower than simply picking up the brick and moving it. Loading the truck with so many bricks so that it is difficult to push and takes a long time is equally not good. The ideal is a balance where 10 or 20 bricks are moved at once in the same time it would take to move one. Don't overload the system with too many interrupts or put too much into the interrupt service routine.

Use handshaking

Just because an interrupt signal has been applied for the correct number of clocks, do not assume that it has been recognised. It might have been masked out or not seen for some other reason. Nearly all processor designs use a handshaking mechanism where the interrupt is maintained until it is explicitly acknowledged and removed. This may be a hardware signal or a constituent of the ISR itself where it will remove the interrupt source by writing to a register in the peripheral.

Control resource sharing

If a resource such as a variable is used by the normal software and within an interrupt routine, care must be taken to prevent corruption.

For example, if you have a piece of C code that modifies a variable *a* as shown in the example, the expected output would be *a=6* if the value of *a* was 3.

```
{
read(a);
a=2*a;
printf("a=", a);
}
```

If variable a was used in an interrupt routine then there is a risk that the original code will fail, e.g. it would print out `a=8`, or some other incorrect value. The explanation is that the interrupt routine was executed in the middle of the original code. This changed the value of a and therefore the wrong value was returned.

```
{
read(a);
                                Interrupt!
                                read(a);
                                Return;
a=2*a;
printf("a=", a);
}
```

Exceptions and interrupts can occur asynchronously and therefore if the system shares any resource such as data, or access to peripherals and so on, it is important that any access is handled in such a way that an interrupt cannot corrupt the program flow. This is normally done by masking interrupts before access and unmasking them afterwards. The example code has been modified to include the mask_int and unmask_int calls. The problem is that while the interrupts are masked out, the interrupt latency is higher and therefore this is not a good idea for all applications.

```
{
mask_int();
read(a);
a=2*a;
printf("a=", a);
unmask_int();
}
```

This problem is often the cause of obscure faults where the system works fine for days and then crashes i.e. the fault only occurs when certain events happen within certain time frames. The best way to solve the problem in the first place is to redesign the software so that the sharing is removed or uses a messaging protocol that copies data to ensure that corruption cannot take place.

Beware false interrupts

Ensure that all the hardware signals and exception routines do not generate false interrupts. This can happen in software when the interrupt mask or the interrupt handler executes the return from interrupt instruction before the original interrupt source is removed.

In hardware, this can be caused by pulsing the interrupt line and assuming that the processor will only recognise the first pulse and mask out the others. Noise and other factors can corrupt the interrupt lines so that the interrupt is not recognised correctly.

Controlling interrupt levels

This was touched on earlier when controlling resources. It is important to assign high priority events to high priority interrupts. If this does not happen then priority inversion can occur where the lower priority event is serviced while higher priority events wait. This is quite a complex topic and is discussed in more detail in the chapter on real-time operating systems.

Controlling stacks

It is important to prevent stacks from overflowing and exceeding the storage space, whether it is external or internal memory. Some software, in an effort to optimise performance, will remove stack frames from the stack so that the return process can go straight back to the initial program. This is common with nested routines and can be a big time saver. However, it can also be a major source of problems if the frames are not correctly removed or if they are when information must be returned. Another common mistake is to assume that all exceptions have the same size stack frames for all exceptions and all processor models within the family. This is not always the case!

7 Real-time operating systems

What are operating systems?

Operating systems are software environments that provide a buffer between the user and the low level interfaces to the hardware within a system. They provide a constant interface and a set of utilities to enable users to utilise the system quickly and efficiently. They allow software to be moved from one system to another and therefore can make application programs hardware independent. Program debugging tools are usually included which speed up the testing process. Many applications do not require any operating system support at all and run direct on the hardware.

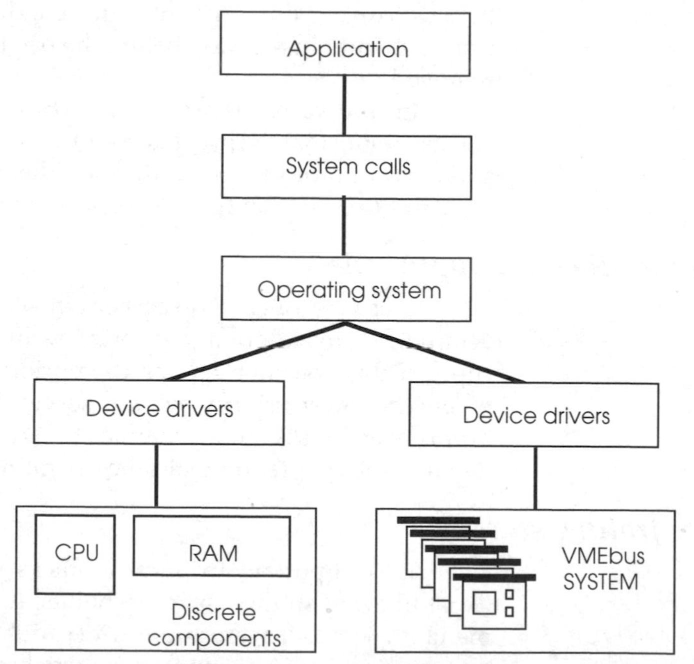

Hardware independence through the use of an operating system

Such software includes its own I/O routines, for example, to drive serial and parallel ports. However, with the addition of mass storage and the complexities of disk access and file structures, most applications immediately delegate these tasks to an operating system.

The delegation decreases software development time by providing system calls to enable application software access to any of the I/O system facilities. These calls are typically made by building a parameter block, loading a specified register with its location and then executing a software interrupt instruction.

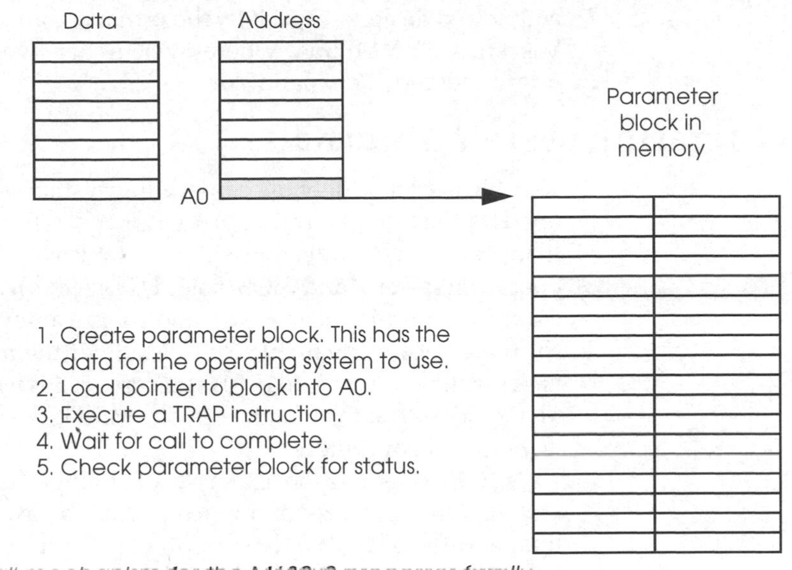

Typical system call mechanism for the M680x0 processor family

The TRAP instruction is the MC68000 family equivalent of the software interrupt and switches the processor into supervisor mode to execute the required function. It effectively provides a communication path between the application and the operating system kernel. The kernel is the heart of the operating system which controls the hardware and deals with interrupts, memory usage, I/O systems etc. It locates a parameter block by using an address pointer stored in a predetermined address register. It takes the commands stored in a parameter block and executes them. In doing so, it is the kernel that drives the hardware, interpreting the commands passed to it through a parameter block. After the command is completed, status information is written back into the parameter block, and the kernel passes control back to the application which continues running in USER mode. The application will find the I/O function completed with the data and status information written into the parameter block. The application has had no direct access to the memory or hardware whatsoever.

These parameter blocks are standard throughout the operating system and are not dependent on the actual hardware performing the physical tasks. It does not matter if the system uses an MC68901 multifunction peripheral or a 8530 serial communication controller to provide the serial ports: the operating system driver software takes care of the dependencies. If the parameter blocks are general enough in their definition, data can be supplied from almost any source within the system, for example a COPY utility could use the same blocks to get data from a serial port and copy it to a parallel port, or for copying data from one file to another. This idea of device independence and unified I/O allows software to be reused rather than rewritten. Software can be easily moved from one system to another. This is important for modular embedded designs, especially those that use an industry standard bus such as VMEbus, where system hardware can easily be upgraded and/or expanded.

Operating system internals

The first widely used operating system was CP/M, developed for the Intel 8080 microprocessor and 8" floppy disk systems. It supported I/O calls by two jump tables — BDOS (basic disk operating system) and BIOS (basic I/O system). It quickly became a standard within the industry and a large amount of application software became available for it. Many of the micro-based business machines of the late 1970s and early 1980s were based on CP/M. Its ideas even formed the basis of MS-DOS, chosen by IBM for its personal computers.

CP/M is a good example of a single tasking operating system. Only one task or application can be executed at any one time and therefore it only supports one user at a time. When an application is loaded, it provides the user-defined part of the total 'CP/M' program.

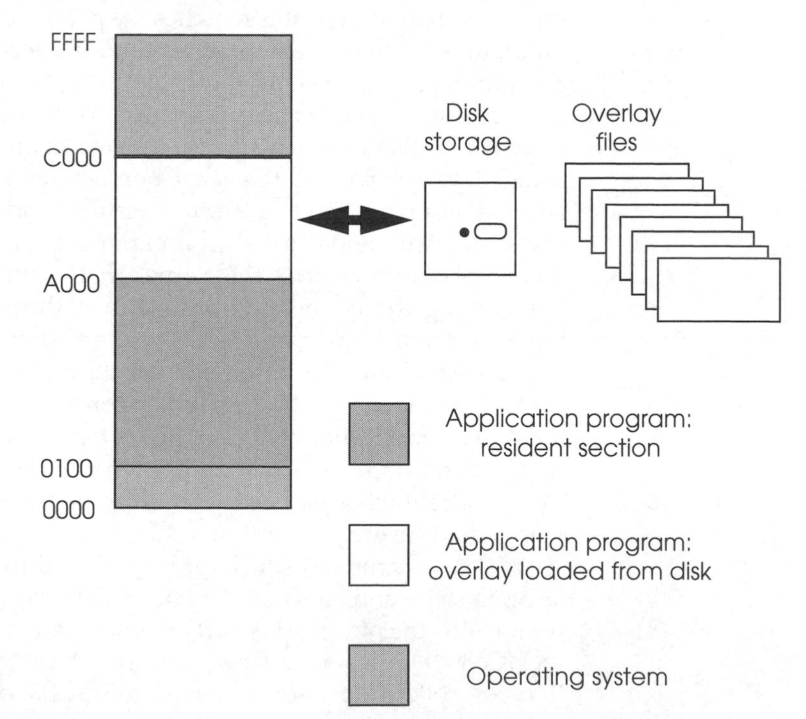

Program overlays

Any application program has to be complete and therefore the available memory often becomes the limiting factor. Program overlays are often used to solve this problem. Parts of the complete program are stored separately on disk and retrieved and loaded over an unused code area when needed. This allows applications larger than the available memory to run, but it places the control responsibility on the application. This is similar to virtual memory schemes where the operating system divides a task's memory into pages and swaps them between memory and mass storage. However, the operating system assumes complete control and such schemes are totally transparent to the user.

With a single tasking operating system, it is not possible to run multiple tasks simultaneously. Large applications have to be run sequentially and cannot support concurrent operations. There is no support for message passing or task control, which would enable applications to be divided into separate entities. If a system needs to take log data and store it on disk and, at the same time, allow a user to process that data using an online database package, a single tasking operating system would need everything to be integrated. With a multitasking operating system, the data logging task can run at the same time as the database. Data can be passed between each element by a common file on disk, and neither task need have any direct knowledge of the other. With a single tasking system, it is likely that the database program would have to be written from scratch. With the multitasking system, a commercial program can be used, and the logging software interfaced to it. These restrictions forced many applications to interface directly with the hardware and therefore lose the hardware independence that the operating system offered. Such software would need extensive modification to port it to another configuration.

Multitasking operating systems

For the majority of embedded systems, a single tasking operating system is too restrictive. What is required is an operating system that can run multiple applications simultaneously and provide intertask control and communication. The facilities once only available to mini and mainframe computer users are now required by 16/32 bit microprocessor users.

A multitasking operating system works by dividing the processor's time into discrete time slots. Each application or task requires a certain number of time slots to complete its execution. The operating system kernel decides which task can have the next slot, so instead of a task executing continuously until completion, its execution is interleaved with other tasks. This sharing of processor time between tasks gives the illusion to each user that he is the only one using the system.

Context switching, task tables, and kernels

Multitasking operating systems are based around a multitasking kernel which controls the time slicing mechanisms. A time slice is the time period each task has for execution before it is stopped and replaced during a context switch. This is periodically triggered by a hardware interrupt from the system timer. This interrupt may provide the system clock and several interrupts may be executed and counted before a context switch is performed.

When a context switch is performed, the current task is interrupted, the processor's registers are saved in a special table for that particular task and the task is placed back on the 'ready' list to await another time slice. Special tables, often called task control blocks, store all the information the system requires about the task, for example its memory usage, its priority level within the system and its error handling. It is this context information that is switched when one task is replaced by another.

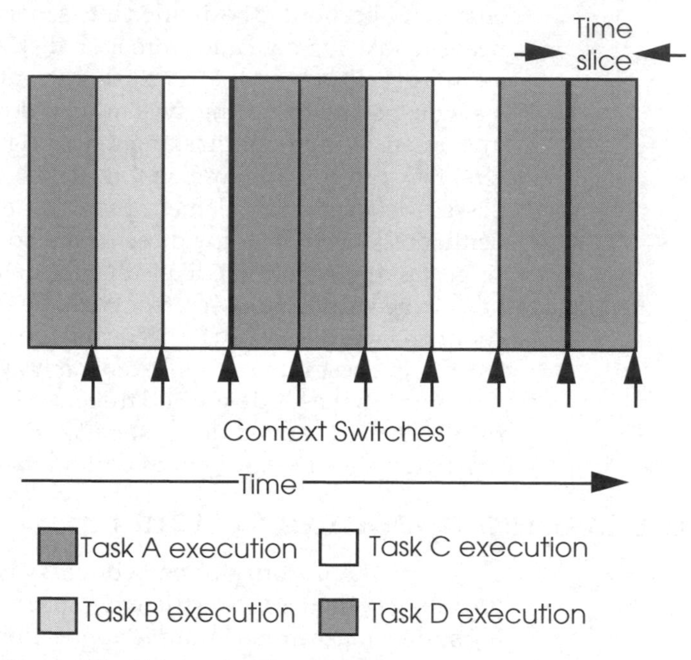

Time slice mechanism for multitasking operating systems

The 'ready' list contains all the tasks and their status and is used by the scheduler to decide which task is allocated the next time slice. The scheduling algorithm determines the sequence and takes into account a task's priority and present status. If a task is waiting for an I/O call to complete, it will be held in limbo until the call is complete.

Once a task is selected, the processor registers and status at the time of its last context switch are loaded back into the processor and the processor is started. The new task carries on as if nothing had happened until the next context switch takes place. This is the basic method behind all multitasking operating systems.

The diagram shows a simplified state diagram for a typical real-time operating system which uses this time slice mechanism. On each context switch, a task is selected by the kernel's scheduler from the 'ready' list and is put into the run state. It is then executed until another context switch occurs. This is normally signalled by a periodic interrupt from a timer. In such cases the task is simply switched out and put back on the 'ready' list, awaiting its next slot. Alternatively, the execution can be stopped by the task executing certain kernel commands. It could suspend itself, where it remains present in the system but no further execution occurs. It could become dormant, awaiting a start command from another task, or even simply waiting for a server task within the operating system to perform a special function for it. A typical example of a server task is a driver performing special screen graphics functions. The most common reason for a task to come out of the run state, is to wait for a message or command, or delay itself for a certain time period. The various wait directives allow tasks to synchronise and control each other within the system. This state diagram is typical of many real-time operating systems.

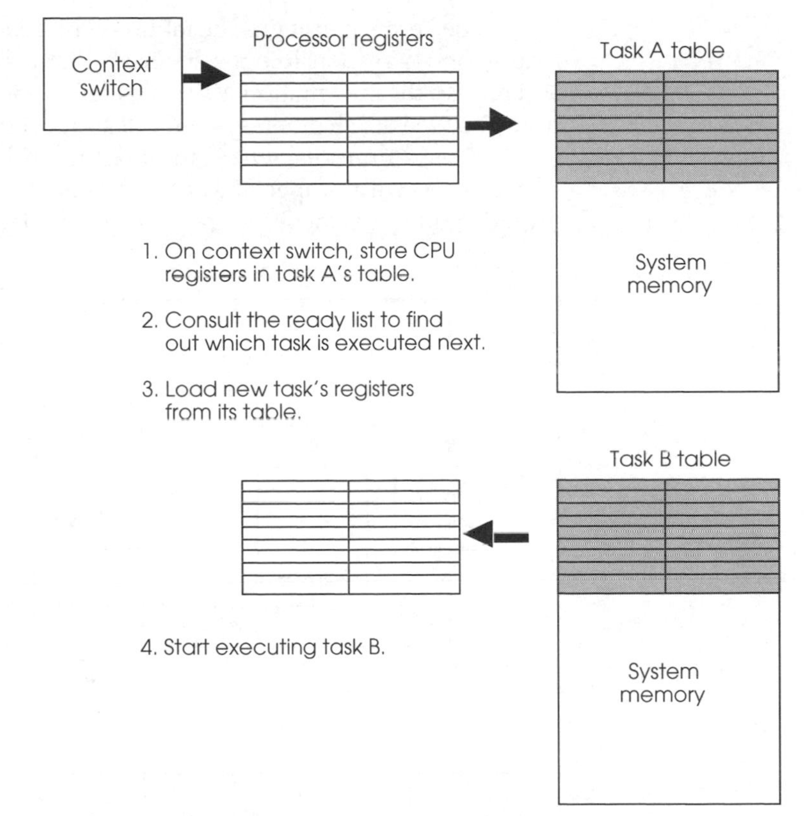

Context switch mechanism

The kernel controls memory usage and prevents tasks from corrupting each other. If required, it also controls memory sharing between tasks, allowing them to share common program modules, such as high level language run-time libraries. A set of memory tables is maintained, which is used to decide if a request is accepted or rejected. This means that resources, such as physical memory and peripheral devices, can be protected from users without using hardware memory management provided the task is disciplined enough to use the operating system and not access the resources directly. This is essential to maintain the system's integrity.

Message passing and control can be implemented in such systems by using the kernel to act as a message passer and controller between tasks. If task A wants to stop task B, then by executing a call to the kernel, the status of task B can be changed and its execution halted. Alternatively, task B can be delayed for a set time period or forced to wait for a message.

With a typical real-time operating system, there are two basic type of messages that the kernel will deal with:

- flags that can control but cannot carry any implicit information — often called semaphores or events and
- messages which can carry information and control tasks — often called messages or events.

The kernel maintains the tables required to store this information and is responsible for ensuring that tasks are controlled and receive the information. With the facility for tasks to communicate between each other, system call support for accessing I/O, loading tasks from disk etc., can be achieved by running additional tasks, with a special system status. These system tasks provide additional facilities and can be included as required.

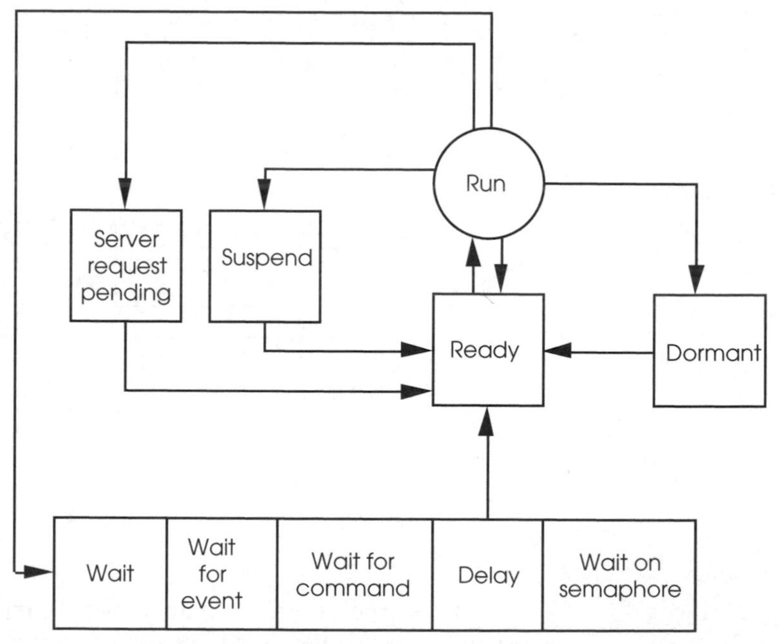

State diagram for a typical real-time kernel

To turn a real-time kernel into a full operating system with file systems and so on, requires the addition of several such tasks to perform I/O services, file handling and file management services, task loading, user interface and driver software. What was about a small <16 kbyte-sized kernel will often grow into a large 120 kbyte operating system. These extra facilities are built up as layers surrounding the kernel. Application tasks then fit around the outside. A typical onion structure is shown as an example. Due to the modular construction, applications can generally access any level directly if required. Therefore, application tasks that just require services provided by the kernel can be developed and debugged under the full environment, and stripped down for integration onto the target hardware.

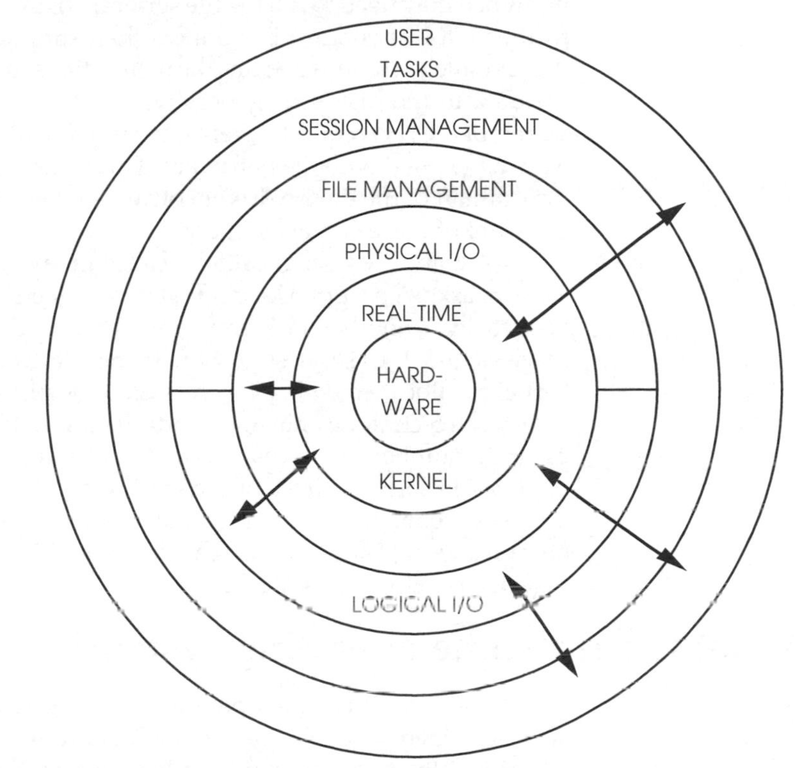

A typical operating system structure

In a typical system, all these service tasks and applications are controlled, scheduled and executed by the kernel. If an application wishes to write some data to a hard disk in the system, the process starts with the application creating a parameter block and asking the file manager to open the file. This system call is normally executed by a TRAP instruction. The kernel then places the task on its 'waiting' list until the file manager had finished and passed the status information back to the application task. Once this event has been received, it wakes up and is placed on the 'ready' list awaiting a time slot.

These actions are performed by the kernel. The next application command requests the file handling services to assign an identifier — often called a logical unit number (LUN) — to the file prior to the actual access. This is needed later for the I/O services call. Again, another parameter block is created and the file handler is requested to assign the LUN. The calling task is placed on the 'waiting' list until this request is completed and the LUN returned by the file handler. The LUN identifies a particular I/O resource such as a serial port or a file without actually knowing its physical characteristics. The device is therefore described as logical rather than physical.

With the LUN, the task can create another parameter block, containing the data, and ask the I/O services to write the data to the file. This may require the I/O services to make system calls of its own. It may need to call the file services for more data or to pass further information on. The data is then supplied to the device driver which actually executes the instructions to physically write the data to the disk. It is generally at this level that the logical nature of the I/O request is translated into the physical characteristics associated with the hardware. This translation should lie in the domain of the device driver software. The user application is unaware of these characteristics.

A complex system call can cause many calls between the system tasks. A program loader that is requested by an application task to load another task from memory needs to call the file services and I/O services to obtain the file from disk, and the kernel to allocate memory for the task to be physically loaded.

The technique of using standard names, files, and/or logical unit numbers to access system I/O makes the porting of application software from one system to another very easy. Such accesses are independent of the hardware the system is running on, and allow applications to treat data received or sent in the same way, irrespective of its source.

What is a real-time operating system?

Many multitasking operating systems available today are also described as 'real-time'. These operating systems provide additional facilities allowing applications that would normally interface directly with the microprocessor architecture to use interrupts and drive peripherals to do so without the operating system blocking such activities. Many multitasking operating systems prevent the user from accessing such sensitive resources. This overzealous caring can prevent many operating systems from being used in applications such as industrial control.

A characteristic of a real-time operating system is its defined response time to external stimuli. If a peripheral generates an interrupt, a real-time system will acknowledge and start to service it within a maximum defined time. Such response times vary from system to system, but the maximum time specified is a worst case figure, and will not be exceeded due to changes in factors such as system workload.

Any system meeting this requirement can be described as real-time, irrespective of the actual value, but typical industry accepted figures for context switches and interrupt response times are about 10 microseconds. This figure gets smaller as processors become more powerful and run at higher speeds. With several processors having the same context switch mechanism, the final context switch time come down to its clock speed and the memory access time.

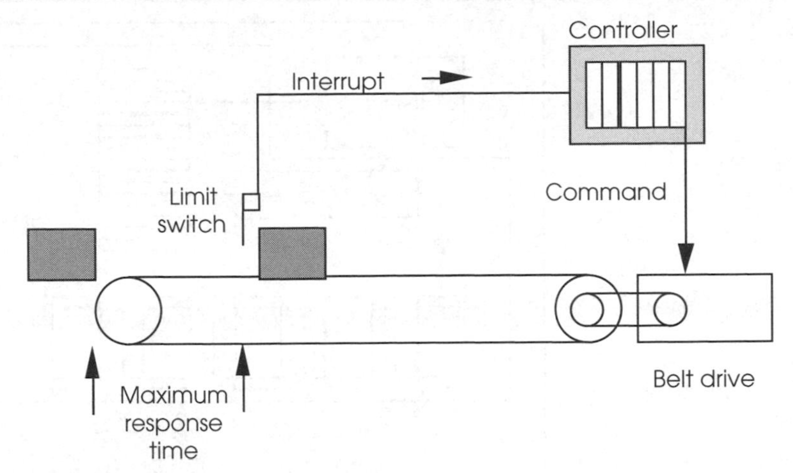

Example of a real-time response

The consequences to industrial control of not having a real-time characteristic can be disastrous. If a system is controlling an automatic assembly line, and does not respond in time to a request from a conveyor belt limit switch to stop the belt, the results are easy to imagine. The response does not need to be instantaneous — if the limit switch is set so that there are 3 seconds to stop the belt, any system with a guaranteed worst case response of less than 3 seconds can meet this real-time requirement.

For an operating system to be real-time, its internal mechanisms need to show real-time characteristics so that the internal processes sequentially respond to external interrupts in guaranteed times.

When an interrupt is generated, the current task is interrupted to allow the kernel to acknowledge the interrupt and obtain the vector number that it needs to determine how to handle it. A typical technique is to use the kernel's interrupt handler to update a linked list which contains information on all the tasks that need to be notified of the interrupt.

If a task is attached to a vector used by the operating system, the system actions its own requirements prior to any further response by the task. The handler then sends an event message to the tasks attached to the vector, which may change their status and completely change the priorities of the task ready list. The scheduler analyses the list, and dispatches the highest priority task to run. If the interrupt and task priorities are high enough, this may be the next time slice.

The diagram depicts such a mechanism: the interrupt handler and linked list searches are performed by the kernel. The first priority is to service the interrupt. This may be from a disk controller indicating that it has completed a data transfer. Once the kernel has satisfied its own needs, the handler will start a linked list search. The list comprises blocks of data identifying tasks that have their own service routines. Each block will contain a reference to the next block, hence the linked list terminology.

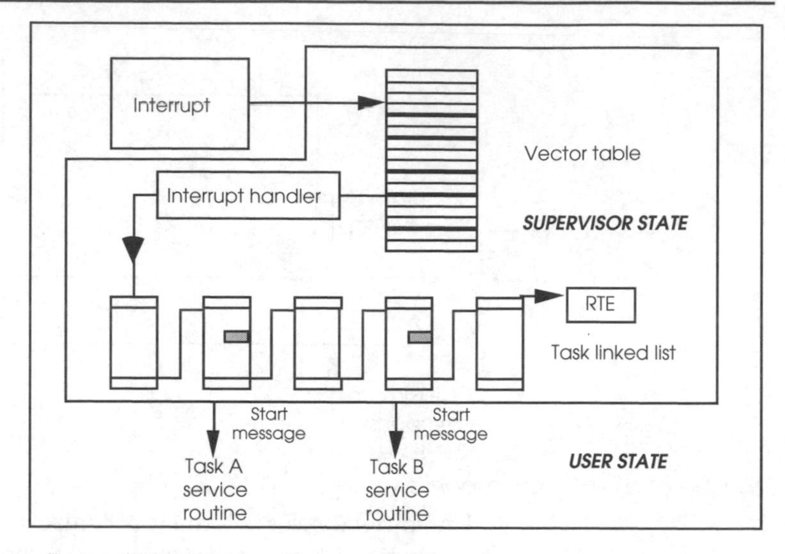

Handling interrupt routines within an operating system

Each identified task is then sent a special message. This will start the task's service routine when it receives its next time slice. The kernel interrupt handler will finally execute an RTE return from the exception instruction which will restore the processor state prior to the interrupt. In such arrangements the task service routines execute in USER mode. The only SUPERVISOR operation is that of the kernel and its own interrupt handler. As can be imagined, this processing can increase the interrupt latency seen by the task quite dramatically. A ten-fold increase is not uncommon.

To be practical, a real-time operating system has to guarantee maximum response times for its interrupt handler, event passing mechanisms, scheduler algorithm and provide system calls to allow tasks to attach and handle interrupts.

With the conveyor belt example above, a typical software configuration would dedicate a task to controlling the conveyor belt. This task would make several system calls on start-up to access the parallel I/O peripheral that interfaces the system to components such as the drive motors and limit switches and tells the kernel that certain interrupt vectors are attached to the task and are handled by its own interrupt handling routine.

Once the task has set everything up, it remains dormant until an event is sent by other tasks to switch the belt on or off. If a limit switch is triggered, it sets off an interrupt which forces the kernel to handle it. The currently executing task stops, the kernel handler searches the task interrupt attachment linked list, and places the controller task on the ready list, with its own handler ready to execute. At the appropriate time slice, the handler runs, accesses the peripheral and switches off the belt. This result may not be normal, and so the task also sends event messages to the others, informing them that it has acted independently and may force other actions. Once this has been done, the task goes back to its dormant state awaiting further commands.

Real-time operating systems have other advantages: to prevent a system from power failure usually needs a guaranteed response time so that the short time between the recognition of and the actual power failure can be used to store vital data and bring the system down in a controlled manner. Many operating systems actually have a power fail module built into the kernel so that no time is lost in executing the module code.

So far in this chapter, an overview of the basics behind a real-time operating system have been explained. There are, however, several variants available for the key functions such as task swapping and so on. The next few sections will delve deeper into these topics.

Task swapping methods

The choice of scheduler algorithms can play an important part in the design of an embedded system and can dramatically affect the underlying design of the software. There are many different types of scheduler algorithm that can be used, each with either different characteristics or different approaches to solving the same problem of how to assign priorities to schedule tasks so that correct operation is assured.

Time slice

Time slicing has been previously mentioned in this chapter under the topic of multitasking and can be used within an embedded system where time critical operations are not essential. To be more accurate about its definition, it describes the task switching mechanism and not the algorithm behind it although its meaning has become synonymous with both.

Time slicing works by making the task switching regular periodic points in time. This means that any task that needs to run next will have to wait until the current time slice is completed or until the current task suspends its operation. This technique can also be used as a scheduling method as will be explained later in this chapter. The choice of which task to run next is determined by the scheduling algorithm and thus is nothing to do with the time slice mechanism itself. It just happens that many time slice-based systems use a round-robin or other fairness scheduler to distribute the time slices across all the tasks that need to run.

For real-time systems where speed is of the essence, the time slice period plus the context switch time of the processor determines the context switch time of the system. With most time slice periods in the order of milliseconds, it is the dominant factor in the system response. While the time period can be reduced to improve the system context switch time, it will increase the number of task switches that will occur and this will reduce the efficiency of the system. The larger the number of switches, the less time there is available for processing.

Pre-emption

The alternative to time slicing is to use pre-emption where a currently running task can be stopped and switched out — pre-empted — by a higher priority *active* task. The *active* qualifier is important as the example of pre-emption later in this section will show. The main difference is that the task switch does not need to wait for the end of a time slice and therefore the system context switch is now the same as the processor context switch.

As an example of how pre-emption works, consider a system with two tasks A and B. A is a high priority task that acts as an ISR to service a peripheral and is activated by a processor interrupt from the peripheral. While it is not servicing the peripheral, the task remains dormant and stays in a suspended state. Task B is a low priority task that performs system housekeeping.

When the interrupt is recognised by the processor, the operating system will process it and activate task A. This task with its higher priority compared to task B will cause task B to be pre-empted and replaced by task A. Task A will continue processing until it has completed and then suspend itself. At this point, task B will context switch task A out because task A is no longer active.

This can be done with a time slice mechanism provided the interrupt rate is less than the time slice rate. If it is higher, this can also be fine provided there is sufficient buffering available to store data without losing it while waiting for the next time slice point. The problem comes when the interrupt rate is higher or if there are multiple interrupts and associated tasks. In this case, multiple tasks may compete for the same time slice point and the ability to run even though the total processing time needed to run all of them may be considerably less than the time provided within a single time slot. This can be solved by artificially creating more context switch points by getting each task to suspend after completion. This may offer only a partial solution because a higher priority task may still have to wait on a lower priority task to complete. With time slicing, the lower priority task cannot be pre-empted and therefore the higher priority task must wait for the end of the time slice or the lower priority task to complete. This is a form of priority inversion which is explained in more detail later.

Most real-time operating systems support pre-emption in preference to time slicing although some can support both methodologies

Co-operative multitasking

This is the mechanism behind Windows 3.1 and while not applicable to real-time operating systems for reasons which will become apparent, it has been included for reference.

The idea of co-operative multitasking is that the tasks themselves co-operate between themselves to provide the illusion of multitasking. This is done by periodically allowing other tasks or applications the opportunity to execute. This requires programming within the application and the system can be destroyed by a single rogue program that hogs all the processing power. This method may be acceptable for a desktop personal computer but it is not reliable enough for most real-time embedded systems.

Scheduler algorithms

So far in this section, the main methods of swapping tasks has been discussed. It is clear that pre-emption is the first choice for embedded systems because of its better system response. The next issue to address is how to assign the task priorities so that the system works and this is the topic that is examined now.

Rate monotonic

Rate monotonic scheduling (RMS) is an approach that is used to assign task priority for a pre-emptive system in such a way that the correct execution can be guaranteed. It assumes that task priorities are fixed for a given set of tasks and are not dynamically changed during execution. It assumes that there are sufficient task priority levels for the task set and that the task set models periodic events only. This means that an interrupt that is generated by a serial port peripheral is modelled as an event that occurs on a periodic rate determined by the data rate, for example. Asynchronous events such as a user pressing a key are handled differently as will be explained later.

The key policy within RMS is that tasks with shorter execution periods are given the highest priority within the system. This means that the faster executing tasks can pre-empt the slower periodic tasks so that they can meet their deadlines. The advantage this gives the system designer is that it is easier to theoretically specify the system so that the tasks will meet their deadlines without overloading the processor. This requires detailed knowledge about each task and the time it takes to execute. This and its periodicity can be used to calculate the processor loading.

To see how this policy works, consider the examples shown in the diagram on the next page. In the diagrams, events that start a task are shown as lines that cross the horizontal time line and tasks are shown as rectangles whose length determines their execution time. Example 1 shows a single periodic task where the task t is executed with a periodicity of time t. The second example adds a second task S where its periodicity is longer than that of task *t*. The task priority shown is with task S having the highest priority. In this case, the RMS policy has not been followed because the longest task has been given a higher priority than the shortest task. However, please note that in this case the system works fine because of the timing of the tasks' periods.

Example 3 shows what can go wrong if the timing is changed and the periodicity for task S approaches that of task t. When t_3 occurs, task t is activated and starts to run. It does not complete because S_2 occurs and task S is swapped-in due to its higher priority. When tasks S completes, task t resumes but during its execution, the event t_4 occurs and thus task t has failed to meets its task 3 deadline. This could result in missed or corrupted data, for example. When task t completes, it is then reactivated to cope with the t_4 event.

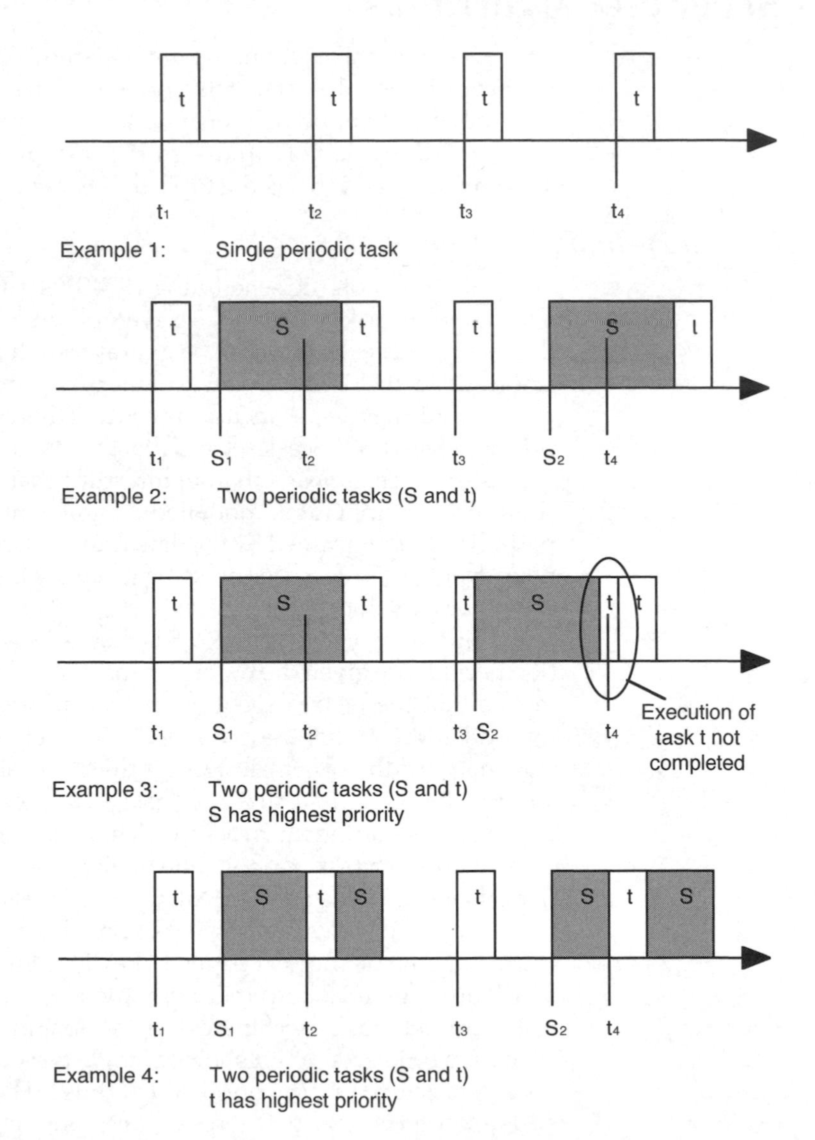

Using RMS policies

Example 4 shows the same scenario with the task priorities reversed so that task t pre-empts task S. In this case, RMS policy has been followed and the system works fine with both tasks meeting their deadlines. This system is useful to understand and does allow theoretical analysis before implementation to prevent a design from having to manually assign task priorities to get it to work using a trial and error approach. It is important to remember within any calculations to take into account the context swapping time needed to pre-empt and resume tasks. The processor utilisation can also be calculated and thus give some idea of the performance needed or how much performance is spare. Typically, the higher the utilisation (>80%), the more chance that the priority assignment is wrong and that the system will fail. If the utilisation is below this, the chances are that the system will run correctly.

This is one of the problems with theoretical analysis of such systems in that the actual design may have to break some of the assumptions on which the analysis is based. Most embedded systems will have asynchronous events and tasks running. While these can be modelled as a periodic task that polls for the asynchronous event and are analysed as if they will run, other factors such as cache memory hit ratios can invalidate the timing by lengthening the execution time of any analysis. Similarly, the act of synchronising tasks by inter-task communication can also cause difficulties within any analysis.

Deadline monotonic scheduling

Deadline monotonic scheduling (DMS) is another task priority policy that uses the nearest deadline as the criterion for assigning task priority. Given a set of tasks, the one with the nearest deadline is given the highest priority. This means that the scheduling or designer must now know when these deadlines are to take place. Tracking and, in fact, getting this information in the first place can be difficult and this is often the reason behind why deadline scheduling is often a second choice compared to RMS.

Priority guidelines

With a system that has a large number of tasks or one that has a small number of priority levels, the general rule is to assign tasks with a similar period to the same level. In most cases, this does not effect the ability to schedule correctly. If a task has a large context, i.e. it has more registers and data to be stored compared to other tasks, it is worth raising its priority to reduce the context switch overhead. This may prevent the system from scheduling properly but can be a worthwhile experiment.

Priority inversion

It is also possible to get a condition called priority inversion where a lower priority task can continue running despite there being a higher priority task active and waiting to pre-empt.

This can occur if the higher priority task is in some way blocked by the lower priority task through it having to wait until a message or semaphore is processed. This can happen for several reasons.

Disabling interrupts

While in the interrupt service routine, all other interrupts are disabled until the routine has completed. This can cause a problem if another interrupt is received and held pending. What happens is that the higher priority interrupt is held pending in favour of the lower priority one — albeit that it occurred first. As a result, priority inversion takes place until interrupts are re-enabled at which point the higher priority interrupt will start its exception processing, thus ending the priority inversion.

One metric of an operating system is the longest period of time that all interrupts are disabled. This must then be added to any other interrupt latency calculation to determine the actual latency period.

Message queues

If a message is sent to the operating system to activate a task, many systems will process the message and then reschedule accordingly. In this way, the message queue order can now define the task priority. For example, consider an ISR that sends an unblocking message to two tasks, A and B, that are blocked waiting for the message. The ISR sends the message for task A first followed by task B. The ISR is part of the RTOS kernel and therefore may be subject to several possible conditions:

- Condition 1

  Although the message calls may be completed, their action may be held pending by the RTOS so that any resulting pre-emption is stopped from switching out the ISR.
- Condition 2

  The ISR may only be allowed to execute a single RTOS call and in doing so the operating system itself will clean up any stack frames. The operating system will then send messages to tasks notifying them of the interrupt and in this way simulate the interrupt signal. This is normally done through a list.

These conditions can cause priority inversion to take place. With condition 1, the ISR messages are held pending and processed. The problem arises with the methodology used by the operating system to process the pending messages. If it processes all the messages, effectively unblocking both tasks before instigating the scheduler to decide the next task to run, all is well. Task B will be scheduled ahead of task A because of its higher priority. The downside is the delay in processing all the messages before selecting the next task.

Most operating systems, however, only have a single call to process and therefore in normal operation do not expect to handle multiple messages. In this case, the messages are handled individually so that after the first message is processed, task A would be unblocked and allowed to execute. The message for task B would either be ignored or processed as part of the housekeeping at the next context switch. This is where priority inversion would occur. The ISR has according to its code unblocked both tasks and thus would expect the higher priority task B to execute. In practice, only task A is unblocked and is running, despite it being at a lower priority. This scenario is a programming error but one that is easy to make.

To get around this issue, some RTOS implementations restrict an ISR to making either one or no system calls. With no system calls, the operating system itself will treat the ISR event as an internal message and will unblock any task that is waiting for an ISR event. With a single system call, a task would take the responsibility for controlling the message order to ensure that priority inversion does not take place.

Waiting for a resource

If a resource is shared with a low priority task and it does not release it, a higher priority task that needs it can be blocked until it is released. A good example of this is where the interrupt handler is distributed across a system and needs to access a common bus to handle the interrupt. This can be the case with a VMEbus system, for example.

VMEbus interrupt messages

VMEbus is an interconnection bus that was developed in the early 1980s for use within industrial control and other real-time applications. The bus is asynchronous and is very similar to that of the MC68000. It comprises of a separate address, data, interrupt and control buses.

If a VMEbus MASTER wishes to inform another that a message is waiting or urgent action is required, a VMEbus interrupt can be generated. The VMEbus supports seven interrupt priority levels to allow prioritisation of a resource.

Any board can generate an interrupt by asserting one of the levels. Interrupt handling can either be centralised, and handled by one MASTER, or can be distributed among many. For multi-processor applications, distributed handling allows rapid direct communication to individual MASTERs by any board in the system capable of generating an interrupt: the MASTER that has been assigned to handle the interrupt requests the bus and starts the interrupt acknowledgement cycle. Here, careful consideration of the arbitration level chosen for the MASTER is required. The interrupt response time depends on the time taken by the handler to obtain the bus prior to the acknowledgement. This has to be correctly assigned to achieve the required performance. If it has a low priority, the overall response time may be more than that obtained for a lower priority interrupt whose handler has a higher arbitration level. The diagrams below show the relationship for both priority and round robin arbitration schemes. Again, as with the case with arbitration schemes, the round robin system has been assumed on average to provide equal access for all the priority levels.

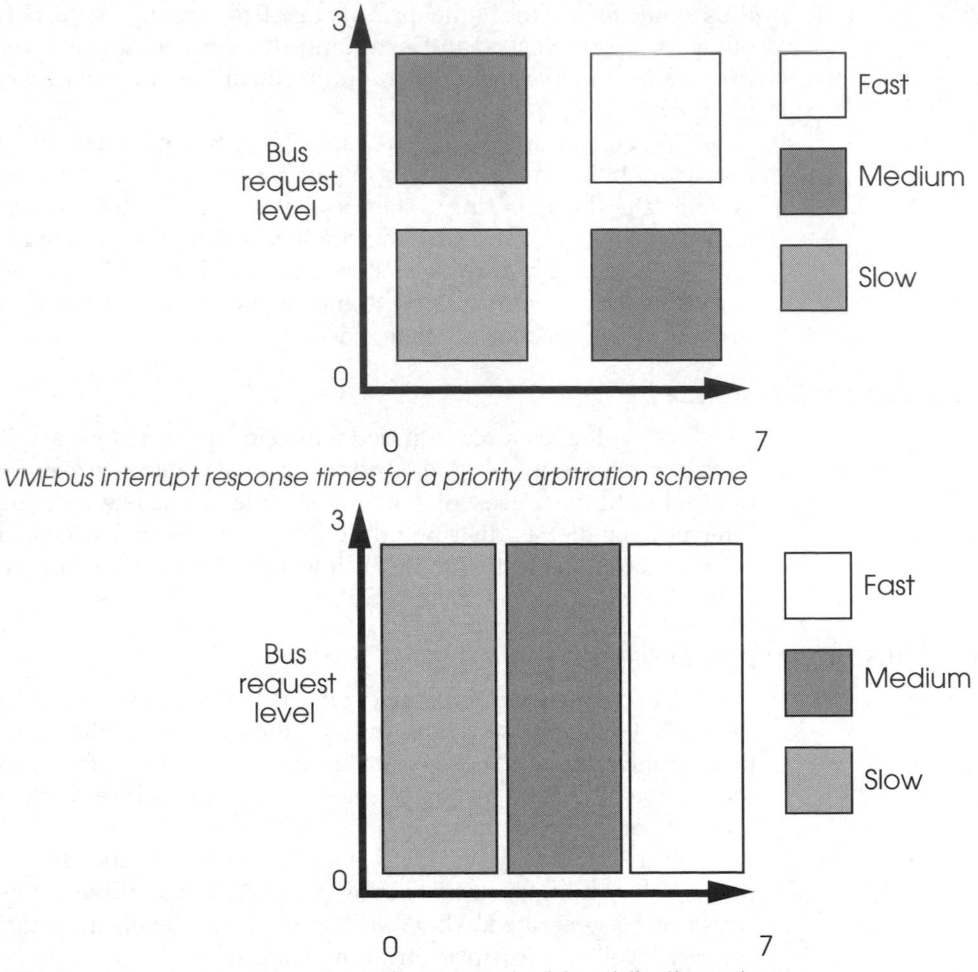

VMEbus interrupt response times for a priority arbitration scheme

VMEbus interrupt response times for a round robin arbitration scheme

Priority inversion can occur if a lower priority interrupt may be handled in preference to a higher priority one, simply because of the arbitration levels. To obtain the best response, high priority interrupts should only be assigned to high arbitration level MASTERs. The same factors, such as local traffic on the VMEbus and access time, increase the response time as the priority level decreases.

VMEbus only allows a maximum of seven separate interrupt levels and this limits the maximum number of interrupt handlers to seven. For systems with larger numbers of MASTERs, polling needs to be used for groups of MASTERs assigned to a single interrupt level. Both centralised and distributed interrupt handling schemes have their place within a multiprocessor system. Distributed schemes allow interrupts to be used to pass high priority messages to handlers, giving a fast response time which might be critical for a real-time application. For simpler designs, where there is a dominant controlling MASTER, one handler may be sufficient.

Fairness systems

There are times when the system requires different characteristics from those originally provided or alternatively wants a system response that is not priority based. This can be achieved by using a fairness system where the bus access is distributed across the requesting processors. There are various schemes that are available such as round-robin where access is simply passed from one processor to another. Other methods can use time sharing where the bus is given to a processor on the understanding that it must be relinquished within a maximum time frame.

These type of systems can affect the interrupt response because bus access is often necessary to service an interrupt.

Tasks, threads and processes

This section of the chapter discusses the nomenclature given to the software modules running within the RTOS environment. Typically, they have been referred to as tasks but other names such as threads and processes are also used to refer to entities within the RTOS. They are sometimes used instead as an interchangeable replacement for the term task. In practice, they refer to different aspects within the system.

So far in this chapter, a task has been used to describe an entity of work within an operating system that has control over resources. When a context switch is performed, it effectively switches to another task which takes over. Strictly speaking the context switch may include additional information that is relevant to the task such as memory management information, which is beyond the simple register swapping that is performed within the processor. As a result, the term process is often used to encompass more than a simple context switch and thus includes the additional information. The problem is that this is very similar to that of a task switch or context switch that the definitions have become blurred and almost interchangeable. A task or process has several characteristics:

- It owns or controls resources, e.g. access to peripherals and so on.
- It has threads of execution. These are paths through the code contained within the task or process. Normally, there is a single thread though this may not always be the case. Multiple threads can be supported if the task or process can maintain separate data areas for each thread. This also requires the code to be written in a re-entrant manner.
- It requires additional information beyond the normal register contents to maintain its integrity, e.g. memory management information, cache flushing and so on. When a new process or task is swapped-in, not only are the processor registers changed but additional work must be done

such as invalidating caches to ensure that the new process or task does not access incorrect information.

A thread has different characteristics:

- It has no additional context information beyond that stored in the processor register set.
- Its ownership of resources is inherited from its parent task or process.

With a simple operating system, there is no difference between the thread context switch and the process level switch. As a result, these terms almost become interchangeable. With a multiuser, multitasking operating system, this is not the case. The process or task is the higher level with the thread(s) the lower level.

Some operating systems take this a stage further and define a three level hierarchy: a process consists of a set of tasks with each task having multiple threads. Be warned! These terms mean many different things depending on who is using them.

Exceptions

With most embedded systems, access to the low level exception handler is essential to allow custom routines to be written to support the system. This can include interrupt routines to control external peripherals, emulation routines to simulate instructions or facilities that the processor does not support — software floating point is a very good example of this — and other exception types.

Some of these exceptions are needed by the RTOS to provide entry points into the kernel and to allow the timers and other facilities to function. As a result, most RTOSs already provide the basic functionality for servicing exceptions and provide access points into this functionality to allow the designer to add custom exception routines. This can be done in several ways:

- Patching the vector table

  This is relatively straight forward if the vector is not used by the RTOS. If it is, then patching will still work but the inserted user exception routine must preserve the exception context and then jump to the existing handler instead of using a return from exception type instruction to restore normal processing. If it is sharing an exception with the RTOS, there must be some form of checking so that the user handler does not prevent the RTOS routine from working correctly.

- Adding user routines to existing exception handlers

  This is very similar to the previous technique in that the user routine is added to any existing RTOS routine. The difference is that the mechanism is more formal and does not require vector table patching or any particular checking by the user exception handler.

- Generating a pseudo exception that is handled by separate user exception handler(s)

  This is even more formal — and slower — and effectively replaces the processor level exception routine with a RTOS level version in which the user creates his own vector table and exception routines. Typically, all this is performed through special kernel calls which register a task as the handler for a particular exception. On completion, the handler uses a special return from the exception call into the RTOS kernel to signify that it has completed.

Memory model

The memory model that the processor offers can and often varies with the model defined by the operating system and is open to the software designer to use. In other words, although the processor may support a full 32 bit address range with full memory mapped I/O and read/write access anywhere in the map at a level of an individual word or map, the operating system's representation of the memory map may only be 28 bits, with I/O access allocated on a 512 byte basis with read only access for the first 4 Mbytes of RAM and so on.

This discrepancy can get even wider, the further down in the levels that you go. For example, most processors that have sophisticated cache memory support use the memory management unit. This then requires fairly detailed information about the individual memory blocks within a system. This information has to be provided to the RTOS and is normally done using various memory allocation techniques where information is provided when the system software is compiled and during operation.

Memory allocation

Most real-time operating systems for processors where the memory map can be configured, e.g. those that have large memory addressing and use memory mapped I/O, get around this problem by using a special file that defines the memory map that the system is expected to use and support. This will define which memory addresses correspond to I/O areas, RAM, ROM and so on. When a task is created, it will be given a certain amount of memory to hold its code area and provide some initial data storage. If it requires more memory, it will request it from the RTOS using a special call such as malloc(). The RTOS will look at the memory request and allocate memory by passing back a pointer to the additional memory. The memory request will normally define the memory characteristics such as read/write access, size, and even its location and attributes such as physical or logical addressing.

The main question that arises is why dynamically allocate memory? Surely this can be done when the tasks are built and included with the operating system? The answer is not straightforward. In simple terms, it is a yes and for many simple embedded systems, this is correct. For more complex systems, however, this static allocation of memory is not efficient, in that memory may be reserved or allocated to a task yet could only be used rarely within the system's operation. By dynamically allocating memory, the total amount of memory that the system requires for a given function can be reduced by reclaiming and reallocating memory as and when needed by the system software. This will be explained in more detail later on.

Memory characteristics

The memory characteristics are important to understand especially when different memory addresses correspond to different physical memory. As a result, asking for a specific block of memory may impact the system performance. For example, consider an embedded processor that has both internal and external memory. The internal memory is faster than the external memory and therefore improves performance by not inserting wait states during a memory access. If a task asks for memory expecting to be allocated internal memory but instead receives a pointer to external memory, the task performance will be degraded and potentially the system can fail. This is not a programming error in the true sense of the word because the request code and RTOS have executed correctly. If the request was not specific enough, then the receiving task should expect the worst case type of memory. If it does not or needs specific memory, this should be specified during the request. This is usually done by specifying a memory address or type that the RTOS memory allocation code can check against the memory map file that was used when the system was built.

- Read/write access

  This is straightforward and defines the access permissions that a task needs to access a memory block.

- Internal/external memory

  This is normally concerned with speed and performance issues. The different types of memory are normally defined not by their speed but indirectly through the address location. As a result, the programmer must define and use a memory map so that the addresses of the required memory block match up the required physical memory and thus its speed. Some RTOSs actually provide simple support flags such as internal/external but this is not common.

- Size

  The minimum and maximum sizes are system dependent and typically are influenced by the page size of any memory management hardware that may be present. Some systems can return with partial blocks, e.g. if the original request was for 8 kbytes, the RTOS may only have 4 kbytes free and instead of returning an error, will return a pointer to the 4 kbytes block instead. This assumes that the requesting task will check the returned size and not simply assume that because there was no error, it has all the 8 kbytes it requested! Check the RTOS details carefully.

- I/O

  This has several implications when using processors that execute instructions out of order to remove pipeline stalls and thus gain performance. Executing instructions that access I/O ports out of sequence can break the program syntax and integrity. The program might output a byte and then read a status register. If this is reversed, the correct sequence has been destroyed and the software will probably crash. By declaring I/O addresses as I/O, the processor can be programmed to ensure the correct sequence whenever these addresses are accessed.

- Cached or non-cachable

  This is similar to the previous paragraph on I/O. I/O addresses should not be cached to prevent data corruption. Shared memory blocks need careful handling with caches and in many cases unless there is some form of bus snooping to check that the contents of a cache is still valid, these areas should also not be cached.

- Coherency policies

  Data caches can have differing coherency policies such as write-through, copy back and so on which are used to ensure the data coherency within the system. Again, the ability to specify or change these policies is useful.

Example memory maps

The first example is that commonly used within a simple microcontroller where its address space is split into the different memory types. The example shows three: I/O devices and peripherals, program RAM and ROM and data RAM. The last two types have then been expanded to show how they could be allocated to a simple embedded system. The program area contains the code for four tasks, the RTOS code and the processor vector table. The data RAM is split into five areas: one for each of the tasks and a fifth area for the stack. In practice, these areas are often further divided into internal and external memory, EPROM and EEPROM, SRAM and even DRAM, depending on the processor architecture and model. This example uses a fixed static memory map where the memory requirements for the whole system are defined at compile and build time. This means that tasks cannot get access to additional memory by using some of the memory allocated to another task. In addition, it should be remembered that although the memory map shows nicely partitioned areas, it does not imply nor should it be assumed that task A cannot access task C's data area, for example. In these simple processors and memory maps, all tasks have the ability to access any memory location and it is only the correct design and programming that ensures that there is no corruption. Hardware can be used to provide this level of protection but it requires some form of memory management unit to check that programs are conforming to their design and not accessing memory that they should not. Memory management is explained in some detail in the next section.

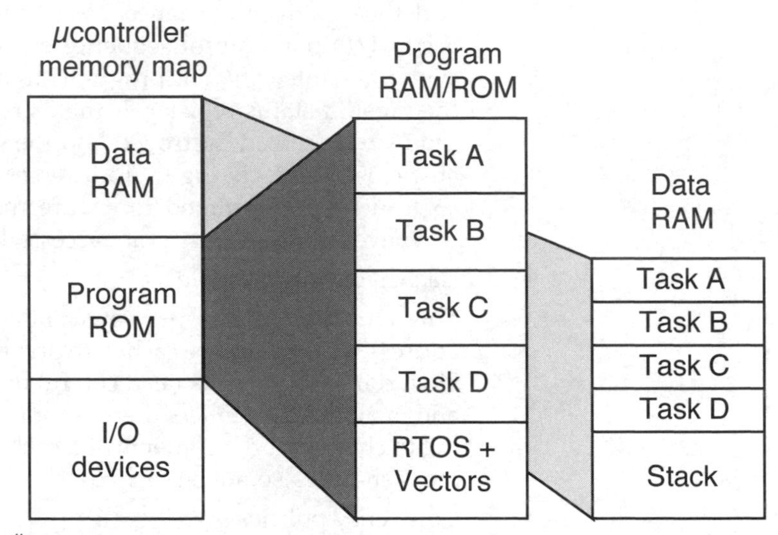

Simple microcontroller memory map

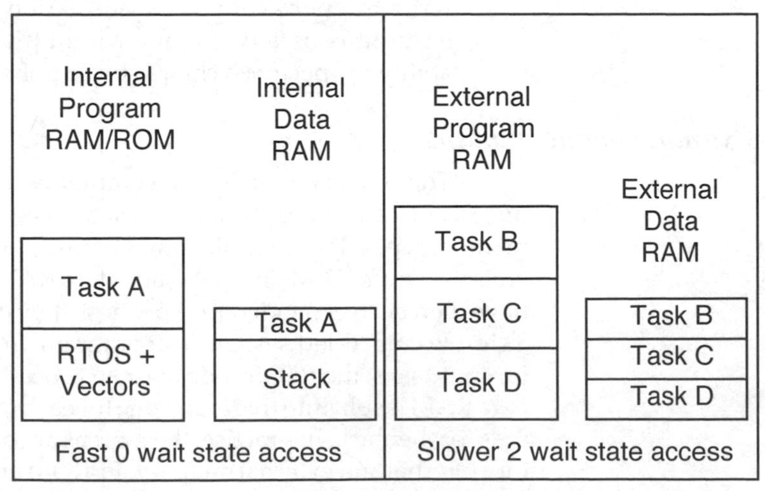

Internal and external memory map

The second example shows a similar system to the first example except that it has been further partitioned into internal and external memory. The internal memory runs faster than the external memory and because it does not have any wait states, its access time is faster and the processor performance does not degrade. The slower external memory has two wait states and with a single cycle processor would degrade performance by 66%

— each instruction would take three clocks instead of one, for example.

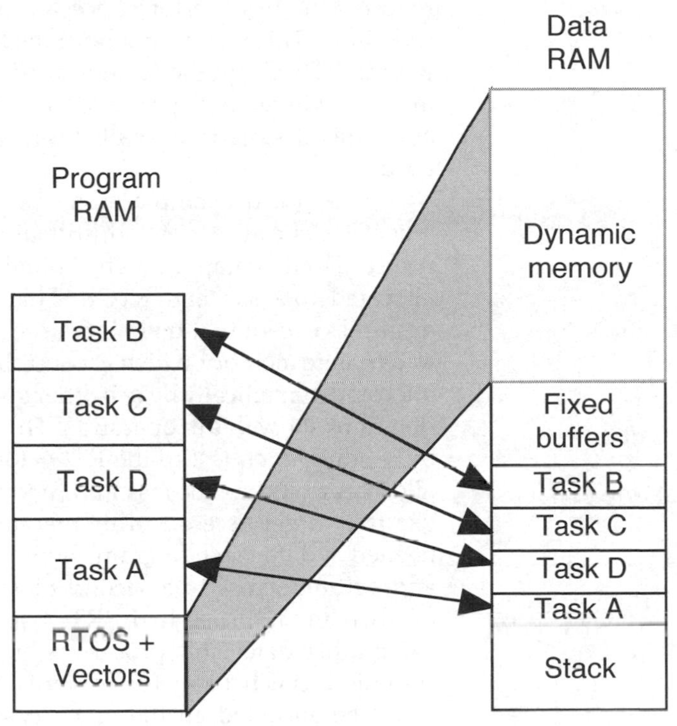

Memory map with dynamic allocation — initial state

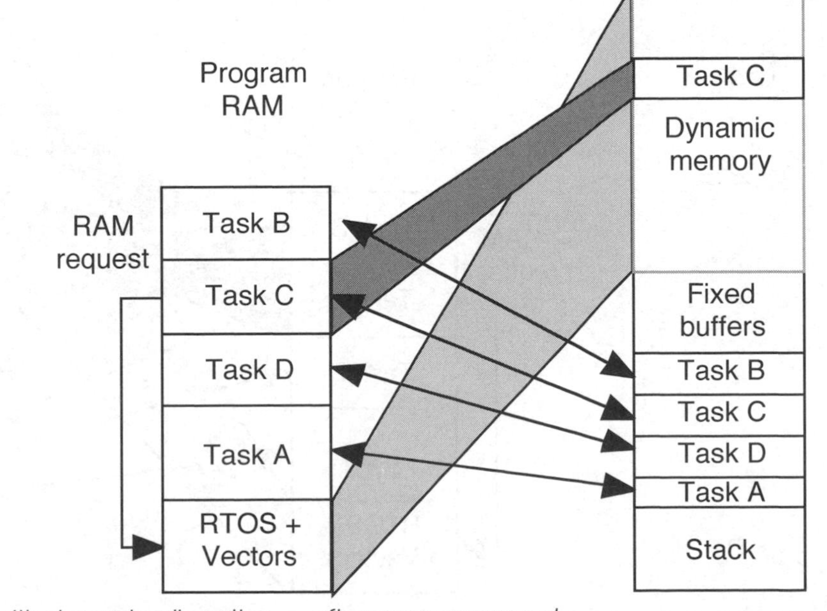

Memory map with dynamic allocation — after memory request

Given this performance difference, it is important that the memory resources are carefully allocated. In the example, task A requires the best performance and the system needs fast task switching. This means that both the task A code and data, along with the RTOS and the system stack, are allocated to the internal memory where it will get the best performance. All other task code and data are stored externally because all the internal memory is used.

The third example shows a dynamic allocation system where tasks can request additional memory as and when they need it. The first map shows the initial state with the basic memory allocated to tasks and RTOS. This is similar to the previous examples except that there is a large amount of memory entitled to dynamic memory which is controlled by the RTOS and can be allocated dynamically by it to other tasks on demand. The next two diagrams show this in operation. The first request by task C starts by sending a request to the RTOS for more memory. The RTOS allocates a block to the task and returns a pointer to it. Task C can use this to get access to this memory. This can be repeated if needed and the next diagram shows task C repeating the request and getting access to a second block. Blocks can also be relinquished and returned to the RTOS for allocation to other tasks at some other date. This process is highly dynamic and thus can provide a mechanism for minimising memory usage. Task C could be allocated all the memory at certain times and as the memory is used and no longer required, blocks can be reallocated to different tasks.

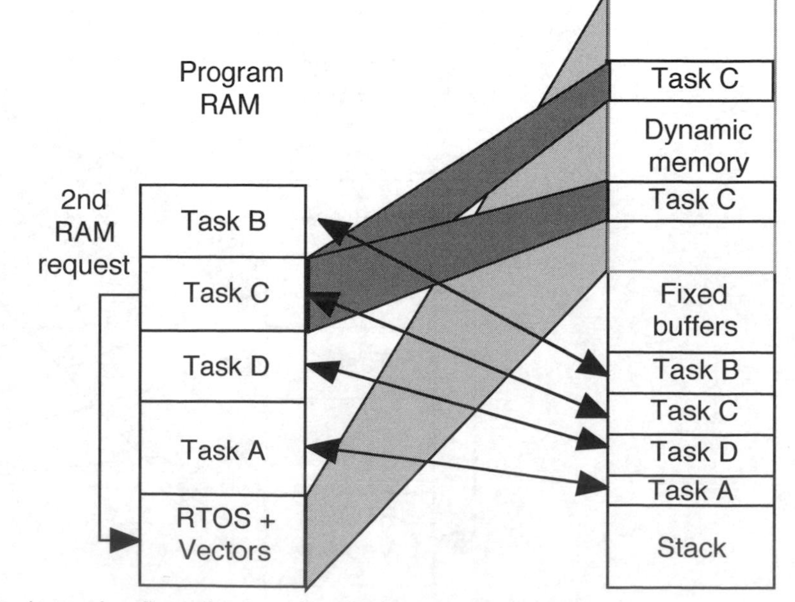

Memory map with dynamic allocation — after 2nd memory request

The problem with this is in calculating the minimum amount of memory that the system will require. This can be difficult to estimate and many designs start with a large amount of memory, get the system running and then find out empirically the minimum amount of required memory.

In this section, the use of memory management within an embedded design has been alluded to in the case of protecting memory for corruption. While this is an important use, it is a secondary advantage compared to its ability to reuse memory through address translation. Before returning to the idea of memory protection, let's consider how address translation works and affects the memory map.

Memory management address translation

While the use of memory management usually implies the use of an operating system to remove the time-consuming job of defining and writing the driver software, it does not mean that every operating system supports memory management.

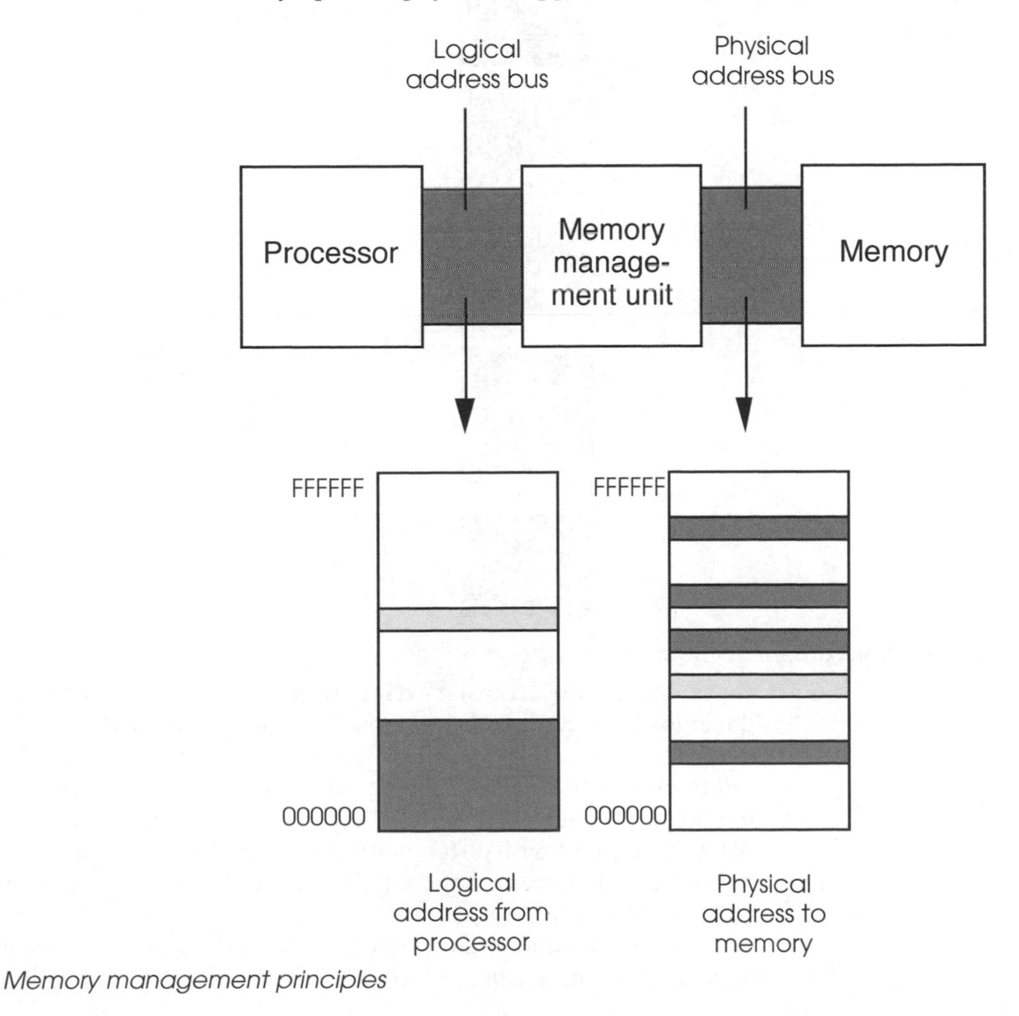

Memory management principles

Many do not or are extremely limited in the type of memory management facilities that they support. For operating systems that do support it, the designer can access standard software that controls the translation of logical addresses to different physical addresses as shown in the diagram.

In this example, the processor thinks that it is accessing memory at the bottom of its memory map, while in reality it is being fetched from different locations in the main memory map. The memory does not even need to be contiguous: the processor's single block or memory can be split into smaller blocks, each with a different translation address.

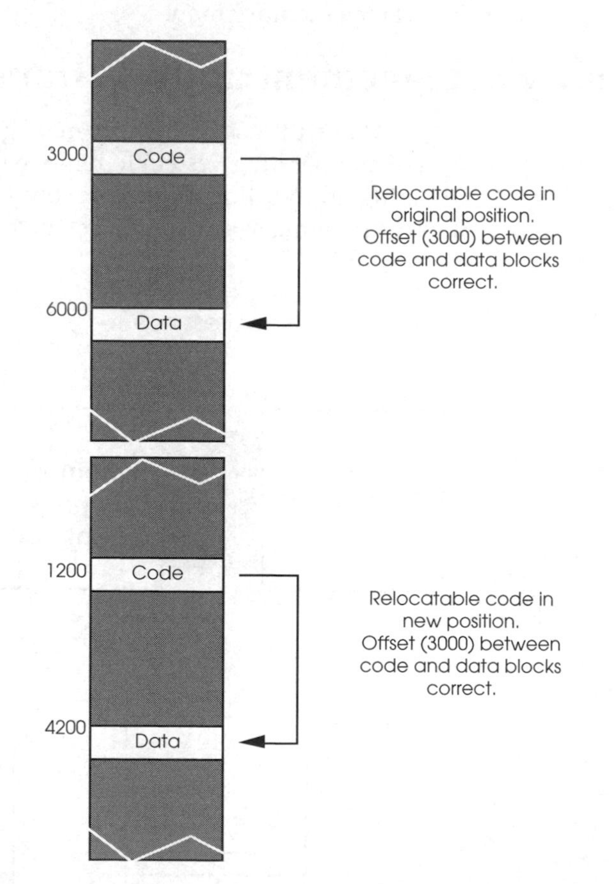

Correct movement of relocatable code

This address translation is extremely powerful and allows the embedded system designer many options to provide greater fault detection and/or security within the system or even cost reduction through the use of virtual memory. The key point is that memory management divides the processor memory map into definable regions with different properties such as read and write only access for one way data transfers and task or process specific memory access.

If no memory management hardware is present, most operating systems can replace their basic address translation facility with a software-based scheme, provided code is written to be position independent and relocatable. The more sophisticated techniques start to impose a large software overhead which in many cases is hard to justify within a simple system. Address translation is often necessary to execute programs in different locations from that in which they were generated. This allows the reuse of existing software modules and facilitates the easy transfer of software from a prototype to a final system.

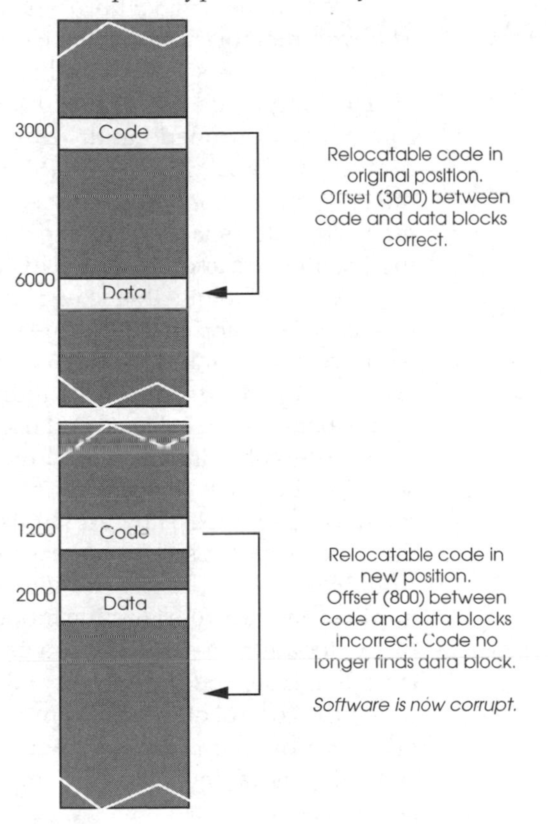

Incorrect movement of relocatable code

The relocation techniques are based on additional software built into the program loader or even into the operating system itself. If the operating system program loader cannot allocate the original memory, the program is relocated into the next available block and the program allowed to execute. Relocatable code does not have any immediate addressing values and makes extensive use of program relative addressing. Data areas or software subroutines are not referenced explicitly but are located by relative addressing modes using offsets:

- Explicit addressing
  e.g. branch to subroutine at address $0F04FF.
- Relative addressing
  e.g. branch to subroutine which is offset from here by $50 bytes.

Provided the offsets are maintained, then the relative addressing will locate data and code wherever the blocks are located in memory. Most modern compilers will use these techniques but do not assume that all of them do.

There are alternative ways of manipulating addresses to provide address translation, however.

Bank switching

When the first 8 bit processors became available with their 64 kbytes memory space, there were many comments concerning whether there was a need for this amount of memory. When the IBM PC appeared with its 640 kbyte memory map, the same comments were made and here we are today using PCs that probably have 32 Mbytes of RAM. The problem faced by many of the early processor architectures with a small 64 kbyte memory map is how the space can be expanded without having to change the architecture and increase register sizes and so on.

One technique that is used is that of bank switching. With this technique, additional bits are used to select different banks of memory. Each bank is the full 64 kbytes in size and is used in the normal way. By writing to the individual selection bits, an individual bank can be selected and used. It could be argued that this is no different from adding additional address bits. Using two selection bits will support four 64 kbyte banks giving a total memory space of 256 kbytes. This is the same amount of memory that can be addressed by increasing the number of address bits from 16 to 18. The difference, however, is that adding address bits implies that the programming model and processor knows about the wider address space and can use it. With bank switching, this is not the case, and the control and manipulation of the banks is under the control of the program that the processor is running. In other words, the processor itself has no knowledge that bank switching is taking place. It simply sees the normal 64 kbyte address space.

This approach is frequently used with microcontrollers that have either small external address spaces or alternatively limited external address buses. The selection bits are created by dedicating bits from the microcontroller's parallel I/O lines and using these to select and switch the external memory banks. The bank switching is controlled by writing to these bits within the I/O port.

This has some interesting repercussions for designs that use a RTOS. The main problem is that the program must understand when the system context is safe enough to allow a bank switch to be made. This means that system entities such as data structures, buffers and anything else that is stored in memory including program software must fit into the boundaries created by the bank switching.

This can be fairly simple but it can also be extremely complex. If the bank switching is used to extend a database, for example, the switching can be easy to control by inserting a check for a memory bank boundary. Records 1–100 could be in bank A, with bank B holding records 101–200. By checking the record number, the software can switch between the banks as needed. Such an implementation could define a subroutine as the access point to the data and within this routine the bank switching is managed so that it is transparent to the rest of the software.

Using bank switching to support large stacks or data structures on the other hand is more difficult because the mechanisms that use the data involve both automatic and software controlled access. Interrupts can cause stacks to be incremented automatically and there is no easy way of checking for an overflow and then incorporating the bank switching and so on needed to use it.

In summary, bank switching is used and there are 8 bit processors that have dedicated bits to support it but the software issues to use it are immense. As a result, it is frequently left for the system designer to figure out a way to use it within a design. As a result, few, if any, RTOS environments support this memory model.

Segmentation

Segmentation can be described as a form of bank switching that the processor architecture does know about! It works by providing a large external address bus but maintaining the smaller address registers and pointers within the original 8 bit architecture. To bridge the gap, the memory is segmented internally into smaller blocks that match the internal addressing and additional registers known as segment registers are used to hold the additional address data needed to complete the larger external address.

Probably the most well-known implementation is the Intel 8086 architecture.

Virtual memory

With the large linear addressing offered by today's 32 bit microprocessors, it is relatively easy to create large software applications which consume vast quantities of memory. While it may be feasible to install 64 Mbytes of RAM in a workstation, the costs are expensive compared with 64 Mbytes of a hard disk. As the memory needs go up, this differential increases. A solution is to use the disk storage as the main storage medium, divide the stored program into small blocks and keep only the blocks in processor system memory that are needed. This technique is known as virtual memory and relies on the presence within the system of a memory management unit.

As the program executes, the MMU can track how the program uses the blocks, and swap them to and from the disk as needed. If a block is not present in memory, this causes a page fault and forces some exception processing which performs the swapping operation. In this way, the system appears to have large amounts of system RAM when, in reality, it does not. This virtual memory technique is frequently used in workstations and in the UNIX operating system. Its appeal in embedded systems is limited because of the potential delay in accessing memory that can occur if a block is swapped out to disk.

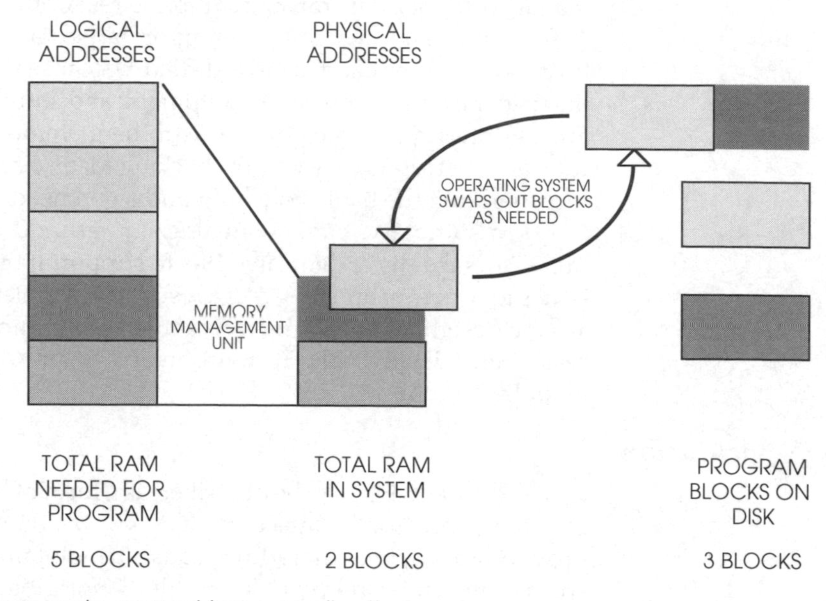

Using virtual memory to support large applications

Choosing an operating system

Comparing an operating system from 10 years ago with one offered today shows how operating system technology has developed over the past years. Although the basic functions provided by the old and the newer operating systems — they all provide multitasking, real-time responses and so on — are still present, there have been some fundamental changes in the improvement in the ease of use, performance and debugging facilities. Comparing a present-day car with one from the 1920s is a good analogy. The basic mechanics and principles have largely remained unchanged — that is, the engine, gearbox, brakes, transmission — but there has been a great improvement in the ease of driving, comfort and facilities. This is similar to what has happened with operating systems. The basic mechanisms of context switches, task control blocks, linked lists and so on are the basic fundamentals of any operating system or kernel.

As a result, it can be quite difficult to select an operating system. To make such a choice, it is necessary to understand the different ways that operating systems have developed over the years and the advantages that this has brought. The rest of this chapter discusses these changes and can help formulate the criteria that can be used to make a decision.

Assembler versus high level language

In the early 1980s, operating systems were developed in response to minicomputer operating systems where the emphasis was on providing the facilities and performance offered by minicomputers. To achieve performance, they were often written in assembler rather than in a high level language such as C or PASCAL. The reason for this was simply one of performance: compiler technology was not advanced enough to provide the compact and fast code needed to run an operating system. For example, many compilers from the early 1980s did not use all the M68000 address and data registers and limited themselves to only one or two. The result was code that was extremely inefficient when compared with hand coded assembler which did use all the processor's registers.

The disadvantage is that assembler code is harder to write and maintain compared to a high level language and is extremely difficult to migrate to other architectures. In addition, the interface between the operating system and a high level language was not well developed and in some cases non-existent! Writing interface libraries was considered part of the software task.

As both processor performance and compiler technology improved, it became feasible to provide an operating system written in a high level language such as C which provided a seamless integration of the operating system interface and application development language.

The choice of using assembler or a high level language with some assembler compared to using an integrated operating system and high level language is fairly obvious. What was acceptable a few years ago is no longer the case and today's successful operating systems are highly integrated with their compiler technology.

ROMable code

With early operating systems, restrictions in the code development often prevented operating systems and compilers from generating code that could be blown into read only memory for an embedded application. The reasons were often historic rather than technical, although the argument that most applications were too big to fit into the relatively small size of EPROM that was available was certainly true for many years. Today, most users declare this requirement as mandatory, and it is a standard offering from compilers and operating system vendors alike.

Scheduling algorithms

One area of constant debate is that of the scheduling algorithms that are used to select which task is to execute next. There are several different approaches which can be used. The first is to switch tasks only at the end of a time slice. This allows a fairer distribution of the processing power across a large number of tasks but at the expense of response time. Another is to take the first approach but allow certain events to force switch a task even if the current one has not used up all its allotted time slice. This allows external interrupts to get a faster response. Another event that can be used to interrupt the task is an operating system call.

Others have implemented priority systems where a task's priority and status within the ready list can be changed by itself, the operating system or even by other tasks. Others have a fixed priority system where the level is fixed when the task is created. Some operating systems even allow different scheduling algorithms to be implemented so that a designer can change them to give a specific response.

Changing algorithms and so on are usually indicative of trying to squeeze the last bit of performance from the system and in such cases it may be better to use a faster processor, or even in extreme cases actually accept that the operating system cannot meet the required performance and use another.

Pre-emptive scheduling

One consistent requirement that has appeared is that of pre-emptive scheduling. This refers to a particular scheduling algorithm where the highest priority task will interrupt or pre-empt a currently executing task irrespective of whether it has used its allotted time slice, and will continue running until a higher level task is ready to. This gives the best response to interrupts and events but can be a little dangerous. If a task is given the highest priority and does not lower its priority or pre-empt itself, then other tasks will not get an opportunity to execute. Therefore the ability to pre-empt is often restricted to special tasks with time critical routines.

Modular approach

The idea of reusing code whenever possible is not a new one but it can be difficult to implement. Obvious candidates with an operating system are device drivers for I/O , and kernels for different processors. The key is in defining a standard interface which allows such modules to be reused without having to alter or change the code. This means that memory maps must not be hardwired, or assumptions made by the driver or operating system. One of the problems with early versions of many operating systems was the fact that it was not until fairly late in their development that a modular approach for device drivers was available. As a result, the standard release included several drivers for the same peripheral chip, depending on which VMEbus board it was located.

Today, this approach is no longer acceptable and operating systems are more modular in their approach and design. The advantages for users are far more compact code, shorter development times and the ability to reuse code. A special driver can be reused without modification. This coupled with the need to keep up with the number of boards that need standard ports has led to the development of automated build systems that can take modular drivers and create a new version extremely quickly.

Re-entrant code

This follows on from the previous topic but has one fundamental difference in that a re-entrant software module can be shared be many tasks and also interrupted at any point and reused without any problems. For example, consider module A which is shared by two tasks B and C. If task B uses module A and exits from it, the module will be left in a known state, ready for another task to use it. If task C starts to use it and in the middle of its execution is switched out in favour of task B, then the problem may appear. If task B starts to use module A, it may experience problems because A will be in an indeterminate state. Even if it does not, other problems may still be lurking. If module A uses global variables, then task B will cause them to be reset. When task C returns to continue execution, its global data will have been destroyed.

A re-entrant module can tolerate such interruptions without experiencing these types of problems. The golden rule is for the module to only access data associated with the task that is using the module code. Variables are stored on stacks, in registers or in memory areas specific to the task and not to the module. If shared modules are not re-entrant, care must be taken to ensure that a context switch does not occur during its execution. This may mean disabling or locking out the scheduler or dispatcher.

Cross-development platforms

Today, most software development is done on cross-development platforms such as Sun workstations, UNIX systems and IBM PCs. This is in direct contrast to early systems which required a dedicated software development system. The degree of platform support and the availability of good development tools which go beyond the standard of symbolic level debug have become a major product selling point.

Integrated networking

This is another area which is becoming extremely important. The ability to use a network such as TCP/IP on Ethernet to control target boards, download code and obtain debugging information is fast becoming a mandatory requirement. It is rapidly replacing the more traditional method of using serial RS232 links to achieve the same end.

Multiprocessor support

This is another area which has changed dramatically. Ten years ago it was possible to use multiple processors provided the developer designed and coded all the inter-processor communication. Now, many of today's operating systems can provide optional modules that will do this automatically. However, multiprocessing techniques are often misunderstood and as this is such a big topic for both hardware and software developers it is treated in more depth later in this text.

Commercial operating systems

pSOS+

pSOS+ is the name of a popular multitasking real-time operating system. Although the name refers to the kernel itself, it is often used in a more generic way to refer to a series of development tools and system components. The best way of looking at the products is to use the overall structure as shown in the diagram. The box on the left is concerned with the development environment while that on the right are the software components that are used in the final target system. The two halves work together via communication links such as serial lines, Ethernet and TCP/IP protocol or even over the VMEbus itself.

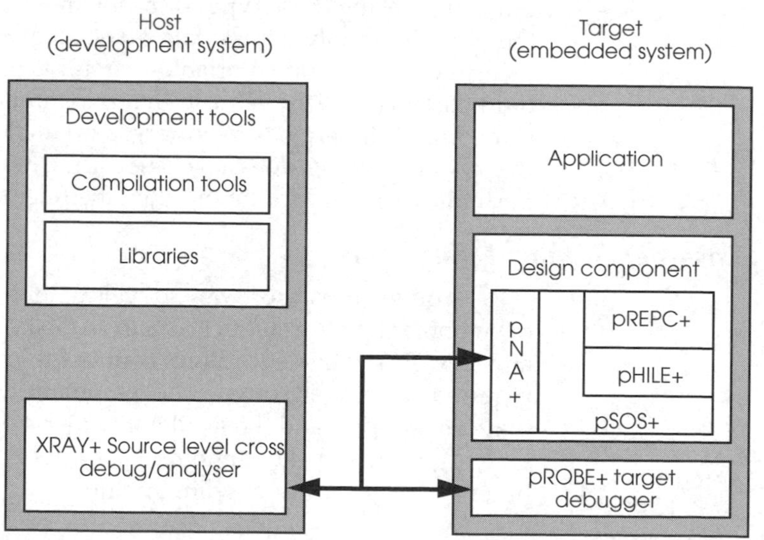

pSOS+ overall structure

pSOS+ kernel

The kernel supports a wide range of processor families like the Motorola M68000 family, the Intel 80x86 range, and the M88000 and i960 RISC processors. It is small in size and typically takes about 15–20 kbytes of RAM, although the final figure will depend on the configuration and processor type.

It supports more than 50000 system objects such as tasks, memory partitions, message queues and so on and will execute time-critical routines consistently irrespective of the application size. In other words, the time to service a message queue is the same irrespective of the size of the message. Note that this will refer to the time taken to pass the message or perform the service only and does not and cannot take into account the time taken by the user to handle messages. In other words, the consistent timing refers to the message delivery and not the actions taken as a result of the message. Worst case figures for interrupt latency and context switch for an MC68020 running at 25 MHz are 6 and 19 μs respectively. Among its 55 service calls, it provides support for:

- Task management
- Message queues
- Event services
- Semaphore services
- Asynchronous services
- Storage allocation services
- Time management and timer services
- I/O supervisor services
- Interrupt management
- Error handling services

pSOS multiprocessor kernel

pSOS+m is the multiprocessing version of the kernel. From an application or task's perspective, it is virtually the same as the single processor version except that the kernel now has the ability to send and receive system objects from other processors within the system. The application or task does not know where other tasks actually reside or the communication method used to link them.

The communication mechanism works in this way. Each processor runs its own copy of the kernel and has a kernel interface to the media used for linking the processors, either a VMEbus backplane or a network. When a task sends a message to a local task, the local processor will handle it. If the message is for task running on another node, the local operating system will look up the recipient's identity and location. The message is then routed across the kernel interface across to the other processor and its operating system where the message is finally delivered. Different kernel interfaces are needed for different media.

pREPC+ runtime support

This is a compiler independent run-time environment for C applications. It is compatible with the ANSI X3J11 technical committee's proposal for C run-time functionality and provides 88 functions that can be called from C programs. Services supported include formatted I/O, file I/O and string manipulation.

pREPC+ is not a standalone product because it uses pSOS+ or pSOS+m for device I/O and task functions and calls pHILE+ for file and disk I/O. Its routines are re-entrant which allows multiple tasks to use the same routine simultaneously.

pHILE$^+$ file system

This product provides file I/O and will support either the MS-DOS file structure or its own proprietary formats. The MS-DOS structure is useful for data interchange while the proprietary format is designed to support the highest data throughput across a wide range of devices. pHILE$^+$ does not drive physical devices directly but provides logical data via pSOS$^+$ to a device driver task that converts this information to physical data and drives the storage device.

pNA$^+$ network manager

This is a networking option that provides TCP/IP communication over a variety of media such as Ethernet and FDDI. It conforms to the UNIX 4.3 BSD socket syntax and approach and is compatible with other TCP/IP–based networking standards such as ftp and NFS.

As a result, pNA$^+$ is used to provide efficient downloading and debugging communication between the target and a host development system. Alternatively, it can be used to provide a communication path between other systems that are also sitting on the same network.

pROBE$^+$ system level debugger

This is the system level debugger which provides the system and low level debugging facilities. With this, system objects can be inspected or even used to act as breakpoints if needed. It can use either a serial port to communicate with the outside world or if pNA$^+$ is installed, use an TCP/IP link instead.

XRAY$^+$ source level debugger

This is a complementary product to pROBE$^+$ as it can use the debugger information and combine it with the C source and other symbolic information on the host to provide a complete integrated debugging environment.

OS-9

OS-9 was originally developed by Microware and Motorola as a real-time operating system for the Motorola MC6809 8 bit processor and it appeared on many 6809-based systems such as the Exorset 165 and the Dragon computer. It provided a true hierarchical filing system and the ability to run multiple tasks. It has since been ported to the M68000 family and the Intel 80x86 processor families. It is best described as a complete operating system with its own user commands, interface and so on. Unlike other products which have concentrated on the central kernel and then built outwards but stopping at below the user and utility level, OS-9 goes from a multi-user multitasking interface with a range of utilities down to the low level kernel. Early on it supported UNIX by using and supporting the same library interface and similar system calls. So much so that one of its strengths was the ability to take UNIX source code, recompile it and then run it.

One criticism has been its poor real-time response although a new version has been released which used a smaller, compact and faster kernel to provide better performance. The full facilities are still provided by the addition of other kernel services around the inner one. It provides more sophisticated support such as multimedia extensions which other operating systems do not, and because of this and its higher level of utilities and expansion has achieved success in the marketplace.

VXWorks

VXWorks has taken another approach to the problem of providing a real-time environment as well as standard software tools and development support. Instead of creating its own or reproducing a UNIX-like environment, it actually has integrated with UNIX to provide its development and operational environment. Through VXWorks' UNIX compatible networking facilities, it can combine with UNIX to form a complete run-time solution as well. The UNIX system is responsible for software development and non real time debugging while the VXWorks kernel is used for testing, debugging and executing the real-time applications, either standalone or part of a network with UNIX.

How does this work? The UNIX system is used as the development host and is used to edit, compile, link and administer the building of real-time code. These modules can then be burned into ROM or loaded from disk and executed standalone under the VXWorks kernel. This is possible because VXWorks can understand the UNIX object module format and has a UNIX compatible software interface. By using the standard UNIX pipe and socket mechanisms to exchange data between tasks and by using UNIX signals for asynchronous events, there is no need for recompilation or any other conversion routines. Instead, the programmer can use the same interface for both UNIX and VXWorks software without having to learn different libraries or programming commands. It supports the POSIX 1003.4 real-time extensions and multiprocessing support for up to 20 processors is offered via another option called VxMP.

The real key to VXWorks is its ability to network with UNIX to allow a hybrid system to be developed or even allow individual modules or groups to be transferred to run in a VXWorks environment. The network can be over an Ethernet or even using shared memory over a VMEbus, for example.

VRTX-32

VRTX-32 from Microtec Research has gained a reputation for being an extremely compact but high performance real-time kernel. Despite being compact — typically about 8 kbytes of code for an MC68020 system — it provides task management facilities, inter-task communication, memory control and allocation, clock and timer functions, basic character I/O and interrupt handling facilities.

Under the name of VRTXvelocity, VRTX-32 systems can be designed using cross-development platforms such as Sun workstations and IBM PCs. The systems can be integrated with the host, usually using an Ethernet, to provide an integrated approach to system design.

IFX

Its associated product IFX (input/output file executive) provides support for more complicated I/O subsystems such as disks, terminals, and serial communications using structures such as pipes, null devices, circular buffers and caches. The file system is MS-DOS compatible although if this is not required, disks can be treated as single partitions to speed up response.

TNX

This is the TCP/IP networking package that allows nodes to communicate with hosts and other applications over the Ethernet. The Ethernet device itself can either be resident on the processor board or accessible across a VMEbus. It supports both stream and datagram sockets.

RTL

This is the run-time library support for Microtec and Sun compilers and provides the library interface to allow C programs to call standard I/O functions and make VRTX-32 calls.

RTscope

This is the real-time multitasking debugger and system monitor that is used to debug VRTX tasks and applications. It operates on two levels: the board level debugger provides the standard features such as memory and register display and modify, software upload and download and so on. In the VRTX-32 system monitor mode, tasks can be interrogated, stopped, suspended and restarted.

MPV

The multiprocessor VRTX-32 extensions allow multiple processors each running their own copy of VRTX to pass messages and other task information from one processor to another and thus create a multiprocessor system. The messages are based across the VMEbus using shared memory although other links such as RS232 or Ethernet are possible.

LynxOS-POSIX conformance

POSIX (IEEE standard portable operating system interface for computer environments) began in 1986 as an attempt to provide an open standard for operating system support. The ideas behind it are to provide vendor independence, protection from technical obsolescence, the availability of standard off-the-shelf applications, the preservation of software investment and to provide connectivity between computers.

It is based on UNIX but has added a set of real-time extensions as defined in the POSIX 1003.4 document. These cover a more sophisticated semaphore system which uses the open() call to create them. This call is more normally associated with opening a file. The facilities include persistent semaphores which retain their binary state after their last use, and the ability to force a task to wait for a semaphore through a locking mechanism.

The extensions also provide a process or task locking mechanism which prevents memory pages associated with the task or process from being swapped out to memory, thus improving the real-time response for critical routines. Shared memory is better supported through the provision of the shmmap() call which will allocate a sheared memory block. Both asynchronous and synchronous I/O and inter-task message passing are supported along with real-time file extensions to speed up file I/O. This uses techniques such as preallocating file space before it is required.

At the time of writing LynxOS is the main real-time product that supports these standards, although many others support parts of the POSIX standard. Indeed, there is an increasing level of support for this type of standardisation.

However, it is not a complete panacea and, while any attempt for standardisation should be applauded, it does not solve all the issues. Real-time operating systems and applications are very different in their needs, design and approach, as can be seen from the diversity of products that are available today. Can all of these be met by a single standard? In addition, the main cost of developing software is not in porting but in testing and documenting the product and this again is not addressed by the POSIX standards. POSIX conformance means that software should be portable across processors and platforms, but it does not guarantee it. With many of today's operating systems available in versions for the major processor families, is the POSIX portability any better? Many of these questions are yet to be answered conclusively by supporters or protagonists.

An alternative way of looking at this problem is: do you assume that a ported real-time product will work because it is POSIX compliant without testing it on the new target? In most cases the answer will be no and that testing will be needed. What POSIX conformance has given is a helping hand in the right direction and this should not be belittled, neither should it be seen as a miracle cure. In the end, the success of the POSIX standards will depend on the market and users seeing benefit in its approach. It is an approach that is gathering pace, and one that the real-time market should be aware of. It is possible that it may succeed where other attempts at a real-time interface standard have failed. Another possibility for POSIX conformance is Windows NT.

Windows NT

Windows NT has been portrayed as many different things during its short lifetime. When it first appeared, it was perceived by many as the replacement for Windows 3.1, an alternative to UNIX, and finally has settled down as an operating system for workstations, servers and power users. This chameleon-like change was not due to any real changes in the product but were caused by a mixture of aspirations and misunderstandings.

Windows NT is rapidly replacing Windows 3.1 and Windows 95 and parts of its technology have already found themselves incorporated into Windows 95 and Windows for Workgroups. Whether the replacement is through a merging of the operating system technologies or through a sharing of common technology, only time will tell. The important message is that the Windows NT environment is becoming prevalent, especially with Microsoft's aim of a single set of programming interfaces that will allow an application to run on any of its operating system environments. Its greater stability and reliability is another feature that is behind its adoption by many business systems in preference over Windows 95. All this is fine, but how does this fit with an embedded system?

There are several reasons why Windows NT is being used in real-time environments. It may not have the speed of a dedicated RTOS but it has the important features and coupled with a fast processor, reasonable performance.

- Portability

  Most PC-based operating systems were written in low-level assembler language instead of a high level language such as C or C++. This decision was taken to provide smaller programs sizes and the best possible performance. The disadvantage is that the operating system and applications are now dependent on the hardware platform and it is extremely difficult to move from one platform to another. MS-DOS is writen in 8086 assembler which is incompatible with the M68000 processors used in the Apple Macintosh. For a software company like Microsoft, this has an additional threat of being dependent on a single processor platform. If the platform changes — who remembers the Z80 and 6502 processors which were the mainstays of the early PCs — then its software technology becomes obsolete.

  With an operating system that is written in a high level language and is portable to other platforms, it allows Microsoft and other application developers to be less hardware dependent.

- True multitasking

  While more performant operating systems such as UNIX and VMS offer the ability to run multiple applications simultaneously, this facility is not really available from the Windows and MS-DOS environments (a full explanation of what they can do and the difference will be offered later in this chapter). This is now becoming a very important aspect for both users and developers alike so that the full performance of today's processors can be utilised.

- Multi-threaded

  Multi-threading refers to a way of creating software that can be reused without having to have multiple copies of the code or memory spaces. This leads to more efficient use of both memory and code.

- Processor independent

  Unlike Windows and MS-DOS which are completely linked to the Intel 80x86 architecture, Windows NT through its portability is processor independent and has been ported to other processor architectures such as Motorola's PowerPC, DEC's Alpha architecture and MIPS RISC processor systems.

- Multiprocessor support

  Windows NT uses a special interface to the processor hardware which makes it independent of the processor architecture that it is running on. As a result, this not only gives processor independence but also allows the operating system to run on multiprocessor systems.

- Security and POSIX support

  Windows NT offers several levels of security through its use of a multi-part access token. This token is created and verified when a user logs onto the system and contains IDs for the user, the group he is assigned to, privileges and other information. In addition, an audit trail is also provided to allow an administrator to check who has used the system, when they used it and what they did. While an overkill for a single user, this is invaluable with a system that is either used by many or connected to a network.

  The POSIX standard defines a set of interfaces that allow POSIX compliant applications to easily be ported between POSIX compliant computer systems.

  Both security and POSIX support are commercially essential to satisfy purchasing requirements from government departments, both in the US and the rest of the world.

Windows NT characteristics

Windows NT is a pre-emptive multitasking environment that will run multiple applications simultaneously and uses a priority based mechanism to determine the running order. It is capable of providing real-time support in that it has a priority mechanism and fast response times for interrupts and so on, but it is less deterministic — there is a wider range of response times — when compared to a real-time operating system such as pSOS or OS-9 used in industrial applications. It can be suitable for many real-time applications with less critical timing characteristics and this is a big advantage over the Windows 3.1 and Windows 95 environments. It is interesting to note that this technology now forms the backbone of all the Windows software environments.

Process priorities

Windows NT calls all applications, device drivers, software tasks and so on processes and this nomenclature will be used from now on. Each process can be assigned one of 32 priority levels which determines its scheduling priority. The 32 levels are divided into two groups called the real-time and dynamic classes.

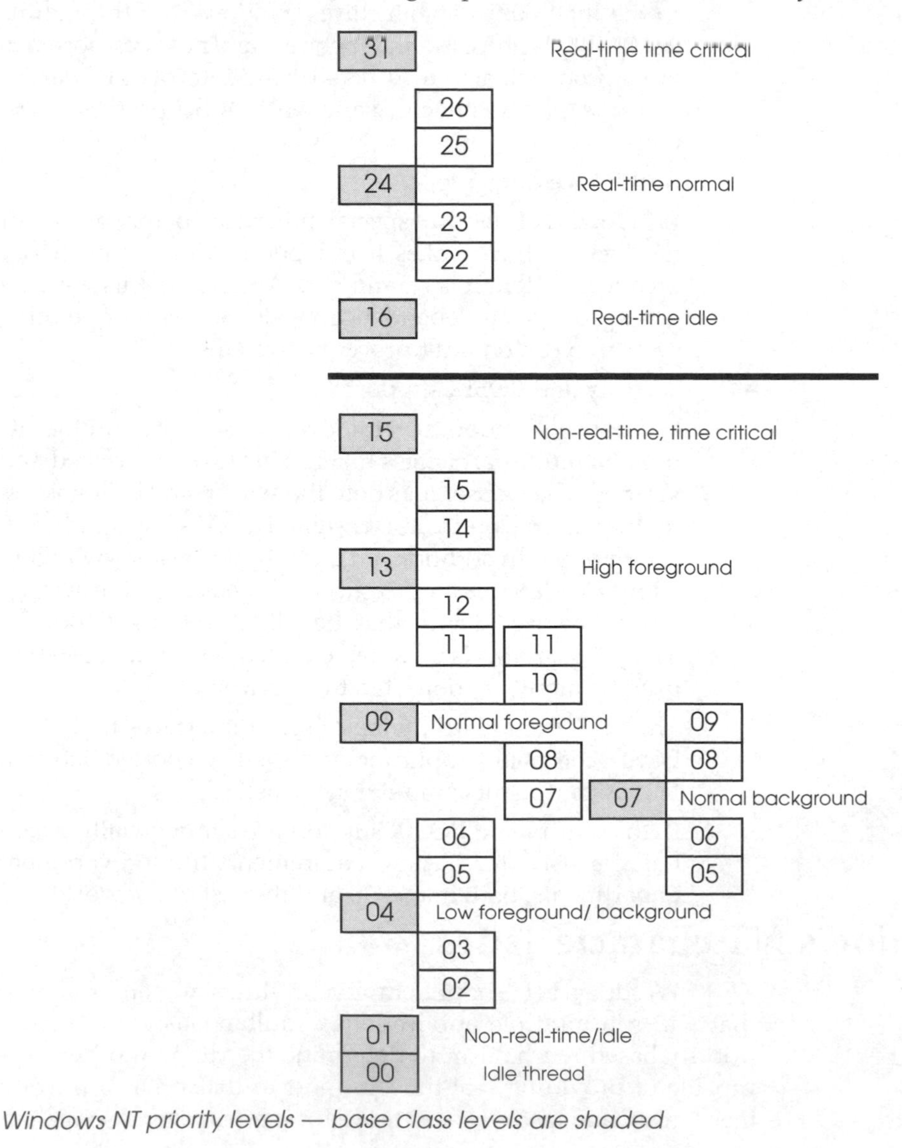

Windows NT priority levels — base class levels are shaded

The real-time classes comprise priority levels 16 through to 31 and the dynamic classes use priority levels 15 to 0. Within these two groups, certain priorities are defined as base classes and processes are allocated a base process. Independent parts of a process — these are called threads — can be assigned their own priority levels which are derived from the base class priority and can be ±2 levels different. In addition, a process cannot move from a real-time class to a dynamic one.

The diagram shows how the base classes are organised. The first point is that within a given dynamic base class, it is possible for a lot of overlap. Although a process may have a lower base class compared to another process, it may be at a higher priority than the other one depending on the actual priority level that has been assigned to it. The real-time class is a little simpler although again there is some possibility for overlap.

User applications like word processors, spread sheets and so on run in the dynamic class and their priority will change depending on the application status. Bring an application from the background to the foreground by expanding the shrunk icon or by switching to that application will change its priority appropriately so that it gets allocated a higher priority and therefore more processing. Real-time processes include device drivers handling the keyboard, cursor, file system and other similar activities.

Interrupt priorities

The concept of priorities is not solely restricted to the preemption priority previously described. Those priorities come into play when an event or series of events occur. The events themselves are also controlled by 32 priority levels defined by the hardware abstraction layer (HAL).

Interrupt	Description
31	Hardware error interrupt
30	Powerfail interrupt
29	Inter-processor interrupt
28	Clock interrupt
27-12	Standard IBM PC AT interrupt levels 0 to 15
11-4	Reserved (not generally used)
3	Software debugger
2-0	Software interrupts for device drivers etc.

Interrupt priorities

The interrupt priorities work in a similar way to those found on a microprocessor: if an interrupt of a higher priority than the current interrupt priority mask is generated, the current processing will stop and be replaced by the associated routines for the new higher priority level. In addition, the mask will be raised to match that of the higher priority. When the higher priority processing has been completed, the previous processing will be restored and allowed to continue. The interrupt priority mask will also be restored to its previous value.

Within Windows NT, the interrupt processing is also subject to the multitasking priority levels. Depending on how these are assigned to the interrupt priority levels, the processing of a high priority interrupt may be delayed until a higher priority process has completed. It makes sense therefore to have high priority interrupts processed by processes with high priority scheduling levels. Comparing the interrupts and the priority levels shows that this maxim has been followed. Software interrupts used to communicate between processes are allocated both low interrupt and scheduling priorities. Time critical interrupts such as the clock and inter-processor interrupts are handled as real-time processes and are allocated the higher real-time scheduling priorities.

The combination of both priority schemes provides a fairly complex and flexible method of structuring how external and internal events and messages are handled.

Resource protection

If a system is going to run multiple applications simultaneously then it must be able to ensure that one application doesn't affect another. This is done through several layers of resource protection. Resource protection within MS-DOS and Windows 3.1 is a rather hit and miss affair. There is nothing to stop an application from directly accessing an I/O port or other physical device and if it did so, it could potentially interfere with another application that was already using it. Although the Windows 3.1 environment can provide some resource protection, it is of collaboration level and not mandatory. It is without doubt a case of self-regulation as opposed to obeying the rules of the system.

Protecting memory

The most important resource to protect is memory. Each process is allocated its own memory which is protected from interference by other processes through programming the memory management unit. This part of the processor's hardware tracks the executing process and ensures that any access to memory that it has not been allocated or given permission to use is stopped.

Protecting hardware

Hardware such as I/O devices are also protected by the memory management unit and any direct access is prevented. Such accesses have to be made through a device driver and in this way the device driver can control who has access to a serial port and so on. A mechanism called a spinlock is also used to control access. A process can only access a device or port if the associated spinlock is not set. If someone else is using it, the process must wait until they have finished.

Coping with crashes

If a process crashes then it is important for the operating system to maintain as much of the system as possible. This requires that the operating system as well as other applications must have its own memory and resources given to it. To ensure this is the case, processes that are specific to user applications are run in a user mode while operating system processes are executed in a special kernel mode. These modes are kept separate from each other and are protected. In addition, the operating system has to have detailed knowledge of the resources used by the crashed process so that it can clean up the process, remove it and thus free up the resources that it used. In some special cases, such as power failures where the operating system may have a limited amount of time to shut down the system in a controlled manner or save as much of the system data as it can, resources are dedicated specifically for this functionality. For example, the second highest interrupt priority is allocated to signalling a power failure.

Windows NT is very resilient to system crashes and while processes can crash, the system will continue. This is essentially due to the use of user and kernel modes coupled with extensive resource protection. Compared to Windows 3.1 and MS-DOS, this resilience is a big advantage.

Multi-threaded software

There is a third difference with Windows NT that many other operating systems do not provide in that it supports multi-threaded processes. Processes can support several independent processing paths or threads. A process may consist of several independent sections and thus form several different threads in that the context of the processing in one thread may be different from that in another thread. In other words, the process has all the resources defined that it will use and if the process can support multi-threaded operations, the scheduler will see multiple threads going through the process. A good analogy is a production line. If the production line is single threaded, it can only produce a single end product at a time. If it is multi-threaded, it separates the production process into several independent parts and each part can work on a different product. As soon as the first operation has taken place, a second thread can be started. The threads do not have to follow the same path and can vary their route through the process.

The diagram shows a simple multi-threaded operation with each thread being depicted by a different shading. As the first thread progresses through, a second thread can be started. As that progresses through, a third can commence and so on. The resources required to process the multiple threads in this case are the same as if only one thread was supported.

The advantage of multi-threaded operation is that the process does not have to be duplicated every time it is used: a new thread can be started. The disadvantage is that the process programming must ensure that there is no contention or conflict between the various threads that it supports. All the threads that exist in the process can access each other's data structures and even files. The operating system which normally polices the environment is powerless in this case. Threads within Windows NT derive their priority from that of the process although the level can be adjusted within a limited range.

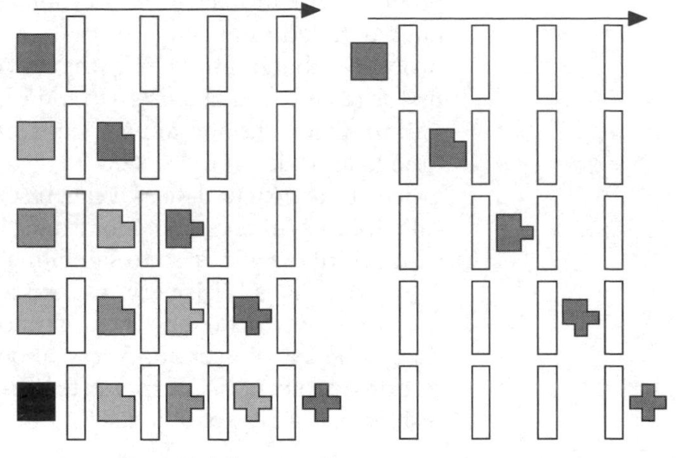

Multi-threaded (left) and single-threaded (right) operations

Addressing space

The addressing space within Windows NT is radically different from that experienced within MS-DOS and Windows 3.1. It provides a 4 Gbyte virtual address space for each process which is linearly addressed using 32 bit address values. This is different from the segmented memory map that MS-DOS and Windows have to use. A segmented memory scheme uses 16 bit addresses to provide address spaces of only 64 kbytes. If addresses beyond this space have to be used, special support is needed to change the segment address to point to a new segment. Gone are the different types of memory such as extended and expanded.

This change towards a large 32 bit linear address space improves the environment for software applications and increases their performance and capabilities to handle large data structures. The operating system library that the applications use is called WIN32 to differentiate it from the WIN16 libraries that Windows 3.1 applications use. Applications that use the WIN32 library are known as 32 bit or even native — this term is also used for Windows NT applications that use the same instruction set as the host processor and therefore do not need to emulate a different architecture.

To provide support for legacy MS-DOS and Windows 3.1 applications, Windows NT has a 16 bit environment which simulates the segmented architecture that these applications use.

Virtual memory

The idea behind virtual memory is to provide more memory than physically present within the system. To make up the shortfall, a file or files are used to provide overflow storage for applications which are to big to fit in the system RAM at one time. Such applications' memory requirements are divided into pages and unused pages are stored on disk.

When the processor wishes to access a page which is not resident in memory, the memory management hardware asserts a page fault, selects the least used page in memory and swaps it with the wanted page stored on disk. Therefore, to reduce the system overhead, fast mass storage and large amounts of RAM are normally required.

Windows NT uses a swap file to provide a virtual memory environment. The file is dynamic in size and varies with the amount of memory that all the software including the operating system, device driver, and applications require. The Windows 3.1 swap file is limited to about 30 Mbytes in size and this effectively limits the amount of virtual memory that it can support.

The internal architecture

The internal architecture is shown in the diagram below and depicts the components that run in the user and kernel modes. Most of the operating system runs in the kernel mode with the exception of the security and WIN32 subsystems.

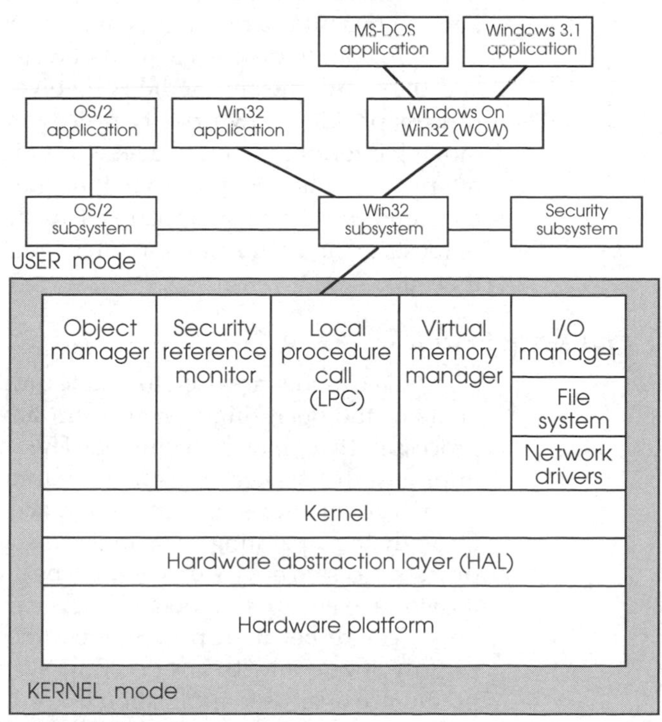

The internal Windows NT architecture

The environments are protected in that direct hardware access is not allowed and that a single application will run in a single environment. This means that some applications or combinations may not work in the Windows NT environment. On the other hand, running them in separate isolated environments does prevent them from causing problems with other applications or the operating system software.

Virtual memory manager

The virtual memory manager controls and supervises the memory requirements of an operating system. It allocates to each process a private linear address space of 4 Gbytes which is unique and cannot be accessed by other processes. This is one reason why legacy software such as Windows 3.1 applications run as if they are the only application running.

With each process running in its own address space, the virtual memory manager ensures that the data and code that the process needs to run is located in pages of physical memory and ensures that the correct address translation is performed so that process addresses refer to the physical addresses where the information resides. If the physical memory is all used up, the virtual memory manager will move pages of data and code out to disk and store it in a temporary file. With some physical memory freed up, it can load from disk previously stored information and make it available for a process to use. This operation is closely associated with the scheduling process which is handled within the kernel. For an efficient operating system, it is essential to minimise the swapping out to disk — each disk swap impacts performance — and the most efficient methods involve a close correlation with process priority. Low priority processes are primary targets for moving information out to disk while high priority processes are often locked into memory so that they offer the highest performance and are never swapped out to disk. Processes can make requests to the virtual memory manager to lock pages in memory if needed.

User and kernel modes

Two modes are used to isolate the kernel and other components of the operating system from any user applications and processes that may be running. This separation dramatically improves the resilience of the operating system. Each mode is given its own addressing space and access to hardware is made through the operating system kernel mode. To allow a user process access, a device driver must be used to isolate and control its access to ensure that no conflict is caused.

The kernel mode processes use the 16 higher real-time class priority levels and thus operating system processes will take preference over user applications.

Local procedure call (LPC)

This is responsible for co-ordinating system calls from an application and the WIN32 subsystem. Depending on the type of call and to some extent its memory needs, it is possible for applications to be routed directly to the local procedure call (LPC) without going through the WIN32 subsystem.

The kernel

The kernel is responsible for ensuring the correct operation of all the processes that are running within the system. It provides the synchronisation and scheduling that the system needs. Synchronisation support takes the form of allowing threads to wait until a specific resource is available such as an object, semaphore, an expired counter or other similar entity. While the thread is waiting it is effectively dormant and other threads and processes can be scheduled to execute.

The scheduling procedures use the 32 level priority scheme previously described in this chapter and is used to schedule threads rather than processes. With a process potentially supporting multiple threads, the scheduling operates on a thread basis and not on a process basis as this gives a finer granularity and control. Not scheduling a multi-threaded process would affect several threads which may not be the required outcome. Scheduling on a thread basis gives far more control.

Interrupts and other similar events also pass through the kernel so that it can pre-empt the current thread and reschedule a higher priority thread to process the interrupt.

File system

Windows NT supports three types of file system and these different file systems can co-exist with each other although there can be some restrictions if they are accessed by non-Windows NT systems across a network, for example.

- FAT

  File allocation table is the file system used by MS-DOS and Windows 3.1 and uses file names with an 8 character name and a 3 character extension. The VFAT system used by Windows 95 and supports long file names is also supported with Windows NT v4 in that it can read Windows 95 long file names.

- HPFS

  High performance file system is an alternative file system used by OS/2 and supports file names with 254 characters with virtually none of the character restrictions that the FAT system imposes. It also uses a write caching to disk technique which stores data temporarily in RAM and writes it to disk at a later stage. This frees up an application from waiting until the physical disk write has completed. The physical disk write is performed when the processor is either not heavily loaded or when the cache is full.

- NTFS

  The NT filing system is Windows NT's own filing system which conforms to various security recommendation and allows system administrators to restrict access to files and directories within the filing system.

All three filing systems are supported — Windows NT will even truncate and restore file names that are not MS-DOS compatible — and are selected during installation.

Network support

As previously stated, Windows NT supports most major networking protocols and through its multi-tasking capabilities can support several simultaneously using one or more network connections. The drivers that do the actual work are part of the kernel and work closely with the file system and security modules.

I/O support

I/O drivers are also part of the kernel and these provide the link between the user processes and threads and the hardware. MS-DOS and Windows 3.1 drivers are not compatible with Windows NT drivers and one major difference between Windows NT and Windows 3.1 is that not all hardware devices are supported. Typically modern devices and controllers can be used but it is wise to check the existence of a driver before buying a piece of hardware or moving from Windows 3.1 to Windows NT.

HAL approach

The hardware abstraction layer (HAL) is designed to provide portability across different processor-based platforms and between single and multi-processor systems. In essence, it defines a piece of virtual hardware that the kernel uses when it needs to access hardware or processor resources. The HAL layer then takes the virtual processor commands and requests and translates them to the actual processor system that it is actually using. This may mean a simple mapping where a Windows NT interrupt level corresponds to a processor hardware interrupt but it can involve a complete emulation of a different processor. Such is the case to support MS-DOS and Windows 3.1 applications where an Intel 80x86 processor is emulated so that an Intel instruction set can be run.

With the rest of Windows NT being written in C, a portable high level language, the only additional work to the recompilation and testing is to write a suitable HAL layer for the processor platform that is being used.

Linux

Linux started as a personal interest project by Linus Torvalds at the University of Helsinki in Finland to produce an operating system that looked and felt like UNIX. It was based on work that he had done in porting Minix , an operating system that had been shipped with a textbook that described its inner workings.

After much discussion via user groups on the Internet, the first version of Linux saw the light of day on the 5 October, 1991. While limited in its abilities — it could run the GNU bash shell and gcc compiler but not much else — it prompted a lot of interest. Inspired by Linus Torvalds' efforts, a band of enthusiasts started to create the range of software that Linux offers today. While this was progressing, the kernel development continued until some 18 months later, when it reached version 1.0. Since then it has been developed further with many ports for different processors and platforms. Because of the large amount of software available for it, it has become a very popular operating system and one that is often thought off as a candidate for embedded systems.

However it is based on the interfaces and design of the UNIX operating system which for various reasons is not considered suitable for embedded design. If this is the case, how is it that Linux is now forging ahead in the embedded world. To answer this question, it is important to understand how it came about and was developed. That means starting with the inspiration behind Linux, the UNIX operating system.

Origins and beginnings

UNIX was first described in an article published by Ken Thompson and Dennis Ritchie of Bell Research Labs in 1974, but its origins owe much to work carried out by a consortium formed in the late 1960s, by Bell Telephones, General Electric and the Massachusetts Institute of Technology, to develop MULTICS — a MULTIplexed Information and Computing Service. Their goal was to move away from the then traditional method of users submitting work in as punched cards to be run in batches — and receiving their results several hours (or days!) later. Each piece of work (or job) would be run sequentially — and this combination of lack of response and the punched card medium led to many frustrations — as anyone who has used such machines can confirm. A single mistake during the laborious task of producing punched cards could stop the job from running and the only help available to identify the problem was often a 'syntax error' message. Imagine how long it could take to debug a simple program if it took the computer several hours to generate each such message!

The idea behind MULTICS was to generate software which would allow a large number of users simultaneous access to the computer. These users would also be able to work interactively and on-line in a way similar to that experienced by a personal computer user today. This was a fairly revolutionary concept. Computers were very expensive and fragile machines that required specially trained staff to keep other users away from and protect *their* machine. However, the project was not as successful as had been hoped and Bell dropped out in 1969. The experienced gained in the project was turned to other uses when Thompson and Ritchie designed a computer filing system on the only machine available — a Digital Equipment PDP-7 mini computer.

While this was happening, work continued on the GE645 computer used in the MULTICS project. To improve performance and save costs (processing time was very expensive), they wrote a very simple operating system for the PDP-7 to enable it to run a space travel game. This operating system, which was essentially the first version of UNIX, included a new filing system and a few utilities.

The PDP-7 processor was better than nothing — but the new software really cried out for a better, faster machine. The problem faced by Thompson and Ritchie was one still faced by many today. It centred on how to persuade management to part with the cash to buy a new computer, such as the newer Digital Equipment Company's PDP-11. Their technique was to interest the Bell legal department in the UNIX system for text processing and use this to justify the expenditure. The ploy was successful and UNIX development moved along.

The next development was that of the C programming language, which started out as an attempt to develop a FORTRAN language compiler. Initially, a programming language called B which was developed, which was then modified into C. The development of C was crucial to the rapid movement of UNIX from a niche within a research environment to the outside world.

UNIX was rewritten in C in 1972 — a major departure for an operating system. To maximise the performance of the computers then available, operating systems were usually written in a low level assembly language that directly controlled the processor. This had several effects. It meant that each computer had its own operating system, which was unique, and this made application programs hardware dependent. Although the applications may have been written in a high level language (such as FORTRAN or BASIC) which could run on many different machines, differences in the hardware and operating systems would frequently prevent these applications from being moved between systems. As a result, many man hours were spent porting software from one computer to another and work around this computer equivalent of the Tower of Babel.

By rewriting UNIX in C, the painstaking work of porting system software to other computers was greatly reduced and it became feasible to contemplate a common operating system running on many different computers. The benefit of this to users was a common interface and way of working, and to software developers, an easy way to move applications from one machine to another. In retrospect, this decision was extremely far sighted.

The success of the legal text processing system, coupled with a concern within Bell about being tied to a number of computer vendors with incompatible software and hardware, resulted in the idea of using the in-house UNIX system as a standard environment. The biggest advantage of this was that only one set of applications needed to be written for use on many different computers. As UNIX was now written in a high level language, it was a lot more feasible to port it to different hardware platforms. Instead of rewriting every application for each computer, only the UNIX operating system would need to be written for each machine — a lot less work. This combination of factors was too good an opportunity to miss. In September 1973, a UNIX Development Support group was formed for the first UNIX applications, which updated telephone directory information and intercepted calls to changed numbers.

The next piece of serendipity in UNIX development was the result of a piece of legislation passed in 1956. This prevented AT&T, who had taken over Bell Telephone, from selling computer products. However, the papers that Thompson and Ritchie had published on UNIX had created a quite a demand for it in academic circles. UNIX was distributed to universities and research institutions at virtually no cost on an 'as is' basis — with no support. This was not a problem and, if anything, provided a motivating challenge. By 1977, over 500 sites were running UNIX.

By making UNIX available to the academic world in this way, AT&T had inadvertently discovered a superb way of marketing the product. As low cost computers became available through the advent of the mini computer (and, later, the microprocessor), academics quickly ported UNIX and moved the rapidly expanding applications from one machine to another. Often, an engineer's first experience of computing was on UNIX systems with applications only available on UNIX. This experience then transferred into industry when the engineer completed training. AT&T had thus developed a very large sales force promoting its products — without having to pay them! A situation that many marketing and sales groups in other companies would have given their right arms for. Fortunately for AT&T, it had started to licence and protect its intellectual property rights without restricting the flow into the academic world. Again, this was either far sighted or simply common sense, because they had to wait until 1984 and more legislation changes before entering the computer market and starting to reap the financial rewards from UNIX.

The disadvantage of this low key promotion was the appearance of a large number of enhanced variants of UNIX which had improved appeal — at the expense of some compatibility. The issue of compatibility at this point was less of an issue than today.

UNIX was provided with no support and its devotees had to be able to support it and its applications from day one. This self sufficiency meant that it was relatively easy to overcome the slight variations between UNIX implementations. After all, most of the application software was written and maintained by the users who thus had total control over its destiny. This is not the case for commercial software, where hard economic factors make the decision for or against porting an application between systems.

With the advent of microprocessors like the Motorola MC68000 family, the Intel 8086 and the Zilog Z8000, and the ability to produce mini computer performance and facilities with low cost silicon, UNIX found itself a low cost hardware platform. During the late 1970s and early 1980s, many UNIX systems appeared using one of three UNIX variants.

XENIX was a UNIX clone produced by Microsoft in 1979 and ported to all three of the above processors. It faded into the background with the advent of MS-DOS, albeit temporarily. Several of the AT&T variants were combined into System III, which, with the addition of several features, was later to become System V. The third variant came from work carried at out at Berkeley (University of California), which produced the BSD versions destined to became a standard for the Digital Equipment Company's VAX computers and throughout the academic world.

Of the three versions, AT&T were the first to announce that they would maintain upward compatibility and start the lengthy process of defining standards for the development of future versions. This development has culminated in AT&T System V release 4, which has effectively brought the System V, XENIX and BSD UNIX environments together.

What distinguishes UNIX from other operating systems is its wealth of application software and its determination to keep the user away from the physical system resources. There are many compilers, editors, text processors, compiler construction aids and communication packages supplied with the basic release. In addition, packages from complete CAD and system modelling to integrated business office suites are available.

The problem with UNIX was that it was quite an expensive operating system to buy. The hardware in many cases was specific to a manufacturer and this restricted the use of UNIX. What was needed was an alternative source of UNIX. With the advent of Linux, this is exactly what happened.

Inside Linux

The key to understanding Linux as an operating system is to understand UNIX and then to grasp how much the operating system protects the user from the hardware it is running on. It is very difficult to know exactly where the memory is in the system, what a disk drive is called and other such information. Many facets of the Linux environment are logical in nature, in that they can be seen and used by the user — but their actual location, structure and functionality is hidden. If a user wants to run a 20 Mbyte program on a system, UNIX will use its virtual memory capability to make the machine behave logically like one with enough memory — even though the system may only have 4 Mbytes of RAM installed. The user can access data files without knowing if they are stored on a floppy or a hard disk — or even on another machine many miles away and connected via a network. UNIX uses its facilities to present a logical picture to the user while hiding the more physical aspects from view.

The Linux file system

Linux like UNIX has a hierarchical filing system which contains all the data files, programs, commands and special files that allow access to the physical computer system. The files are usually grouped into directories and subdirectories. The file system starts with a root directory and divides it into subdirectories. At each level, there can be subdirectories that continue the file system into further levels and files that contain data. A directory can contain both directories and files. If no directories are present, the file system will stop at that level for that path.

A file name describes its location in the hierarchy by the path taken to locate it, starting at the top and working down. This type of structure is frequently referred to as a tree structure which, if turned upside down, resembles a tree by starting at a single root directory — the trunk — and branching out.

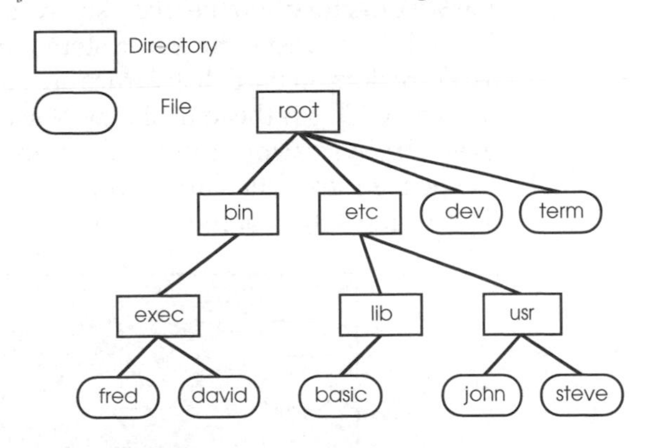

The Linux file system

The full name, or path name, for the file *steve* located at the bottom of the tree would be */etc/usr/steve*. The / character at the beginning is the symbol used for the starting point and is known as root or the root directory. Subsequent use within the path name indicates that the preceding file name is a directory and that the path has followed down that route. The / character is in the opposite direction to that used in MS-DOS systems: a tongue in cheek way to remember which slash to use is that MS-DOS is backward compared with Linux — and thus its slash character points backward.

The system revolves around its file structure and all physical resources are also accessed as files. Even commands exist as files. The organisation is similar to that used within MS-DOS — but the original idea came from UNIX, and not the other way around. One important difference is that with MS-DOS, the top of the structure is always referred to by the name of the hard disk or storage medium. Accessing an MS-DOS root directory C:\ immediately tells you that drive C holds the data. Similarly, A:\ and B:\ normally refer to floppy disks. With UNIX, such direct references to hardware do not exist. A directory is simply present and rarely gives any clues as to its physical location or nature. It may be a floppy disk, a hard disk or a disk on another system that is connected via a network.

All Linux files are typically one of four types, although it can be extremely difficult to know which type they are without referring to the system documentation. A *regular* file can contain any kind of data and is not restricted in size. A *special* file represents a physical I/O device, such as a terminal. *Directories* are files that hold lists of files rather than actual data and *named pipes* are similar to regular files but restricted in size.

The physical file system

The physical file system consists of mass storage devices, such as floppy and hard disks, which are allocated to parts of the logical file system. The logical file system (previously described) can be implemented on a system with two hard disks by allocating the *bin* directory and the filing subsystem below it to hard disk no. 2 — while the rest of the file system is allocated to hard disk no. 1. To store data on hard disk 2, files are created somewhere in the *bin* directory. This is the logical way of accessing mass storage. However, all physical input and output can be accessed by sending data to special files which are normally located in the */dev* directory. This organisation of files is shown.

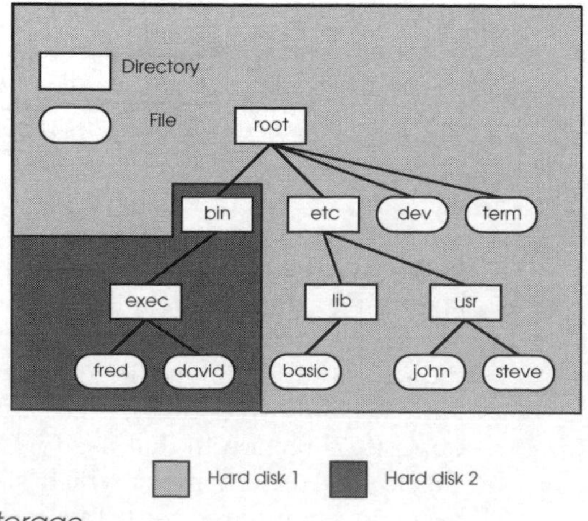

The file system and physical storage

This can create a great deal of confusion: one method of sending data to a hard disk is by allocating it to part of the logical file system and simply creating data files. The second method involves sending the data directly to the special */dev* file that represents the physical disk drive — which itself exists in the logical file system!

This conflict can be explained by an analogy using bookcases. A library contains many bookcases where many books are stored. The library represents the logical file system and the bookcases the physical mass storage. Books represent the data files. Data can be stored within the file system by putting books into the bookcases. Books can be grouped by subject on shelves within the bookcases — these represent directories and subdirectories. When used normally, the bookcases are effectively transparent and the books are located or stored depending on their subject matter. However, there may be times when more storage is needed or new subjects created and whole bookcases are moved or cleared. In these cases, the books are referred to using the bookcase as the reference — rather than subject matter.

The same can occur within Linux. Normally, access is via the file system, but there are times when it is easier to access the data as complete physical units rather than lots of files and directories. Hard disk copying and the allocation of part of the logical file system to a floppy disk are two examples of when access via the special */dev* file is used. Needless to say, accessing hard disks directly without using the file system can be extremely dangerous: the data is simply accessed by block numbers without any reference to the type of data that it contains. It is all too easy to destroy the file system and the information it contains. Another important difference between the access methods is that direct access can be performed at any time and with the mass storage in any state. To access data via the logical file system, data structures must be present to control the file structure. If these are not present, logical access is impossible.

Building the file system

When a Linux system is powered up, its system software boots the Linux kernel into existence. One of the first jobs performed is the allocation of mass storage to the logical file system. This process is called mounting and its reverse, the de-allocation of mass storage, is called unmounting. The *mount* command specifies the special file which represents the physical storage and allocates it to a target directory. When *mount* is complete, the file system on the physical storage media has been added to the logical file system. If the disk does not have a filing system, i.e. the data control structures previously mentioned do not exist, the disk cannot be successfully mounted.

The *mount* and *umount* commands can be used to access removable media, such as floppy disks, via the logical file system.

The disk is mounted, the data accessed as needed and the disk unmounted before physically removing it. All that is needed for this access to take place is the name of the special device file and the target directory. The target directory normally used is */mnt* but the special device file name varies from system to system. The *mount* facility is not normally available to end users for reasons that will become apparent later in this chapter.

The file system

Files are stored by allocating sufficient blocks of storage to contain all the data they contain. The minimum amount of storage that can be allocated is determined by the block size, which can range from 512 bytes to 8 kbytes in more recent systems. The larger block size reduces the amount of control data that is needed — but can increase the storage wastage. A file with 1,025 bytes would need two 1,024 byte blocks to contain it, leaving 1,023 bytes allocated and therefore not accessible to store other files. End of file markers indicate where the file actually ends within a block. Blocks are controlled and allocated by a superblock, which contains an inode allocated to each file, directory, subdirectory or special file. The inode describes the file and where it is located.

di_mod	File type, flags, access permission
di_nlink	Number of inode directory references
di_uid	File owner user id
di_gid	File owner group id
di_size	File size
di_addr • • • • • • • di_addr	13 address fields for data block allocation
di_atim	Last time data was read
di_mtime	Last time data was modified
di_ctime	Last time inode was modified

The inode structure

Using the library and book analogy, the *superblock* represents the library catalogue which is used to determine the size and location of each book. Each book has an entry — an *inode* — within the catalogue.

The example *inode* below, which is taken from a Motorola System V/68 computer, contains information describing the file type, status flags and access permissions (read, write and execute) for the three classifications of users that may need the file: the owner who created the file originally, any member of the owner's group and, finally, anyone else. The owner and groups are identified by their identity numbers, which are included in the *inode*. The total file size is followed by 13 address fields, which point to the blocks that have been used to store the file data. The first ten point directly to a block, while the other three point indirectly to other blocks to effectively increase the number of blocks that can be allocated and ultimately the file size. This concept of direct and indirect pointers is analogous to a library catalogue system: the *inode* represents the reference card for each book or file. It would have sufficient space to describe exactly where the book was located, but if the entry referred to a collection, the original card may not be able to describe all the books and overflow cards would be needed. The *inode* uses indirect addresses to point to other data structures and solve the overflow problem.

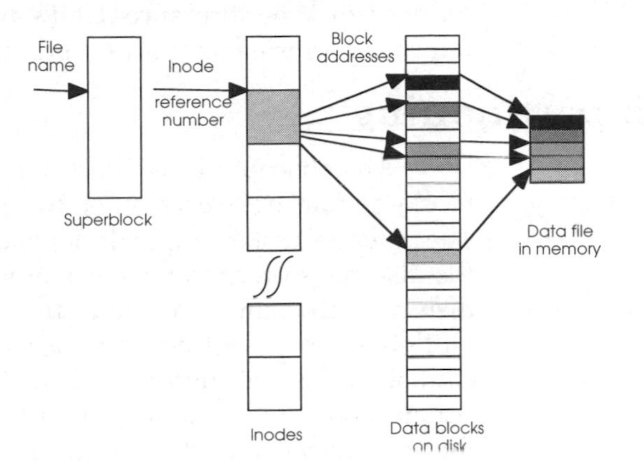

File access mechanism

Why go to these lengths when all that is needed is the location of the starting block and storage of the data in consecutive blocks? This method reduces the amount of data needed to locate a complete file, irrespective of the number of blocks the file uses. However, it does rely on the blocks being available in *contiguous* groups, where the blocks are consecutively ordered. This does not cause a problem when the operating system is first used, and all the files are usually stored in sequence, but as files are created and deleted, the free blocks become fragmented and intermingled with existing files. Such a fragmented disk may have 20 Mbytes of storage free, but would be unable to create files greater than the largest contiguous number of blocks — which could be 10 or 20 times smaller. There is little more frustrating than being told there is insufficient storage available when the same system reports that there are many megabytes free. Linux is more efficient in using the mass storage — at the expense of a more complicated directory control structure. For most users, this complexity is hidden from view and is not relevant to their use of the file system.

So what actually happens when a user wants a file or executes a command? In both cases, the mechanism is very similar. The operating system takes the file or command name and looks it up within the *superblock* to discover the *inode* reference number. This is used to locate the *inode* itself and check the access permissions before allowing the process to continue. If permission is granted, the *inode* block addresses are used to locate the data blocks stored on hard disk. These blocks are put into memory to reconstitute the file or command program. If the data file represents a command, control is then passed to it, and the command executed.

The time taken to perform file access is inevitably dependant on the speed of the hard disk and the time it takes to access each individual block. If the blocks are consecutive or close to each other, the total access time is much quicker than if they are dispersed throughout the disk. Linux also uses mass storage as a replacement for system memory by using memory management techniques and its system response is therefore highly dependant on hard disk performance. UNIX uses two techniques to help improve performance: partitioning and data caching.

Disk partitioning

The concept of disk partitioning is simple: the closer the blocks of data are to each other, the quicker they can be accessed. The potential distance apart is dependant on the number of blocks the disk can store, and thus its storage capacity. Given two hard disks with the same access time, the drive with the largest storage will give the slowest performance, on average. The principle is similar to that encountered when shopping in a small or large supermarket. It takes longer to walk around the larger shop than the smaller one to fetch the same goods.

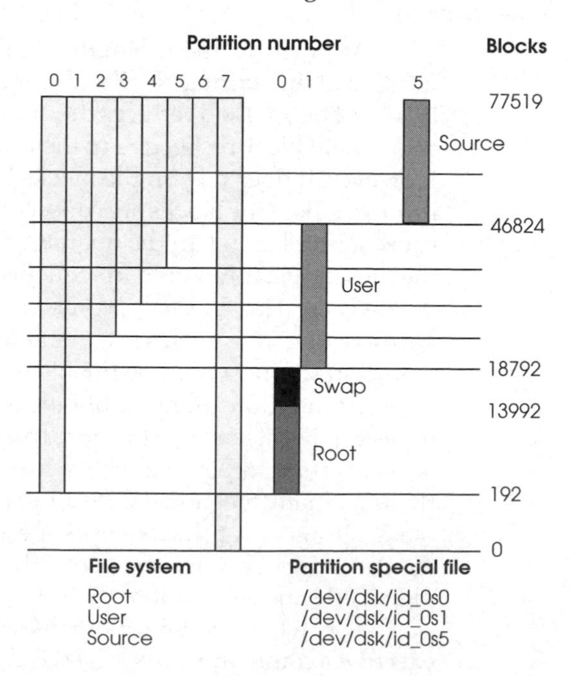

Example Linux partitioning

Linux has the option of partitioning large hard disks so the system sees them as a set of smaller disks. This greatly reduces the amount of searching required and considerably improves overall access times. Each partition (or slice, as it sometimes called) is created by allocating a consecutive number of blocks to it. The partition is treated exactly as if it is a separate mass storage device and can be formatted, have a file system installed and mounted, if required. The partitions can be arranged so they either overlap or are totally separate. In both cases, an installed file system cannot exceed the partition size (i.e. the number of blocks allocated to it) and the lower boundaries of the file system and partition are the same. With non-overlapped partitions, the file system cannot be changed so it overlaps and destroys the data of an adjacent partition. With an overlapped arrangement, this is possible. Changing partition dimensions requires, at best, the reinstallation of the operating system or other software and, at worst, may need them to be completely rebuilt. Fortunately, this only usually concerns the system administrator who looks after the system. Users do not need to worry about these potential problems.

The Motorola System V/68 implementation uses such techniques as shown. The standard hard disk has 77,519 physical 512 byte blocks, which are allocated to 8 overlapping partitions. The whole disk can be accessed using partition 7, the lower 192 blocks of which are reserved for the boot software, which starts UNIX when the system is powered on. Partition 0 has root, the main file system, installed in blocks 192 to 13,991. Blocks 13,992 to 18,791 are used as a swap area for the virtual memory implementation and do not have a file system as such. Partition 1 is used to implement a User file system as far as block 46,823. This could not have been implemented using partitions 2, 3 or 4 without creating a gap — and effectively losing storage space. A third file system, Source, is implemented in partition 5 to use the remaining blocks for data storage.

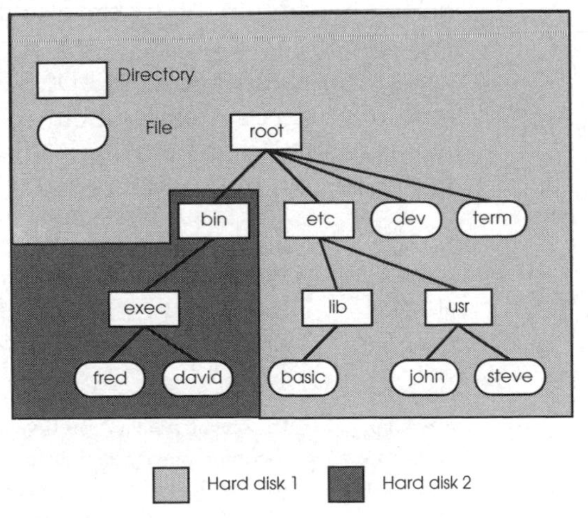

The file system and physical storage

Partitioning provides several other advantages. It allows partitions to be used exclusively by Linux or another operating system, such as MS-DOS, and it reduces the amount of data backup needed to maintain a system's integrity. Most Linux implementations running on an IBM PC allocate partitions to either MS-DOS or Linux and the sizes of these partitions are usually decided when the Linux software is first installed. Some implementations use the same idea, but create large MS-DOS files, which are used as UNIX partitions. In both cases, partitioning effectively divides the single physical hard disk into several smaller logical ones and allows the relatively easy transfer from Linux to MS-DOS, and vice versa. The same principles are also used if Linux is running on an Apple PowerMAC as well.

Most tape backup systems access hard disks directly and not through the file system, so whole disks can be quickly backed up onto tape. In practice, it is common for only parts of the file system to require backing up, such as user files and data, and this is more efficient if the backup process is restricted to these specific parts of the file system. This is easily done using partitions which are used by different parts of the file system. To back up specific parts, the special */dev* file for that partition is used. With the file structure shown below, copying partition 2 to tape would back up all the files and subdirectories in the */root/bin* directory. Copying partition 1 would copy everything excluding the */root/bin* directory.

The Linux disk partitioning

The Linux operating system uses partitions to allow it to co-exist with MS-DOS files and disks as used on IBM PCs. Any IBM PC disk can be partitioned using the MS-DOS FDISK command to create separate partitions. These partitions can then be assigned to Linux and thus support both MS-DOS (and Windows) as well as Linux. The partition naming follows a simple syntax as shown below. In addition, Linux can also directly read and write to MS-DOS disks.

This ability allows MS-DOS disks and thus files to co-exist within a Linux system without the need for special utilities to mount MS-DOS hard and floppy disks and then transfer files from MS-DOS and Linux and vice versa.

Name	Description
/dev/fd0	The first floppy disk drive (A:).
/dev/fd1	The second floppy disk drive (B:).
/dev/hda	The ***whole*** first disk drive (IDE or BIOS compatible disk drive, e.g. ESDI, ST506 and so on).
/dev/hda1	The first primary partition on the first drive.
/dev/hda2	The second primary partition on the first drive.
/dev/hda3	The third primary partition on the first drive.
/dev/hda4	The fourth primary partition on the first drive.
/dev/hdb	The ***whole*** second disk drive.
/dev/hdb1	The first primary partition on the second drive.
/dev/hdb2	The second primary partition on the second drive.
/dev/hdb3	The third primary partition on the second drive.
/dev/hdb4	The fourth primary partition on the second drive.
/dev/hdc	The ***whole*** third disk drive.
/dev/hdc1	The first primary partition on the third drive.
/dev/hdc2	The second primary partition on the third drive.

/dev/hdc3	The third primary partition on the third drive.
/dev/hdc4	The fourth primary partition on the third drive.
/dev/hdd	The ***whole*** fourth disk drive.
/dev/hdd1	The first primary partition on the fourth drive.
/dev/hdd2	The second primary partition on the fourth drive.
/dev/hdd3	The third primary partition on the fourth drive.
/dev/hdd4	The fourth primary partition on the fourth drive.
/dev/sda	The ***whole*** first disk drive (LUN 0) on the first SCSI controller.
/dev/sda1	The first primary partition on the first drive.

The /proc *file system*

Linux has an additional special file system called */proc*. This is not a file system in the true sense of the term but a simple method of getting information about the system using the normal tools and utilities. As will be shown, this is not always the case and there is at least one special utility commonly used with this file system. It is useful in making sure that all the drivers and other system components you expected to be installed are actually there. To access the */proc* file system, it must be built into the kernel. Most, if not all, standard Linux kernels do this to provide debugging information at the very least.

Data caching

One method of increasing the speed of disk access is to keep copies of the most recently used data in memory so it can be fetched without having to keep accessing the slower electro-mechanical disk. The first time the data is needed, it is read from disk and is copied into the cache memory. The next time this data is required, it comes directly from cache memory — without using the disk. This access can be up to 1,000 times faster — which greatly improves system performance. The amount of improvement depends on the amount of cache memory present and the quantity of data needed from disk. If cache memory exceeds the required amount of data, the maximum performance improvement is gained — all the data is read once and can be completely stored in cache memory. If the amount of data is larger than the amount of cache memory, that the actual disk has been updated. The system frequently caches the new data — so the *only* copy is in cache memory. As this memory is volatile, if the machine is switched off the data is lost. If this information also includes superblock and inode changes, the file system will have been corrupted and, at best, parts of it will have been destroyed. At worst, the whole file system can be lost by switching the power off without executing a power down sequence. Most times, an accidental loss of power will not cause any real damage — but it is playing Russian roulette with the system.

The user can force the system to update the disk by executing the sync command as required. This is a well recommended practice.

Multi-tasking systems

Most operating systems used on PCs today, such as MS-DOS, can only execute one application at a time. This means that only one user can use the computer at any time, with the further limitation that only one application can run at a time. While a spreadsheet is executing, the PC can only wait for commands and data from the keyboard. This is a great waste of computer power because the PC could be executing other programs or applications or, alternately, allow other users to run their software on it. The ability to support multiple users running multiple applications is called multi-user multi-tasking. This is a feature of Linux — and is one of the reasons for its rapid adoption. The multi-tasking techniques are where standard Linux falls down in that they use a time slice mechanism (as explained earlier in this chapter) and this is not real-time. As a result, the initial use of Linux into the embedded market has been restricted because of this and the amount of resources such as memory that it needs to function. This has prompted the development of embedded Linux (eLinux) that will be explained later in this chapter.

Multi-user systems

Given a multi-tasking operating system, it is easy to create a multi-user environment, where several users can share the same computer. This is done by taking the special interface program that provides the command line and prompts, and running multiple copies of it as separate processes. When a user logs into the computer, a copy of the program is automatically started. In the UNIX environment, this is called the shell, and there are several different versions available. The advantages of multi-user systems are obvious — powerful computer systems can be shared between several users, rather than each having a separate system. With a shared system, it can also be easier to control access and data, which may be important for large work groups.

With any multi-user system, it is important to prevent users from corrupting eachothers work, or gaining access to sensitive data. To facilitate this, Linux allocates each user a password protected login name, which uniquely identifies him. Each user is normally allocated his own directory within the file system and can configure his part of the system as needed. Users can be organised into groups and every file within the system is given access permissions controlling which user or group can read, write or execute it. When a file is accessed, the requesting user's identity (or ID) is checked against that of the file. If it matches, the associated permissions are checked against the request. The file may be defined as read only, in which case a request to modify it would not be allowed — even if the request came from the user who created it in the first place. If the user ID does not match, the group IDs are checked. If these match, the group permissions are used to judge the validity of the request. If neither IDs match, a third set of permissions, known as others, are checked as the final part of this process.

These permissions can be changed as required by the system administrator, who must set up the Linux system and control how much or how little access each user has to the system and its facilities. This special user (or superuser) has unlimited access by being able to assume any user and/or group identity. This allows an organised structure to be easily implemented and controlled.

Linux software structure

The software structure used within UNIX is very modular, despite its long development history. It consists of several layers starting with programming languages, command scripts and applications, shells and utilities, which provide the interface software or application the user sees. In most cases, the only difference between the three types of software is their interaction and end product, although the more naive user may not perceive this.

The most common software used is the shell, with its commands and utilities. The shell is used to provide the basic login facilities and system control. It also enables the user to select application software, load it and then transfer control to it. Some of its commands are built into the software but most exist as applications, which are treated by the system in the same way as a database or other specialised application.

Programming languages, such as C and Fortran, and their related tools are also applications but are used to write new software to run on the system.

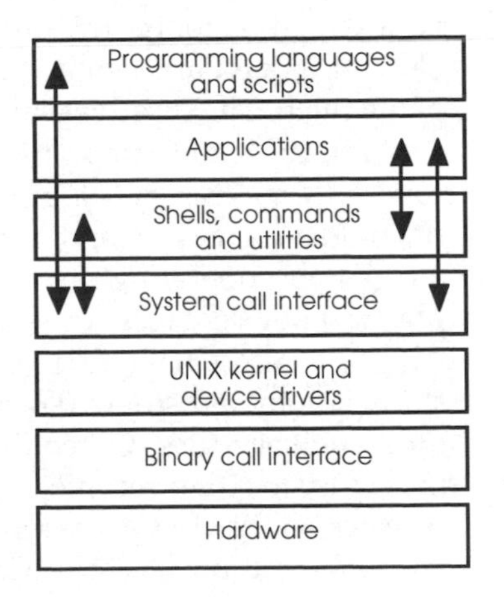

Example Linux software structure

As shown in the diagram, all these layers of software interface with the rest of the operating system via the system call interface. This provides a set of standard commands and services for the software above it and enables their access to the hardware, filing systems, terminals and processor. To read data from a file, a set of system calls are carried out which locate the file, open it and transfer the required data to the application needing it. To find out the time, another call is used, and so on. Having transferred the data, system calls are used to call the UNIX kernel and special software modules, known as device drivers, which actually perform the work required.

Up to this point, the software is essentially working in a standard software environment, where the actual hardware configuration (processor, memory, peripherals and so on) is still hidden. The hardware dependant software which actually drives the hardware is located at the binary call interface.

Of all the layers, the kernel is the heart of the operating system and it is here that the multi-tasking and multi-user aspects of Linux and memory control are implemented. Control is achieved by allocating a finite amount of processor time to each process — an application, a user shell, and so on.

When a process starts executing, it will either be stopped involuntarily (when its CPU time has been consumed) or, if it is waiting for another service to complete, such as a disk access. The next process is loaded and its CPU time starts. The scheduler decides which process executes next, depending on how much CPU time a process needs, although the priority can be changed by the user. It should be noted that often the major difference between UNIX variants and/or implementations is the scheduling algorithm used.

Processes and standard I/O

One problem facing multi-user operating systems is that of access to I/O devices, such as printers and terminals. Probably the easiest method for the operating system is to address all peripherals by unique names and force application software to directly name them. The disadvantage of this for the application is that it could be difficult to find out which peripheral each user is using, especially if a user may login via a number of different terminals. It would be far easier for the application to use some generic name and let the operating system translate these logical names to physical devices. This is the approach taken by Linux.

Processes have three standard files associated with them: stdin, stdout and stderr. These are the default input, output and error message files. Other files or devices can be reassigned to these files to either receive or provide data for the active process. A list of all files in the current directory can be displayed on the terminal by typing ls<cr> because the terminal is automatically assigned to be stdout. To send the same data to a file, ls > filelist<cr> is entered instead. The extra > filelist redirects the output of the ls command. In the first case, ls uses the normal stdout file that is assigned to the terminal and the directory information appears on the terminal screen. In the second example, stdout is temporarily assigned to the file filelist. Nothing is sent to the terminal screen — the data is stored within the file instead.

Data can be fed from one process directly into another using a 'pipe'. A process to display a file on the screen can be piped into another process which converts all lower case characters to upper case. It can then pipe its data output into another process, which pages it automatically before displaying it. This command line can be written as a data file or 'shell script', which provides the commands to the user interface or shell. Shell scripts form a programming language in their own right and allow complex commands to be constructed from simple ones. If the user does not like the particular way a process presents information, a shell script can be written which executes the process, edits and reformats the data and then presents it. There are two commonly used shell interfaces: the standard Bourne shell and the 'C' shell. Many application programs provide their own shell which hide the Linux operating system completely from the user.

Executing commands

After a user has logged onto the system, Linux starts to run the shell program by assigning stdin to the terminal keyboard and stdout to the terminal display. It then prints a prompt and waits for a command to be entered. The shell takes the command line and deciphers it into a command with options, file names and so on. It then looks in a few standard directories to find the right program to execute or, if the full path name is used, goes the directory specified and finds the file. If the file cannot be found, the shell returns an error message.

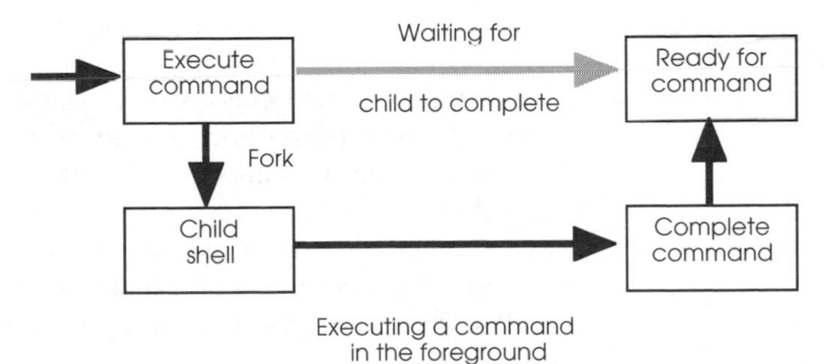

Executing a command in the foreground

Foreground execution

The next stage appears to be a little strange. The shell forks i.e. it creates a duplicate of itself, with all its attributes. The only difference between this new child process and its parent is the value returned from the operating system service that performs the creation. The parent simply waits for the child process to complete. The child starts to run the command and, on its completion, it uses the operating system exit call to tell the kernel of its demise. It then dies. The parent is woken up, it issues another prompt, and the mechanism repeats. Running programs in this way is called executing in the foreground and while the child is performing the required task, no other commands will be executed (although they will still be accepted from the keyboard!).

An alternative way of proceeding is to execute the command in background. This is done by adding an ampersand to the end of the command. The shell forks, as before, with each command as it is entered — but instead of waiting for the child to complete, it prompts for the next command. The example below shows three commands executing in background. As they complete, their output appears on stdout. (To prevent this disrupting a command running in foreground, it is usual to redirect stdout to a file.) Each child is a process and is scheduled by the kernel, as necessary.

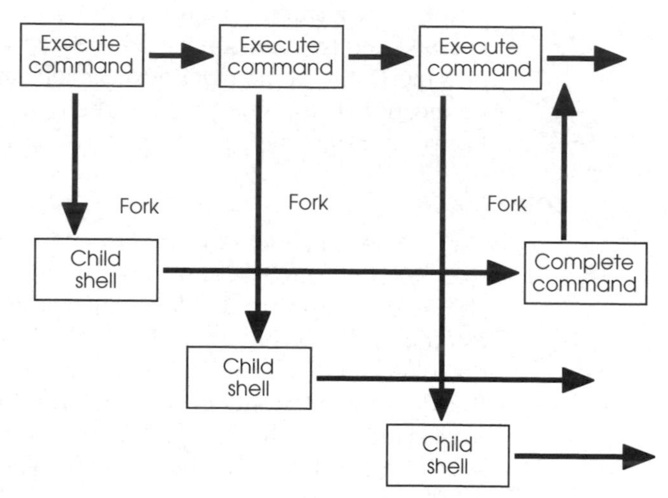

Background execution

Physical I/O

There are two classes of physical I/O devices and their names describe the method used to collect and send data to them. All are accessed by reading and writing to special files, usually located in the */dev* directory, as previously explained. Block devices transfer data in multiples of the system block size. This method is frequently used for disks and tapes. The other method is called character I/O and is used for devices such as terminals and printers.

Block devices use memory buffers and pools to store data. These buffers are searched first to see if the required data is present, rather than going to the slow disk or tape to fetch it. This gives some interesting user characteristics. The first is that frequently used data is often fetched from memory buffers rather than disk, making the system response apparently much better. This effect is easily seen when an application is started a second time. On its first execution, the system had to fetch it in from disk but now the information is somewhere in memory and can be accessed virtually instantaneously. The second characteristic is the performance improvement often seen when the system RAM is increased. With more system memory, more buffers are available and the system spends less time accessing the disk.

It is possible to access some block devices directly, without using memory buffers. These are called raw accesses and are used, for example, when copying disks.

For devices such as terminals and printers, block transfers do not make much sense and here character transfers are used. Although they do not use memory buffers and pools in the same way as block devices, the Linux kernel buffers requests for single characters. This explains why commands must be terminated with a carriage return — to tell the kernel that the command line is now complete and can be sent to the shell.

Memory management

With many users running many different processes, there is a great potential for problems concerning memory allocation and the protection of users from accesses to memory and peripherals that are currently being used by other users of the system. This is especially true when software is being tested which may attempt to access memory that does not exist or is already being used. To solve this sort of problem, Linux depends on a memory management unit (MMU) — hardware which divides all the memory and peripherals into sections which are marked as read only, read or write, operating system accesses only, and so on. If a program tries to write to a read only section, an error occurs which the kernel can handle. In addition, the MMU can translate the memory addresses given by the processor to a different address in memory.

Linux limitations

Unfortunately, Linux is not the utopian operating system for all applications. Its need for memory management, many megabytes of RAM and large mass storage (>40 Mbytes) immediately limits the range of hardware platforms it can successfully run on. Mass storage is not only used for holding file data. It also provides, via its virtual operating system and memory management scheme, overflow storage for applications which are to big to fit in the system RAM all at once. Its use of a non-real-time scheduler, which gives no guarantee as to when a task will complete, further excludes UNIX from many applications.

Through its use of memory management to protect its resources, the simple method of writing an application task which drives a peripheral directly via its physical memory is rendered almost impossible. Physical memory can be accessed via the slow '*/dev/mem*' file technique or by incorporating a shared memory driver, but these techniques are either very slow or restrictive. There is no straightforward method of using or accessing the system interrupts and this forces the user to adopt polling techniques.

In addition, there can be considerable overheads in managing all the look up tables and checking access rights etc. These overheads appear on loading a task, during any memory allocation and when any virtual memory system needs to swap memory blocks out to disk. The required software support is usually performed by an operating system. In the latter case, if system memory is very small compared with the virtual memory size and application, the memory management driver will consume a lot of processing power and time in simply moving data to and from the disk. In extreme cases, this overhead starts to dominate the system — which is working hard but achieving very little. The addition of more memory relieves the need to swap and releases more of the processing power to execute the application.

Finally, the system makes extensive use of disk caching techniques, which use RAM buffers to hold recent data in memory for faster access. This helps to reduce the system performance degradation, particularly when used with a combination of external page swapping and slow mass storage. The system does not write data immediately to disk but stores it in a buffer. If a power failure occurs, the data may only be memory resident and is therefore lost. As this can include directory structures from the superblock, it can corrupt or destroy files, directories or even entire systems! Such systems cannot be treated with the contempt other, more resilient, operating systems can tolerate — Linux systems have to be carefully started, shut down, administered and backed up.

One of the more interesting things about the whole Linux movement is that give the developers a problem and someone somewhere will find a way round the problem and come up with a new version. Given that Linux in its initial form is not ideal for embedded systems, can an embedded real-time version be created that would allow the wealth of Linux software to be executed and reused? The answer has been yes.

eLinux

While UNIX is a wonderful operating system for workstations, desktops and servers, it suffers from several restrictions that prevented it from being used in the embedded system environment. It was large in terms of memory requirements both as main memory and virtual memory where hard disk storage is used to extend the amount of memory the system thinks it has. It makes a lot of assumptions about its environment that may not be true in an embedded system. How many embedded systems do you see with terminals and hard disks? These problems can be overcome but the real issue is the characteristics shown by the kernel. Yes the kernel is multi-tasking and yes it will support multiple users if needed but it does not support real-time operation. As a result, the embedded system community has largely excluded it from consideration. There were also problems with access to code, licensing, royalties that didn't help its cause either.

Linux then appears and as it developed under the various licences that required easy access to the source code, it started to do several things. Firstly, it attracted a large applications base who ended up writing applications and drivers that were freely available. At this point several people started to play with the kernel and its internals. Access was as simple as downloading the code from a web site and the idea of a version to support embedded designers started to come together.

The first thing that has to be remembered that while most real-time applications are embedded systems, not all embedded systems need real-time support. Taking this one step further, many designs actually need the ability to complete critical operations within a certain time frame and need a "fast enough" system and not necessarily a real-time one. This is important as the processing power available to designers today is getting faster and faster and this means that the need to chase every last bit of processing power from the system is no longer needed and instead a faster processor and memory can be used. In this case, the margin between the time taken to complete the operation and when it needs to be completed is so great that there is no longer any need to have a real-time system. This has led to the introduction of non-real-time operating systems with fast processors as a viable alternative for some embedded system designs. So the mantra of "it is embedded and therefore must need a real-time operating system" has been shattered. It is no longer such a universal truth and many designs can use non-real-time operating systems for an embedded design providing the software is designed to cope with the characteristics that the system provides. Indeed the last chapter goes through a design that does exactly that and uses MS-DOS with no multitasking or real-time capabilities as a real-time data logger.

This change coupled with the growth and confidence in Linux-based applications and systems has encouraged the use of Linux in embedded design. This has meant that embedded Linux has taken two development directions: the first concerns adapting the operating system to fit in a constrained hardware system with reduced amounts of memory and peripherals. This is based on stripping out as much of the software that can be done while maintaining the required functionality and this has dramatically reduced the amount of memory that is needed to run the operating system. Add to that support for RAM disks and other solid state memory technology such as flash and an ability to boot up system tasks without the need for a terminal connection and these basic problems are addressed. However, the next issue is not so easy and is concerned with modifying the Linux kernel to provide real-time support.

The standard kernel is multi-tasking and uses a sophisticated fairness scheme that will try and give a fair distribution and sharing of the processing time to the different tasks and threads that are running at the time. It is not pre-emptive in that a currently running task or thread cannot be shut down and replaced as soon as a higher priority task needs to run. So how can Linux be made to run real-time? There are three main methods.

1. Run the standard kernel on a fast enough hardware. This can achieve some impressive figures for task switching (60 microseconds is not uncommon) but it should be remembered that a fast task switch does not mean that there is not some thread or task in the system that might block the kernel and prevent the system from working. This approach requires making some risk assessments over the potential for this to happen. It is fair to say that Linux drivers and code tends to be consistently written following the recommended practices and that this gives a high level of confidence. However, in the same way that MS-DOS can be used for real-time embedded systems by carefully designing how the software is written and runs, the same can be done for Linux without changing the kernel.

   Care needs to be taken however to ensure that there are no hidden problems and in particular tasks and drivers that can hog the execution time and block operations. As drivers are normally supplied in source code, they can be inspected (and modified if necessary) to prevent this. This does require some level of expertise and understanding of how the driver was written, which can be a daunting prospect at first.

2. Replace the standard kernel with a real-time version. This is a little strange in that one of the reasons why Linux is popular is because of the stability of its kernel and the fact that it is freely available. Yet this route advocates replacing the kernel with a real-time version. Several proposals have been made and have included the idea of using a standard RTOS and wrapping it so that it looks like a normal Linux kernel from a software perspective.
3. Enhance the standard kernel with pre-emptive scheduling. In this case, the standard kernel is modified to allow blocking tasks to be pre-empted and removed from execution.

Now the joy of working with Linux is that there are many keen developers ready to meet the challenge but this can lead to many different implementations. It is fair to say that there are several types of embedded Linux implementations available now that use different techniques to provide embedded support.

One method that is used with the TLinux / TAI releases is to use a second kernel that has real-time characteristics and thus the operating system becomes a hybrid with normal Linux software running under the Linux kernel and real-time tasks running under the second real-time kernel. Communication between the Linux and RTOS worlds is performed using shared memory. This does work but is not as elegant as some purists would like in that why have the Linux kernel there in the first place and also it forces the software developer to classify software into the two camps. It is a suitable solution for very time critical applications where the

Linux components are not critical and are restricted to housekeeping and other activities.

The alternative to a second kernel is to make the kernel real-time itself. It turns out that there is real-time support deep down in the kernel e.g. the SCHED_FIFO and SCHED_ calls that support up to 100 priority levels. The idea is that these priority levels are serviced first and then if there are no such tasks that need execution then control can be passed to normal Linux tasks and threads via the standard scheduler. This provides a priority scheme but there are some restrictions. While the calls are present in the interfaces, this does not mean that the implementations actually support or enforce them. In addition, there is no pre-emption which means that a lower priority task or thread can still prevent a higher priority one from pre-empting and gaining control. However, providing the implementation does support these schemes, this can provide an improved real-time Linux environment. Couple it with fast hardware and good interrupt latencies can be obtained.

So ideally, a pre-emptive version of the kernel is needed. It turns out that the standard SMP (Symmetric Multi Processing) version of Linux does just that and by modifying it slightly to support a single processor, it can become a transparent priority-based pre-emptive eLinux kernel. This is the approach that Monta Vista has taken with its eLinux support.

In summary, the restrictions that prevented eLinux from becoming a mainstream embedded RTOS have by and large been removed and eLinux is poised to become a dominant player in the RTOS market.

8 Writing software for embedded systems

There are several different ways of writing code for embedded systems depending on the complexity of the system and the amount of time and money that can be spent. For example, developing software for ready-built hardware is generally easier than for discrete designs. Not only are hardware problems removed — or at least they should have been — but there is more software support available to overcome the obstacles of downloading and debugging code. Many ready-built designs provide libraries and additional software support which can dramatically cut the development time.

The traditional method of writing code has centred on a two pronged approach based on the use of microprocessor emulation. The emulator would be used by the hardware designer to help debug the board before allowing the software engineer access to the prototype hardware. The software engineer would develop his code on a PC, workstation or development system, and then use the emulator as a window into the system. With the emulator, it would be possible to download code, and trace and debug it.

This approach can still be used but the ever increasing cost of emulation and the restrictions it can place on hardware design, such as timing and the physical location of the CPU and its signals, coupled with the fact that ready-built boards are proven designs, prompted the development of alternative techniques which did not need emulation. Provided a way could be found to download code and debug it on the target board, the emulator could be dispensed with. The initial solution was the addition and development of the resident onboard debugger. This has been developed into other areas and includes the development of C source and RTOS aware software simulators that can simulate both software and hardware on a powerful workstation. However, there is more to writing software for microprocessor-based hardware than simply compiling code and downloading it. Debugging software is covered in the next chapter.

The compilation process

When using a high level language compiler with an IBM PC or UNIX system, it is all too easy to forget all the stages that are encountered when source code is compiled into an executable file. Not only is a suitable compiler needed, but the appropriate run-time libraries and linking loader to combine all the modules are also required. The problem is that these may be well integrated for the native system, PC or workstation, but this may not be the case for a VMEbus system, where the hardware configuration may well be unique. Such cross-compilation methods, where software for another processor or target is generated on a different machine, are attractive if a suitable PC or workstation is available, but can require work to create the correct software environment. However, the popularity of this method, as opposed to the more traditional use of a dedicated development system, has increased dramatically. It is now common for operating systems to support cross-compilation directly, rather than leaving the user to piece it all together.

Compiling code

Like many compilers, such as PASCAL or C, the high level language only generates a subset of its facilities and commands from built-in routines and relies on libraries to provide the full range of functions. These libraries use the simple commands to create well-known functions, such as `printf` and `scanf` from the C language, which print and interpret data. As a result, even a simple high level language program involves several stages and requires access to many special files.

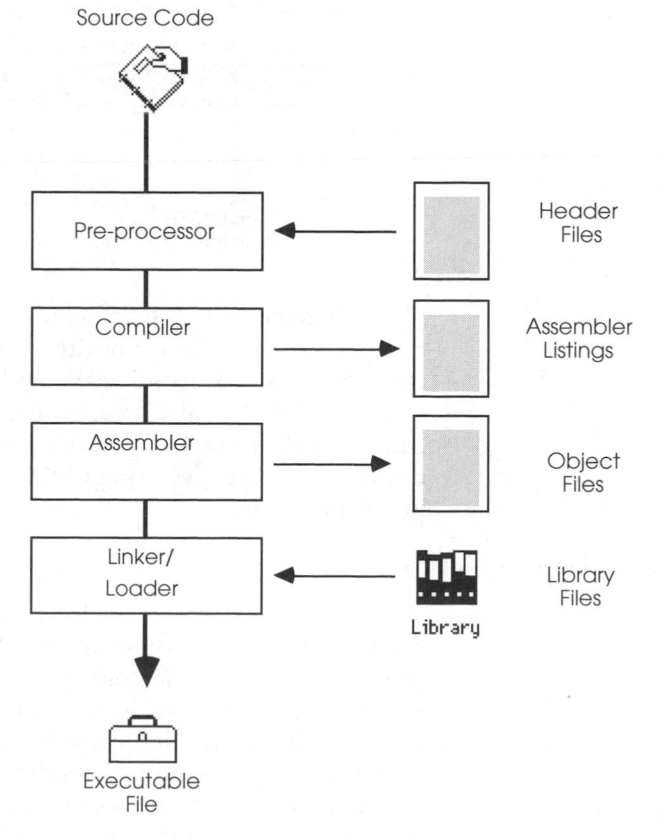

The compilation process

The first stage involves pre-processing the source, where include files are added to it. These files define constants, standard functions and so on. The output of the pre-processor is fed into the compiler, where it produces an assembler file using the native instruction codes for the processor. This file may have references to other software files, called libraries. The assembler file is next assembled and converted into an object file.

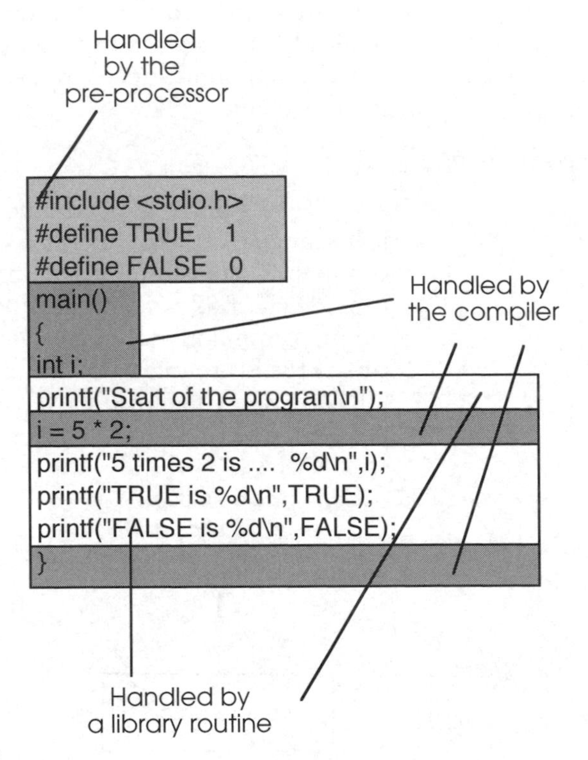

Example C source program

This contains the hexadecimal coding for the instructions, except that memory addresses and file references are not completed; these are resolved by the loader (sometimes known as a linker) that finally creates an executable file. The loader calculates all the memory addresses and takes software routines from library files to supply the standard functions called by the program.

The pre-processor

The pre-processor, as its name suggests, processes the source code before it goes through the compiler. It allows the programmer to define constants, variable types and other information. It also includes other files (*include* files) and combines them into the program source. These tasks can be conditionally performed, depending on the value of constants, and so on. The pre-processor is programmed using one of five basic commands which are inserted into the C source.

#define

#define identifier string

This statement replaces all occurrences of identifier with string. The normal convention is to put the identifier in capital letters so it can easily be recognised as a pre-processor statement. In this example it has been used to define the values of TRUE and FALSE. The main advantage of this is usually the ability to make C code more readable by defining names to be certain values. Statements like `if i == 1` can be replaced in the code by `i == TRUE` which makes their meaning far easier to understand. This technique is also used to define constants, which also make the code easier to understand.

One important point to remember is that the substitution is literal, i.e. the identifier is replaced by the string, irrespective of whether the substitution makes sense. While this is not usually a problem with constants, some programs use *#define* to replace part or complete program lines. If the wrong substitution or definition is made, the resulting program line may cause errors which are not immediately apparent from looking at the program lines. This can also cause problems with different compiler syntax where the definition is valid and accepted by one compiler but rejected by another. This problem can be solved by using the *#define* to define different versions. This is usually done with using the *#if* def variation of the *#define* statement.

It is possible to supply definitions from the C compiler command line direct to the pre-processor, without having to edit the file to change the definitions, and so on. This often allows features for debugging to be switched on or off, as required. Another use for this command is with macros.

#define MACRO() statement

#define MACRO() statement

It is possible to define a macro which is used to condense code either for space reasons or to improve its legibility. The format is *#define*, followed by the macro name and the arguments, within brackets, that it will use in the statement. There should be no space between the name and the brackets. The statement follows the bracket. It is good practice to put each argument within the statement in brackets, to ensure that no problems are encountered with strange arguments.

```
#define SQ(a) ((a)*(a))
#define MAX(i,j) ((i) > ( j) ? (i) : (j))
...
...
x = SQ(56);
z = MAX(x,y);
```

#include

#include "filename"
#include <filename>

This statement takes the contents of a file name and includes it as part of the program code. This is frequently used to define standard constants, variable types, and so on, which may be used either directly in the program source or are expected by any library routines that are used. The difference between the two forms is in the file location. If the file name is in quotation marks, the current directory is searched, followed by the standard directory — usually */usr/include*. If angle brackets are used instead, only the standard directory is searched.

Included files are usually called header files and can themselves have further *#include* statements. The examples show what happens if a header file is not included.

#ifdef

#ifdef identifier
code
#else
code
#endif

This statement conditionally includes code, depending on whether the identifier has been previously defined using a *#define* statement. This is extremely useful for conditionally altering the program, depending on definitions. It is often used to insert machine dependent software into programs. In the example, the source was edited to comment out the CPU_68000 definition so that cache control information was included and a congratulations message printed. If the CPU_68040 definition had been commented out and the CPU_68000 enabled, the reverse would have happened — no cache control software is generated and an update message is printed. Note that *#ifndef* is true when the identifier does not exist and is the opposite of *#ifdef*. The *#else* and its associated code routine can be removed if not needed.

```
#define CPU_68040
/*define CPU_68000 */
#ifdef CPU_68040
/* insert code to switch on caches */
else
/* Do nothing !    */
#endif
#ifndef CPU_68040
printf("Considered upgrading to an MC68040\n");
#else
printf("Congratulations !\n");
#endif
```

#if

```
#if expression
code
#else
code
#endif
```

This statement is similar to the previous *#ifdef,* except that an expression is evaluated to determine whether code is included. The expression can be any valid C expression but should be restricted to constants only. Variables cannot be used because the pre-processor does not know what values they have. This is used to assign values for memory locations and for other uses which require constants to be changed. The total memory for a program can be defined as a constant and, through a series of *#if* statements, other constants can be defined, e.g. the size of data arrays, buffers and so on. This allows the pre-processor to define resources based on a single constant and using different algorithms — without the need to edit all the constants.

Compilation

This is where the processed source code is turned into assembler modules ready for the linker to combine them with the run-time libraries. There are several ways this can be done. The first may be to generate object files directly without going through a separate assembler stage. The usual approach is to create an assembler source listing which is then run through an assembler to create an object file. During this process, it is sometimes possible to switch on automatic code optimisers which examine the code and modify it to produce higher performance.

The standard C compiler for UNIX systems is called *cc* and from its command line, C programs can be pre-processed, compiled, assembled and linked to create an executable file. Its basic options shown below have been used by most compiler writers and therefore are common to most compilers, irrespective of the platform. This procedure can be stopped at any point and options given to each stage, as needed. The options for the compiler are:

-c	Compiles as far as the linking stage and leaves the object file (suffix .o). This is used to compile programs to form part of a library.
-p	Instructs the compiler to produce code which counts the number of times each routine is called. This is the profiling option which is used with the *prof* utility to give statistics on how many subroutines are called. This information is extremely useful for finding out which parts of a program are consuming most of the processing time.

-f Links the object program with the floating point software rather than using a hardware processor. This option is largely historic as many processors now have floating point co-processors. If the system does not, this option performs the calculations in software — but more slowly.

-g Generates symbolic debug information for debuggers like *sdb*. Without this information, the debugger can only work at assembler level and not print variable values and so on. The symbolic information is passed through the compilation process and is stored in the executable file it produces.

-O Switch on the code optimiser to optimise the program and improve its performance. An environment variable OPTIM controls which of two levels is used. If OPTIM=HL (high level), only the higher level code is optimised. If OPTIM=BOTH, the high level and object code optimisers are both invoked. If OPTIM is not set, only the object code optimiser is used. This option cannot be used with the -g flag.

-W*c,args* Passes the arguments *args* to the compiler process indicated by c, where c is one of *p012al* and stands for pre-processor, compiler first pass, compiler second pass, optimiser, assembler and linker, respectively.

-S Compiles the named C programs and generates an assembler language output file only. This file is suffixed .s. This is used to generate source listings and allows the programmer to relate the assembler code generated by the compiler back to the original C source. The standard compiler does not insert the C source into assembler output, it only adds line references.

-E Only runs the pre-processor on the named C programs and sends the result to the standard output.

-P Only runs the pre-processor on the named C programs and puts the result in the corresponding files suffixed .i.

-D*symbol* Defines a symbol to the pre-processor. This mechanism is useful in defining a constant which is then evaluated by the pre-processor, without having to edit the original source.

-U*symbol* Undefine symbol to the pre-processor. This is useful in disabling pre-processor statements.

-I*dir* Provides an alternative directory for the pre-processor to find #include files. If the file name is in quotes, the pre-processor searches the current directory first, followed by *dir* and finally the standard directories.

Here is an example C program and the assembler listing it produced on an MC68010-based UNIX system. The assembler code uses M68000 UNIX mnemonics.

```
$cat math.c
main()
{
int a,b,c;
a=2;
b=4;
c=b-a;
b=a-c;
exit();
}
```

```
$cat math.s
file      "math.c"
   data   1
   text
   def    main; val   main; scl   2;    type  044;  endef
   global      main
main:
   ln     1
   def    ~bf;  val   ~;    scl   101;  line  2;    endef
   link.l      %fp,&F%1
#movm.l   &M%1,(4,%sp)
#fmovm    &FPM%1,(FPO%1,%sp)
   def    a;    val   -4+S%1;     scl   1;    type  04;
endef
   def    b;    val   -8+S%1;     scl   1;    type  04;
endef
   def    c;    val   -12+S%1;    scl   1;    type  04;
endef
   ln     4
   mov.l  &2,((S%1-4).w,%fp)
   ln     5
   mov.l  &4,((S%1-8).w,%fp)
   ln     6
   mov.l  ((S%1-8).w,%fp),%d1
  sub.l    ((S%1-4).w,%fp),%d1
   mov.l  %d1,((S%1-12).w,%fp)
   ln     7
   mov.l  ((S%1-4).w,%fp),%d1
  sub.l    ((S%1-12).w,%fp),%d1
   mov.l  %d1,((S%1-8).w,%fp)
   ln     8
   jsr    exit
L%12:
   def    ~ef;  val   ~;    scl   101;  line  9;    endef
   ln     9
#fmovm    (FPO%1,%sp),&FPM%1
#movm.l   (4,%sp),&M%1
   unlk   %fp
   rts
   def    main; val   ~;    scl   -1;   endef
   set    S%1,0
   set    T%1,0
   set    F%1,-16
   set    FPO%1,4
   set    FPM%1,0x0000
   set    M%1,0x0000
   data   1
$
```

as assembler

After the compiler and pre-processor have finished their passes and have generated an assembler source file, the assembler is used to convert this to hexadecimal. The UNIX assembler differs from many other assemblers in that it is not as powerful and does not have a large range of built-in macros and other facilities. It also frequently uses a different op code syntax from that normally used or specified by a processor manufacturer. For example, the Motorola MC68000 *MOVE* instruction becomes *mov* for the UNIX assembler. In some cases, even source and destination operand positions are swapped and some instructions are not supported. The assembler has several options:

-o objfile	Puts the assembler output into file *objfile* instead of replacing the input file's .s suffix with .o.
-n	Turns off long/short address optimisation. The default is to optimise and this causes the assembler to use short addressing modes whenever possible. The use of this option is very machine dependent.
-m	Runs the m4 macro pre-processor on the source file.
-V	Writes the assembler's version number on standard error output.

Linking and loading

On their own, object files cannot be executed as the object file generated by the assembler contains the basic program code but is not complete. The linker, or loader as it is also called, takes the object file and searches library files to find the routines it calls. It then calculates all the address references and incorporates any symbolic information. Its final task is to create a file which can be executed. This stage is often referred to as linking or loading. The linker gives the final control to the programmer concerning where sections are located in memory, which routines are used (and from which libraries) and how unresolved references are reconciled.

Symbols, references and relocation

When the compiler encounters a `printf()` or similar statement in a program, it creates an external reference which the linker interprets as a request for a routine from a library. When the linker links the program to the library file, it looks for all the external references and satisfies them by searching either default or user defined libraries. If any of these references cannot be found, an error message appears and the process aborts. This also happens with symbols where data types and variables have been used but not specified. As with references, the use of undefined symbols is not detected until the linker stage, when any unresolved or multiply defined symbols cause an error message. This situation is similar to a partially complete jigsaw, where there are pieces missing which represent the object file produced by the assembler. The linker supplies the missing pieces, fits them and makes sure that the jigsaw is complete.

The linker does not stop there. It also calculates all the addresses which the program needs to jump or branch to. Again, until the linker stage, these addresses are not calculated because the sizes of the library routines are not known and any calculations performed prior to this stage would be incorrect. What is done is to allocate enough storage space to allow the addresses to be inserted. Although the linker normally locates the program at $00000000 in memory, it can be instructed to relocate either the whole or part of the code to a different memory location. It also generates symbol tables and maps which can be used for debugging.

As can be seen, the linker stage is not only complicated but can also be extremely complex. For most compilations, the defaults used by the compiler are more than adequate.

ld linker/loader

As explained earlier, an object file generated by the assembler contains the basic program code but is not complete and cannot be executed. The command *ld* takes the object file and searches library files to find the routines it calls. It calculates all the address references and incorporates any symbolic information. Its final task is to create a COFF (common object format file) file which can be executed. This stage is often referred to as linking or loading and *ld* is often called the linker or loader. *ld* gives the final control to the programmer concerning where sections are located in memory, which routines are used (and from which libraries) and how unresolved references are reconciled. Normally, three sections are used — *.text* for the actual code, and *.data* and *.bss* for data. Again, there are several options:

-a	Produces an absolute file and gives warnings for undefined references. Relocation information is stripped from the output object file unless the option is given. This is the default if no option is specified.
-e *epsym*	Sets the start address for the output file to *epsym*.
-f *fill*	Sets the default fill pattern for holes within an output section. This is space that has not been used within blocks or between blocks of memory. The argument *fill* is a 2 byte constant.
-l*x*	Searches library libx.a, where *x* contains up to seven characters. By default, libraries are located in */lib* and */usr/lib*. The placement of this option is important because the libraries are searched in the same order as they are encountered on the command line. To ensure that an object file can extract routines from a library, the library must be searched after the file is given to the linker. Common values for *x* are *c*, which searches the standard C library and *m*, which accesses the maths library.
-m	Produces a map or listing of the input/ output sections on the standard output. This is useful when debugging.
-o *outfile*	Produces an output object file called *outfile*. The name of default object file is *a.out*.
-r	Retains relocation entries in the output object file. Relocation entries must be saved if the output file is to become an input file in a subsequent *ld* session.
-s	Strips line number entries and symbol table information from the output file — normally to save space.
-t	Turns off the warning about multiply-defined symbols that are not of the same size.
-u*symname*	Enters *symname* as an undefined symbol in the symbol table.
-x	Does not preserve local symbols in the output symbol table. This option reduces the output file size.

-L*dir*	Changes the library search order so libx.a looks in *dir* before */lib* and */usr/lib*. This option needs to be in front of the -l option to work!
-N	Puts the data section immediately after the text in the output file.
-V	Outputs a message detailing the version of *ld* used.
-VS *num*	Uses *num* as a decimal version stamp to identify the output file produced.

Native versus cross-compilers

With a native compiler, all the associated run-time libraries, default memory locations and loading software are supplied, allowing the software engineer to concentrate on writing software. It is possible to use a native compiler to write software for a target board, provided it is running the same processor. For example, it is possible to use an IBM PC compiler to write code for an embedded 80386 design or an Apple MAC compiler to create code for an M68000 target. The problem is that all the support libraries and so on must be replaced and this can be a considerable amount of work.

This is beginning to change and many compiler suppliers have realised that it is advantageous to provide many different libraries or the ability to support their development through the provision of a library source. For example, the MetroWorks compilers for the MC68000 and PowerPC for the Apple MAC support cross-compilation for Windows, Windows 95 and Windows NT environments as well as embedded systems.

Run-time libraries

The first problem for any embedded design is that of run-time libraries. These provide the full range of functions that the high level language offers and can be split into several different types, depending on the functionality that they offer and the hardware that they use. The problem is that with no such thing as an embedded design, they often require some modification to get them to work.

Processor dependent

The bulk of a typical high level language library simply requires the processor and memory to execute. Mathematical functions, string manipulation, and so on, all use the processor and do not need to communicate with terminals, disk controllers and other peripherals. As a result these libraries normally require no modification. There are some exceptions concerning floating point and instruction sets. Some processors, such as the MC68020 and MC68030, can use an optional floating point co-processor while others, such as the MC68000 and MC68010, cannot. Further complications can arise between processor variants such as the MC68040 family where some have on-chip floating point, while others do not. Running floating point instructions without the hardware support can generate unexpected processor exceptions and cause the system to crash. Instruction sets can also vary, with the later generations of M68000 processors adding new codes to their predecessor's instruction set. To overcome these differences, compilers often have software switches which can be set to select the appropriate run-time to match the processor configuration.

I/O dependent

If a program does not need any I/O at all, it is very easy to move from one machine to another. However, as soon as any I/O is needed, this immediately defines the hardware that the software needs to access. Using a `printf` statement calls the `printf` routine from the appropriate library which, in turn, either drives the hardware directly or calls the operating system to perform the task of printing data to the screen. If the target hardware is different from the native target, then the `printf` routine will need to be rewritten to replace the native version. Any attempt to use the native version will cause a crash because either the hardware or the operating system is different.

System calls

This is a similar problem to that of I/O dependent calls. Typical routines are those which dynamically allocate memory, task control commands, use semaphores, and so on. Again, these need to be replaced with those supported by the target system.

Exit routines

These are often neglected but are essential to any conversion. With many executable files created by compilers, the program is not simply downloaded into memory and the program counter set to the start of the module. Some systems attach a module header to the file which is then used by the operating system to load the file correctly and to preload registers with address pointers to stack and heap memory and so on. Needless to say, these need to be changed or simulated to allow the file to execute on the target. The start-up routine is often not part of a library and is coded directly into the module.

Similar problems can exist with exit routines used to terminate programs. These normally use an `exit()` call which removes the program and frees up the memory. Again, these need to be replaced. Fortunately, the routines are normally located in the run-time library rather than being hard coded.

Writing a library

For example, given that you have an M68000 compiler running on an IBM PC and that the target is an MC68040 VMEbus system, how do you modify or replace the runtime libraries? There are two generic solutions: the first is to change the hardware design so that it looks like the hardware design that is supported by the run-time libraries. This can be quite simple and involve configuring the memory map so that memory is located at the same addresses. If I/O is used, then the peripherals must be the same — so too must their address locations to allow the software to be used without modification. The second technique is to modify the libraries so that they work with the new hardware configuration. This will involve changing the memory map, adding or changing drivers for new or different peripherals and in some cases even porting the software to a new processor or variant.

The techniques used depend on how the run-time libraries have been supplied. In some cases, they are supplied as assembler source modules and these can simply be modified. The module will have three sections: an entry and exit section to allow data to be passed to and from the routine and, sandwiched between them, the actual code that performs the requested command. It is this middle section that is modified.

Other compilers supply object libraries where the routines have already been assembled into object files. These can be very difficult to patch or modify, and in such cases the best approach is to create an alternative library.

Creating a library

The first step is to establish how the compiler passes data to routines and how it expects information to be returned. This information is normally available from the documentation or can be established by generating an assembler listing during the compilation process. In extreme cases, it may be necessary to reverse engineer the procedure using a debugger. A break point is set at the start of the routine and the code examined by hand.

The next problem is concerned with how to tell the compiler that the routine is external and needs to be specially handled. If the routine is an addition and not a replacement for a standard function, this is normally done by declaring the routines to be external when they are defined. To complement this, the routines must each have an external declaration to allow the linker to correctly match the references.

With replacements for standard library functions, the external declaration from within the program source is not needed, but the one within the replacement library routine is. The alternative library is accessed first to supply the new version by setting the library search list used by the linker.

To illustrate these procedures, consider the following PASCAL example. The first piece of source code is written in PASCAL and controls a semaphore in a typical real-time operating system, which is used to synchronise other tasks. The standard PASCAL did not have any run-time support for operating system calls and therefore a library needed to be created to supply these. The data passing mechanism is typical of most high level languages, including C and FORTRAN, and the trap mechanism, using directive numbers and parameter blocks, is also common to most operating systems.

The PASCAL program declares the operating system calls as external procedures by defining them as procedures and marking them as FORWARD. This tells the compiler and linker that they are external references that need to be resolved at the linker stage. As part of the procedure declaration, the data types that are passed to the procedure have also been defined. This is essential to force the compiler to pass the data to the routine — without it, the information will either not be accepted or the routine will misinterpret the information. In the example, four external procedures are declared: `delay`, `wtsem`, `sgsem` and `atsem`. The procedure `delay` takes an integer value while the others pass over a four character string — described as a packed array of char. Their operation is as follows:

delay delays the task by a number of milliseconds.

atsem creates and attaches the task to a semaphore.

wtsem causes the task to wait for a semaphore.

sgsem signals the semaphore.

```
program timer(input,output);
type
         datatype = packed array[1..4] of char;
var
         msecs:integer;
         name :datatype;
         i :integer;

procedure delay( msecs:integer); FORWARD;
procedure wtsem( var name:datatype); FORWARD;
procedure sgsem( var name:datatype); FORWARD;
procedure atsem( var name:datatype); FORWARD;

         begin
                  name:= '1sec';
                  atsem(name);
                  delay(10000);
                  sgsem(name);
                  for i := 1 to 10 do begin;
                                 wtsem(name);
                                 delay(10000);
                                 sgsem(name);
                  end;
         end.
```

PASCAL source for the program 'TIMER'

The program TIMER works in this way. When it starts, it assigns the identity 1sec to the variable name. This is then used to create a semaphore called 1sec using the `atsem` procedure. The task now delays itself for 10000 milliseconds to allow a second task to load itself, attach itself to the semaphore 1sec and wait for its signal. The signal comes from the `sgsem` procedure on the next line. The other task receives the signal, TIMER goes into a loop where it waits for the 1sec semaphore, delays itself for 10000 milliseconds and then signals with the 1sec semaphore. The other task complements this operation by signalling and then waiting, using the 1sec semaphore.

The end result is that the program TIMER effectively controls and synchronises the other task through the use of the semaphore.

The run-time library routines for these procedures were written in MC68000 assembler. Two of the routines have been listed to illustrate how integers and arrays are passed across the stack from a high level language — PASCAL in this case — to the routine. In C, the assembler routines would be declared as functions and the assembler modules added at link time. Again, it should be remembered that this technique is common to most compilers.

```
DELAY    IDNT  1,0
* ++++++++++++++++++++++++++++++++++++++++++++++++++++
* ++++++++++++++++++++++++++++++++++++++++++++++++++++
* ++++ ++++
* ++++ Runtime procedure call for PASCAL ++++
* ++++ ++++
* ++++ Version 1.0 ++++
* ++++ ++++
* ++++ Steve Heath - Motorola Aylesbury ++++
* ++++ ++++
* ++++++++++++++++++++++++++++++++++++++++++++++++++++
* ++++++++++++++++++++++++++++++++++++++++++++++++++++
*
* PASCAL call structure:
*
* procedure delay(msecs:integer);FORWARD
*
* This routine calls the delay directive of the OS
* and delays the task for a number of ms.
* The number is passed directly on the stack
*

         XDEF DELAY

         SECTION 9

DELAY    EQU        *
         MOVE.L     (A7)+,A4    Save return address
         MOVE.L     (A7)+,A0    Load time delay into A0

         MOVE.L     A3,-(A7)    Save A3 for PASCAL
         MOVE.L     A5,-(A7)    Save A5 for PASCAL
         MOVE.L     A6,-(A7)    Save A6 for PASCAL
```

```
EXEC      MOVE.L   #21,D0      Load directive number 21
          TRAP     #1          Execute OS command
          BNE      ERROR       Error handler if problem

POP       MOVE.L   (A7)+,A6    Restore saved values
          MOVE.L   (A7)+,A5    Restore saved values
          MOVE.L   (A7)+,A3    Restore saved values

          JMP      (A4)        Jump back to PASCAL

ERROR     MOVE.L   #14,D0      Load abort directive no.
          TRAP     #1          Abort task

          END
```

Assembler listing for the delay call

The code is divided into four parts: the first three correspond with the entry, execution and exit stages previously mentioned. A fourth part that handles any error conditions has been added.

The routine is identified to the linker as the delay procedure by the XDEF delay statement. The section 9 command instructs the linker to insert this code in the program part of the file. Note how there are no absolute addresses or address references in the source. The actual values are calculated and inserted by the linker during the linking stage.

The next few instructions transfer the data from PASCAL to the assembler routine. The return address is taken from the stack followed by the time delay. These values are stored in registers A4 and A0, respectively. Note that the stack pointer A7 is incremented after the last transfer to effectively remove the passed parameters. These are not left on the stack. The next three instructions save the address registers A3, A5 and A6 onto the stack so that they are preserved. This is necessary to successfully return to PASCAL. If they are corrupted, then the return to PASCAL will either not work or will cause the program to crash at a later point. With some compilers, more registers may need saving and it is a good idea to save all registers if it is not clear which ones must be preserved. With this example, only these three are essential.

The next part of the code loads the directive number into the right register and executes the system call using the TRAP #1 instruction. The directive needs the delay value in A0 and this is loaded earlier from the stack.

If the system call fails, the condition code register is returned with a non-zero setting. This is tested by the BNE ERROR instruction. The error routine simply executes a termination or abort system call to halt the task execution.

The final part of the code restores the three address registers and uses the return address in A4 to return to the PASCAL program. If the procedure was expecting a returned value, this would be placed on the stack using the same technique used to place the data on the stack. A common fault is to use the wrong method or fail to clear the stack of old data.

The next example routine executes the `atsem` directive which creates the semaphore. The assembler code is a little more complex because the name is passed across the stack using a pointer rather than the actual value and, secondly, a special parameter block has to be built to support the system call to the operating system.

```
ATSEM    IDNT     1,0
* ++++++++++++++++++++++++++++++++++++++++++++++++++
* ++++++++++++++++++++++++++++++++++++++++++++++++++
* ++++ ++++
* ++++ Runtime procedure call for PASCAL ++++
* ++++ ++++
* ++++ Version 1.0 ++++
* ++++ ++++
* ++++ Steve Heath - Motorola Aylesbury ++++
* ++++ ++++
* ++++++++++++++++++++++++++++++++++++++++++++++++++
* ++++++++++++++++++++++++++++++++++++++++++++++++++
*
*        PASCAL call structure:
*
*        type
*                 datatype = packed array[1..4] of char
*
*        procedure atsem(var name:datatype);FORWARD
*
*        This routine calls the OS and creates a
*        semaphore. Its name is passed across on the
*        stack using an address pointer.
*

         XDEF ATSEM

         SECTION 9

DELAY    EQU                    *
         MOVE.L   (A7)+,A4      Save return address
         MOVE.L   (A7)+,A0      Get pointer to the name
         LEA      PBL(PC),A1    Load the PBL address
         MOVE.L   (A0),(A1)     Move the name into PBL
         MOVE.L   A3,-(A7)      Save A3 for PASCAL
         MOVE.L   A5,-(A7)      Save A5 for PASCAL
         MOVE.L   A6,-(A7)      Save A6 for PASCAL

EXEC     MOVE.L   #21,D0        Load directive number 21
         LEA      PBL(PC),A0    Load the PBL address
         TRAP     #1            Execute OS command
         BNE      ERROR         Error handler if problem

POP      MOVE.L   (A7)+,A6      Restore saved values
         MOVE.L   (A7)+,A5      Restore saved values
         MOVE.L   (A7)+,A3      Restore saved values

         JMP      (A4)          Jump back to PASCAL
```

```
ERROR    MOVE.L   #14,D0      Load abort directive no.
         TRAP                 #1      Abort task

         SECTION 15

PBL      EQU                  *
         DC.L                 ' '     Create space for
name
         DC.L                 0       Semaphore key
         DC.B                 0       Initial count
         DC.B                 1       Semaphore type

         END
```

Assembler listing for the atsem call

The name is passed via a pointer on the stack. The pointer is fetched and then used to point to the packed array that contains the semaphore name. Normally, each byte is taken in turn by using the pointer and moving it on to the next location until it points to a null character, i.e. hexadecimal 00. Instead of writing a loop to perform this task, a short cut was taken by assuming that the name is always 4 bytes and by transferring the four characters as a single 32 bit long word.

The address of the parameter block PBL is obtained using the PC relative addressing mode. Again, the reason for this is to allow the linker freedom to locate the parameter block wherever it wants to, without the need to specify an absolute address. The address is calculated and transferred to register A1 using the load effective address instruction, LEA.

The parameter block is interesting because it has been put into section 15 as opposed to the code which is located in section 9. Both of these operations are carried out by the appropriate SECTION command. The reason for this is to ensure that the routines work in all target types, irrespective of whether there is a memory management unit present or the code is in ROM. With this compiler and linker, two sections are used for any program: section 9 is used to hold the code while section 15 is used for data. Without the section 15 command, the linker would put the parameter block immediately after the code routine somewhere in section 9. With a target with no memory management, or with it disabled, this would not cause a problem — provided the code was running in RAM. If the memory management declares the program area as read only — standard default for virtually all operating systems — or the code is in ROM, the transfer of the semaphore name would fail as the parameter block was located in read only memory. By forcing it into section 15, the block is located correctly in RAM and will work correctly, whatever the system configuration.

These routines are extremely simple and quick to create. By using a template, it is easy to modify them to create new procedure calls. More sophisticated versions could transfer all the data to build the parameter block rather than just the name, as in these examples. The procedure could even return a completion code back to the PASCAL program, if needed. In addition, register usage in these examples is not very efficient and again could be improved. However, the important point is that the amount of sophistication is dependent on what the software engineer requires.

Device drivers

This technique is not just restricted to creating run-time libraries for operating systems and replacement I/O functions. The same technique can even be used to drive peripherals or access special registers. This method creates a pseudo device driver which allows the high level language access to the lower levels of the hardware, while not going to the extreme of hard coding or in-lining assembler. If the application is moved to a different target, the pseudo device driver is changed and the application relinked with the new version.

Debugger supplied I/O routines

I/O routines which read and write data to serial ports or even sectors to and from disk can be quite time consuming to write. However, such routines already exist in the onboard debugger which is either shipped with a ready built CPU board or can be obtained for them.

```
* Output a character to console
*
* The character to be output is passed to
* this routine on the stack as byte 5 with
* reference to A7.
*
* A TRAP #14 call to the debugger does the actual work
* Tabs are handled separately

putch
 move.b 5(A7),D0  Get char from stack
 cmp #09,D0       Is it tab character?
 beq _tabput      Yes,go to tab routine
 trap #14         Call debugger I/O
 dc.w 1           Output char in D0.B
 rts
```

An example putchar routine for C using debugger I/O

Many suppliers provide a list of basic I/O commands which can be accessed by the appropriate trap calls. The mechanism is very similar to that described in the previous examples: parameter block addresses are loaded into registers, the command number loaded into a data register and a trap instruction executed. The same basic technique template can be used to create replacement I/O libraries which use the debugger rather than an operating system.

Run-time libraries

The example assembler routines simply use the predefined stack mechanisms to transfer data to and from PASCAL. At no point does the routine actually know that the data is coming from a high level language as opposed to an assembler routine — let alone differentiate between C and PASCAL. If a group of high level languages have common transfer mechanisms, it should be possible to share libraries and modules between them, without having to modify them or know how they were generated. Unfortunately, this utopia has not quite been realised, although some standards have been put forward to implement it.

Using alternative libraries

Given that the new libraries have been written, how are they built into the program? This is done by the linker. The program and assembler routines are compiled and assembled into object modules. The object modules are then linked together by the linker to create the final executable program. The new libraries are incorporated using one of two techniques. The actual details vary from linker to linker and will require checking in the documentation.

Linking additional libraries

This is straightforward. The new libraries are simply included on the command line with all the other modules or added to the list of libraries to search.

Linking replacement libraries

The trick here is to use the search order so that the replacement libraries are used first instead of the standard ones. Some linkers allow you to state the search order on the command line or through a command file. Others may need several link passes, where the first pass disables the automatic search and uses the replacement library and the second pass uses the automatic search and standard library to resolve all the other calls.

Using a standard library

The reason that porting software from one environment to another is often complicated and time consuming is the difference in run-time library support. If a common set of system calls were available and only this set was used by the compiler to interface to the operating system, it would be very easy to move software across from one platform to another — all that would be required would be a simple recompilation. In addition, using a common library would take advantage of common knowledge and experience.

If these improvements are so good, why is this not a more common approach? The problem is in defining the set of library calls and interface requirements. While some standards have appeared and are used, such as UNIX System V interface definition (SVID), they cannot provide a complete set for all operating system environments. Other problems can also exist with the interpretation and operation of the library calls. A call may only work with a 32 bit integer and not with an 8 or 16 bit one. Others may rely on undocumented or vaguely specified functions which may vary from one system to another, and so on. Even after taking these considerations into account, the ability to provide some standard library support is a big advantage. With it, a real-time operating system can support SVID calls and thus allow UNIX software to be transferred through recompilation with a minimum of problems.

There have been several attempts to go beyond the SVID type library definitions and provide a system library that truly supports the real-time environment. Both the VMEexec and ORKID specifications tried to implement a real-time library that was kernel independent with the plan of allowing software that used these definitions to be moved from one kernel to another. Changing kernels would allow application software to be reused with different operating system characteristics, without the need to rewrite libraries and so on. The POSIX real-time definitions are another example of this type of approach.

It can be very dangerous to pin too much hope on these types of standards. The first problem is that they are source code definitions and are therefore subject to misinterpretation not only by the user, but also by the compiler, its run-time libraries and the response from the computer system itself. All of these can cause software to exhibit different behaviour. It may work on one machine but not on another. As a result, the use of a standard library does not in itself guarantee that software will work after recompilation and that it will not require major engineering effort to make it do so. What it does do, however, is provide a better base to work from and such work should be encouraged.

Porting kernels

So far, it has been assumed that the operating system or real-time kernel is already running on the target board. While this is sometimes true, it is not always the case. The operating system may not be available for the target board or the hardware may be a custom design.

Board support

One way to solve this problem is to buy the operating system software already configured for a particular board.

Many software suppliers have a list of supported platforms — usually the most popular boards from the top suppliers — where their software has been ported to and is available off the shelf. For many projects, this is a very good way to proceed as it removes one more variable from the development chain. Not only do you have tested hardware, but you also have preconfigured and tested software.

Rebuilding kernels for new configurations

What happens if you cannot use a standard package or if you need to make some modifications? These changes can be made by rebuilding the operating system or kernel. This is not as difficult as it sounds and is more akin to a linking operation, where the various modules that comprise the operating system are linked together to form the final version.

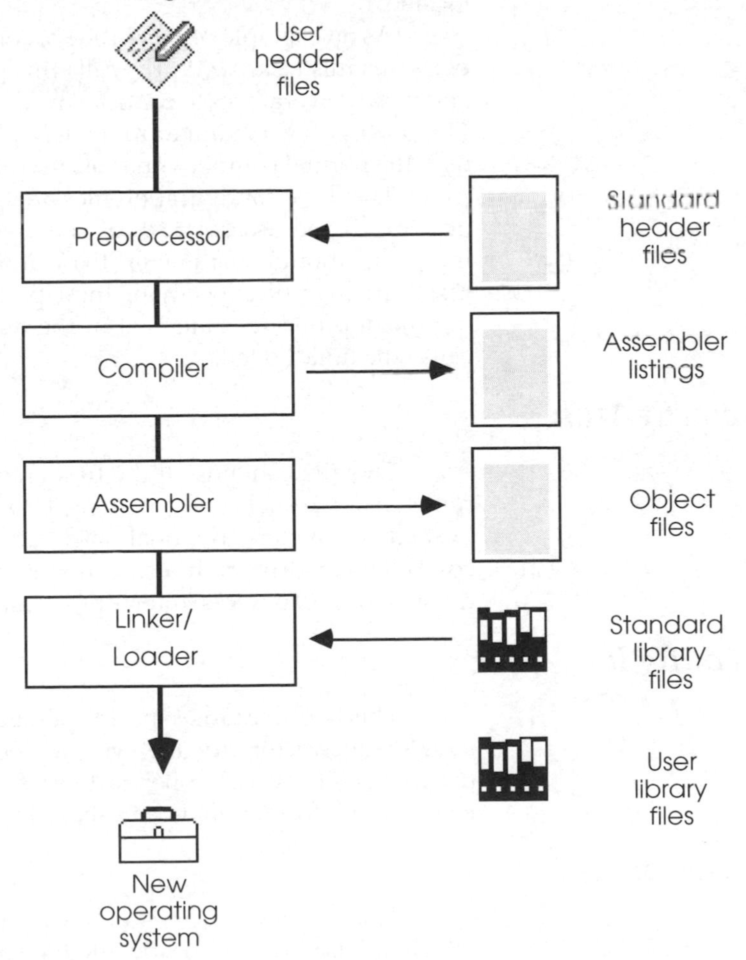

The configuration process

This was not always the case. Early versions of operating systems offered these facilities but took several hours to complete and involved the study of tens of pages of tables to set various switches to include the right modules. Those of you who remember the SYSGEN command within VersaDOS will understand the problem. It did not simply link together modules, it often created them by modifying source code files and patching object files! A long and lengthy process and extremely prone to errors.

This procedure has not gone away but has become quicker and easier to manage. Through the use of reusable modules and high level languages, operating systems are modified and built using a process which is similar to compilation. User created or modified modules are compiled and then linked with the basic operating system to form the final version. The various parameters and software switches are set by various header files — similar to those used with C programs — and these control exactly what is built and where it is located.

As an example of this process, consider how VXWorks performs this task. VXWorks calls this process configuration and it uses several files to control how the kernel is configured. The process is very similar to the UNIX make command and uses the normal compilation tools used to generate tasks.

The three configuration files are called `configAll.h`, `config.h` and `usrConfig.c`. The first two files are header files, which supply parameters to the modules specified in the `usrConfig.c` file. Specifying these parameters without adding the appropriate statement in the `usrConfig.c` file will cause the build to fail.

configAll.h

This file contains all the fundamental options and parameters for kernel configurations, I/O and Networking File System parameters, optional software modules and device controllers or drivers. It also contains cache modes and addresses for I/O devices, interrupt vectors and levels.

config.h

This is where target specific parameters are stored, such as interrupt vectors for the system clock and parity errors, target specific I/O controller addresses, interrupt vectors and levels, and information on any shared memory.

usrConfig.c

This contains a series of software include statements which are used to omit or include the various software modules that the operating system may need. This file would select which Ethernet driver to use or which serial port driver was needed. These modules use parameters from the previous two configuration files within the rebuilding process.

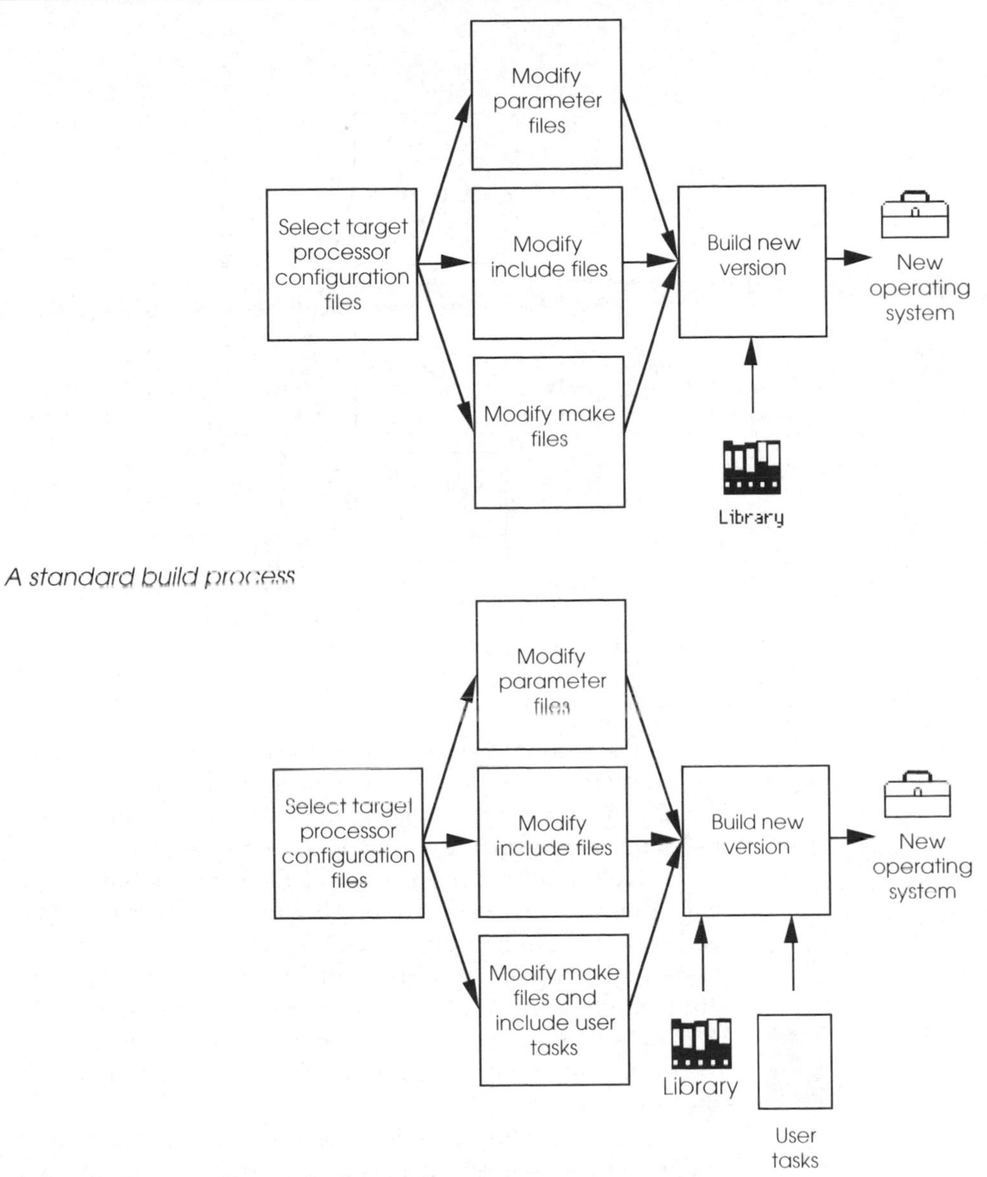

A standard build process

Building tasks into the operating system

Several standard make files are supplied which will create bootable, standalone versions of the operating system, as well as others that will embed tasks into the standalone version. These are used by the compiler and linker to control the building process. All the requisite files are stored in several default library directories but special directories can also be used by adding further options to the make file.

The diagrams show the basic principles involved. A standard build process usually involves modification of the normal files and a simple rebuild. New modules are extracted from the library files or directories as required to build the new version of the operating system.

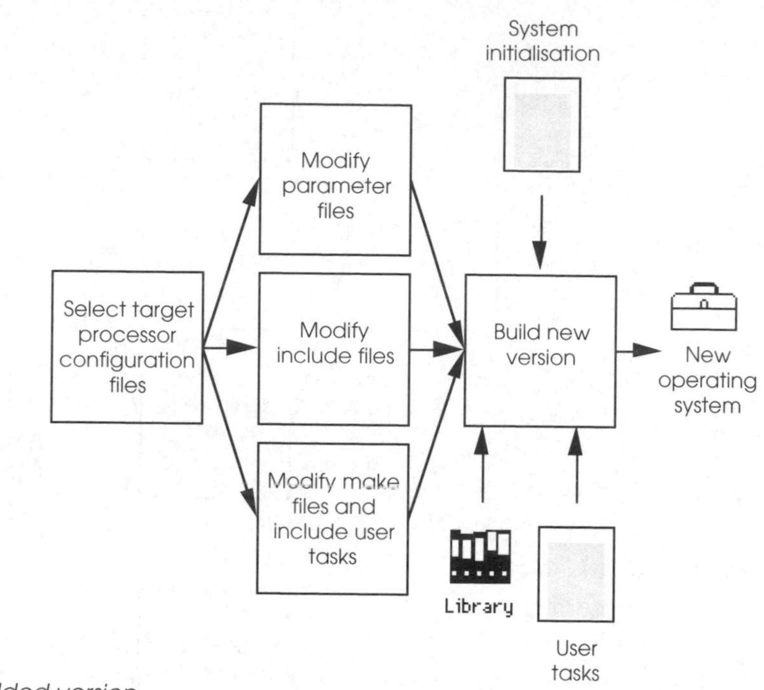

Building an embedded version

The second diagram shows the basic principles behind including user tasks into the operating system. User tasks are usually included at the make or link level and are added to the list of object files that form the operating system.

Note that this process is not the same as building an embedded standalone version. Although the tasks have been embedded, the initialisation code is still the standard one used to start up the operating system only. The tasks may be present, but the operating system is not aware of their existence. This version is often used as an intermediate stage to allow the tasks to be embedded but started under the control of a debugging shell or task.

To create a full embedded version, the user must supply some initialisation routines as well as the tasks themselves. This may involve changing the operating system start point to that of the user's own routine rather than the default operating system one. This user routine must also take care of any variable initialisation, setting up of the vector table, starting tasks in the correct order and allocating memory correctly.

Other options that may need to be included are the addition of any symbol tables for debugging purposes, multi-processor communication and networking support.

pSOSystem+

Rebuilding operating systems is not difficult, once a basic understanding of how the process works and what needs to be changed is reached. The biggest problem faced by the user and by software suppliers is the sheer number of different parameters and drivers that are available today. With hundreds of VMEbus processor boards and I/O modules available, it is becoming extremely difficult to keep up with new product introductions. In an effort to reduce this problem, more user friendly rebuilding systems, such as pSOSystem+, are becoming available which provide a menu driven approach to the problem. The basic options are presented and the user chooses from the menu. The program then automatically generates the changes to the configuration file and builds the new version automatically.

C extensions for embedded systems

Whenever writing in a high level language, there are always times when there is a need to go down to assembler level coding, either for speed reasons or because it is simpler to do so. Accessing memory ports is another example of where this is needed. C in itself does not necessarily support this. Assembler routines can be written as library routines and included at link time — a technique that has been explained and used in this and later chapters. It is possible to define access a specific memory mapped peripheral by defining a pointer and then assigning the peripheral's memory address to the pointer. Vector tables can be created by an array of pointers to functions. While these techniques work in many cases, they are susceptible to failing.

Many compilers provide extensions to the compilers that allow embedded system software designers facilities to help them use low level code. These extensions are compiler specific and may need changing or not be supported if a different compiler is substituted. Many of these extensions are supplied as additional *#pragma* definitions that supply additional information to the compiler on how to handle the routines. These routines may be in C or in assembler and the number and variety will vary considerably. It is worth checking out the compiler documentation to see what it does support.

#pragma interrupt `func2`

This declares the function `func2` as an interrupt function and therefore will ensure that the compiler will save all registers so that the function code does not need to do this. It also instructs the compiler that the return mechanism is different — with a PowerPC instruction a special assembler level instruction has to be used to synchronise and restart the processor.

#pragma pure_function **func2**

This declares that the function `func2` does not use or modify any global or static data and that it is a pure function. This can be used to identify assembler-based routines that configure the processor without accessing any data. This could be to change the cache control, disable or enable interrupts.

#pragma no_side_effects **func2**

This declares that the function `func2` does not modify any global or static data and that it has no side effects. This could be used in preference to the pure_function option to allow access to data to allow an interrupt mask to be changed depending on a global value, for example.

#pragma no_return **func2**

This declares that the function `func2` does not return and therefore the normal preparation of retaining the subroutine return address can be dispensed with. This is used when an exit or abort function is used. Jumps can also be better implemented using this as the stack will be correctly maintained and not filled with return addresses that will never be used. This can cause stack overflows.

#pragma mem_port **int2**

This declares that the variable `int2` is a value of a specific memory address and therefore should be treated accordingly. This is normally used with a definition that defines where the address is.

asm and __asm

The asm and __asm directives — note that the number of underlines varies from compiler to compiler — provide a way to generate assembly code from a C program. Both usually have similar functionality that allows assembler code to be directly inserted in the middle of C without having to use the external routine and linking technique. In most cases, the terms are interchangeable, but beware since this is not always the case. Care must also be taken with them as they break the main standards and enforcing strict compatibility with the compiler can cause them to either be flagged up as an error or simply ignored.

There are two ways of using the asm/__asm directives. The first is a simple way to pass a string to the assembler, an asm string. The second is an advanced method to define an asm macro that in-lines different assembly code sections, depending on the type of arguments given. The examples shown are based on the Diab PowerPC compiler.

asm strings

An asm string can be specified wherever a statement or an external declaration is allowed. It must have exactly one argument, which should be a string constant to be passed to the assembly output. Some optimisations will be turned off when an asm string statement is encountered.

```
int f() { /* returns value at $$address */
        asm(" addis r3,r0,$$address)@ha");
        asm(" lwz r3,r3,$$address@l");
```

This technique is very useful for executing small functions such as enabling and disabling interrupts, flushing caches and other processor level activities. With the code directly inlined into the assembler, it is very quick with little or no overhead.

asm macros

An asm macro definition looks like a function definition in that the body of the function is replaced with one or more assembly code sequences. The compiler chooses one of these sequences depending on which types of arguments are provided when using the asm macro, e.g.

```
asm int busy_wait(char *addr)
        { % reg addr; lab loop;
              addi  r4,r0,1
        loop:                       # label is replaced by
compiler
              lwarx r5,r0,addr  # argument is forced to
register
              cmpi cr0,r5,0
              bne loop
              stwcx. r4,r0,addr
              bae loop
        }

extern char *sem
fn(char *addr) {
              busy_wait(addr); /* wait for semaphore */
              busy_wait(sem) ; /* wait for semaphore */
           }
```

The first part of the source defines the assembler routine that waits for the semaphore or event to change. The second part of the source calls this assembler function twice with the event name as its parameter.

```
        addi  r4,r0,1
   .L11:              # label is replaced by compiler
        lwarx         r5,r0,r31 # argument is forced to
register
        cmpi          cr0,r5,0
        bne           .L11
        stwcx.        r4,r0,r31
        bne           .L11
        addis         r3,r0,sem@ha
        lwz           r3,sem@l(r3)
        addi          r4,r0,1
```

```
     .L12:                # label is replaced by compiler
            lwarx         5,r0,r3      # argument is forced to
register
            cmpl          cr0,r5,0
            bne           .L12
            stwcx.        r4,r0,r3
            bne           .L12
```

Downloading

Having modified libraries, linked modules together and so on, the question arises of how to get the code down to the target board. There are several methods available to do this.

Serial lines

Programs can be downloaded into a target board using a serial comms port and usually an onboard debugger. The first stage is to convert the executable program file into an ASCII format which can easily be transmitted to the target. This is done either by setting a switch on the linker to generate a download format or by using a separate utility. Several formats exist for the download format, depending on which processor family is being used. For Motorola processors, this is called the S-record format because each line commences with an S. The format is very simple and comprises of an S identifier, which describes the record type and addressing capability, followed by the number of bytes in the record and the memory address for the data. The last byte is a checksum for error checking.

```
S0080000612E6F757410
S223400600480EFFFFFFEC42AEFFF00CAE00002710FFF06C0000322D7C00000002FFFC2D27
S22340061F7C00000004FFF8222EFFF892AEFFFC2D41FFF4222EFFFC92AEFFF42D41FFF83A
S21E40063E52AEFFF06000FFC64EB9004006504E5E4E754E4B000000004E71E5
S9030000FC
```

An example S-record file

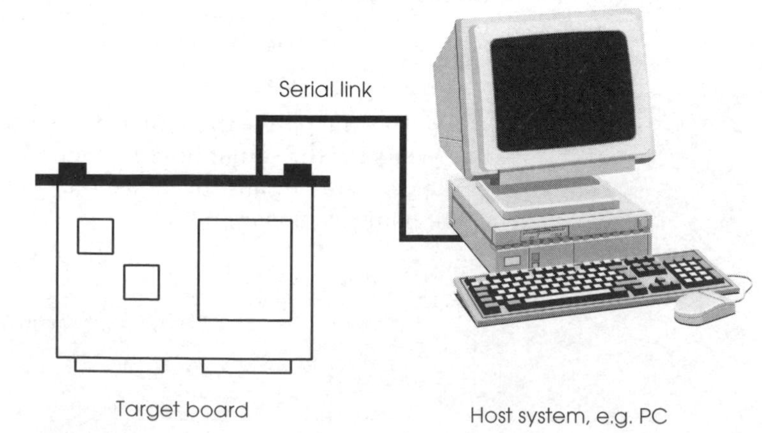

Downloading via a serial link

The host is then connected to the target, invokes the download command on the target debugger and sends the file. The target debugger then converts the ASCII format back into binary and loads at the correct location. Once complete, the target debugger can be used to start the program.

This method is simple but slow. If large programs need to be moved it can take all day — which is not only an efficiency problem but also leads to the practice of patching rather than updating source code. Faced with a three hour download, it is extremely tempting to go in and patch a variable or routine, rather than modify the program source, recompile and download. In practice, this method is only really suitable for small programs.

EPROM and FLASH

An alternative is to burn the program into EPROM, or some other form of non-volatile memory such as FLASH or battery backed-up SRAM, insert the memory chips into the target and start running the code. This can be a lot quicker than downloading via a serial line, provided the link between the development system and the PROM programmer is not a serial link itself!

There are several restrictions with this. The first is that there may not be enough free sockets on the target to accept the ROMs and second, modifications cannot be made to read only memory which means that patching and setting breakpoints will not function. If the compiler does not produce ROMable code or, for some reason, data structures have been included in the code areas, again the software may not run.

There are some solutions to this. The code in the ROMs can be block transferred to RAM before execution. This can either be done using a built-in block move command in the onboard debugger or with a small 4 or 5 line program.

Parallel ports

This is similar to the serial line technique, except that data is transferred in bytes rather than bits using a parallel interface — often a reprogrammed Centronics printer port. While a lot faster, it does require access to parallel ports which tend to be less common than serial ones.

From disk

This is an extremely quick and convenient way of downloading code. If the target is used to develop the code then this is a very easy way of downloading. If the target VMEbus board can be inserted into the development host, the code can often be downloaded directly from disk into the target memory. This technique is covered in more detail later on.

Downloading from disk can even be used with cross-compilation systems, provided the target can read floppy disks. Many target operating systems are file compatible with MS-DOS systems and use the IBM PC as their development host. In such cases, files can be transferred from the PC to the target using floppy disk(s).

Ethernet

For target systems that support networking, it is possible to download and even debug using the Ethernet and TCP/IP as the communications link. This method is very common with development hosts that use UNIX and is used widely in the industry. It does require an Ethernet port though. Typically, the target operating system will have a communications module which supports the TCP/IP protocols and allows it to replace a serial line for software downloading. The advantage is one of far greater transfer rates and ease of use, but it does rely on having this support available within the operating system. VXWorks, VMEexec, and pSOS+ can all cover these types of facilities.

Across a common bus

An ideal way of downloading code would be to go across a data bus such as PCI or VMEbus. This general method has already been briefly explained using an extra memory board to connect ROMs to the bus and transfer data, and the idea of adding the target boards to the host to allow the host to download directly into the target. Some operating systems can already provide this mechanism for certain host configurations. For those that do not, the methods are very simple, provided certain precautions are taken.

The first of these concerns how the operating system sees the target board. Unless restricted or told otherwise, the operating system may automatically use the memory on the target board for its own uses.

This may appear to be exactly what is required as the host can simply download code into this memory. On the other hand, the operating system may use the target memory for its own software and free up memory elsewhere in the system. Even if the memory is free, there is often no guarantee that the operating system will not overwrite the target memory.

To get around this problem, it may be necessary to physically limit the operating system so that it ignores the target memory. This can cause problems with memory management units, which will not allow access to what the operating system thinks is non-existent memory. The solution is either to disable the memory management, or to use an operating system call to get access to the physical memory to access and reserve it.

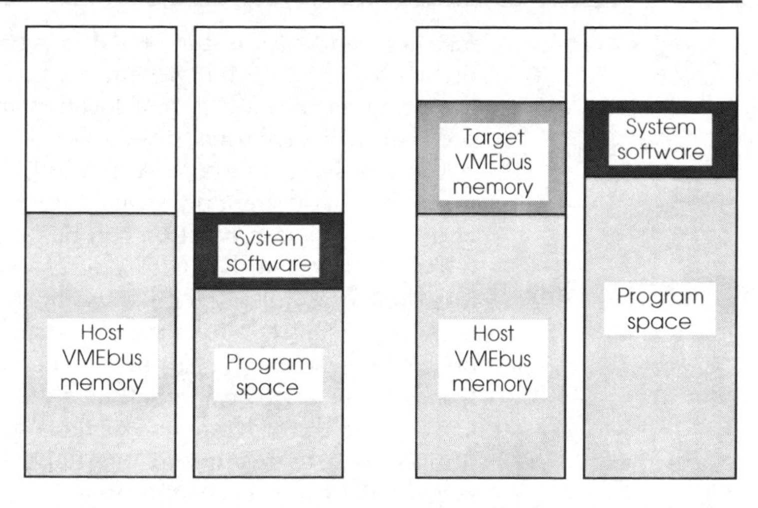

Target memory by the operating system

With real-time-based operating systems, I would recommend disabling the MMU, thus allowing the host to access any physical memory location that the processor generates. Without the MMU, the CPU can access any address that it is instructed to, even if this address is outside those used by the operating system. This is normally not good practice but in this case the benefits justify its use. With UNIX, the best way is to declare the target memory as a shared memory segment or to access it via the device /dev/mem.

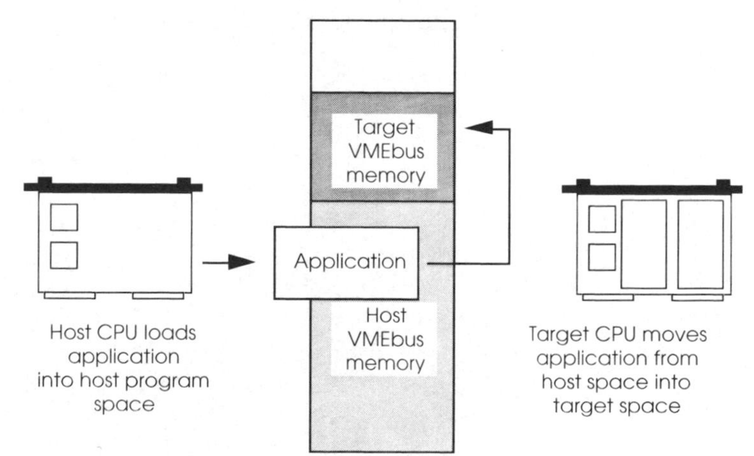

Using the target CPU to move an application to the right memory

There are occasions when even these solutions are not feasible. However, it is still possible to download software using a modification to the technique. The program is loaded into the host memory by the host. The target then moves the code across the VMEbus from the host memory space to its own memory space. The one drawback with this is the problem of memory conflicts. The target VMEbus memory must not conflict with the host VMEbus memory, and so the host must load the program into a different location to that intended. The program will have been linked to the target address, which is not recognised by the host. As a result, the host must load the program at a different physical address to that intended. This address translation function can be performed by an MMU which can translate the logical addresses of the target memory to physical addresses within the host memory space. It is usually possible to obtain the physical address by calling the operating system. This address is then used by the target to transfer the data. The alternative to this is to write a utility to simply load the program image into a specified memory location in the host memory map. After this has been done, the target CPU can transfer the program to its correct memory location.

One word of warning. It is important that there are no conflicting I/O addresses, interrupt levels, resets and so on that can cause a conflict. For example, the reset button on the target should only generate a local reset and not a VMEbus one, so that the downloading system will not see it and immediately start its reset procedure. Similarly, the processor boards should not be set up to respond to the same interrupt level or memory addresses.

This method of downloading is very quick and versatile once it has been set up. Apart from its use in downloading code during developments, the same techniques are also applicable to downloading code in multiprocessor designs.

9 Emulation and debugging techniques

Debugging techniques

The fundamental aim of a debugging methodology is to restrict the introduction of untested software or hardware to a single item. This is good practice and of benefit to the design and implementation of any system, even those that use emulation early on in the design cycle.

It is to prevent this integration of two unknowns that simulation programs to simulate and test software, hardware or both can play a critical part in the development process.

High level language simulation

If software is written in a high level language, it is possible to test large parts of it without the need for the hardware at all. Software that does not need or use I/O or other system dependent facilities can be run and tested on other machines, such as a PC or a engineering workstation. The advantage of this is that it allows a parallel development of the hardware and software and added confidence, when the two parts are integrated, that it will work.

Using this technique, it is possible to simulate I/O using the keyboard as input or another task passing input data to the rest of the modules. Another technique is to use a data table which contains data sequences that are used to test the software.

This method is not without its restrictions. The most common mistake with this method is the use of non-standard libraries which are not supported by the target system compiler or environment. If these libraries are used as part of the code that will be transferred, as opposed to providing a user interface or debugging facility, then the modifications needed to port the code will devalue the benefit of the simulation.

The ideal is when the simulation system is using the same library interface as the target. This can be achieved by using the target system or operating system as the simulation system or using the same set of system calls. Many operating systems support or provide a UNIX compatible library which allows UNIX software to be ported using a simple recompilation. As a result, UNIX systems are often employed in this simulation role. This is an advantage which the POSIX compliant operating system Lynx offers.

This simulation allows logical testing of the software but rarely offers quantitative information unless the simulation environment is very close to that of the target, in terms of hardware and software environments.

Low level simulation

Using another system to simulate parts of the code is all well and good, but what about low level code such as initialisation routines? There are simulation tools available for these routines as well. CPU simulators can simulate a processor, memory system and, in some cases, some peripherals and allow low level assembler code and small HLL programs to be tested without the need for the actual hardware. These tools tend to fall into two categories: the first simulate the programming model and memory system and offer simple debugging tools similar to those found with an onboard debugger. These are inevitably slow, when compared to the real thing, and do not provide timing information or permit different memory configurations to be tested. However, they are very cheap and easy to use and can provide a low cost test bed for individuals within a large software team. There are even shareware simulators for the most common processors such as the one from the University of North Carolina which simulates an MC68000 processor.

```
<D0> =00000000 <D4> =00000000 <A0> =00000000 <A4> =00000000
<D1> =00000000 <D5> =0000abcd <A1> =00000000 <A5> =00000000
<D2> =00000000 <D6> =00000000 <A2> =00000000 <A6> =00000000
<D3> =00000000 <D7> =00000000 <A3> =00000000 <A7> =00000000
trace: on     sstep: on     cycles:    416    <A7'>= 00000f00
         cn tr st rc          T S  INT   XNZVC  <PC> = 00000090
  port1  00 00 82 00  SR = 1010101111011111
--------------------------------------------------
executing a ANDI         instruction at location   58
executing a ANDI         instruction at location   5e
executing a ANDI         instruction at location   62
executing a ANDI_TO_CCR  instruction at location   68
executing a ANDI_TO_SR   instruction at location   6c
executing a OR           instruction at location   70
executing a OR           instruction at location   72
executing a OR           instruction at location   76
executing a ORI          instruction at location   78
executing a ORI          instruction at location   7e
executing a ORI          instruction at location   82
executing a ORI_TO_CCR   instruction at location   88
executing a ORI_TO_SR    instruction at location   8c
TRACE exception occurred at location   8c.
Execution halted
```

Example display from the University of North Carolina 68k simulator

The second category extends the simulation to provide timing information based on the number of clock cycles. Some simulators can even provide information on cache performance, memory usage and so on, which is useful data for making hardware decisions. Different performance memory systems can be exercised using the simulator to provide performance data. This type of information is virtually impossible to obtain without using such tools. These more powerful simulators often require very powerful hosts with large amounts of memory. SDS provide a suite of such tools that can simulate a processor and memory and with some of the integrated processors that are available, even emulate onboard peripherals such as LCD controllers and parallel ports.

Simulation tools are becoming more and more important in providing early experience of and data about a system before the hardware is available. They can be a little impractical due to their performance limitations — one second of processing with a 25 MHz RISC processor taking 2 hours of simulation time was not uncommon a few years ago — but as workstation performance improves, the simulation speed increases. With instruction level simulators it is possible with a top of the range workstation to get simulation speeds of 1 to 2 MHz.

Onboard debugger

The onboard debugger provides a very low level method of debugging software. Usually supplied as a set of EPROMs which are plugged into the board or as a set of software routines that are combined with the applications code, they use a serial connection to communicate with a PC or workstation. They provide several functions: the first is to provide initialisation code for the processor and/or the board which will normally initialise the hardware and allow it to come up into a known state. The second is to supply basic debugging facilities and, in some cases, allow simple access to the board's peripherals. Often included in these facilities is the ability to download code using a serial port or from a floppy disk.

```
>TR

PC=000404 SR=2000 SS=00A00000 US=00000000          X=0
A0=00000000 A1=000004AA A2=00000000 A3=00000000 N=0
A4=00000000 A5=00000000 A6=00000000 A7=00A00000 Z=0
D0=00000001 D1=00000013 D2=00000000 D3=00000000 V=0
D4=00000000 D5=00000000 D6=00000000 D7=00000000 C=0
---------->LEA      $000004AA,A1

>TR

PC=00040A SR=2000 SS=00A00000 US=00000000          X=0
A0=00000000 A1=000004AA A2=00000000 A3=00000000 N=0
A4=00000000 A5=00000000 A6=00000000 A7=00A00000 Z=0
D0=00000001 D1=00000013 D2=00000000 D3=00000000 V=0
D4=00000000 D5=00000000 D6=00000000 D7=00000000 C=0
---------->MOVEQ    #19,D1

>
```

Example display from an onboard M68000 debugger

When the board is powered up, the processor fetches its reset vector from the table stored in EPROM and then starts to initialise the board. The vector table is normally transferred from EPROM into a RAM area to allow it to be modified, if needed. This can be done through hardware, where the EPROM memory address is temporarily altered to be at the correct location for power-on, but is moved elsewhere after the vector table has been copied. Typically, a counter is used to determine a preset number of memory accesses, after which it is assumed that the table has been transferred by the debugger and the EPROM address can safely be changed.

The second method, which relies on processor support, allows the vector table to be moved elsewhere in the memory map. With the later M68000 processors, this can also be done by changing the vector base register which is part of the supervisor programming model.

The debugger usually operates at a very low level and allows basic memory and processor register display and change, setting RAM-based breakpoints and so on. This is normally performed using hexadecimal notation, although some debuggers can provide a simple disassembler function. To get the best out of these systems, it is important that a symbol table is generated when compiling or linking software, which will provide a cross-reference between labels and symbol names and their physical address in memory. In addition, an assembler source listing which shows the assembler code generated for each line of C or other high level language code is invaluable. Without this information it can be very difficult to use the debugger easily. Having said that, it is quite frustrating having to look up references in very large tables and this highlights one of the restrictions with this type of debugger.

While considered very low level and somewhat limited in their use, onboard debuggers are extremely useful in giving confidence that the board is working correctly and working on an embedded system where an emulator may be impractical. However, this ability to access only at a low level can also place severe limitations on what can be debugged.

The first problem concerns the initialisation routines and in particular the processor's vector table. Breakpoints use either a special breakpoint instruction or an illegal instruction to generate a processor exception when the instruction is executed. Program control is then transferred to the debugger which displays the breakpoint and associated information. Similarly, the debugger may use other vectors to drive the serial port that is connected to the terminal.

This vector table may be overwritten by the initialisation routines of the operating system which can replace them with its own set of vectors. The breakpoint can still be set but when it is reached, the operating system will see it instead of the debugger and not pass control back to it. The system will normally crash because it is not expecting to see a breakpoint or an illegal instruction!

To get around this problem, the operating system may need to be either patched so that its initialisation routine writes the debugger vector into the appropriate location or this must be done using the debugger itself. The operating system is single stepped through its initialisation routine and the instruction that overwrites the vector simply skipped over, thus preserving the debugger's vector. Some operating systems can be configured to preserve the debugger's exception vectors, which removes the need to use the debugger to preserve them.

A second issue is that of memory management where there can be a problem with the address translation. Breakpoints will still work but the addresses returned by the debugger will be physical, while those generated by the symbol table will normally be logical. As a result, it can be very difficult to reconcile the physical address information with the logical information.

The onboard debugger provides a simple but sometimes essential way of debugging VMEbus software. For small amounts of code, it is quite capable of providing a method of debugging which is effective, albeit not as efficient as a full blown symbolic level debugger — or as complex or expensive. It is often the only way of finding out about a system which has hung or crashed.

Task level debugging

In many cases, the use of a low level debugger is not very efficient compared with the type of control that may be needed. A low level debugger is fine for setting a breakpoint at the start of a routine but it cannot set them for particular task functions and operations. It is possible to set a breakpoint at the start of the routine that sends a message, but if only a particular message is required, the low level approach will need manual inspection of all messages to isolate the one that is needed — an often daunting and impractical approach!

To solve this problem, most operating systems provide a task level debugger which works at the operating system level. Breakpoints can be set on system circumstances, such as events, messages, interrupt routines and so on, as well as the more normal memory address. In addition, the ability to filter messages and events is often included. Data on the current executing tasks is provided, such as memory usage, current status and a snapshot of the registers.

Symbolic debug

The ability to use high level language instructions, functions and variables instead of the more normal addresses and their contents is known as symbolic debugging. Instead of using an assembler listing to determine the address of the first instruction of a C function and using this to set a breakpoint, the symbolic debugger allows the breakpoint to be set by quoting a line reference or the function name. This interaction is far more efficient than working at the assembler level, although it does not necessarily mean losing the ability to go down to this level if needed.

The reason for this is often due to the way that symbolic debuggers work. In simple terms, they are intelligent front ends for assembler level debuggers, where software performs the automatic look-up and conversion between high level language structures and their respective assembler level addresses and contents.

```
12 int prime,count,iter;
13
14 for (iter = 1;iter<=MAX_ITER;iter++)
15 {
16 count = 0;
17 for(i = 0; i<MAX_PRIME; i++)
18 flags[i] = 1;
19 for(i = 0; i<MAX_PRIME; i++)
20 if(flags[i])
21 {
22 prime = i + i + 3;
23 k = i + prime;
24 while (k < MAX_PRIME)
25 {
26 flags[k] = 0;
27 k += prime;
28 }
29 count++;
```

Source code listing with line references

```
000100AA 7C01 MOVEQ #$1,D6
000100AC 7800 MOVEQ #$0,D4
000100AE 7400 MOVEQ #$0,D2
000100B0 207C 0001 2148 MOVEA.L #$12148,A0 000100B6
11BC 0001 2000 MOVE.B #$1,($0,A0,D2.W)
000100BC 5282 ADDQ.L #$1,D2
000100BE 7011 MOVEQ #$11,D0
000100C0 B082 CMP.L D2,D0
000100C2 6EEC BGT.B $100B0
000100C4 7400 MOVEQ #$0,D2 000100C6 207C 0001 2148
MOVEA.L #$12148,A0 000100CC 4A30 2000 TST.B
($0,A0,D2.W) 000100D0 6732 BEQ.B $10104 000100D2
2A02 MOVE.L D2,D5 000100D4 DA82 ADD.L D2,D5 000100D6
5685 ADDQ.L #$3,D5
```

Assembler listing

```
›>> 12 int prime,count,iter;
›>> 13
›— 14 => for (iter = 1;<=iter<=MAX_ITER;iter++)
› 000100AA 7C01 MOVEQ #$1,D6
›>> 15 {
›>> 16 count = 0;
› 000100AC 7800 MOVEQ #$0,D4
›— 17 => for(i = 0;<= i<MAX_PRIME; i++)› > 000100AE
7400 MOVEQ #$0,D2
›>> 18 flags[i] = 1;
› 000100B0 207C 0001 2148 MOVEA.L #$12148,A0 {flags}
› 000100B6 11BC 0001 2000 MOVE.B #$1,($0,A0,D2.W)
```

```
›– 17 for(i = 0; i<MAX_PRIME; => i++)<=
› 000100BC 5282 ADDQ.L #$1,D2
›– 17 for(i = 0; => i<MAX_PRIME;<=i++)
› 000100BE 7011 MOVEQ #$11,D0
› 000100C0 B082 CMP.L D2,D0
› 000100C2 6EEC BGT.B $100B0
```

Assembler listing with symbolic information

The key to this is the creation of a symbol table which provides the cross-referencing information that is needed. This can either be included within the binary file format used for object and absolute files or, in some cases, stored as a separate file. The important thing to remember is that symbol tables are often not automatically created and, without them, symbolic debug is not possible.

When the file or files are loaded or activated by the debugger, it searches for the symbolic information which is used to display more meaningful information as shown in the various listings. The symbolic information means that break-points can be set on language statements as well as individual addresses. Similarly, the code can be traced or stepped through line by line or instruction by instruction.

This has several repercussions. The first is the number of symbolic terms and the storage they require. Large tables can dramatically increase file size and this can pose constraints on linker operation when building an application or a new version of an operating system. If the linker has insufficient space to store the symbol tables while they are being corrected — they are often held in RAM for faster searching and update — the linker may crash with a symbol table overflow error. The solution is to strip out the symbol tables from some of the modules by recompiling them with symbolic debugging disa-bled or by allocating more storage space to the linker.

The problems may not stop there. If the module is then embedded into a target and symbolic debugging is required, the appropriate symbol tables must be included in the build and this takes up memory space. It is not uncommon for the symbol tables to take up more space than the spare system memory and prevent the system or task from being built or running correctly. The solution is to add more memory or strip out the symbol tables from some of the modules.

It is normal practice to remove all the symbol table information from the final build to save space. If this is done, it will also remove the ability to debug using the symbol informa-tion. It is a good idea to have at least a hard copy of the symbol table to help should any debugging be needed.

Emulation

Even using the described techniques, it cannot be stated that there will never be a need for additional help. There will be times when instrumentation, such as emulation and logic analysis, are necessary to resolve problems within a design quickly. Timing and intermittent problems cannot be easily solved without access to further information about the processor and other system signals. Even so, the recognition of a potential problem source, such as a specific software module or hardware, allows more productive use and a speedier resolution. The adoption of a methodical design approach and the use of ready built boards as the final system, at best remove the need for emulation and, at worst, reduce the amount of time required to debug the system.

There are some problems with using emulation within a board-based system or any rack mounted system. The first is how to get the emulation or logic analysis probe onto the board in the first place. Often the gap between the processor and adjacent boards is too small to cope with the height of the probe. It may be possible to move adjacent boards to other slots, but this can be very difficult or impossible in densely populated racks. The answer is to use an extender board to move the target board out of the rack for easier access. Another problem is the lack of a socketed processor chip which effectively prevents the CPU from being removed and the emulator probe from being plugged in. With the move towards surface mount and high pin count packages, this problem is likely to increase. If you are designing your own board, I would recommend that sockets are used for the processor to allow an emulator to be used. If possible, and the board space allows it, use a zero insertion force socket. Even with low insertion force sockets, the high pin count can make the insertion force quite large. One option that can be used, but only if the hardware has been designed to do so, is to leave the existing processor *in situ* and tri-state all its external signals. The emulator is then connected to the processor bus via another connector or socket and takes over the processor board.

The second problem is the effect that large probes can have on the design especially where high speed buses are used. Large probes and the associated cabling create a lot of additional capacitance loading which can prevent an otherwise sound electronic design from working. As a result, the system speed very often must be downgraded to compensate. This means that the emulator can only work with a slower than originally specified design. If there is a timing problem that only appears while the system is running at high speed, then the emulator is next to useless in providing any help. We will come back to emulation techniques at the end of this chapter.

Optimisation problems

The difficulties do not stop with hardware mechanical problems. Software debugging can be confused or hampered by optimisation techniques used by the compiler to improve the efficiency of the code. Usually set by options from the command line, the optimisation routines examine the code and change it to improve its efficiency, while retaining its logical design and context. Many different techniques are used but they fall into two main types: those that remove code and those that add code or change it. A compiler may remove variables or routines that are never used or do not return any function. Small loops may be unrolled into straight line code to remove branching delays at the expense of a slightly larger program. Floating point routines may be replaced by inline floating point instructions. The net result is code that is different from the assembler listing produced by the compiler. In addition, the generated symbol table may be radically different from that expected from the source code.

These optimisation techniques can be ruthless; I have known whole routines to be removed and in one case a complete program was reduced to a single NOP instruction! The program was a set of functions that performed benchmark routines but did not use any global information or return any values. The optimiser saw this and decided that as no data was passed to it and it did not modify or return any global data, it effectively did nothing and replaced it with a NOP. When benchmarked, it gave a pretty impressive performance of zero seconds to execute several million calculations.

```
/* sieve.c – Eratosthenes Sieve prime number
calculation */
/* scaled down with MAX_PRIME set to 17 instead of
8091 */

#define MAX_ITER     1
#define MAX_PRIME    17

char      flags[MAX_PRIME];

main ()
{
          register int i,k,l,m;
          int   prime,count,iter;

          for (iter = 1;iter<=MAX_ITER;iter++)
                {
                count = 0;
/* redundant code added here */
                for(l = 0; l < 200; l++ );
                for(m = 128; l > 1; m– );
/* redundant code ends here */
                for(i = 0; i<MAX_PRIME; i++)
                      flags[i] = 1;
                for(i = 0; i<MAX_PRIME; i++)
                      if(flags[i])
                            {
                            prime = i + i + 3;
                            k = i + prime;
                            while (k < MAX_PRIME)
                                  {
```

```
                        flags[k] = 0;
                        k += prime;
                        }
                    count++;
                    printf(" prime %d =
%d\n", count, prime);
                    }
            }
        printf("\n%d primes\n",count);
}
```

Source listing for optimisation example

```
        file    "ctmlAAAa00360"

        def     autl.,32
        def     argl.,64
        text
        global  _main
_main:
        subu    r31,r31,argl.
        st      rl,r31,argl.-4
        st      r19,r31,autl.+0
        st      r20,r31,autl.+4
        st      r21,r31,autl.+8
        st      r22,r31,autl.+12
        st      r23,r31,autl.+16
        st      r24,r31,autl.+20
        st      r25,r31,autl.+24
        or      r19,r0,1
        br      @L25
@L26:
        or      r20,r0,r0
        or      r23,r0,r0
        br      @L6
@L7:
        addu    r23,r23,1
@L6:
        cmp     r13,r23,200
        bbl     lt,r13,@L7
        or      r22,r0,128
        br      @L10
@L11:
        subu    r22,r22,1
@L10:
        cmp     r13,r23,1
        bbl     gt,r13,@L11
        or      r25,r0,r0
        br      @L14
@L15:
        or.u    r13,r0,hi16(_flags)
        or      r13,r13,lo16(_flags)
        or      r12,r0,1
        st.b    r12,r13,r25
        addu    r25,r25,1
@L14:
        cmp     r13,r25,17
        bbl     lt,r13,@L15
        or      r25,r0,r0
        br      @L23
@L24:
        or.u    r13,r0,hi16(_flags)
        or      r13,r13,lo16(_flags)
        ld.b    r13,r13,r25
        bcnd    eq0,r13,@L17
        addu    r13,r25,r25
        addu    r21,r13,3
        addu    r24,r25,r21
        br      @L19
@L20:
        or.u    r13,r0,hi16(_flags)
        or      r13,r13,lo16(_flags)
        st.b    r0,r13,r24
        addu    r24,r24,r21
@L19:
        cmp     r13,r24,17
        bbl     lt,r13,@L20
        addu    r20,r20,1
        or.u    r2,r0,hi16(@L21)
        or      r2,r2,lo16(@L21)
        or      r3,r0,r20
        or      r4,r0,r21
        bsr     _printf
@L17:
        addu    r25,r25,1
@L23:
        cmp     r13,r25,17
        bbl     lt,r13,@L24
        addu    r19,r19,1
@L25:
        cmp     r13,r19,1
        bbl     le,r13,@L26
        or.u    r2,r0,hi16(@L27)
        or      r2,r2,lo16(@L27)
        or      r3,r0,r20
        bsr     _printf
        ld      r19,r31,autl.+0
        ld      r20,r31,autl.+4
        ld      r21,r31,autl.+8
        ld      r22,r31,autl.+12
        ld      r23,r31,autl.+16
        ld      r24,r31,autl.+20
        ld      r25,r31,autl.+24
        ld      rl,r31,argl.-4
        addu    r31,r31,argl.
        jmp     rl
```

No optimisation

```
        file    "ctmlAAAa00355"

        def     autl.,32
        def     argl.,56
        text
        global  _main
_main:
        subu    r31,r31,argl.
        st      rl,r31,argl.-4
        st.d    r20,r31,autl.+0
        st.d    r22,r31,autl.+8
        st      r25,r31,autl.+16
        or      r20,r0,1
@L26:
        or      r21,r0,r0
        or      r25,r0,r0
@L7:
        addu    r25,r25,1
        cmp     r13,r25,200
        bbl     lt,r13,@L7
        br.n    @L28
        or      r2,r0,128
@L11:
        subu    r2,r2,1
@L28:
        cmp     r13,r25,1
        bbl     gt,r13,@L11
        or      r25,r0,r0
        or.u    r22,r0,hi16(_flags)
        or      r22,r22,lo16(_flags)
@L15:
        or      r13,r0,1
        st.b    r13,r22,r25
        addu    r25,r25,1
        cmp     r12,r25,17
        bbl     lt,r12,@L15
        or      r25,r0,r0
@L24:
        ld.b    r12,r22,r25
        bcnd    eq0,r12,@L17
        addu    r12,r25,r25
        addu    r23,r12,3
        addu    r2,r25,r23
        cmp     r12,r2,17
        bbl     ge,r12,@L18
@L20:
        st.b    r0,r22,r2
        addu    r2,r2,r23
        cmp     r13,r2,17
        bbl     lt,r13,@L20
@L18:
        addu    r21,r21,1
        or.u    r2,r0,hi16(@L21)
        or      r2,r2,lo16(@L21)
        or      r3,r0,r21
        bsr.n   _printf
        or      r4,r0,r23
@L17:
        addu    r25,r25,1
        cmp     r13,r25,17
        bbl     lt,r13,@L24
        addu    r20,r20,1
        cmp     r13,r20,1
        bbl     le,r13,@L26
        or.u    r2,r0,hi16(@L27)
        or      r2,r2,lo16(@L27)
        bsr.n   _printf
        or      r3,r0,r21
        ld.d    r20,r31,autl.+0
        ld      rl,r31,argl.-4
        ld.d    r22,r31,autl.+8
        ld      r25,r31,autl.+16
        jmp.n   rl
        addu    r31,r31,argl.
```

Full optimisation

Assembler listings for optimised and non-optimised compilation

To highlight how optimisation can dramatically change the generated code structure, look at the C source listing for the Eratosthenes Sieve program and the resulting M88000 assembler listings that were generated by using the default non-optimised setting and the full optimisation option. The immediate difference is in the greatly reduced size of the code and the use of the .n suffix with jump and branch instructions to make use of the delay slot. This is a technique used on many RISC processors to prevent a pipeline stall when changing the program flow. If the instruction has a .n suffix, the instruction immediately after it is effectively executed with the branch and not after it, as it might appear from the listing!

In addition, the looping structures have been reorganised to make them more efficient, although the redundant code loops could be encoded simply as a loop with a single branch. If the optimiser is that good, why has it not done this? The reason is that the compiler expects loops to be inserted for a reason and usually some form of work is done within the loop which may change the loop variables. Thus the compiler will take the general case and use that rather than completely remove it or rewrite it. If the loop had been present in a dead code area — within a conditional statement where the conditions would never be met — the compiler would remove the structure completely.

The initialisation routine `_main` is different in that not all the variables are initialised using a store instruction and fetching their values from a stack. The optimised version uses the faster 'or' instruction to set some of the variables to zero.

These and other changes highlight several problems with optimisation. The obvious one is with debugging the code. With the changes to the code, the assembler listing and symbol tables do not match. Where the symbols have been preserved, the code may have dramatically changed. Where the routines have been removed, the symbols and references may not be present. There are several solutions to this. The first is to debug the code with optimisation switched off. This preserves the symbol references but the code will not run at the same speed as the optimised version, and this can lead to some timing problems. A second solution is becoming available from compiler and debugger suppliers, where the optimisation techniques preserve as much of the symbolic information as possible so that function addresses and so on are not lost.

The second issue is concerned with the effect optimisation may have on memory mapped I/O. Unless the optimiser can recognise that a function is dealing with memory mapped I/O, it may not realise that the function is doing some work after all and remove it — with disastrous results. This may require declaring the I/O addresses as a global variable, returning a value at the function's completion or even passing the address to the function itself, so that the optimiser can recognise its true role. A third complication can arise with optimisations such as unrolling loops and software timing. It is not uncommon to use instruction sequences to delay certain accesses or functions. A peripheral may require a certain number of clock cycles to respond to a command. This delay can be accomplished by executing other instructions, such as a loop or a divide instruction. The optimiser may remove or unroll such loops and replace the inefficient divide instruction with a logical shift. While this does increase the performance, that is not what was required and the delayed peripheral access may not be long enough — again with disastrous results.

Such software timing should be discouraged not only for this but also for portability reasons. The timing will assume certain characteristics about the processor in terms of processing speed and performance which may not be consistent with other faster board designs or different processor versions.

Xray

It is not uncommon to use all the debugging techniques that have been described so far at various stages of a development. While this itself is not a problem, it has been difficult to get a common set of tools that would allow the various techniques to be used without having to change compilers or libraries, learn different command sets, and so on. The ideal would be a single set of compiler and debugger tools that would work with a simulator, task level debugger, onboard debugger and emulator. This is exactly the idea behind Microtec's Xray product.

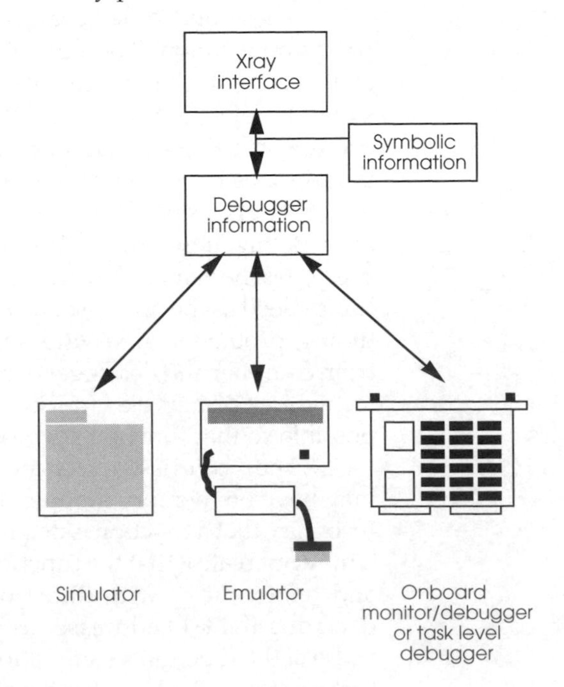

Xray structure

```
DATA                                        12     STACK               14
1                                                  0000FFE4=00000000
2                                                  0000FFE0=00000000
3                                                  0000FFDC=00000000
4                                                  0000FFD8=00000000
5                                                SP->0000FFD4=00000000

CODE                                        11     REGISTERS           13
--    17  =>              for(i = 0;<=  i<MAX_PRIME;  PC=000100AE pi=000100AC
*000100AE 7400              MOVEQ    #$0,D2          D0=000122BC A0=000122C4
>>    18            flags[i] = 1;                     D1=FFFFFFFF A1=00012084
 000100B0 207C 0001 2148    MOVEA.L #$12148,A0       D2=00000000 A2=0001215C
 000100B6 11BC 0001 2000    MOVE.B  #$1,($0,A0,D2.W) D3=00000000 A3=00000000
--    17              for(i = 0; i<MAX_PRIME; => i+   D4=00000000 A4=00000000
 000100BC 5282              ADDQ.L  #$1,D2           D5=00000000 A5=00000000
--    17              for(i = 0; => i<MAX_PRIME;<=    D6=00000001 A6=00000000
 000100BE 7011              MOVEQ   #$11,D0          D7=00000000 A7=0000FFD4
 000100C0 B082              CMP.L   D2,D0            SR=0010011100010100
 000100C2 6EEC              BGT.B   $100B0             T S  III   XNZVC

Command  ↑ ↓    68000   MODULE: SIEVE          BREAK #: 1 HELP=F5   MRI 2.2A
COMMAND                                                                  10
> go
  Break # 1 on instr module SIEVE line 17
>
```

```
BREAK                                                                    25
 #     ADDRESS          MOD/FNCT       LINE         TYPE      COMMAND ARGUMENT
 1 000100AE             SIEVE        #17:1         INST/H    #17
 2 000100D8             SIEVE          #23         INST/H    #23

CODE                                                                      2
*    17               for(i = 0; i<MAX_PRIME; i++)
     18                       flags[i] = 1;
     19               for(i = 0; i<MAX_PRIME; i++)
     20                       if(flags[i])
     21                               {
     22                               prime = i + i + 3;
*    23                               k = i + prime;
     24                               while (k < MAX_PRIME)
     25                                       {
     26                                       flags[k] = 0;

Command          68000   MODULE: SIEVE          BREAK #: 1 HELP=F5   MRI 2.2A
COMMAND                                                                   1
> go
  Break # 1 on instr module SIEVE line 17
>
```

Xray screen shots

Xray consists of a consistent debugger system that can interface with a simulator, emulator, onboard debugger or operating system task level debugger. It provides a consistent interface which greatly improves the overall productivity because there is no relearning required when moving from one environment to another. It obtains its debugging information from a variety of sources, depending on how the target is being accessed. With the simulator, the information is accessed directly. With an emulator or target hardware, the link is via a simple serial line, via the Ethernet or directly across a shared memory interface. The data is then used in conjunction with symbolic information to produce the data that the user can see and control on the host machine.

The interface consists of a number of windows which display the debugging information. The windows consist of two types: those that provide core information, such as breakpoints and the processor registers and status. The second type are windows concerned with the different environments, such as task level information. Windows can be chosen and displayed at the touch of a key.

The displays are also consistent over a range of hosts, such as Sun workstations, IBM PCs and UNIX platforms. Either a serial or network link is used to transfer information from the target to the debugger. The one exception is that of the simulator which runs totally on the host system.

So how are these tools used? Xray comes with a set of compiler tools which allows software to be developed on a host system. This system does not have to use the same processor as the target. To execute the code, there is a variety of choices. The simulator is ideal for debugging code at a very early stage, before hardware is available, and allows software development to proceed in parallel with hardware development. Once the hardware is available, the Xray interface can be taken into the target through the use of an emulator or a small onboard debug monitor program. These debug monitors are supplied as part of the Xray package for a particular processor. They can be easily modified to reflect individual memory maps and have drivers from a large range of serial communications peripherals.

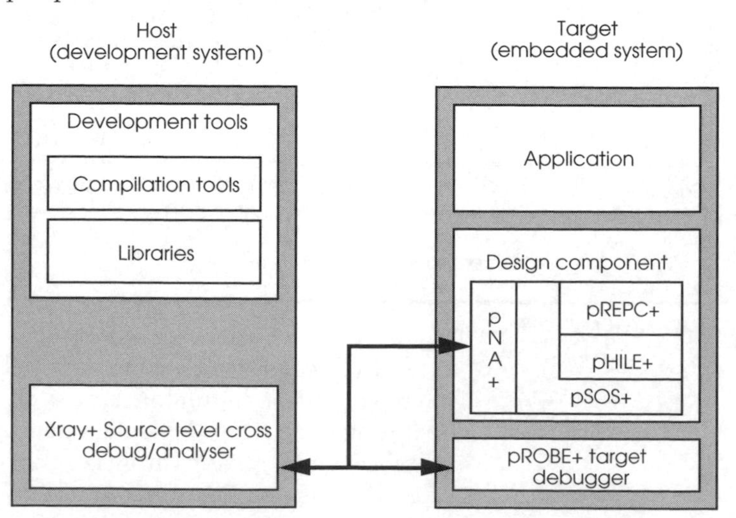

pSOS+ debugging

With Xray running in the target, the hardware and initial software routines can be debugged. The power of Xray can be further extended by having an Xray interface from the operating system debugger. pSOS+ uses this method to provide its debugging interface. This allows task level information to be used to set breakpoints, and so on, while still preserving the lower level facilities. This provides an extremely powerful and flexible way of debugging a target system. Xray has become a *de facto* standard for debugging tools within the real-time and VMEbus market. This type of approach is also being adopted by many other software suppliers.

The role of the development system

An alternative environment for developing software for embedded systems is to use the final operating system as the development environment either on the target system itself or on a similar system. This used to be the traditional way of developing software before the advent of cheap PCs and workstations and integrated cross-compilation.

It still offers distinct advantages over the cross-compilation system. Run-time library support is integrated because the compilers are producing native code. Therefore the run-time libraries that produce executable code on the development system will run unmodified on the target system. The final software can even be tested on the development system before moving it to the target. In addition, the full range of functions and tools can be used to debug the software during this testing phase, which may not be available on the final target. For example, if a target simply used the operating system kernel, it would not have the file system and terminal drivers needed to support an onscreen debugger or help download code quickly from disk to memory. Yet a development system running the full version of the operating system could offer this and other features, such as downloading over a network. However, there are some important points to remember.

Floating point and memory management functions

Floating point co-processors and memory management units should be considered as part of the processor environment and need to be considered when creating code on the target. For example, the development system may have a floating point unit and memory management, while the target does not. Code created on the development system may not run on the target because of these differences. Executing floating point instructions would cause a processor exception while the location of code and data in the memory may not be compatible.

This means that code created for the development system may need recompiling and linking to ensure that the correct run-time routines are used for the target and that they are located correctly. This in turn may mean that the target versions may not run on the development system because its resources do not match up. This raises the question of the validity of using a development system in the first place. The answer is that the source code and the bulk of the binary code does not need modifying. Calling up a floating point emulation library instead of using floating point instructions will not affect any integer or I/O routines. Linking modules to a different address changes the addresses, not the instructions and so the two versions are still extremely similar. If the code works on the development system, it is likely that it will work on the target system.

While the cross-compilation system is probably the most popular method used today to develop software for embedded systems — due to the widespread availability of PCs and workstations and the improving quality of software tools — it is not the only way. Dedicated development systems can offer faster and easier software creation because of the closer relationship between the development environment and the end target.

Emulation techniques

In-circuit emulation (ICE) has been the traditional method employed to emulate a processor inside an embedded design so that software can be downloaded and debugged *in situ* in the end application. For many processors this is still an appropriate method for debugging embedded systems but the later processors have started to dispense with the emulator as a tool and replace it with alternative approaches.

The main problem is concerned with the physical issues associated with replacing the processor with a probe and cable. These issues have been touched on before but it is worth revisiting them. The problems are:

- Physical limitation of the probe

  With high pin count and high density packages that many processors now use such as quad flat packs, ball grid arrays and so on, the job of getting sockets that can reliably provide good electrical contacts is becoming harder. This is starting to restrict the ability of probe manufacturers to provide headers that will fit these sockets, assuming that the sockets are available in the first place.

  The ability to get several hundred individual signal cables into the probe is also causing problems and this has meant that for some processors, emulators are no longer a practical proposition.

- Matching the electrical characteristics

  This follows on from the previous point. The electrical characteristics of the probe should match that of the device the emulator is emulating. This includes the electrical characteristics of the pins. The difficulty is that the probe and its associated wiring make this matching very difficult indeed and in some cases, this imposes speed limits on the emulation or forces the insertion of wait states. Either way, the emulation is far from perfect and this can cause restrictions in the use of emulation. In some cases, where speed is of the essence, emulation can prevent the system from working at the correct design speed.

- Field servicing

  This is an important but often neglected point. It is extremely useful for a field engineer to have some form of debug access to a system in the field to help with fault identification and rectification. If this relies on an emulator, this can pose problems of access and even power supplies if the system is remote.

So, faced with these difficulties, many of the more recent processors have adopted different strategies to provide emulation support without having to resort to the traditional emulator and its inherent problems.

The basic methodology is to add some debugging support to the processor that enables a processor to be single stepped and breakpointed under remote control from a workstation or host. This facility is made possible through the provision of dedicated debug ports.

JTAG

JTAG ports were originally designed and standardised to provide a way of taking over the pins of a device to allow different bit patterns to be imposed on the pins allowing other devices within the circuit to be tested. This is important to implement boundary scan techniques without having to remove a processor. It allows access to all the hardware within the system.

The system works by using a serial port and clocking data into a large shift register inside the device. The outputs from the shift register are then used to drive the pins under control from the port.

OnCE

OnCE or on-chip emulation is a debug facility used on Motorola's DSP 56x0x family of DSP chips. It uses a special serial port to access additional registers within the device that provide control over the processor and access to its internal registers. The advantage of this approach is that by bringing out the OnCE port to an external connector, every system can provide its own in circuit emulation facilities by hooking this port to an interface port in a PC or workstation. The OnCE port allows code to be downloaded and single stepped, breakpoints to be set and the display of the internal registers, even while operating. In some cases, small trace buffers are available to capture key events.

BDM

BDM or background debug mode is provided on Motorola's MC683xx series of processors as well as some of the newer 8 bit microcontrollers such as the MC68HC12. It is similar in concept to OnCE, in that it provides remote control and access over the processor, but the way that it is done is slightly different. The processor has additional circuitry added which provides a special background debug mode where the processor does not execute any code but is under the control of the remote system connected to its BDM port. The BDM state is entered by the assertion of a BDM signal or by executing a special BDM instruction. Once the BDM mode has been entered, low level microcode takes over the processor and allows breakpoints to be set, registers to be accessed and single stepping to take place and so on, under command from the remote host.

10 Buffering and other data structures

This chapter covers the aspects of possibly the most used data structure within embedded system software. The use and understanding behind buffer structures is an important issue and can greatly effect the design of software.

What is a buffer?

A buffer is as its name suggests an area of memory that is used to store data, usually on a temporary basis prior to processing it. It is used to compensate for timing problems between software modules or subsystems that cannot always guarantee to process every piece of data as it becomes available. It is also used as a collection point for data so that all the relevant information can be collected and organised before processing.

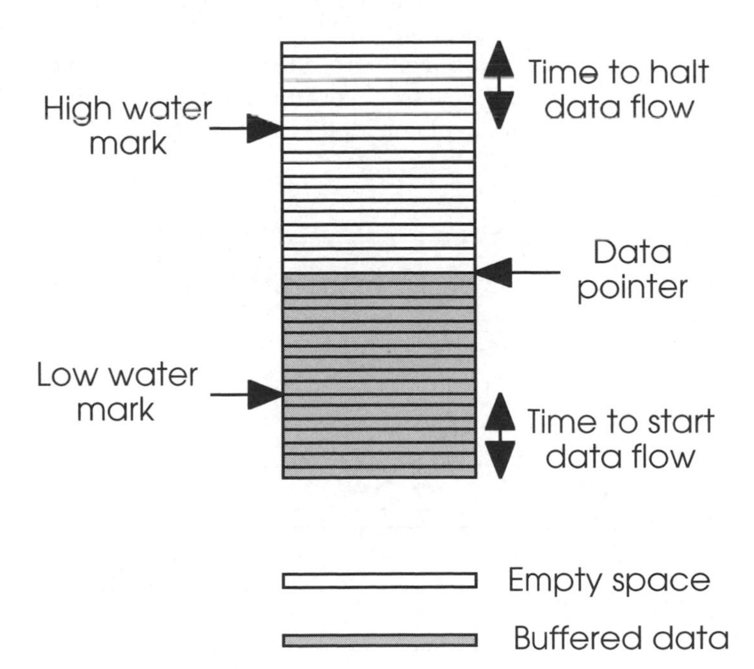

A basic buffer structure

The diagram shows the basic construction of a buffer. It consists of a block of memory and a pointer that is used to locate the next piece of data to be accessed or removed from the buffer. There are additional pointers which are used to control the buffer and prevent data overrun and underrun. An overrun occurs when the buffer cannot accept any more data. An underrun is caused when it is asked for data and cannot provide it.

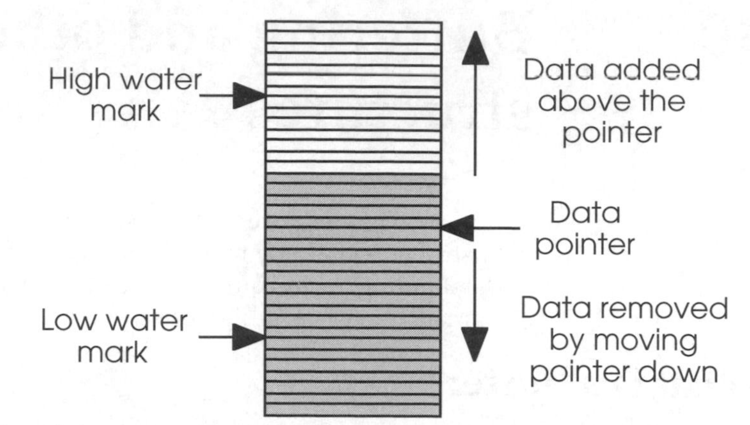

Adding and removing data

Data is removed by using the pointer to locate the next value and moving the data from the buffer. The pointer is then moved to the next location by incrementing its value by the number of bytes or words that have been taken. One common programming mistake is to confuse words and bytes. A 32 bit processor may access a 32 bit word and therefore it would be logical to think that the pointer is incremented by one. The addressing scheme may use bytes and therefore the correct increment is four. Adding data is the opposite procedure. The details on exactly how these procedures work determine the buffer type and its characteristics and are explained later in this chapter.

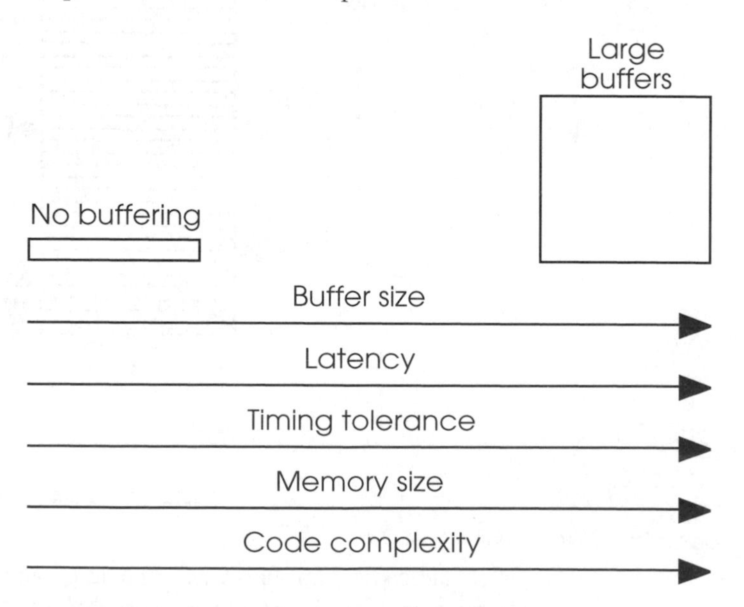

Buffering trade-offs

However, while buffering does undoubtedly offer benefits, they are not all for free and their use can cause problems. The diagram shows the common trade-offs that are normally encountered with buffering.

Latency

If data is stored in a buffer, then there is normally a delay before the first and subsequent data is received from the buffer. This delay is known as the buffer latency and in some systems, it can be a big advantage. In others, however, its effect is the opposite.

For a real-time system, the buffer latency defines the earliest that information can be processed and therefore any response to that information will be delayed by the latency irrespective of how fast or efficient the processing software and hardware is. If data is buffered so that eight samples must be received before the first is processed, the real-time response is now eight times the data rate for a single sample plus the processing time. If the first sample was a request to stop a machine or ignore the next set of data, the processing that determines its meaning would occur after the event it was associated with. In this case, the data that should have been ignored is now in the buffer and has to be removed.

Latency can also be a big problem for data streams that rely on real-time to retain their characteristics. For example, digital audio requires a consistent and regular stream of data to ensure accurate reproduction. Without this, the audio is distorted and can become unintelligible in the case of speech. Buffering can help by effectively having some samples in reserve so that the audio data is always there for decoding or processing. This is fine except that there is now an initial delay while the buffer fills up. This delay means an interaction with the stream is difficult as anyone who has had an international call over a satellite link with the large amount of delay can vouch for. In addition some systems cannot tolerate delay. Digital telephone handsets have to demonstrate a very small delay in the audio processing path which limits the size of any buffering for the digital audio data to less than four samples. Any higher and the delay caused by buffer latency means that the phone will fail its type approval.

Timing tolerance

Latency is not all bad, however, and used in the right amounts can provide a system that is more tolerant and resilient than one that is not. The issue is based around how time critical the system is and perhaps more importantly how deterministic is it.

Consider a system where audio is digitally sampled, filtered and stored. The sampling is performed on a regular basis and the filtering takes less time than the interval between samples. In this case, it is possible to build a system that does not need buffering and will have a very low latency. As each sample is received, it is processed and stored. The latency is the time to take a single sample.

If the system has other activities and especially if those involve asynchronous events such as the user pressing a button on the panel, then the guarantee that all the processing can be completed between samples may no longer be true. If this deadline is not made, then a sample may be lost. One solution to this — there are others such as using a priority system as supplied by a real-time operating system — is to use a small buffer to temporarily store the data so that it is not lost. By doing this the time constraints on the processing software are reduced and are more tolerant of other events. This is, however, at the expense of a slightly increased latency.

Memory size

One of the concerns with buffers is the memory space that they can take. With a large system this is not necessarily a problem but with a microcontroller or a DSP with on-chip memory, this can be an issue when only small amounts of RAM are available.

Code complexity

There is one other issue concerned with buffers and buffering technique and that is the complexity of the control structures needed to manage them. There is a definite trade-off between the control structure and the efficiency that the buffer can offer in terms of memory utilisation. This is potentially more important in the region of control and interfacing with interrupts and other real-time events. For example, a buffer can be created with a simple area of memory and a single pointer. This is how the frequently used stack is created. The control associated with the memory — or buffer which is what the memory really represents — is a simple register acting as an address pointer. The additional control that is needed to remember the stacking order and the frame size and organisation is built into the frame itself and is controlled by the microprocessor hardware. This additional level of control must be replicated either in the buffer control software or by the tasks that use the buffer. If a single task is associated with a buffer, it is straightforward to allow the task to implement the control. If several tasks use the same buffer, then the control has to cope with multiple and, possibly, conflicting accesses and while this can be done by the tasks, it is better to nominate a single entity to control the buffer. However, the code complexity associated with the buffer has now increased.

The code complexity is also dependent on how the buffer is organised. It is common for multiple pointers to be used along with other data such as the number of bytes stored and so on. The next section in this chapter will explain the commonly used buffer structures.

Linear buffers

The term linear buffer is a generic reference to many buffers that are created with a single piece of linear contiguous memory that is controlled by pointers whose address increments linearly. The examples so far discussed are all of linear buffers.

The main point about them is that they will lose data when full and fail to provide data when empty. This is obvious but as will be shown, the way in which this happens with linear buffers compared to circular ones is different. With a linear buffer, it loses incoming data when full so that the data it does contain becomes older and older. This is the overrun condition. When it is empty, it will provide old data, usually the last entry, and so the processor will continue to process potentially incorrect data. This is the underrun condition.

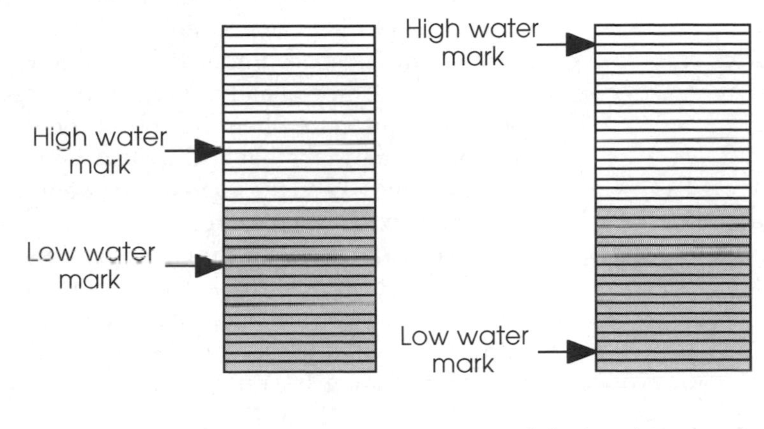

Adjusting the water marks

Within a real-time system, these conditions are often but not always considered error conditions. In some cases, the loss of data is not critical but with any data processing that is based on regular sampling, it will introduce errors. There are further complications concerning how these conditions are prevented from occurring. The solution is to use a technique where the pointers are checked against certain values and the results used to trigger an action such as fetching more data and so on. These values are commonly referred to as high and low water marks, so named because they are similar to the high and low water marks seen at the coast that indicate the minimum and maximum levels that tidal water will fall and rise.

The number of entries below the low water mark determine how many entries the buffer still has and thus the amount of time that is available to start filling the buffer before the buffer empties and the underrun condition exists. The number of empty entries in the buffer above the high water mark determines the length of time that is available to stop the further filling of the buffer and thus prevent data loss through overrun. By comparing the various input and output pointers with these values, events can be generated to start or stop filling the buffer. This could simply take the form of jumping to a subroutine, generating a software interrupt or within the context of an operating system posting a message to another task to fill the buffer.

Directional buffers

If you sit down and start playing with buffers, it quickly becomes apparent that there is more to buffer design than first meets the eye. For example, the data must be kept in the same order in which it was placed to preserve the chronological order. This is especially important for signal data or data that is sampled periodically. With an infinitely long buffer, this is not a problem. The first data is placed at the top of the buffer and new data is inserted underneath. The data in and out pointers then simply move up and down as needed. The order is preserved because there is always space under the existing data entries for more information. Unfortunately, such buffers are rarely practical and problems can occur when the end of the buffer is reached. The previous paragraphs have described how water marks can be used to trigger when these events are approaching and thus give some time to resolve the situation.

The resolution is different depending on how the buffer is used, i.e. is it being used for inserting data, extracting data or both. The solutions are varied and will lead onto the idea of multiple buffers and buffer exchange. The first case to consider is when a buffer is used to extract and insert data at the same time.

Single buffer implementation

In this case the buffer is used by two software tasks or routines to insert or extract information. The problem with water marks is that they have data above or below them but the free space that is used to fill the buffer does not lie in the correct location to preserve the order.

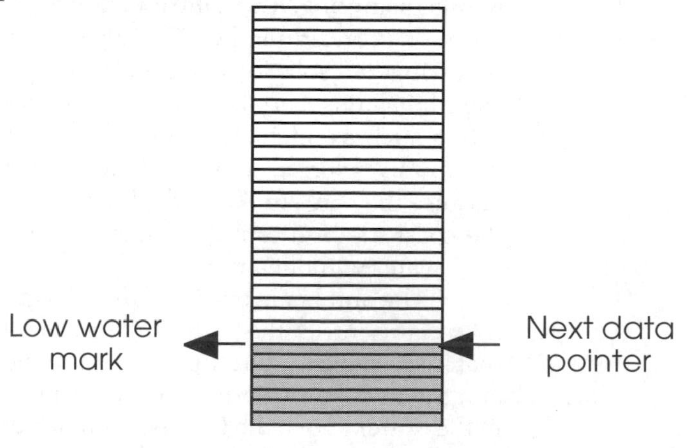


```
If next_data_pointer = low_water
      then
            copy current samples to top of buffer
            set next_data_pointer to top of buffer
            fill rest of buffer with new samples
      endif
```

Single buffer implementation

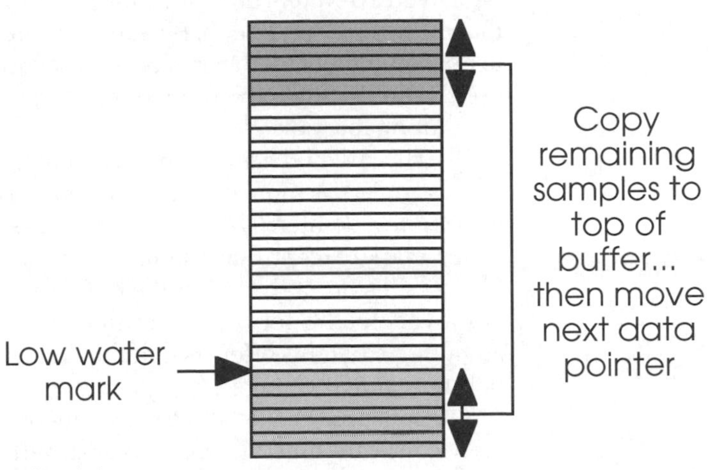

Moving the samples and pointer

One solution to this problem is to copy the data to a new location and then continue to back-fill the buffer. This is the method shown in the next three diagrams.

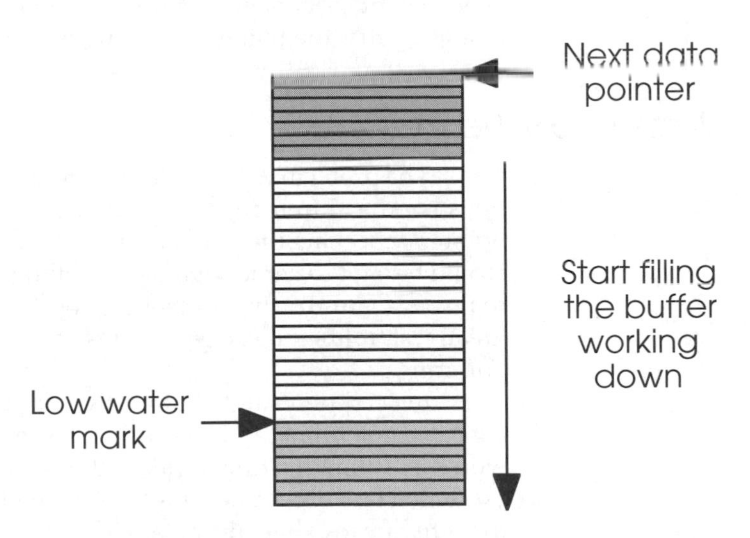

Back-filling the buffer

It uses a single low water mark and a next data pointer. The next data pointer is used to access the next entry that should be extracted. Data is not inserted into the buffer until the next data pointer hits the low water mark. When that happens, the data below the low water mark is copied to the top of the buffer and the next data pointer moved to the top of the buffer. A new pointer is then created whose initial value is that of the first data entry below the copied data. This is chronologically the correct location and thus the buffer can be filled by using this new pointer. The original data at the bottom of the buffer can be safely overwritten because the data was copied to the top of the buffer. Data can still be extracted by the next data pointer. When the temporary pointer reaches the end of the buffer, it stops filling. The low water mark

— or even a different one — can be used to send a message to warn that filling must stop soon. By adding more pointers, it is possible to not completely fill the area below the low water mark and then use this to calculate the number of entries to move and thus the next filling location.

This method has a problem in that there is a delay while the data is copied. A more efficient alternative to copying the data, is to copy the pointer. This approach works by still using the low water mark, except that the remaining data is not copied. The filling will start at the top of the buffer and the next data pointer is moved to the top of the buffer when it hits the end. The advantage that this offers is that the data is not copied and only a pointer value is changed.

Both approaches allow simultaneous filling and extraction. However, care must be taken to ensure that the filling does not overwrite the remaining entries at the bottom of the buffer with the pointer copying technique, and that extracting does not access more data than has been filled. Additional pointer checking may be needed to ensure this integrity in all circumstances and not leave the integrity dependent on the dynamics of the system, i.e. assuming that the filling/extracting times will not cause a problem.

Double buffering

The problem with single buffering is that there is a tremendous overhead in managing the buffer in terms of maintaining pointers, checking the pointers against water marks and so on. It would be a lot easier to separate the filling from the extraction. It removes many of the contention checks that are needed and greatly simplifies the design. This is the idea behind double buffering.

Instead of a single buffer, two buffers are used with one allocated for filling and the second for extraction. The process works by filling the first buffer and passing a pointer to it to the extraction task or routine. This filled buffer is then simply used by the software to extract the data. While this is going on, the second buffer is filled so that when the first buffer is emptied, the second buffer will be full with the next set of data. This is then passed to the extraction software by passing the pointer. Many designs will recycle the first buffer by filling it while the second buffer is emptied. The process will add delay into the system which will depend on the time taken to fill the first buffer.

Care must be taken with the system to ensure that the buffer swap is performed correctly. In some cases, this can be done by passing the buffer pointer in the time period between filling the last entry and getting the next one. In others, water marks can be used to start the process earlier so that the extraction task may be passed to the second buffer pointer before it is completely filled. This allows it the option of accessing data in the buffer if needed instead of having to wait for the buffer to complete filling. This is useful when the extraction timing is not consistent and/or requires different amounts of data. Instead of making the buffers the size of the largest data structure, they can be smaller and the double buffering used to ensure that data can be supplied. In other words, the double buffering is used to give the appearance of the presence of a far bigger buffer than is really there.

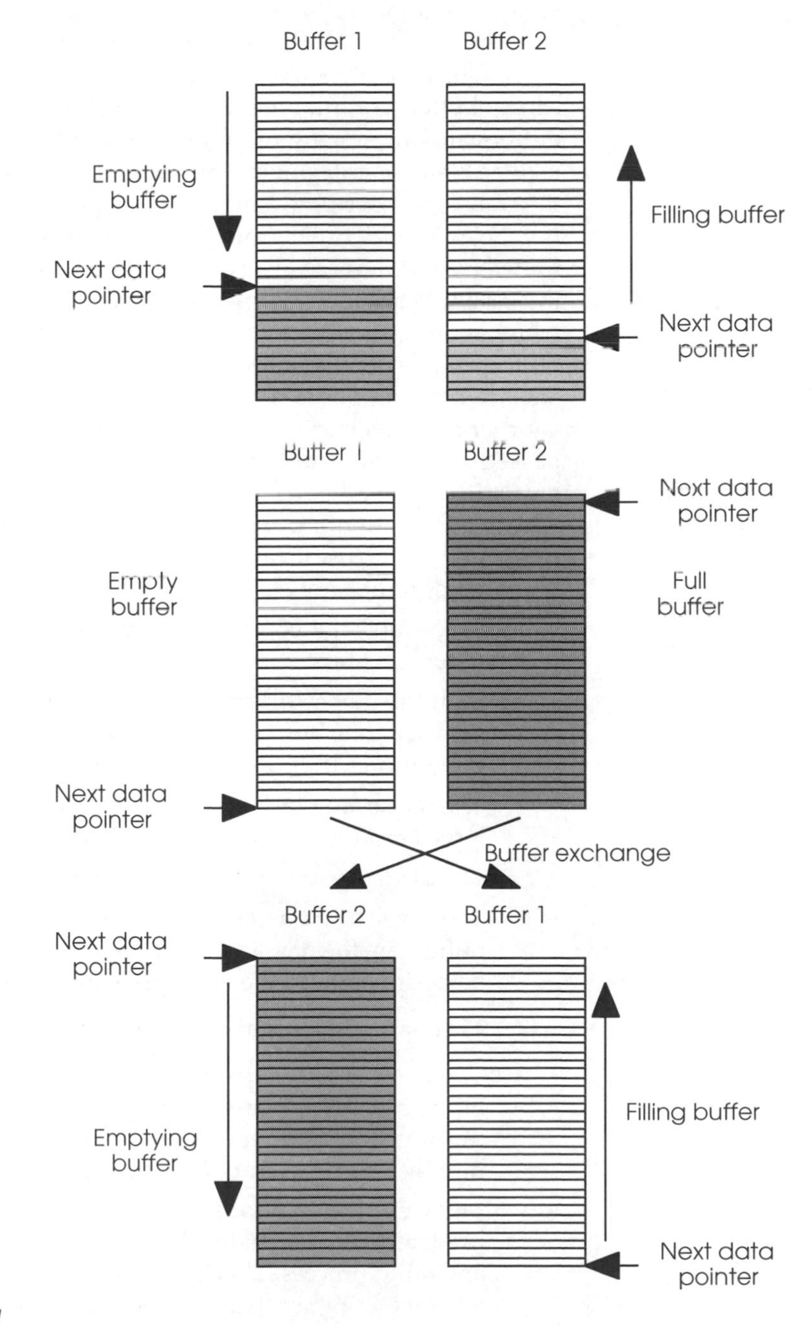

Double buffering

Buffer exchange

Buffer exchange is a technique that is used to simplify the control code and allow multiple tasks to process data simultaneously without having to have control structures to supervise access. In many ways it is a variation of the double buffering technique.

This type of mechanism is common to the SPOX operating system used for DSP processors and in these types of embedded systems it is relatively simple to implement.

The main idea of the system is the concept of exchanging empty buffers for full ones. Such a system will have at least two buffers although many more may be used. Instead of normally using a read or write operation where the data to be used during the transfer is passed as a parameter, a pointer is sent across that points to a buffer. This buffer would contain the data to be transferred in the case of a write or simply be an empty buffer in the case of a read. The command is handled by a device driver which returns another pointer that points to a second buffer. This buffer would contain data with a read or be empty with a write. In effect what happens is that a buffer is passed to the driver and another received back in return. With a read, an empty buffer is passed and a buffer full of data is returned. With a write, a full buffer is passed and an empty one is received. It is important to note that the buffers are different and that the driver does not take the passed buffer, use it and then send it back. The advantages that this process offers are:

- The data is not copied between the device driver and the requesting task.
- Both the device driver and the requesting task have their own separate buffer area and there is thus no need to have semaphores to control any shared buffers or memory.
- The requesting task can use multiple buffers to assimilate large amounts of data before processing.
- The device driver can be very simple to write.
- The level of inter-task communication to indicate when buffers are full or ready for collection can be varied and thus be designed to fit the end application or system.

There are some disadvantages however:

- There is a latency introduced dependent on the size of the buffer in the worst case. Partial filling can be used to reduce this if needed but requires some additional control to signify the end of valid data within a buffer.
- Many implementations assume a fixed buffer size which is predetermined, usually during the software compilation and build process. This has to be big enough for the largest message but may therefore be very inefficient in terms of

memory usage for small and simple data. Variable size buffers are a solution to this but require more complex control to handle the length of the valid data. The buffer size must still be large enough for the biggest message and thus the problem of buffer size granularity may come back again.

- The buffers must be accessible by both the driver and requesting tasks. This may seem to be very obvious but if the device driver is running in supervisor mode and the requesting task is in the user mode, the memory management unit or address decode hardware may prevent the correct access. This problem can also occur with segmented architectures like the 8086 where the buffers are in different segments.

Linked lists

Linked lists are a way of combining buffers in a regular and methodical way using pointers to point to the next entry in the list. The linking is maintained by adding an entry to a buffer which contains the address to the next buffer or entry in the list. Typically, other information such as buffer size may be included as well as allowing the list to support different size entries. Each buffer will also include information for its control such as next data pointers and water marks depending on their design or construction.

With a single linked list, the first entry will use the pointer entry to point to the location of the second entry and so on. The last entry will have a special value which indicates that the entry is the last one.

New entries are added by breaking the link, extracting the link information and then inserting the new entry and remaking the link. When appending the entry to the end, the new entry pointer takes the value of the original last entry — the end of list special value in this case. The original last entry will update its link to point to the new entry and in this way the link is created.

The process for inserting in the middle of the list is shown in the diagram and follows the basic principles.

Linked lists have some interesting properties. It is possible to follow all the links down to a particular insertion point. Please note that this can only be done from the first entry down without storing additional information. With a single linked list like the one shown, there is no information in each entry to show where the link came from, only where it goes to. This can be overcome by storing the link information as you traverse down the list but this means that any search has to start at the top and work through, which can be a tiresome process.

The double linked list solves this by using two links. The original link is used to move down and the new link contains the missing information to move back up. This provides a lot more flexibility and efficiency when searching down the list to determine an entry point. If a list is simply used to concatenate buffers or structures together, then the single link list is more than adequate. If the ability to search up and down the list to reorder and sort the list or find different insertion points, then the double linked list is a better choice to consider.

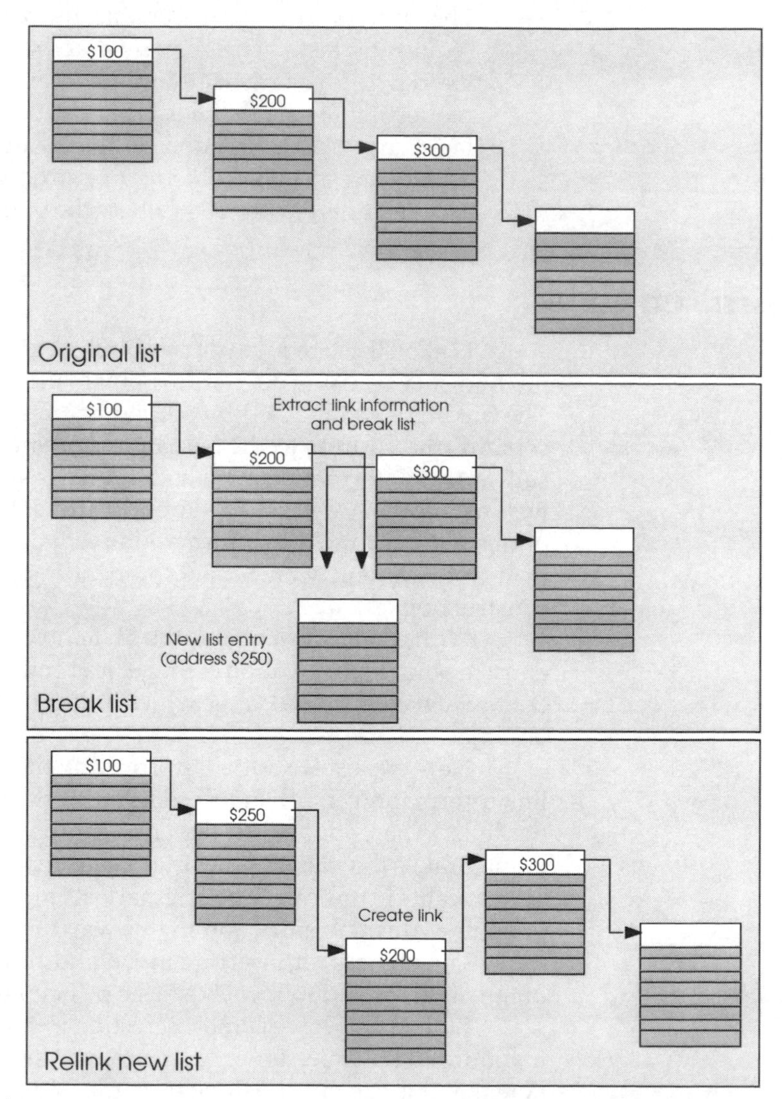

Inserting a new entry

FIFOs

FIFOs or first in, first out are a special form of buffer that uses memory to form an orderly queue to hold information. Its most important attribute is that the data can be extracted in the same way as it was entered. These are used frequently within serial comms chips to hold data temporarily while the processor is busy and cannot immediately service the peripheral. Instead of losing data, it is placed in the FIFO and extracted later. Many of the buffers described so far use a FIFO architecture.

Their implantation can be done either in software or more commonly with special memory chips that automatically maintain the pointers that are needed to control and order the data.

Circular buffers

Circular buffers are a special type of buffer where the data is circulated around a buffer. In this way they are similar to a single buffer that moves the next data pointer to the start of the buffer to access the next data. In this way the address pointer circulates around the addresses. In that particular case, care was taken so that no data was lost. It is possible to use such a buffer and lose data to provide a different type of buffer structure. This is known as a circular buffer where the input data is allowed to overwrite the last data entries. This keeps the most recent data at the expense of losing some of the older data. This is useful for capturing trace data where a trace buffer may be used to hold the last *n* data samples where *n* is the size of the buffer. By doing this, the buffer updating can be frozen at any point and the last *n* samples can be captured for examination. This technique is frequently used to create trace and history buffers for test equipment and data loggers.

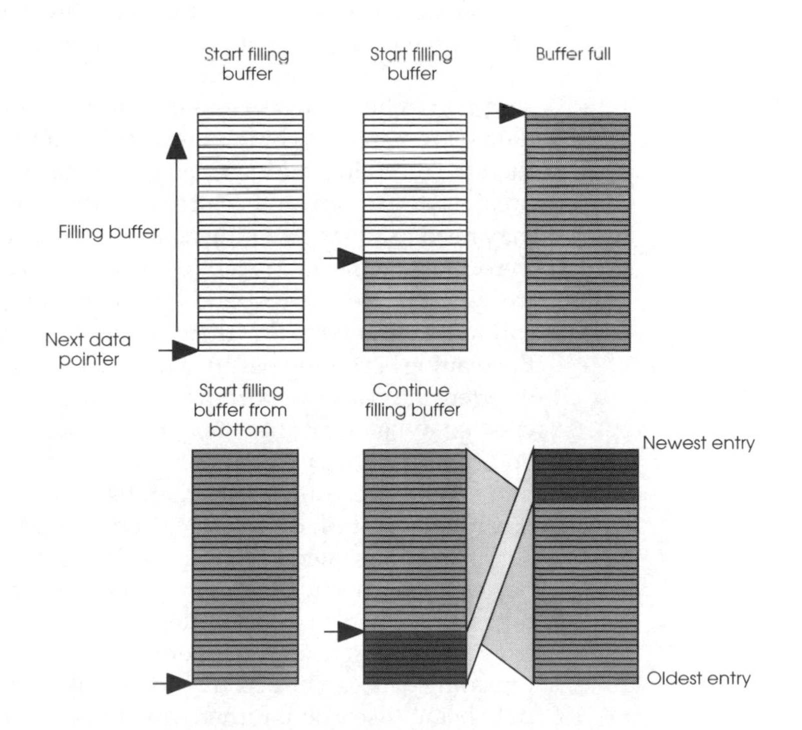

Circular buffers

This circular structure is also a very good match for coefficient tables used in digital signal processing where the algorithm iterates around various tables performing arithmetic operations to perform the algorithm.

The only problem is that the next data pointer must be checked to know when to reset it to the beginning of the buffer. This provides some additional overhead. Many DSPs provide a special modulo addressing mode that will automatically reset the address for a particular buffer size. This buffer size is normally restricted to a power of two.

Buffer underrun and overrun

The terms overrun and underrun have been described throughout this chapter and have been portrayed as things to be avoided when designing buffers and the system that supports them. While this is generally true, it is not always the case and there are times where they do not constitute an error but indicate some other problem or state in the system.

When a buffer underruns, it indicates that there is no more data in the buffer and that further processing should be stopped. This may indicate an error if the system is designed so that it would never run out of data. If it can happen in normal operation then the data underrun signal indicates a state and not an error. In both cases, a signal is needed to recognise this point. This can be done by comparing the buffer pointer to the buffer memory range. If the pointer value is outside of this range, it indicates that an underrun or overrun has occurred and this can redirect flow to the appropriate routine to decide what to do. Valid underrun conditions can occur when incoming data is buffered and not continuously supplied. A break in transmission or the completion of sending a data packet are two common examples. If this break is expected, then the receiving software can complete and go dormant. If the break is due to some other problem, then the receiving software may need to adopt a waiting action to see if other data comes into the buffer and the data reception continues. To prevent the system freezing up and not responding, the waiting action needs to be carefully designed so that there is some kind of timeout to prevent the waiting becoming permanent and locking up the system. This is often the cause of a hung up system that appears to have crashed. What in fact has happened is that it is waiting for data or an event that will not happen.

Data overrun is more than likely caused by some kind of error that results in data being lost because it cannot be accepted. Again in some systems this may not be an error and the system can be designed to carry on. Again pointer comparison can be used to determine when an overrun has occurred. In some cases this may trigger a request to allocate more memory to the buffer so that the incoming data can be accommodated. It may simply result in the data being discarded. Either way, the control software that surrounds the buffer can quickly become quite complicated depending on what desired outcome and behaviour is required. In other words, while buffers are simple to construct and use, designing the control software for buffers that are tolerant and can cope with underrun and overrun conditions is more complex. If this is not done then it can lead to many different problems.

If the pointers that are used to define and create buffers are used with no range checking i.e. they are always used on the assumption that the values are within the buffer range and are correct, then there is always the risk that an error condition may cause the pointers to go out of range. If a buffer underruns and the buffer pointer now points to the next memory location outside of the buffer and no checking is done, incorrect data will be supplied and the pointer incremented to the next location. If these locations are also data structures then no real problem will occur providing the locations are read and not written to. As soon as new data arrives for the buffer, it will be stored in the first available location outside the buffer which will overwrite the contents and destroy the data structure. This pointer corruption can quickly start corrupting data structures which will eventually reveal themselves as corrupt data and strange system behaviour such as crashes and system hang ups. These types of problems are notoriously difficult to solve as the resulting symptoms may have little to do with the actual cause.

It is recommended that care is taken with buffer design to ensure that both underrun and overrun are handled correctly without causing these type of errors. A little bit of care and attention can reap big dividends when the system is actually tested.

Allocating buffer memory

Buffer memory can be allocated in one of two generic ways: statically at build time by allocating memory through the linker and dynamically during run time by calling an operating system to allocate / program memory access.

Static allocation requires the memory needs to be defined before building the application and allocating memory either through special directives at the assembler level or through the definition of global, static or local variables within the various tasks within an application. This essentially declares variables which in turn allocate storage space for the task to use. The amount of storage that is allocated depends on the variable type and definition. Strings and character arrays are commonly used.

malloc()

`malloc()` and its counterpart `unmalloc()` are system calls that were originally used within UNIX environments to dynamically allocate memory and return it. Their popularity has meant that these calls are supported in many real-time operating systems. The call works by passing parameters such as memory size and type, starting address and access parameters for itself and other tasks that need to access the buffer or memory that will be returned. This is a common programming error. In reply, the calling task receives a pointer that points at the start of the memory if the call was successful, or an error code if it was not. Very often, the call is extended to support partial allocation where the system will still return a pointer to the start of the memory along with an error/status message stating that not all the memory is available and that the *x* bytes were allocated. Some other systems would class this partial allocation as an error and return an error message indicating failure.

Memory that was allocated via `malloc()` calls can be returned to the memory pool by using the `unmalloc()` call along with the appropriate pointer. This allows the memory to be recycled and used/allocated to other tasks. This recycling allows memory to be conserved but at the expense of processing the `malloc()` and `unmalloc()` calls. Some software finds this overhead unacceptable and allocates memory statically at build time. It should be noted that in many cases, the required memory may require a different design strategy. For memory efficient designs, requesting and recycling memory may be the best option. For speed, where the recycling overhead cannot be tolerated, static allocation may be the best policy.

Memory leakage

Memory leakage is a term that is used to describe a bug that gradually uses all the memory within a system until such point that a request to use or access memory that should succeed, fails. The term leakage is analogous to a leaking bucket where the contents gradually disappear. The contents within an embedded system are memory. This is often seen as a symptom to buffer problems where data is either read or written using locations outside the buffer.

The common symptoms are stack frame errors caused by the stack overflowing its allocated memory space and malloc() or similar calls to get memory failing. There are several common programming faults that cause this problem.

Stack frame errors

It is common within real-time systems, especially those with nested exceptions, to use the exception handler to clean up the stack before returning to the previous executing software thread or to a generic handler. The exception context information is typically stored on the stack either automatically or as part of the initial exception routine. If the exception is caused by an error, then there is probably little need to return execution to the point where the error occurred. The stack, however, contains a frame with all this return information and therefore the frames need to be removed by adjusting the stack pointer accordingly. It is normally this adjustment where the memory leakage occurs.

- Adjusting the stack for the wrong size frame. If the adjustment is too large, then other stack frames can be corrupted. If it is too small, then at best some stack memory can be lost and at worst the previous frame can be corrupted.
- Adjusting the stack pointer by the wrong value, e.g. using the number of words in the frame instead of the number of bytes.
- Setting the stack pointer to the wrong address so that it is on an odd byte boundary, for example.

Failure to return memory to the memory pool

This is a common cause of bus and memory errors. It is caused by tasks requesting memory and then not releasing it when their need for it is over. It is good practice to ensure that when a routine uses malloc() to request memory that it also uses unmalloc() to return it and make it available for reuse. If a system has been designed with this in mind, then there are two potential scenarios that can occur that will result in a memory problem. The first is that the memory is not returned and therefore subsequent malloc() requests cannot be serviced when they should be. The second is similar but may only occur in certain circumstances. Both are nearly always caused by failure to return memory when it is finished, but the error may not occur until far later in time. It may be the same task asking for memory or another that causes the problem to arise. As a result, it can be difficult to detect which task did not return the memory and is responsible for the problem.

In some cases where the task may return the memory at many different exit points within its code — this could be deemed as bad programming practice and it would be better to use a single exit sub-routine for example — it is often a programming omission at one of these points that stops the memory recycling.

It is difficult to identify when and where memory is allocated unless some form of record is obtained. With memory management systems, this can be derived from the descriptor tables and address translation cache entries and so on. These can be difficult to retrieve and decode and so a simple transaction record of all malloc() and unmalloc() calls along with a time stamp can prove invaluable. This code can be enabled for debugging if needed by passing a DEBUG flag to the pre-processor. Only if the flag is true will it compile the code.

Housekeeping errors

- Access rights not given

  This is where a buffer is shared between tasks and only one has the right access permission. The pointer may be passed correctly by using a mailbox or message but any access would result in a protection fault or if mapped incorrectly in accessing the wrong memory location.

- Pointer corruption

  It is very easy to get pointers mixed up and to use or update the wrong one and thus corrupt the buffer.

- Timing problems with high and low water marks

  Water marks are used to provide early warning and should be adjusted to cope with worst case timings. If not, then despite their presence, it is possible to get data overrun and underrun errors.

Wrong memory specification

This can be a very difficult problem to track down as it is can be very temporal in nature. It may only happen in certain situations which can be hard to reproduce. The problem is caused by a programming error essentially where it is assumed that any type of success message that malloc() or similar function returns actually means that all the memory requested is available. The software then starts to use what it thinks it has been allocated only to find that has not with disastrous results.

This situation can occur when porting software from one system to another where the `malloc()` call is used but has different behaviour and return messages. In one system it may return an error if the complete memory specification can be met while in another it will return the number of bytes that are allocated which may be less than the total requested. In one situation, an error message is returned and in another a partial success is reported back. These are very different and can cause immense problems.

Other errors can occur with non-linear addressing processors which may have restrictions on the pointer types and addressing range that is supported that is not there with a linear addressing architecture. This can be very common with 80x86 architectures and can cause problems or require a large redesign of the software.

11 Memory and performance trade-offs

This chapter describes the trade-offs made when designing an embedded system to cope with the speed and performance of the processor in doing its tasks. The problem faced by many designers is that the overall design requires a certain performance level in terms of processing or data throughput which on first appearance is satisfied by a processor. However, when the system is actually implemented, its performance is lacking due to the performance degradation that the processor can suffer as a result of its memory architecture, its I/O facilities and even the structure and coding techniques used to create the software.

The effect of memory wait states

This is probably the most important factor to consider as it can have the biggest impact on the performance of any system. With most high performance CPUs such as RISC and DSP processors offering single cycle performance where one or more instructions are executed on every clock edge, it is important to remember the conditions under which this is permitted:

- Instruction access is wait state free

  To achieve this, the instructions are fetched either from internal on-chip memory (usually wait state free but not always), or from internal caches. The problem with caches is that they only work with loops so that the first time through the loop, there is a performance impact while the instructions are fetched from external memory. Once the instructions have been fetched, they are then available for any further execution and it is here that the performance starts to improve.

- Data access is wait state free

  If an instruction manipulates data, then the time taken to execute the instruction must include the time needed to store the results of any data manipulation or access data from memory or a peripheral I/O port. Again, if the processor has to wait — known as stalling — while data is stored or read, then performance is lost. If an instruction modifies some data and it takes five clocks to store the result, this potentially can cause processing power to be lost. In many cases, processor architectures make the assumption that there is wait state free data access by either using local memory, registers or cache memory to hold the information.

- There are no data dependencies outstanding

  This leads on from the previous discussion and concerns the ability of an instruction to immediately use the result from a previous instruction. In many cases, this is only permitted if there is a delay to allow the processor to synchronise itself. As a result, the single cycle delay has the same result as a single cycle wait state and thus the performance is degraded.

As a result of all these conditions, it should not be assumed that a 80 MHz single cycle instruction processor such as a DSP- or a RISC- based machine can provide 80 MIPs of processing power. It can provided the conditions are correct and there are no wait states, data dependencies and so on. If there are, then the performance must be degraded. This problem is not unrecognised and many DSP and processor architectures utilise a lot of silicon in providing clever mechanisms to reduce the performance impact. However, the next question that must be answered is how do you determine the performance degradation and how can you design the code to use these features to minimise any delay?

Scenario 1 — Single cycle processor with large external memory

In this example, there is a single cycle processor that has to process a large external table by reading a value, processing it and writing it back. While the processing level is small — essentially a data fetch, three processing instructions and a data store — the resulting execution time can be almost three times longer than expected. The reason is stalling due to data latency. The first figure shows how the problem can arise.

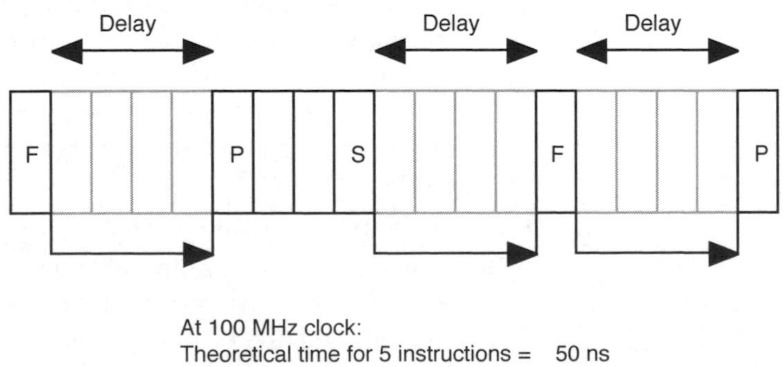

Stalling due to data latency

The instruction sequence consists of a data fetch followed by three instructions that process the information before storing the result. The data access goes to external memory where there are access delays and therefore the first processing instruction must wait until the data is available. In this case, the instruction execution is stalled and thus the fetch instruction takes the equivalent of five cycles instead of the expected single cycle. Once the data has been received, it can be processed and the processing instructions will continue, one per clock. The final instruction stores the end result and again this is where further delays can be experienced. The store instruction experiences the same external memory delays as the fetch instruction. However, its data is not required by the next set of instructions and therefore the rest of the instruction stream should be able to continue. This is not the case. The next instruction is a fetch and has to compete with the external bus interface which is still in use by the preceding store. As a result, it must also wait until the transaction is completed.

The next processing instruction now cannot start until the second fetch instruction is completed. These delays mean that the total time taken at 100 MHz for the five instructions (1 fetch + 3 processing + 1 store) is not 50 ns but 130 ns — an increase of 2.6 times.

The solution to this involves reordering the code so that the delays are minimised by overlapping operations. This assumes that the processor can do this, i.e. the instructions are stored in a separate memory space that can be accessed simultaneously with data. If not, then this conflict can create further delays and processor stalls. The basic technique involves moving the processing segment of the code away from the data access so that the delays do not cause processing stalls because the data dependencies have been removed. In other words, the data is already available before the processing instructions need it.

Moving dependencies can be achieved by restructuring the code so that the data fetch preceding the processing fetches the data for the next processing sequence and not the one that immediately follows it.

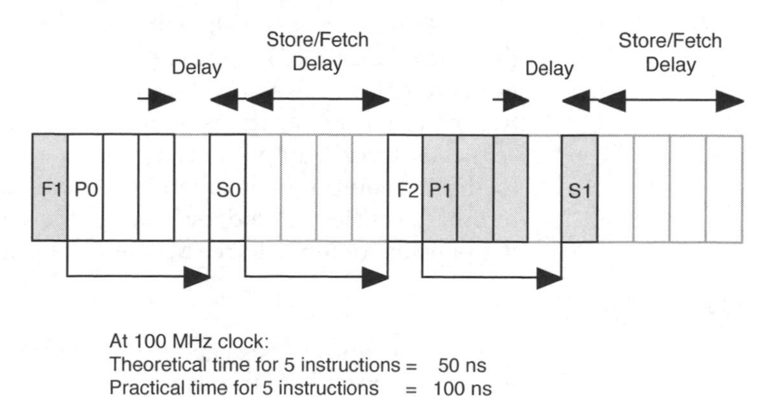

Removing stalling

The diagram above shows the general approach. The fetch instruction is followed by the processing and storage instruction for the preceding fetch. This involves using an extra register or other local storage to hold the sample until it is needed but it removes the data dependency. The processing instructions P0 onward that follow the fetch instruction F1 do not have any data dependency and thus can carry on processing. The storage instruction S0 has to wait one cycle until F0 has completed and similarly the fetch instruction F2 must wait until S0 has finished. These delays are still there because of the common resource that the store and fetch instructions use, i.e. the external memory interface. By reordering in this way, the five instruction sequence is completed twice in every 20 clocks giving a 100 ns timing which is a significant improvement.

This example also shows that the task in this case is I/O bound in that the main delays are caused by waiting for data to be fetched or stored. The processing load could almost be doubled and further interleaved with the store operations without changing or delaying the data throughput of the system. What would happen, however, is an increase in the processing load that the system could handle.

The delays that have been seen are frequently exploited by optimising routines within many modern compilers. These compilers know from information about the target processor when these types of delays can occur and how to reschedule instructions to make use of them and regain some of the lost performance.

Scenario 2 — Reducing the cost of memory access

The preceding scenario shows the delays that can be caused by accessing external memory. If the data is accessible from a local register the delay and thus the performance loss is greatly reduced and may be zero. If the data is in local on-chip memory or in an on-chip cache, the delay may only be a single cycle. If it is external DRAM, the delay may be nine or ten cycles. This demonstrates that the location of data can have a dramatic effect on any access delay and the resultant performance loss.

A good way of tackling this problem is to create a table with the storage location, its storage capability and speed of access in terms of clock cycles and develop techniques to move data between the various locations so that it is available when the processor needs it. For example, moving the data into registers compared to direct manipulation externally in memory can reduce the number of cycles needed, even when the saving and restoring of the register contents to free up the storage is taken into account.

Using registers

Registers are the fastest access storage resource available to the processor and are the smallest in size. As a result they are an extremely scarce resource which has to be used and managed carefully. The main problem is in deciding what information to store and when. This dilemma comes from the fact that there is frequently insufficient register space to store all the data all of the time. As a result, registers are used to store temporary values before updating main memory with a final result, to hold counter values for loop constructions and so on and for key important values. There are several approaches to doing this and exploiting their speed:

- Let the compiler do the work

  Many compilers will take care of register management automatically for you when it is told to use optimisation techniques. For many applications that are written in a high level language such as C, this is often a good choice.

- Load and store to provide faster local storage

  In this case, variables stored in external memory or on a stack are copied to an internal register, processed and then the result is written back out. With RISC processors that do not support direct memory manipulation, this is the only method of manipulating data. With CISC processors, such as the M68000 family, there is a choice. By writing code so that data is brought in to be manipulated, instead of using addressing modes that operate directly on external memory, the impact of slow memory access can be minimised.

- Declaring register-based variables

  By assigning a variable to a register during its initial declaration, the physical access to the variable will be quicker. This can be done explicitly or implicitly. Explicit declarations use special attributes that the programmer uses in the declaration, e.g. reg. An implicit declaration is where the compiler will take a standard declaration such as global or static and implicitly use this to allocate a register if possible.

- Context save and restore

  If more variables are assigned to registers than there are registers within the processor, the software may need to perform a full save and restore of the register set before using or accessing a register to ensure that these variables are not corrupted. This is an accepted process when multitasking so that the data of one task that resides in the processor registers is not corrupted. This procedure may need to be used at a far lower level in the software to prevent inadvertent register-based data corruption.

Using caches

Data and instruction caches work on the principle that both data and code are accessed more than once. The cache memory will store the information as it is fetched from the main memory so that any subsequent access is from the faster cache memory. This assumption is important because straight line code without branches, loops and so on will not benefit from a cache.

The size and organisation of the cache is important because it determines the efficiency of the overall system. If the program loops will fit into the cache memory, the fastest performance will be realised and the whole of the external bus bandwidth will be available for data movement. If the data cache can hold all the data structures, then the data access will be the fastest possible. In practice, the overall performance gain is less than ideal because inevitably access to the external memory will be needed either to fetch the code and data the first time around, when the cache is not big enough to contain all the instructions or data, or when the external bus must be used to access an I/O port where data cannot be cached. Interrupts and other asynchronous events will also compete for the cache and can cause instructions and data that has been cached prior to the event to be removed, thus forcing external memory accesses when the original program flow is continued.

Preloading caches

One trick that can be used with caches is to preload them so that a cache miss is never encountered. With normal operation, a cache miss will force an external memory access and while this is in progress, the processor is stalled waiting for the information — data or instruction — to be returned. In many code sequences, this is more likely to happen with data, where the first time that the cache and external bus are used to access the data is when it is needed. As described earlier with scenario 1, this delay occurs at an important point in the sequence and the delay prevents the processor from continuing.

By using the same technique as used in scenario 1, the data cache can be preloaded with information for the next processing iteration before the current one has completed. The PowerPC architecture provides special instructions that allow this to be performed. In this way, the slow data access is overlapped with the processing and data access from the cache and does not cause any processor stalls. In other words, it ensures that the cache always continues to have the data that the instruction stream needs to have.

By the very nature of this technique, it is one that is normally performed by hand and not automatically available through the optimisation techniques supplied with high level language compilers.

It is very useful with large amounts of data that would not fit into a register. However, if the time taken to fetch the data is greater than the time taken to process the previous block, then the processing will stall.

Caches also have one other interesting characteristic in that they can make it very difficult to predict how long a particular operation will take to execute. If everything is in the cache, the time will be short. If not then it will be long. In practice, it will be somewhere in between. The problem is that the actual time will depend on the number of cache hits and misses which will depend in turn on the software that has run before which will have overwritten some of the entries. As a result, the actual timing becomes more statistical in nature and in many cases the worst case timing has to be assumed, even though statistically the routine will execute faster 99.999% of the time!

Using on-chip memory

Some microcontrollers and DSP chips have local memory which can be used to store data or instructions and thus offers fast local storage. Any access to it will be fast and thus data and code structures will always gain some benefit if they are located here. The problem is that to gain the best benefit, both the code and data structures must fit in the on-chip memory so that no external accesses are necessary. This may not be possible for some programs and therefore decisions have to be made on which parts of the code and data structures are allocated this resource. With a real-time operating system, local on-chip memory is often used to gain the best context switching time. This memory requirement now has to compete with algorithms that need on-chip storage to meet the performance requirements by minimising any processor stalls.

One good thing about using on-chip memory is that it makes performance calculations easier as the memory accesses will have a consistent access time.

Using DMA

Some microcontrollers and DSPs have on-chip DMA controllers which can be used in conjunction with local memory to create a sort of crude but efficient cache. In reality, it is more like a buffering technique with an intelligent agent filling up and controlling the buffers in parallel with the processing.

The basic technique defines the local memory into two or more buffers, and programs the DMA controller to transfer data from the external memory to the local on-chip memory buffer while the data in the other buffer is processed. The overlapping of the DMA data transfer and the processing means that the data processing has local access to its data instead of having to wait for far slower memory access.

The buffering technique can be made more sophisticated by incorporating additional DMA transfers to move data out of the local memory back to the external memory. This may require the use of many more smaller buffers with different DMA characteristics. Constants could be put into one buffer which are read in but not read out. Variables can be stored in another where the information is written out to external memory.

Making the right decisions

The main problems faced by designers with these techniques is in knowing which one(s) should be used. The problem is that they involve a high degree of knowledge about the processor and the system characteristics. While a certain amount of information can be obtained from a documentation-based analysis, the use of simulation tools to run through code sequences and provide information concerning cache hits ratios, processor stalls and so on is a vital part in obtaining the optimum solution. Because of this, many cycle level processor simulation tools are becoming available which help provide this level of information.

12 Software examples

Benchmark example

The difficulty faced here appears to be a very simple one, yet actually poses an interesting challenge. The goal was to provide a simple method of testing system performance of different VMEbus processor boards to enable a suitable board to be selected. The problem was not how to measure the performance — there were plenty of benchmark routines available — but how to use the same compiler to create code that would run on several different systems with the minimum of modification. The idea was to generate some code which could then be tested on several different VMEbus target systems to obtain some relative performance figures. The reason for using the compiler was that typical C routines could be used to generate the test code.

The first decision made was to restrict the C compiler to non-I/O functions so that a replacement I/O library was not needed for each board. This still meant that arithmetic operations and so on could be performed but that the ubiquitous `printf` statements and disk access would not be supported. This decision was more to do with time constraints than anything else. Again for time reasons, it was decided to use the standard UNIX-based M680x0 cc compiler running on a UNIX system. The idea was not to test the compiler but to provide a vehicle for testing relative performance. Again, for this reason, no optimisation was done.

A simple C program was written to provide a test vehicle as shown. The `exit()` command was deliberately inserted to force the compiler to explicitly use this function. UNIX systems normally do not need this call and will insert the code automatically. This can cause difficulties when trying to examine the code to see how the compiler produces the code and what is needed to be modified.

```
main()
{
int a,b,c;

a=2;
b=4;
c=b-a;
b=a-c;
exit();
}
```

The example C program

The next stage was to look at the assembler output from the compiler. The output is different from the more normal M68000 assembler printout for two reasons. UNIX-based assemblers use different mnemonics compared to the standard M68000 ones and, secondly, the funny symbols are there to prompt the linker to fill in the addresses at a later stage.

The appropriate assembler source for each line is shown under the line numbers. The code for line 4 of the C source appears in the section headed `ln 4` and so on. Examining the code shows that some space is created on the stack first using the `link.l` instruction. Lines 4 and 5 load the values 2 and 4 into the variable space on the stack. The next few instructions perform the subtraction before the jump to the exit subroutine.

```
file  "math.c"
      data  1
      text
      def   main; val   main; scl   2;    type  044;  endef
      global      main
main:
      ln    1
      def   ~bf;  val   ~;    scl   101;  line  2;    endef
      link.l      %fp,&F%1
#movm.l     &M%1,(4,%sp)
#fmovm      &FPM%1,(FPO%1,%sp)
      def   a;    val   -4+S%1;     scl   1;    type  04;
endef
      def   b;    val   -8+S%1;     scl   1;    type  04;
endef
      def   c;    val   -12+S%1;    scl   1;    type  04;
endef
      ln    4
      mov.l &2,((S%1-4).w,%fp)
      ln    5
      mov.l &4,((S%1-8).w,%fp)
      ln    6
      mov.l ((S%1-8).w,%fp),%d1
      sub.l   ((S%1-4).w,%fp),%d1
      mov.l %d1,((S%1-12).w,%fp)
      ln    7
      mov.l ((S%1-4).w,%fp),%d1
      sub.l   ((S%1-12).w,%fp),%d1
      mov.l %d1,((S%1-8).w,%fp)
      ln    8
      jsr   exit
L%12:
      def   ~ef;  val   ~;    scl   101;  line  9;    endef
      ln    9
#fmovm      (FPO%1,%sp),&FPM%1
#movm.l     (4,%sp),&M%1
      unlk  %fp
      rts
      def   main; val   ~;    scl   -1;   endef
      set   S%1,0
      set   T%1,0
      set   F%1,-16
      set   FPO%1,4
      set   FPM%1,0x0000
      set   M%1,0x0000
      data  1
```

The resulting assembler source code

This means that provided the main entry requirements are to set-up the stack pointer to a valid memory area, the code located at a valid memory address and the exit routine replaced with one more suitable for the target, the code should execute correctly. The first point can be solved during the code downloading. The other two require the use of the linker and replacement run-time routine for exit. All the target boards have an onboard debugger which provides a set of I/O functions including a call to restart the debugger. This would be an ideal way of terminating the program as it would give a definite visual signal of the termination of the software. So what was required was a routine that executed this debugger call. The routine for a Flight MC68020 evaluation board (EVM) is shown. This method is generic for M68000-based VMEbus boards. The other routines were very similar and basically used a different trap call number, e.g. TRAP #14 and TRAP #15 as opposed to TRAP #11. The global statement defines the label exit as an external reference so that the linker can recognise it. Note also the slightly different syntax used by the UNIX assembler. The byte storage command inserts zeros in the following long word to indicate that this is a call to restart the debugger.

```
exit:
global exit
      trap  &11
      byte  0,0,0,0
```

The exit() routine for the MC68020 EVM

This routine was then assembled into an object file and linked with the C source module using the linker. By including the new exit module on the command line with the C source module, it was used instead of the standard UNIX version. If this version was executed on the UNIX machine, it caused a core dump because a TRAP #11 system call is not normal.

```
SECTIONS
{
      GROUP 0x400600:
      {
            .text :{}
            .data :{}
            .bss :{}
      }
}
```

The MC68020 EVM linker command file

The next issue was to relocate the code into the correct memory location. With a UNIX system, there are three sections that are used to store code and data, called .text, .data and .bss. Normally these are located serially starting at the address $00000000. UNIX with its memory management system will translate this address to a different physical address so that the code can execute correctly, instead of corrupting the M68000 vector table which is physically located at this address. With the target boards, this was not possible and the software had to be linked to a valid absolute address.

This was done by writing a small command file with SECTIONS and GROUP commands to instruct the linker to locate the software at a particular absolute address. The files for the MC68020 EVM and for the VMEbus board are shown. This file is included with the other modules on the command line.

```
SECTIONS
{
	GROUP 0x10000:
	{
		.text :{}
		.data :{}
		.bss :{}
	}
}
```

The VMEbus board linker command file

To download the files, the resulting executable files were converted to S-records and downloaded via a serial port to the respective target boards. Using the debugger, the stack pointer was correctly set to a valid area and the program counter set to the program starting address. This was obtained from the symbol table generated during the linking process. The program was then executed and on completion, returned neatly to the debugger prompt, thus allowing time measurements to be made. With the transfer technique established, all that was left was to replace the simple C program with more meaningful code.

To move this code to different M68000-based VMEbus processors is very simple and only the `exit()` routine with its `TRAP` instruction needs to be rewritten. To move it to other processors would require a change of compiler and a different version of the `exit()` routine to be written. By adding some additional code to pass and return parameters, the same basic technique can be extended to access the onboard debugger I/O routines to provide support for `printf()` statements and so on. Typically, replacement `putchar()` and `getchar()` routines are sufficient for terminal I/O.

Creating software state machines

With many real-time applications, a common approach taken with software is first to determine how the system must respond to an external stimulus and then to create a set of state diagrams which correspond with these definitions. With a state diagram, a task or module can only exist in one of the states and can only make a transition to another state provided a suitable event has occurred. While these diagrams are easy to create, the software structure can be difficult.

One way of creating the equivalent of software state diagrams is to use a modular approach and message passing. Each function or state within the system — this can be part of or a whole state diagram — is assigned to a task. The code for the task is extremely simple in that it will do nothing and will wait for a message to arrive. Upon receipt of the message, it will decode it and use the data to change state. Once this has been completed, the task will go back to waiting for further input. The processing can involve other changes of state as well. Referring back to the example, the incoming interrupt will force the task to come out of its waiting state and read a particular register. Depending on the contents of that register, one of two further states can be taken and so on until the final action is to wait for another interrupt.

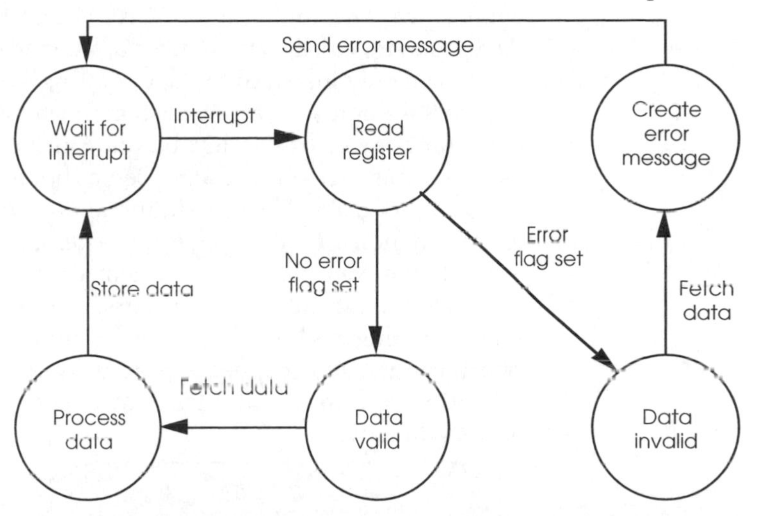

A state diagram

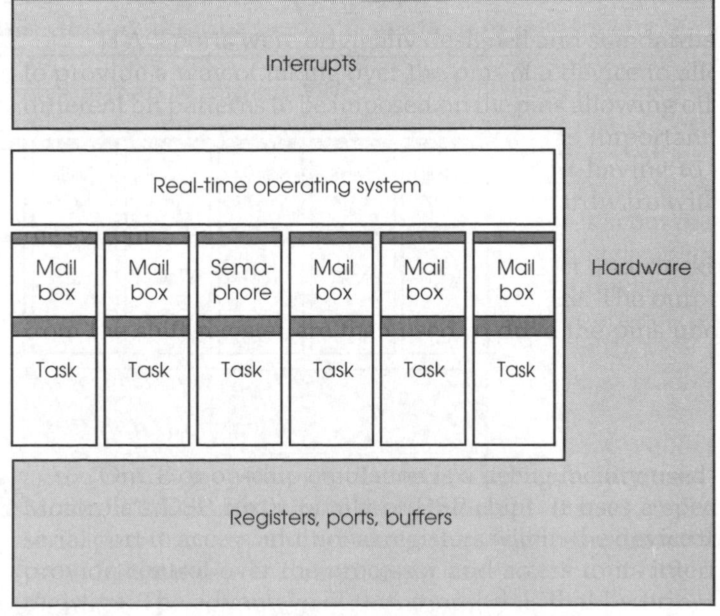

Basic system organisation

This type of code can be written without using an operating system, so what are the advantages? With a multitasking real-time operating system, other changes of state can happen in parallel. By allocating a task to each hardware interrupt, multiple interrupts can easily be handled in parallel. The programmer simply codes each task to handle the appropriate interrupt and the operating system takes care of running the multiple tasks. In addition, the operating system can easily resolve which interrupt or operation will get the highest priority. With complex systems, the priorities may need to change dynamically. This is an easy task for an operating system to handle and is easier to write compared to the difficulty of writing straight line code and coping with the different pathways through the software. The end result is easier code development, construction and maintenance.

The only interface to the operating system is in providing the message and scheduling system. Each task can use messages or semaphores to trigger its operation and during its processing, generate messages and toggle semaphores which will in turn trigger other tasks. The scheduling and time sharing of the tasks are also handled by the operating system.

In the example shown overleaf, there are six tasks with their associated mailboxes or semaphore. These interface to the real-time operating system which handles message passing and semaphore control. Interrupts are routed from the hardware via the operating system, but the tasks can access registers, ports and buffers directly.

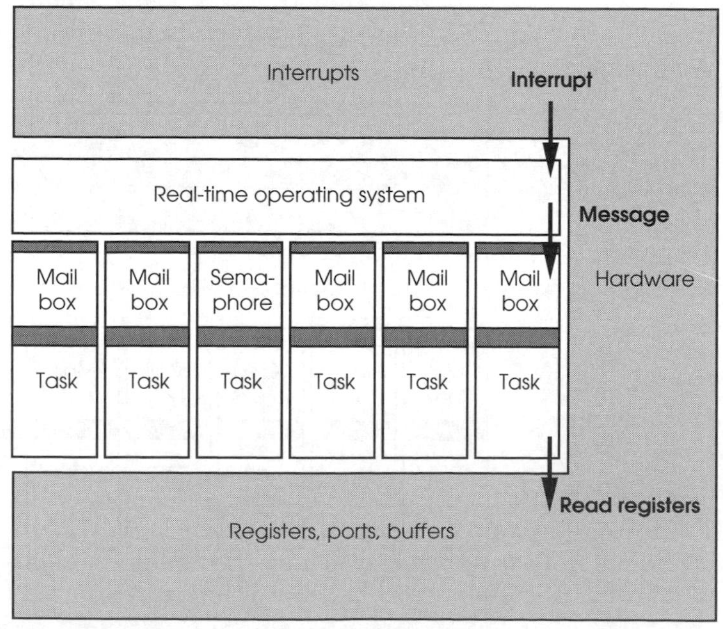

Handling an interrupt

If the hardware generates an interrupt, the operating system will service it and then send a message to the sixth task to perform some action. In this case, it will read some registers. Once read, the task can now pass the data on to another task for processing. This is done via the operating system. The task supplies the data either directly or by using a memory pointer to it and the address of the receiving mail box. The operating system then places this message into the mail box and the receiving task is woken up. In reality, it is placed on the operating system scheduler ready list and allowed to execute.

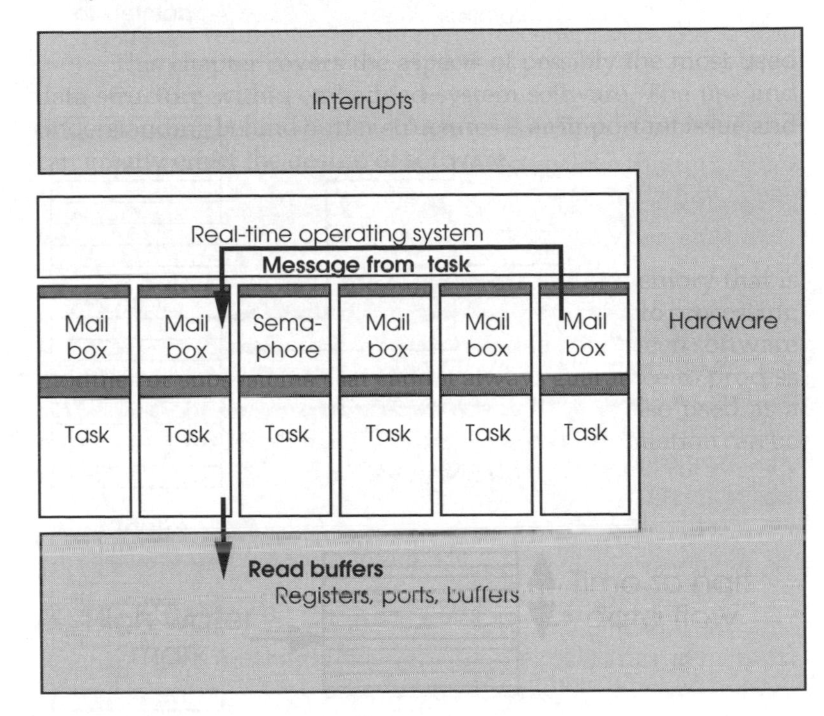

Handling a message from a task

Once woken up, the receiving task can then accept the message and process it. The message is usually designed to contain a code. This data may be an indication of a particular function that the task is needed to perform. Using this value, it can check to see if this function is valid given its current state and, if so, execute it. If not, the task can return an error message back via the operating system.

```
for_ever
      {
      Wait_for_ message();
      Process message();
      Discard_message();
      }
```

Example task software loop

Coding this is relatively easy and can be done using a simple skeleton program. The mechanism used to select the task's response is via two pointers. The first pointer reflects the current state of the task and points to an array of functions. The second pointer is derived from the message and used to index into the array of functions to execute the appropriate code. If the message is irrelevant, then the selected function may do nothing. Alternatively, it may process information, send a message to other tasks or even change its own current state pointer to select a different array of functions. This last action is synonymous to changing state.

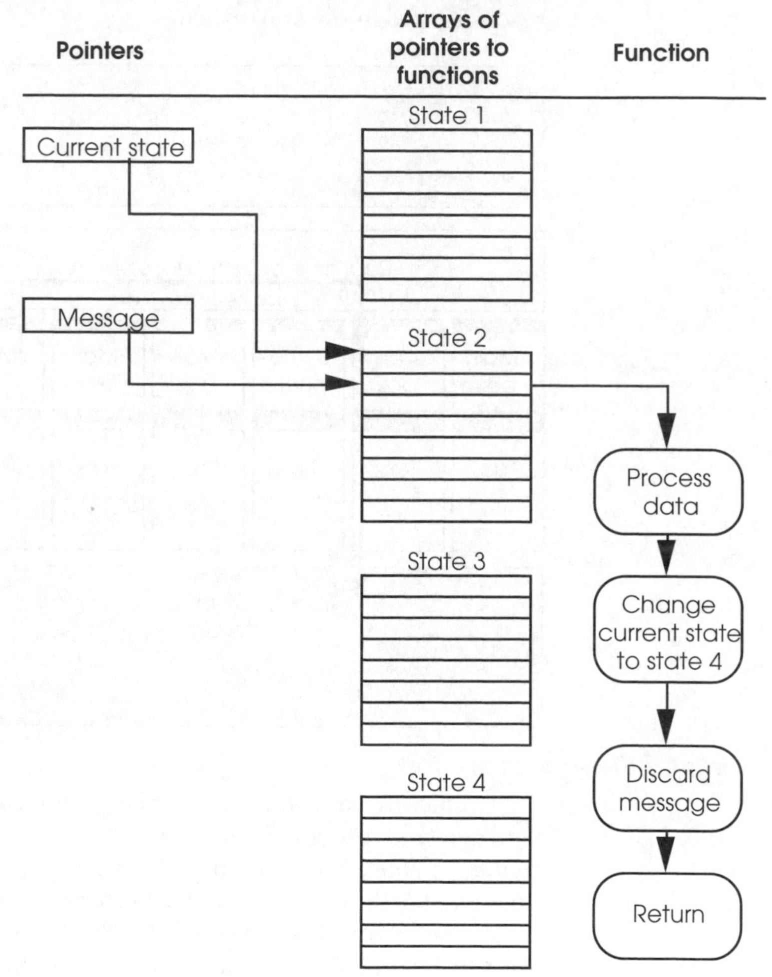

Responding to a message

Priority levels

So far in this example, the tasks have been assumed to have equal priority and that there is no single task that is particularly time critical and must be completed before a particular window expires. In real applications, this is rarely the case and certain routines or tasks are critical to meeting the system requirements. There are two basic ways of ensuring this depending on the facilities offered by the operating system. The first is to set the time critical tasks and routines as the highest priority. This will normally ensure that they will complete in preference to others. However, unless the operating system is pre-emptive and can halt the current task execution and swap it for the time critical one, a lower priority task can still continue execution until the end of its time slot. As a result, the time critical task may have to wait up to a whole time slice period before it can start. In such worse cases, this additional delay may be too long for the routine to complete in response to the original demand or interrupt. If the triggers are asynchronous, i.e. can happen at any time and are not particularly tied to any one event, then the lack of pre-emption can cause a wide range of timings. Unfortunately for real-time systems, it is the worst case that has to be assumed and used in the design.

An alternative approach offered by some operating systems is the idea of an explicit lock where the task itself issues directives to lock itself into permanent execution. It will continue executing until it removes the lock. This is ideal for very time critical routines where the process cannot be interrupted by a physical interrupt or a higher priority task switch. The disadvantage is that it can lead to longer responses by other tasks and in some extreme cases system lock-ups when the task fails to remove the explicit lock. This can be done either with the technique of special interrupt service routines or through calls to the operating system to explicitly lock execution and mask out any other interrupts. Real-time operating systems usually offer at least one or other of these techniques.

Explicit locks

With this technique, the time critical software will make a system call prior to execution which will tell the operating system to stop it from being swapped out. This can also include masking out all maskable interrupts so that only the task itself or a non-maskable interrupt can interrupt the proceedings. The problem with this technique is that it greatly affects the performance of other tasks within the system and if the lock is not removed can cause the task to hog all of the processing time. In addition, it only works once the time critical routine has been entered. If it has to wait until another task has finished then the overall response time will be much lower.

Interrupt service routines

Some operating systems, such as pSOS+, offer the facility of direct interrupt service routines or ISRs where time critical code is executed directly. This allows critical software to execute before other priority tasks would switch out the routines as part of a context switch. It is effectively operating at a very low level and should not be confused with tasks that will activate or respond to a message, semaphore or event. In these cases, the operating system itself is working at the lower level and effectively supplies its own ISR which in turn passes messages, events and semaphores which activate other tasks.

The ISR can still call the operating system, but it will hold any task switching and other activities until the ISR routine has completed. This allows the ISR to complete without interruption.

It is possible for the ISR to send a message to its associated task to start performing other less time critical functions associated with the interrupt. If the task was responsible for reading data from a port, the ISR would read the data from the port and clear the interrupt and send a message to its task to process the data further. After completing, the task would be activated and effectively continue the processing started by the ISR. The only difference is that the ISR is operating at the highest possible priority level while the task can operate at whatever level the system demands.

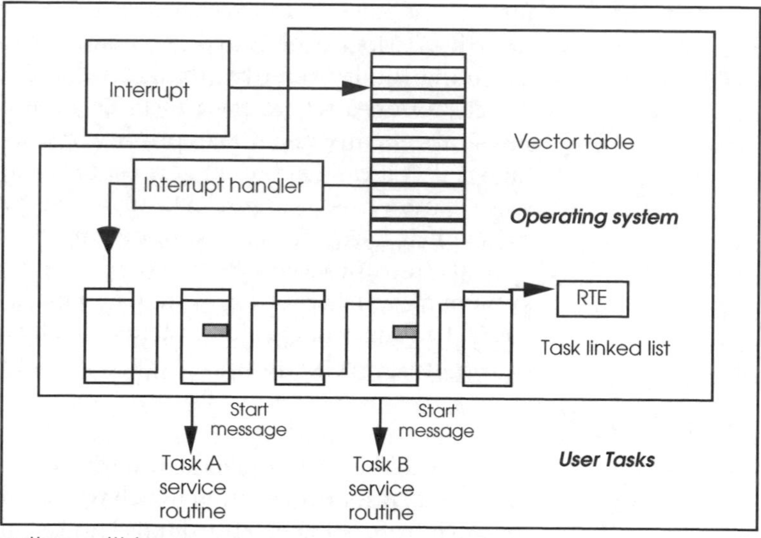

Handling interrupt routines within an operating system

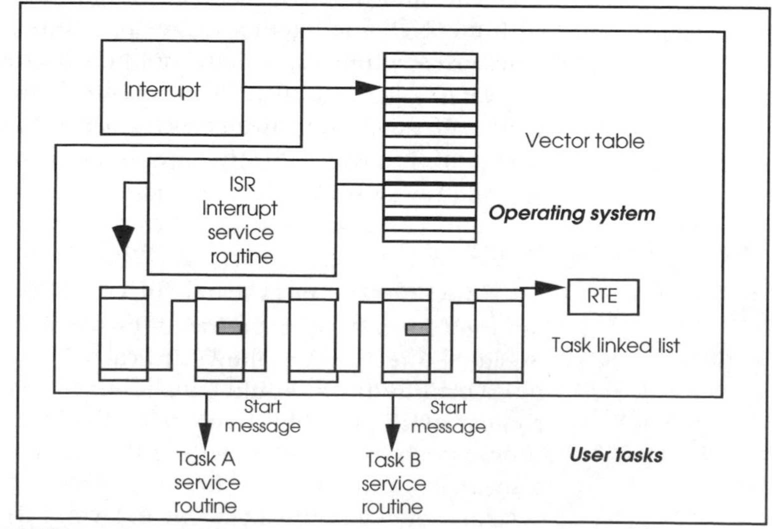

Handling interrupt routines using an ISR

Setting priorities

Given all these different ways of synchronising and controlling tasks, how do you decide which ones to use and how to set them up? There is no definitive answer to this as there are many solutions to the same problem, depending on the detailed characteristics that the system needs to exhibit. The best way to illustrate this is to take an example system and examine how it can be implemented.

The system shown in the diagram below consists of three main tasks. Task A receives incoming asynchronous data and passes this information onto task B which processes it. After processing, task C takes the data and transmits it synchronously as a packet. This operation by virtue of the processing and synchronous nature cannot be interrupted. Any incoming data can fortunately be ignored during this transmission.

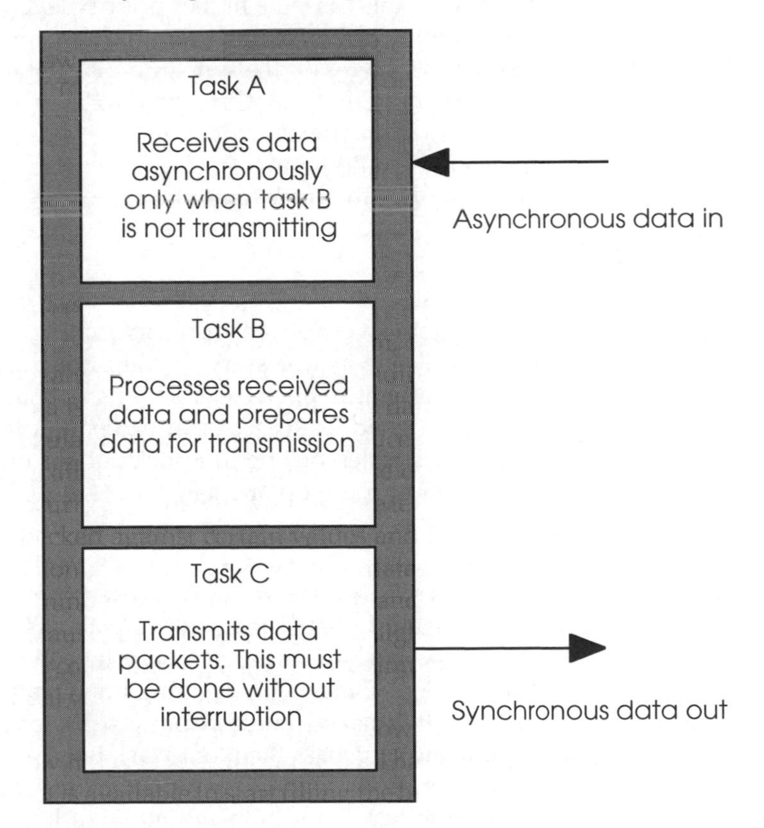

Example system

Task A highest priority

In this implementation, task priorities are used with A having the highest followed by C and finally B. The reasoning behind this is that although task C is the most time critical, the other tasks do not need to execute while it is running and therefore can simply wait until C has completed the synchronous transmission. When this has finished C can wake them up and make itself dormant until it is next required. Task A with its higher priority is then able to pre-empt B when an interrupt occurs, signalling the arrival of some incoming data.

However, this arrangement does require some careful consideration. If task A was woken up while C was transmitting data, task A would replace C by virtue of its higher priority. This would cause problems with the synchronous transmission. Task A could be woken up if it uses external interrupts to know when to receive the asynchronous data. So the interrupt level used by task A must be masked out or disabled prior to moving into a waiting mode and allowing task C to transfer data. This also means that task A should not be allocated a non-maskable interrupt.

Task C highest priority

An alternative organisation is to make task C the highest priority. In this case, the higher priority level will prevent task A from gaining any execution time and thus prevent any interrupt from interfering with the synchronous operation. This will work fine providing that task C is forced to be in a waiting mode until it is needed to transmit data. Once it has completed the data transfer, it would remove itself from the ready list and wait, thus allowing the other tasks execution time for their own work.

Using explicit locks

Task C would also be a candidate for using explicit locks to enable it to override any priority scheme and take as much execution time as necessary to complete the data transmission. The key to using explicit locks is to ensure that the lock is set by all entry points and is released on all exit points. If this is not done, the locks could be left on and thus lock out any other tasks and cause the system to hang up or crash.

Round-robin

If a round-robin system was used, then the choice of executing task would be chosen by the tasks themselves and this would allow task C to have as much time as it needed to transfer its data. The problem comes with deciding how to allocate time between the other two tasks. It could be possible for task A to receive blocks of data and then pass them onto task B for processing. This gives a serial processing line where data arrives with task A, is processed with task B and transmitted by task C. If the nature of the data flow matches this scenario and there is sufficient time between the arrival of data packets at task A to allow task B to process them, then there will be no problem. However, if the arriving data is spread out, then task B's execution may have to be interleaved with task A and this may be difficult to schedule, and be performed better by the operating system itself.

Using an ISR routine

In the scenarios considered so far, it has been assumed that task A does not overlap with task C and they are effectively mutually exclusive. What happens if this is not the case? It really depends on the data transfer rates of both tasks. If they are slow, i.e. the need to send or receive a character is slower than the context switching time, then the normal priority switching can be used with one of the tasks allocated the highest priority. With its synchronous communication, it is likely that it would be task C.

The mechanism would work as follows. Both tasks would be allocated their own interrupt level with C allocated the higher priority to match that of its higher task priority. This is important otherwise it may not get an opportunity to respond to the interrupt routine itself. Its higher level hardware interrupt may force the processor to respond faster but this good work could be negated if the operating system does not allocate a time slice because an existing higher priority task was already running.

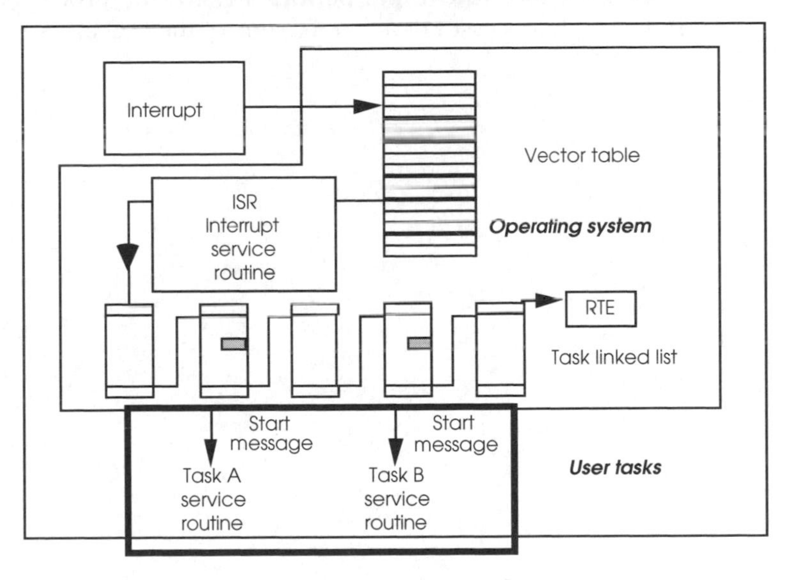

Responding to interrupts

If task A was servicing its own interrupt and a level C interrupt was generated, task A would be pre-empted, task C would start executing and on completion put itself back into waiting mode and thus allow task A to complete. With a pre-emptive system, the worst case latency for task C would be the time taken by the processor to recognise the external task C interrupt, the time taken by the operating system to service it and finally the time taken to perform a context switch. The worst case latency for task A is similar, but with the addition of the worst case value for task C plus another context switch time. The total value is the time taken by the processor to recognise the external task A interrupt, the time taken by the operating system to service it and, finally, the time taken to perform a context switch. The task C latency and context switch time must be added because the task A sequence could be interrupted and replaced at any time by task C. The extra context switch time must be included for when task C completes and execution is switched back to task A.

Provided these times and the time needed to execute the appropriate response to the interrupt fit between the time critical windows, the system will respond correctly. If not then time must be saved.

The diagram shows the mechanism that has been used so far which relies on a message to be sent that will wake up a task. It shows that this operation is at the end of a complex chain of events and that using an ISR, a lot of time can be saved.

The interrupt routines of tasks A and C would be defined as ISRs. These would not prevent context switches but will reduce the decision-making overhead to an absolute minimum and is therefore more effective.

If the time windows still cannot be met, the only solution is to improve the performance of the processor or use a second processor dedicated to one of the I/O tasks.

13 Design examples

Burglar alarm system

This example describes the design and development of an MC68008-based burglar alarm with particular reference to the software and hardware debugging techniques that were used. The design and debugging was performed without the use of an emulator, the traditional development tool, and used cross-compilers and evaluation boards instead. The design process was carefully controlled to allow the gradual integration of the system with one unknown at a time. The decision process behind the compiler choice, the higher level software development and testing, software porting, hardware testing and integration are fully explained.

Design goals

The system under design was an MC68008-based intelligent burglar alarm which scanned and analysed sensor inputs to screen out transient and false alarms. The basic hardware consisted of a processor, a 2k × 8 static RAM, a 32k × 8 EPROM and three octal latches (74LS373) to provide 16 sensor and data inputs and 8 outputs.

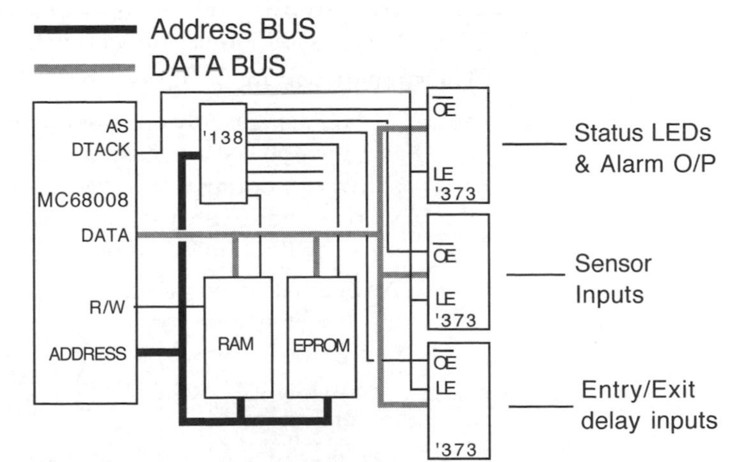

The simplified target hardware

A 74LS138 was used to generate chip selects and output enables for the memory chips and latches from three higher order address lines. Three lines were left for future expansion. The sirens etc., were driven via 5 volt gate power MOSFETs. The controlling software was written in C and controlled the whole timing and response of the alarm system. Interrupts were not used and the power on reset signal generated using a CR network and a Schmidt gate.

Development strategy

The normal approach would be to use an in-circuit emulator to debug the software and target hardware, but it was decided at an early stage to develop the system without using an emulator except as a last resort. The reasoning was simple:

- The unit was a replacement for a current analogue system, and the physical dimensions of the case effectively prevented the insertion of an emulation probe. In addition, the case location was very inaccessible.
- The hardware design was a minimum system which relied on the MC68008 signals to generate the asynchronous handshakes automatically, e.g. the address strobe is immediately routed back to generate a DTACK signal. This configuration reduces the component count but any erroneous accesses are not recognised. While these timings and techniques are easy to use with a processor, the potential timing delays caused by an emulator could cause problems which are exhibited when real silicon is used.
- The software development was performed in parallel with the hardware development and it was important that the software was tested in as close an environment as possible to a debugged target system early on in the design. While emulators can provide a simple hardware platform, they can have difficulties in coping with power-up tests and other critical functions.

The strategy was finally based on several policies:

- At every stage, only one unknown would be introduced to allow the fast and easy debugging of the system, e.g. software modules were developed and tested on a known working hardware platform, cross-compiled and tested on an evaluation board etc.
- An evaluation board would be used to provide a working target system for the system software debugging. One of the key strategies in this approach for the project was to ensure the closeness of this environment to the target system.
- Test modules would be written to test hardware functionality of the target system, and these were tested on the evaluation board.
- The system software would only be integrated on the target board if the test modules executed correctly.

Software development

The first step in the development of the software was to test the logic and basic software design using a workstation. A UNIX workstation was initially used and this allowed the bulk of the software to be generated, debugged and functionally tested within a known working hardware environment, thus keeping with the single unknown strategy. This restricts any new unknown software or hardware to a single component and so makes debugging easier to perform.

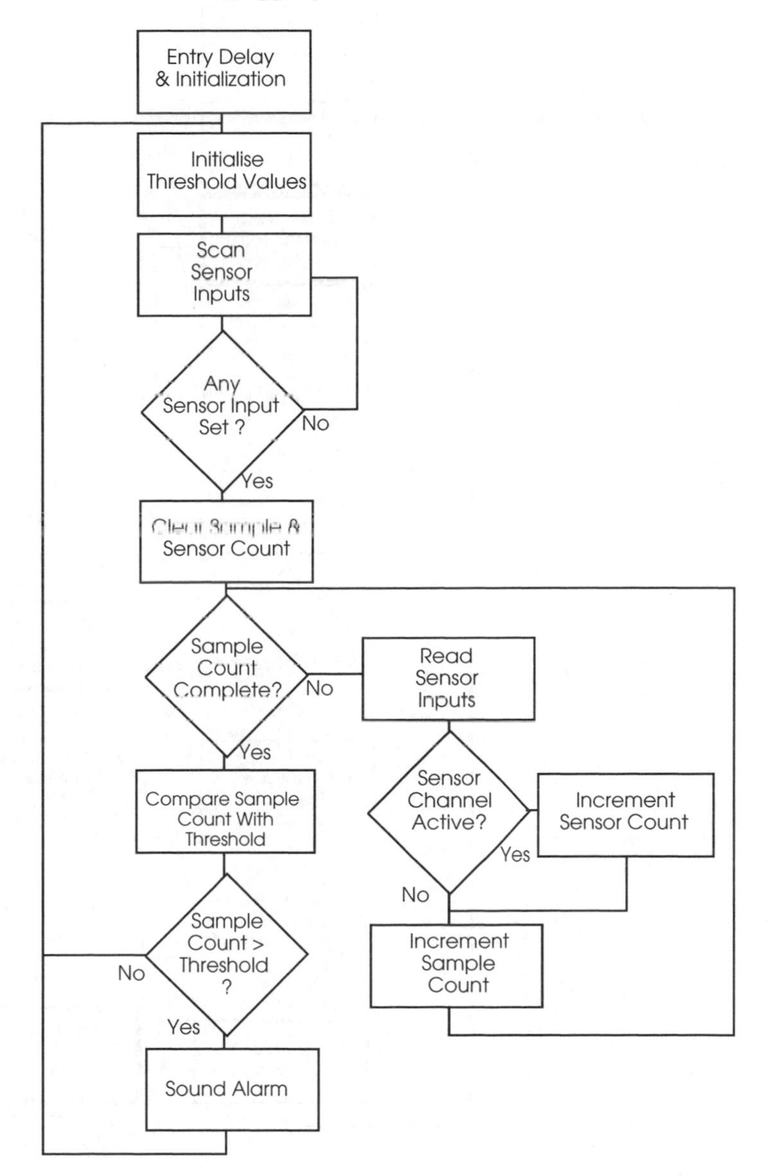

The main software flow diagram

Sensor inputs were simulated using `getchar()` to obtain values from a keyboard, and by using the multitasking signalling available with a UNIX environment. As a result, the keyboard could be used to input values with the hexadecimal value representing the input port value. Outputs were simulated using a similar technique using `printf()` to display the information on the screen. Constants for software delays etc., were defined using `#define C` pre-processor statements to allow their easy modification. While the system could not test the software in real-time, it does provide functional and logical testing.

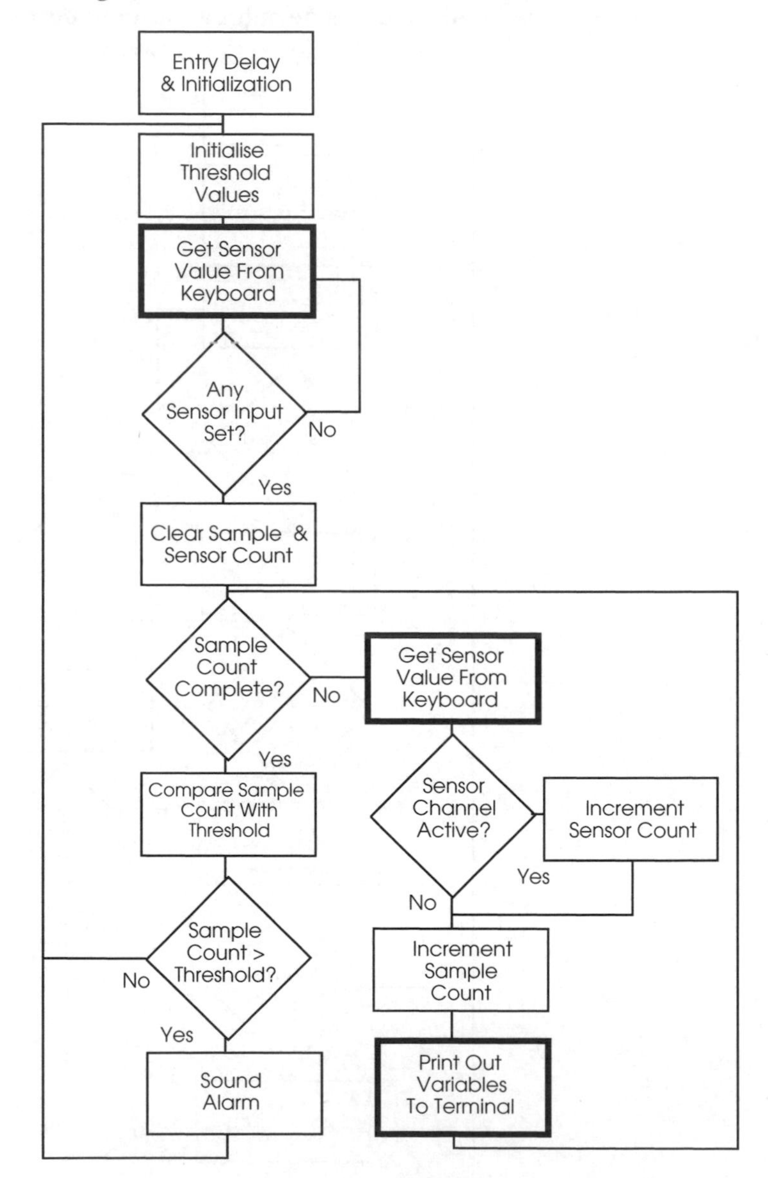

The modified software flow diagram

While it is easy to use the `getchar()` routine to generate an integer value based on the ASCII value of the key that has been pressed, there are some difficulties. The first problem that was encountered was that the UNIX system uses buffered input for the keyboard. This meant that the return key had to be pressed to signal the UNIX machine to pass the data to the program. This initially caused a problem in that it stopped the polling loop to wait for the keyboard every time it went round the loop. As a result, the I/O simulation was not very good or representative.

In the end, two solutions were tried to improve the simulation. The simplest one was to use the waiting as an aid to testing the software. By having to press the return key, the execution of the polling loop could be brought under the user's control and by printing out the sample and sensor count variables, it was possible to step through each loop individually. By effectively executing discrete polling loops and have the option of simulating a sensor input by pressing a key before pressing return, other factors like the threshold values and the disabling of faulty sensors could be exercised.

The more complex solution was to use the UNIX multitasking and signalling facilities to set-up a second task to generate simple messages to simulate the keyboard input. While this allowed different sensor patterns to be sent to simulate various false triggering sequences without having the chore of having to calculate the key sequences needed from the keyboard, it did not offer much more than that offered by the simple solution.

The actual machine used for this was an OPUS personal mainframe based on a MC88100 RISC processor running at 25 MHz and executing UNIX System V/88. The reasons behind this decision were many:

- It provided an extremely fast system, which dramatically reduced the amount of time spent in compilation and development.
- The ease of transferring software source to other systems for cross-compilation. The OPUS system can directly read and write MS-DOS files and switch from the UNIX to PC environment at a keystroke.
- The use of a UNIX- based C compiler, along with other utilities such as lint, was perceived to provide a more standard C source than is offered by other compilers.

This was deemed important to prevent problems when cross-compiling. Typical errors that have been encountered in the past are: byte ordering, variable storage sizes, array and variable initialisation assumed availability of special library routines etc.

Cross-compilation and code generation

Three MC68000 compilation environments were available: the first was a UNIX C compiler running on a VMEbus system, the second was a PC-based cross-compiler supplied with the Motorola MC68020 evaluation board, while a third option of another PC-based cross compiler was a possibility.

The criteria for choosing the cross-compilation development were:

- The ease of modifying run-time libraries to execute standalone on the MC68020 evaluation board and, finally the target system.
- The quality of code produced.

The second option was chosen primarily for the ease with which the run-time libraries could be modified. As standard, full run-time support was available for the evaluation board, and these modules, including the all-important initialisation routines, were supplied in assembler source and are very easy to modify. Although the code quality was not as good as the other options, it was adequate for the design and the advantage of immediate support for the software testing on the evaluation board more than compensated. This support fitted in well with the concept of introducing only a single unknown.

If the software can be tested in as close an environment as possible to the target, any difficulties should lie with the hardware design. With this design, the only differences between the target system configuration and the evaluation board is a different memory map. Therefore, by testing the software on an evaluation board, one of the unknowns can be removed when the software is integrated with the target system. If the run-time libraries are already available, this further reduces the unknowns to that of just the system software.

The C source was then transferred to a PC for cross-compilation. The target system was a Flight MC68020 evaluation board which provides a known working MC68xxx environment and an on-board debugger. The code was cross-compiled, downloaded via a serial link and tested again.

The testing was done in two stages: the first simply ran the software that had been developed using the OPUS system without modifying the code. This involved using the built-in `getchar()` system calls and so on within the Flight board debugger. It is at this point that any differences between the C compilers would be found such as array initialisation, bit and byte alignment etc. It is these differences that can prevent software being ported from one system to another. These tests provided further confidence in the software and its functionality before further modification.

The second stage was to replace these calls by a pointer to a memory location. This value would be changed through the use of the onboard debugger. The evaluation board abort button is pressed to stop the program execution and invoke the debugger. The corresponding memory location is modified and the program execution restarted. At this point, the software is virtually running as expected on the target system. All that remains to do is to take into account the target hardware memory map and initialisation.

Porting to the final target system

The next stage involved porting the software to the final target configuration. These routines allocate the program and stack memory, initialise pointers etc., and define the sensor and display locations within the memory map. All diagnostic I/O calls were removed. The cross-compiler used supplies a start-up assembly file which performs these tasks. This file was modified and the code recompiled all ready for testing.

Generation of test modules

Although the target hardware design was relatively simple, it was thought prudent to generate some test modules which would exercise the memory and indicate the success by lighting simple status LEDs. Although much simpler than the controlling software, these go-nogo tests were developed using the same approach: written and tested on the evaluation board, changed to reflect the target configuration and then blown into EPROM.

The aim of these tests was to prove that the hardware functioned correctly: the critical operations that were tested included power-up reset and initialisation, reading and writing to the I/O functions, and exercising the memory.

These routines were written in assembler and initially tested using the Microtec Xray debugger and simulator before downloading to the Flight board for final testing.

Target hardware testing

After checking the wiring, the test module EPROM was installed and the target powered up. Either the system would work or not. Fortunately, it did! With the hardware capable of accessing memory, reading and writing to the I/O ports, the next stage was to install the final software.

While the system software logically functioned, there were some timing problems associated with the software timing loops which controlled the sample window, entry/exit time delays and alarm duration. These errors were introduced as a direct result of the software methodology chosen: the delay values would obviously change depending on the hardware environment, and while the values were defined using `#define` pre-processor statements and were adjusted to reflect the processing power available, they were inaccurate. To solve this problem, some additional test modules were written, and by using trial and error, the correct values were derived. The system software was modified and installed.

Future techniques

The software loop problem could have been solved if an MC68000 software simulator had been used to execute, test and time the relevant software loops. This would have saved a day's work.

If a hardware simulator had been available, it could have tested the hardware design and provided additional confidence that it was going to work.

Relevance to more complex designs

The example described is relatively simple but many of the techniques used are extremely relevant to more complex designs. The fundamental aim of a test and development methodology, which restricts the introduction of untested software or hardware to a single item, is a good practice and of benefit to the design and implementation of any system, even those that use emulation early on in the design cycle.

The use of evaluation boards or even standalone VME or Multibus II boards can be of benefit for complex designs. The amount of benefit is dependent of the closeness of the evaluation board hardware to the target system. If this design had needed to use two serial ports, timers and parallel I/O, it is likely that an emulator would still not have been used provided a ready built board was available which used the same peripheral devices as the target hardware. The low level software drivers for the peripherals could be tested on the evaluation board and these incorporated into the target test modules for hardware testing.

There are times, however, when a design must be tested and an emulator is simply not available. This scenario occurs when designing with a processor at a very early stage of product life. There is inevitably a delay between the appearance of working silicon and instrumentation support. During this period, similar techniques to those described are used and a logic analyser used instead of an emulator to provide instrumentation support, in case of problems. If the processor is a new generation within an existing family, previous members can be used to provide an interim target for some software testing. If the design involves a completely new processor, similar techniques can be applied, except at some point untested software must be run on untested hardware.

It is to prevent this integration of two unknowns that simulation software to either simulate and test software, hardware or both can play a critical part in the development process.

The need for emulation

Even using the described techniques, it cannot be stated that there will never be a need for additional help. There will be times when instrumentation, such as emulation and logic analysis, is necessary to resolve problems within a design quickly. Timing and intermittent problems cannot be easily solved without access to further information about the processor and other system signals. Even so, the recognition of a potential problem source such as a specific software module or hardware allows a more constructive use and a speedier resolution. The adoption of a methodical design approach and the use of ready built boards as test vehicles may, at best, remove the need for emulation and, at worst, reduce the amount of time debugging the system.

Digital echo unit

This design example follows the construction of a digital echo unit to provide echo and reverb effects.

With sound samples digitally recorded, it is possible to use digital signal processing techniques to create far better and more flexible effects units (or sound processors, as they are more commonly called). Such units comprise a fast digital signal processor with A to D and D to A converters and large amounts of memory. An analogue signal is sent into the processor, converted into the digital domain, processed using software running on the processor to create filters, delay, reverb and other effects before being converted back into an analogue signal and being sent out.

They can be completely software based, which provides a lot of flexibility, or they can be pre-programmed. They can take in analogue or, in some cases, digital data, and feed it back into other units or directly into an amplifier or audio mixing desk, just like any other instrument.

Creating echo and reverb

Analogue echo and reverb units usually rely on an electromechanical method of delaying an audio signal to create reverberation or echo. The WEM Copycat used a tape loop and a set of tape heads to record the signal onto tape and then read it from the three or more tape heads to provide three delayed copies of the signal.

The delay was a function of the tape speed and the distance between the recording and read tape heads. This provides a delay of up to 1 second. Spring line delays used a transducer to send the audio signal mechanically down a taut spring where the delayed signal would be picked up again by another transducer.

Bucket brigade devices have also been used to create a purely electronic delay. These devices take an analogue signal and pass it from one cell to another using a clock. The technique is similar to passing a bucket of water by hand down a line of men. Like the line of men, where some water is inevitably lost, the analogue signal degrades — but it is good enough to achieve some good effects.

With a digitised analogue signal, creating delayed copies is easy. The samples can be stored in memory in a buffer and later retrieved. The advantage this offers is that the delayed

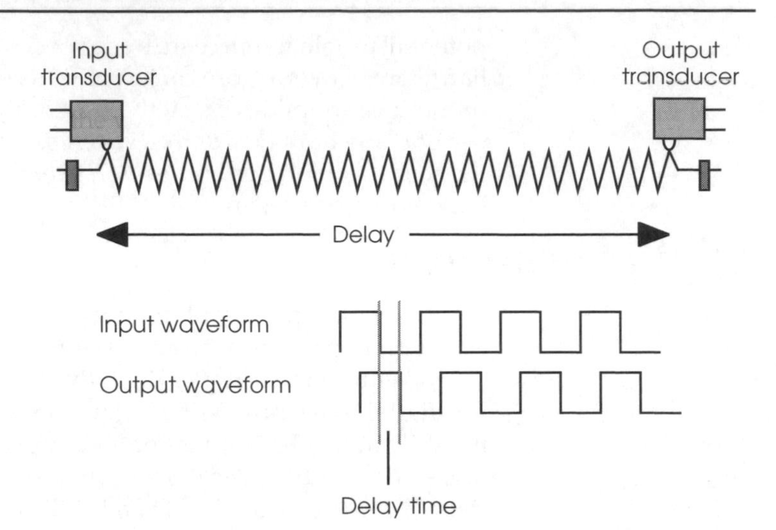

Spring line delay

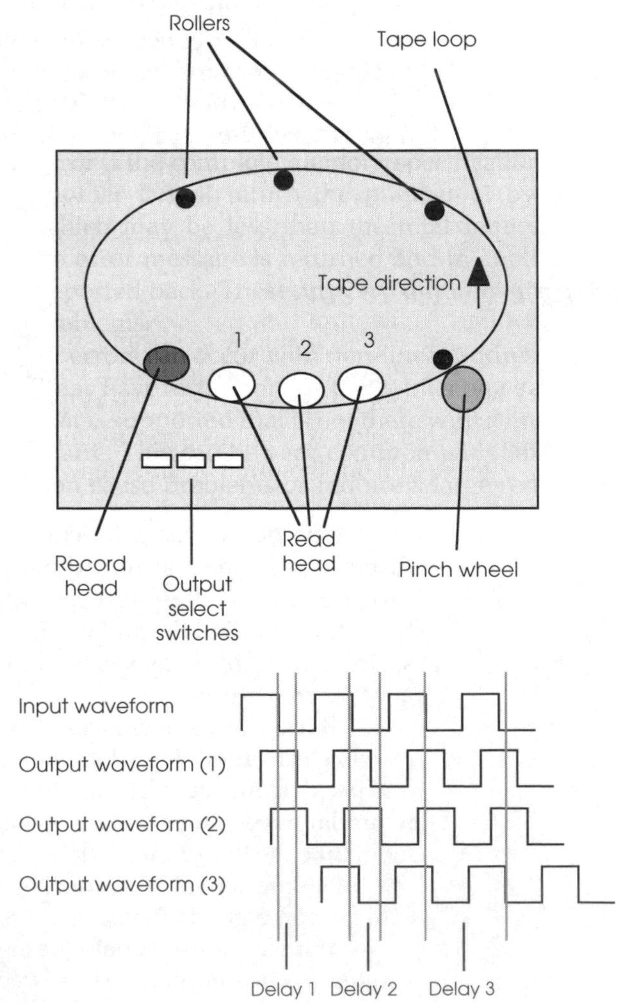

A tape loop-based delay unit

sample is an exact copy of the original sound and, unlike the techniques previously described, has not degraded in quality or had tape noise introduced. The number of delayed copies is dependent on the number of buffers and hence the amount of memory that is available. This ability, coupled with a signal processor allows far more accurate and natural echoes and reverb to be created.

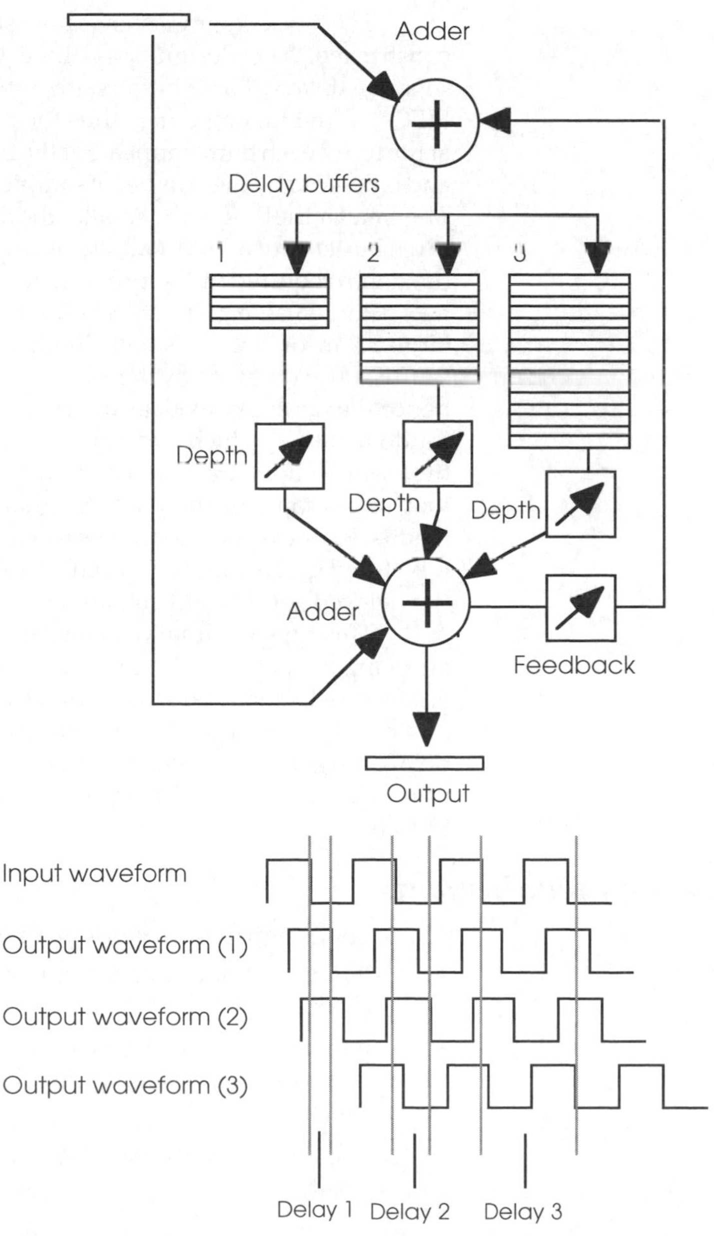

A digital echo/reverb unit

The problem with many analogue echo and reverb units is that they simplify the actual reverb and echo. In natural conditions, such as a large concert hall, there are many delay sources as the sound bounces around and this cannot be reproduced with only two or three voices which are independently mixed together with a bit of feedback. The advantage of the digital approach is that as many delays can be created as required and the signal processor can combine and fade the different sources as needed to reproduce the environment required.

The block diagram shows how such a digital unit can be constructed. The design uses three buffers to create three separate delays. These buffers are initially set to zero and are FIFOs — first in, first out — thus the first sample to be placed at the top of each buffer appears at the bottom at different times and is delayed by the number of samples that each buffer holds. The smaller the buffer, the smaller the delay. The outputs of the three buffers are all individually reduced in size according to the depth required or the prominence that the delayed sound has. A large value gives an echoing effect, similar to that of a large room or hall. A small depth reduces it. The delayed samples are combined by the adder with the original sample — hence the necessity to clear the buffer initially to ensure that random values, which add noise, do not get added before the first sample appears — to create the final effect. A feedback loop takes some of the output signal, as determined by the feedback control, and combines it with the original sample as it is stored in the buffers. This effectively controls the decay of the delayed sounds and creates a more natural effect.

This type of circuit can become more sophisticated by adapting the depth with time and having separate independent feedback loops and so on. This circuit can also be the basis of other effects such as chorus, phasing and flanging where the delayed signal is constantly delayed but varies. This can be done by altering the timing of the sample storage into the buffers.

Design requirements

The design requirements for the echo unit are as follows:

- It must provide storage for at least one second on all its channels.
- It must provide control over the echo length and depth.
- It must take analogue signals in and provide analogue signals out.
- The audio quality must be good with a 20 kHz bandwidth.

Designing the codecs

The first decision concerns the A to D and D to A codec design. Many lower specification units use 8 bit A to D and D to A units to digitise and convert the delayed analogue signal. This signal does not need to be such good quality as the original and using an 8 bit resolution converter saves on cost and reduces the amount of memory needed to store the delayed signal. Such systems normally add the delayed signal in the analogue domain and this helps to cover any quality degradation.

With this design, the quality requirement precludes the use of 8 bit converters and effectively dictates that a higher quality codec is used. With the advent of the Compact Disc, there are now plenty of high quality audio codecs available with sample sizes of 12 or more bits. A top end device would use 16 bit conversion and this would fit nicely with 16 bit memory. This is also the sample size used with Compact Disc.

The next consideration is the conversion rate. To achieve a bandwidth of 20 kHz, a conversion rate of 40 kHz is needed. This has several knock-on effects: it determines the number of samples needed to store one second of digital audio and hence the amount of memory. It also defines the timing that the system must adhere to remove any sampling errors. The processor must be able to receive the digitised audio, store it and copy it as necessary, retrieve the output samples, combine them and convert them to the analogue signals every 25 μs.

Designing the memory structures

In examining the codec design, some of the memory requirements have already started to appear. The first requirement is the memory storage for the digital samples. For a single channel of delay where only a single delayed audio signal is combined with the original signal, the memory storage is the sample size multiplied by the sample rate and the total storage time taken. For a 16 bit sample and a 40 kHz rate, 80000 bytes of storage needed. Rounding up, this is equivalent to just over 78 kbytes of storage (a kbyte of memory is 1024 bytes and not 1000).

This memory needs to be organised as a by 16 structure which means that the final design will need 40 k by 16 words of memory per second of audio. For a system with three delayed audio sources, this is about 120 k words which works out very nicely at two 128k by 8 RAM chips. The spare 8 kbytes in each RAM chip can be used by the supervisor software that will run on the control processor.

Now that the amount of memory is known, then the memory type and access speed can be worked out. DRAM is applicable in this case but requires refresh circuitry and because it is very high density may not be cost effective. If 16 Mb

DRAM is used then with a by 16 organisation, a single chip would provide 1 Mbyte of data storage which is far too much for this application. The other potential problem is the effect of refresh cycles which would potentially introduce sampling errors. This means that static RAM is probably the best solution.

To meet the 25 μs cycle time which includes a minimum of a data read and a data write, this means that the overall access time must be significantly less than half of the cycle time, i.e. less than 12.5 μs. This means that almost any memory is capable of performing this function.

In addition, some form of non-volatile memory is needed to contain the control software. This would normally be stored in an EPROM. However, the EPROM access times are not good and therefore may not be suitable for running the software directly. If the control program is small enough, then it could be transferred from the EPROM to the FSRAM and executed from there.

The software design

The software design is relatively simple and treats the process as a pipeline. While the A to D is converting the next sample, the previous sample is taken and stored in memory using a circular buffer to get the overall delay effect. The next sample for D to A conversion and output is retrieved from the buffer and sent to the converter. The circular buffer pointers are then updated, including checking for the end of the buffer.

This sequence is repeated every 25 μs. While the processor is not performing this task, it can check and maintain the user controls. As stated previously, circular buffers are used to hold the digitised data. A buffer is used with two pointers: one points to the next storage location for the incoming data and a second pointer is used to locate the delayed data. The next two diagrams show how this works. Each sample is stored consecutively in memory and the two pointers are separated by a constant value which is equivalent to the number of samples delay that is required. In the example shown, this is 16 samples. This difference is maintained so that when a new sample is inserted, the corresponding old value is removed as well and then both pointers are updated.

When either pointer reaches the end of the data block, its value is changed to point to the next location. In the example shown, the `New_data` pointer is reset to point at the first location in the buffer which held the first sample. This sample is no longer needed and its value can be overwritten. By changing the difference between the two pointers, the time delay can be changed. In practice, the pointers are simply memory addresses and every time they are updated, they should be checked and if necessary reset to the beginning of the table. This form of addressing is known as modulo addressing and some DSP processors support it directly and therefore do not need to check the address.

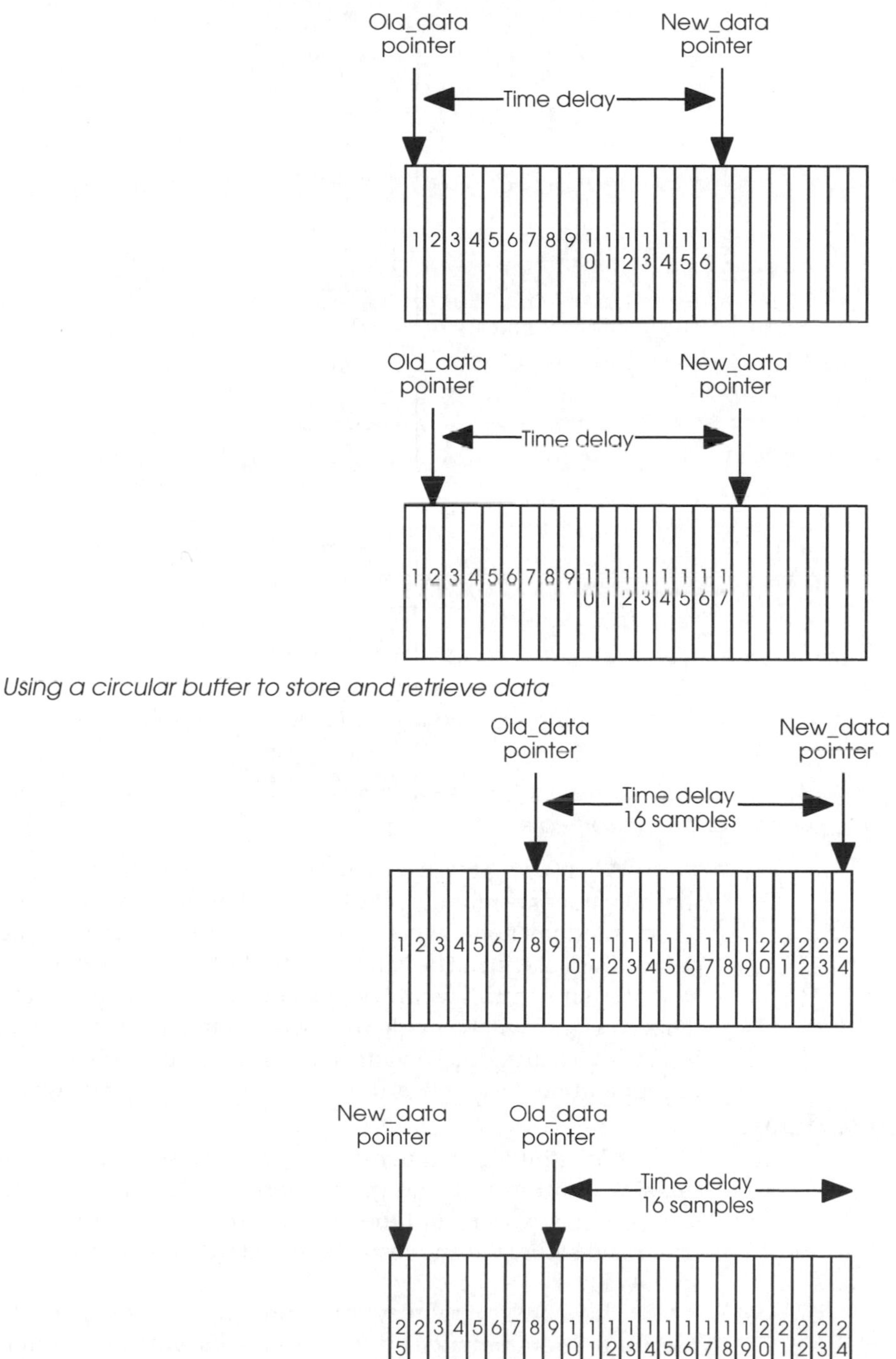

Using a circular buffer to store and retrieve data

Implementing modulo addressing

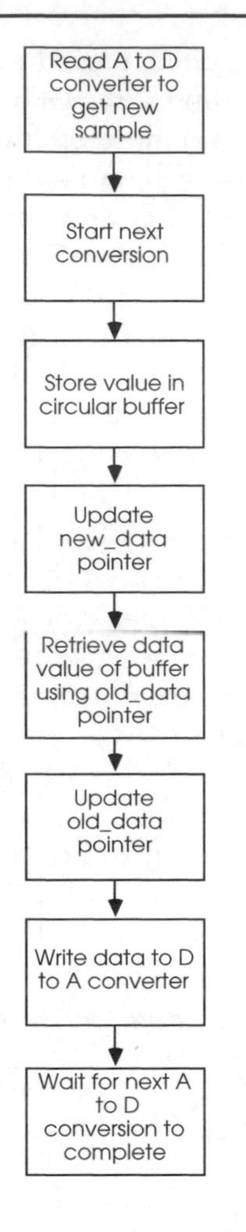

The basic pipeline flow for the software

When using these structures, it is important to ensure that all values are initially set to zero so that the delayed signal is not random noise when the system first starts up. The delayed signal will not be valid until the buffer has filled. In the examples shown, this would be 16 samples. Until this point, the delayed signal will be made from the random contents of the buffer. By clearing these values to zero, silence is effectively output and no noise is heard until the correct delayed signal.

Multiple delays

With a multiple source system, the basic software design remains intact except that the converted data is copied into several delay buffers and the outputs from these buffers are combined before the end result is converted into the analogue signal.

There are several ways of setting this up. The first is to use multiple buffers and copy each new value into each buffer. Each buffer then supplies its own delayed output which can be combined to create the final effect. A more memory efficient system is to use a single buffer but add additional `old_data` pointers with different time delays to create the different delay length outputs.

The overhead in doing this is small. There is the maintenance of the pointers to be done and the combination of the delay values to create the final output for the D to A converter. This can be quite complex depending on the level of sophistication needed.

Digital or analogue adding

There are some options depending on the processing power available. With a real echo or reverb, the delayed signals need to be gradually attenuated as the signals die away and therefore, the delayed signal must be attenuated. This can be done either digitally or in the analogue domain. With a single source, the analogue implementation is easy. The delayed signal is converted and an analogue mixer is used to attenuate and combine the delayed signal with the original to create the reverb or echo effect. An analogue feedback bath can also be created.

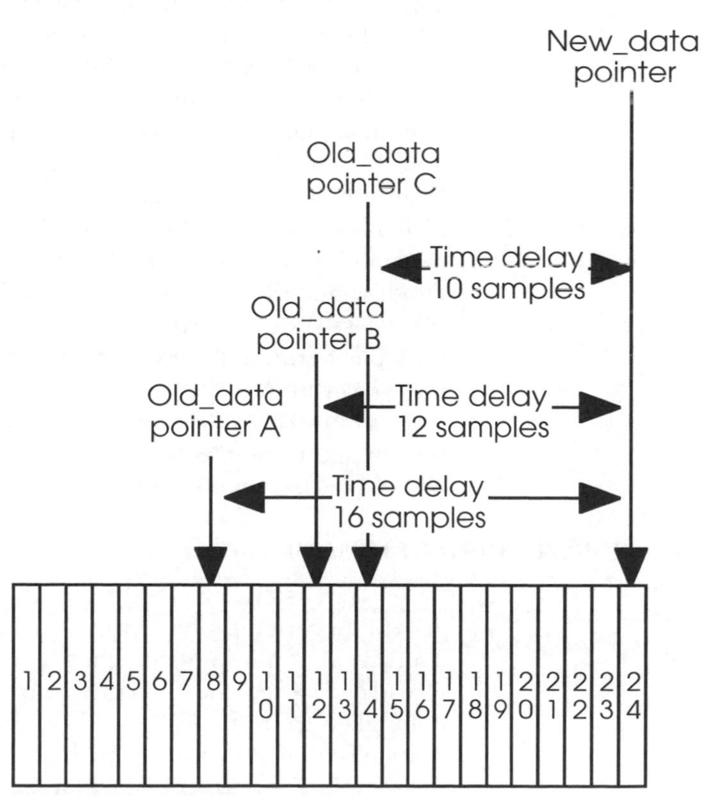

Using a single buffer with multiple pointers to create multiple delays

The multiple delayed source design can use this same analogue method but requires a separate D to A converter for each delayed signal. This can be quite expensive. Instead, the processor can add the signals together, along with attenuation factors to create a combined delay signal that can be sent to the

D to A converter for combination with the original analogue signal. It is therefore possible to perform all this, including the combination with the original signal in the digital domain, and then simply output the end value to the D to A converter. With this version, the attenuation does not need to be constant and can be virtually any type of curve.

The disadvantage is the computation that is needed. The arithmetic that is required is saturation arithmetic which is a little more than simply adding two values together. This is needed to ensure that the combined value only provides a peak value and not cause an overflow error. In addition, all the calculations must be done within 25 μs to meet the sampling rate criteria and this can be pushing the design a little with many general-purpose processors.

Microprocessor selection

The choice of microprocessor is dependent on several factors. It must have an address range of greater than 64 kbytes and have a 16 bit data path. It must be capable of performing 16 bit arithmetic and thus this effectively rules out 8 bit microprocessors and microcontrollers.

In terms of architecture, multiple address pointers that auto-increment would make the circular buffer implementations very efficient and therefore some like a RISC processor or a fast MC68000 would be suitable. Other architectures can certainly do the job but their additional overhead may reduce their ability to perform all the processing within the 25 μs window that there is. One way of finding this out is to create some test code and run it on a simulator or emulator, for example, to find out how many clocks it does take to execute these key routines.

A low cost DSP processor is also quite attractive in this type of application, especially if it supports modulo addressing and saturation arithmetic.

The overall system design

The basic design for the system uses a hardware timer to generate a periodic interrupt every 25 μs. The associated interrupt service routine is where the data from the A to D converter is read and stored, the next conversion started and the delayed data taken from the buffer and combined. The pointers are updated before returning from the service routine. In this way, the sampling is done on a regular basis and is given a higher priority than the background processing.

While the processor is not servicing the interrupt, it stays in a forever loop, polling the user interface for parameters and commands. The delay times are changed by manipulating the pointers. It is possible to do this by changing the sampling rate instead but the audio quality does not stay constant.

The system initialises by clearing the RAM to zero and using some of it to hold the program code which is copied from EPROM. If a battery backed SRAM is used instead, then used defined parameters and settings could be stored here as well and retained when the system is switched off.

14 Real-time without a RTOS

This chapter describes the design and development of a real-time data logger that is used to collate information from various data logging sources in a race car. It brings together many of the topics discussed in previous chapters into a real world project. The project's goal was to design a real-time data logger. Its function was to fetch data samples from the various systems in the car and store them locally for display once the car has finished competing. The car is fitted with several computer based control systems, including a engine management unit (EMU) and traction control system that would provide a snapshot of the current input data (all four wheel speeds, engine revs and traction control intervention) when prompted. The EMU communicates using a serial port at 19.2 kbaud. The system comes with some basic data logging software that would run on a PC laptop but this was not very reliable at time stamping. The timing would depend on the performance of the laptop that received the data which made 'before and after' comparisons very difficult to make. What was needed was an embedded system that could periodically request a data sample and then store it in a format that the standard display software could use. Simple: well, as with most designs, the reality was slightly different.

Choosing the software environment

So where do you start with a project like this? Most designers will often start by defining the hardware and then rely on the software to either cope or fit with the hardware selection. This is often adopted because of the so-called flexibility that software offers. It's only software, it be changed, rewritten, and so on. While there is some truth in this, it should be understood that many software components such as the operating system and compiler cannot be modified and that it can be as fixed as hardware. In practice, these decisions have to be taken in conjunction and the design based on a system approach that takes into account both hardware and software issues. However, you have to start somewhere and in this case, the software environment was given the highest priority.

The first question to be answered was that of which operating system. In practice there were three candidates: a real-time operating system of some kind, MS-DOS or Windows. The design was definitely a real-time one with a deadline to meet to maintain the correct sampling. It probably needed to provide access to the hardware directly due to the slightly different use that the parallel ports in particular were going to be put to. It also needed to be simple and stable. For most engineers, the immediate reaction would be to use a real-time OS. The challenge here is that these are expensive (compared to a copy of MS-DOS or Windows) and it would be difficult to use the target system directly for the development. MS-DOS was easily available and permitted easy access to hardware and could be both the target and development environment. Windows was ruled out because of its size, lack of direct hardware access and its complexity. A data logger did not need to display different backgrounds for a start. In addition, Windows does impose a huge overhead on the hardware resources which again was not attractive.

The decision was taken to use MS-DOS. The Borland C compiler for MS-DOS can be downloaded from Borland's website free of charge and is a fast compact compiler. It might be old but it is more than adequate for the job. This complemented the idea of using the laptop target system for both the actual implementation as well as the final target. The compiler also provides extensive library support for low-level direct access to hardware and BIOS routines. These would provide a rich source of function with which to control the data logging without the need for assembler programs.

Does this decision mean that MS-DOS is a real-time operating system? Well it all depends. It is possible to design real-time systems without the need for a real-time operating system, providing the system designer understands the constraints imposed on the design when doing this. With MS-DOS, this invariably means a single thread of execution with a single task running and performing all the work. This may have procedure calls and so on and be a structured modular program but there is only one thread of execution running. If the design required multiple threads then this can be difficult in MS-DOS as it does not support this functionality. This doesn't mean it cannot be done but it requires the 'multitasking' to be embedded into the single task. In effect the multitasking support is taken out of the operating system and built into the program. This can quickly get very complex and cumbersome to write and maintain and this is where the true multitasking operating system can come to the fore. As soon as this step is taken, there is usually an increased need for a real-time operating system. Windows is multitasking but it is not real-time. It can do real-time work providing the rest of the software running in the system co-operates and shares processing time with the other tasks and threads running in the system. If this co-operation is not maintained, the system can appear to hang up because another that is not co-operating and hogging the CPU blocks the waiting task that needs to meet a deadline. The operating system is powerless in this situation to do anything about this. With a real-time operating system, the waiting task can pre-empt the hogging task if it has a high enough priority.

Given that MS-DOS was to be used, how were the real-time aspects of the design to be implemented, such as the data sampling scheduling and perhaps more importantly, how the system would respond when no data was received? This would be the next set of design decisions to be made.

Deriving real-time performance from a non-real-time system

Before this can be tackled, the first step is to quantify the real-time performance that is needed and from that determine the level of processing power needed. Real-time means guaranteeing that work will be completed within a certain time period. That time period could be seconds or minutes but as long as the work requested is completed within a given time period, the system is a real-time one. The key time critical function is sampling the data. The time frame for processing this is determined by the interval between the samples which in turn depends on the sampling rate. This in turn also determines the minimum processing needed to perform these tasks.

The display software imposes several constraints: its maximum sampling rate is 30Hz and the data file should be less than 350 kbytes in size. Each sample contains 8 bytes and each second generates 240 bytes of data. This means that a 350 kbytes file would contain about 1493 seconds or over 24 minutes of data. This is more than enough as the longest sessions that the car experiences are about 15—24 minutes. It was decided to not limit the sampling data and if longer periods were needed, the sampling rate would be reduced or the resulting data file sub-sampled to reduce its size at the expense of timing resolution. The serial data link is fixed at 19.2 kbaud. As a rough rule of thumb, dividing the baud rate by 10 gives the maximum number of bytes that can be transferred. This works out at just under 2 kbytes per second. This is eight times the required data rate for the 30 Hz sampling rate. This indicates that the serial link is more than capable of supplying the required data and perhaps more importantly, indicates the level or CPU performance needed. A fast 386 or entry level 486 is quite capable of transferring 2 kbytes of data per second over a serial link. Indeed faster rates are often achieved with higher serial link speeds during file transfer. Bearing in mind that all the data logger has to do in real-time is receive some data and store it, then almost any laptop with a higher speed 386 or entry level 486 should be capable of meeting the processing load.

Another way of looking at the same problem is to consider how many instructions the processor can execute between the samples. A 30 MHz 386 with a data sampling rate of 30 Hz can provide 1 MHz of CPU processing per sample. If it takes 10 clocks per instruction (a very conservative figure), that means that 100,000 instructions can be executed between samples. This is several orders of magnitude higher than the number of instructions actually needed to perform the task. Again, this is good information to help back-up the conclusion that CPU performance was not going to be an issue.

Choosing the hardware

The next task was to choose the hardware that was needed. Always a difficult problem but this turned out to be quite simple. Simple microcontrollers were ruled out because of their limited functionality. They have small amounts of memory that would restrict the amount of data that could be sampled and stored. This constrained both the data sampling rate and the total time that the sampling could take place. In addition, the data had to be transferred to the laptop for display and in addition, writing the display and analysis software would have been a major undertaking.

The next hardware candidate was a single board computer. These have more memory and often have disk drive support so that the data could be transferred using a floppy disk. A serial port could also be used. They have a downside in that they require external power which is not something that is easy to provide cleanly from a car system, let alone a competition car. Then there was the level of software support. This needed to be a quick development because it had to be in place for the start of the car's testing dates so that its information could be used to set up the car prior to the start of the season.

The next candidate was a laptop PC itself. This seems to be a bit extreme but has several advantages that make it very attractive:

- It has a built-in battery independent of the car's power supplies and systems.
- It can be used to immediately display the data as well as storing it on disk or removable media like a PC memory card.
- It can be used as the development equipment and thus allow the software to be changed in the pit lane if needed.
- There is a wealth of software available.
- Low cost. This may sound strange but second-user laptops are inexpensive and the project does not require the latest all-singing all-dancing hardware. A fast 486 or entry-level Pentium is more than capable of meeting the processing needs.
- Allows migration to low cost PC-based single board computer if needed.

The disadvantage is that they can be a little fragile and the environment inside a competition car can be a little extreme with a lot of physical shocks and jolts. However, this is similar to the treatment that is handed out to a laptop used by a road warrior. My own experience is that laptops are pretty well indestructible providing they are not dropped on to hard surfaces from great height. By placing the unit in a padded jacket, it would probably be fine. The only way to find out would be to try it! If there were problems then a single board PC in an industrial case could be used instead.

Scheduling the data sampling

For any data logging system the ability to sample data on a regular basis with consistent time intervals between samples is essential for any time-based system. The periodicity needs to be consistent so that the time aspect of the data is not skewed. For example if the logged data was wheel speed then this data is only accurate when the time period between the samples is known. If this is not known, or is subject to variation then the resulting derived information such as distance travelled or acceleration will also be inaccurate.

The easiest way of collating data is let the data collection routines to free run and then calibrate the time interval afterwards against a known time period. The periodicity of the sampling rate is now determined by how long it takes to collect the data, which in turn is determined by the ability of the source to send the data. While this is simple to engineer, it is subject to several errors and problems:

- The time period will vary from system to system depending on the system's processing and which serial port controller is used.
- The time taken to execute the sample collection may vary from sample to sample due to caching and other hard to predict effects. This means that the periodicity will show jitter which will affect the accuracy of any derived data.
- It is difficult for any data display software to display the sampling rate without being supplied with some calibration data.

For a system that provides data about how a car accelerates and behaves over time, these restrictions are not really acceptable. They will give an approximate idea of what is going on but it is not consistent. This also causes problems when data sequences are compared with each other: is the difference due to actual variation in the performance or due to variation in the timing?

Given these problems, then some form of time reference is needed. The PC provides a real-time clock that can be accessed by a system call. This returns the time in microseconds and this can be used to provide a form of software timing loop. The loop reads the system time, compares it the previous version and works out how long has elapsed. If the elapsed time is less than the sampling time interval, the loop is repeated. If it is equal, the program control jumps out of the loop and collects the data. The current time is stored as the previous time and when the data collection is completed, the timing loop is re-entered and the elapsed time continually checked.

This sounds very promising at first. Although timing loops are notorious for being inaccurate and inconsistent from machine to machine due to the different time taken to execute the routines, this software loop is simply measuring the elapsed time from a hardware-based real-time clock. As a result, the loop should synchronise to the hardware derived timing and be consistently accurate from system to system. The data sampling processing time is automatically included in the calculations and so this is a perfect way of simply synchronising the data sampling.

While in general, these statements are true, in practice they are a very simple approximation of what goes on in a PC. The first problem is the assumption that the system time is accurate down to microseconds. While the system call will return the time in microseconds, you do not get microsecond accuracy. In other words, the granularity of any change is far higher. While the design might expect that the returned time would only vary by a few microseconds if the call was consecutively called, in practice the time resolution is typically in seconds. So if the test condition in the software loop relies on being able to detect an exact difference of 100 microseconds, it may miss this as the returned system time may not be exactly 100 microseconds. It might be 90 on the first loop and 180 on the second and so on. None of these match the required 100 microsecond interval and the loop becomes infinite and the system will appear to hang.

This problem can be overcome by changing the condition to be equal to or greater than 100 microseconds. With this, the first time round will cause the loop to repeat since 90 is less than 100. On the second time round, the returned difference of 180 will be greater than 100, the exit condition will be met and the loop exited and the data sample taken. The previous value will be updated to 180 and the loop repeated. The next value will be 270 which is less than the 100 difference (180 + 100 = 280) so the loop will repeat. The next returned value will be 360 which exceeds the 100 microsecond difference and cause the loop to be exited and the data sample to be taken.

At this point note that when the data samples are taken, the required time interval is 100 microseconds. The first sample is taken at 180 microseconds followed by a second sample at 360 microseconds giving a sample interval of 180 microseconds. Compare this with what was intended with the software timing. The software timing was designed to generate a 100 microsecond interval but in practice, it is generating ones with a 180 microsecond interval. If the calculations are carried through what does happen is that one of the intervals becomes very short so that over a given time period, the possible error with the correct number of samples and the respective intervals between them will reduce.

However, this timing cannot be described as meeting the periodicity requirements that the design should meet. This approach may give the appearance of being synchronised but the results are far from this.

So while the system time is not a good candidate because of its poor resolution, there are other time-related functions that are better alternatives. One of these are the oft neglected INT 21 calls which allow a task to go to sleep for some time and then wake up and continue execution or wait for a certain time period to expire. The time resolution is far better than used by the system clock and there is no need for a software loop. The process is simple: collect the data sample and then make the INT 21 call to force the process to wait or sleep for the required time period.

The value for the time period needs to be calculated for each system. The sampling time is now part of the interval timing and this will vary from system to system. This provides us with the first problem with this technique in that the wait period will also vary from system to system. The second issue is that the time to collect the samples may also vary depending on the number of cache hits, or when a DRAM refresh cycle occurs and so on. So again, what appears to be a good solution suffers from the same type of problem. The sampling rate will suffer from jitter and will require calibration from system to system.

This system calibration problem is made worse when the software is run on laptop PCs that incorporate power saving modes to control power consumption. This will alter the clock frequency and even power down parts of the circuitry. This results in a variation in the time taken to collect the data samples. Any variation from the value used to calculate the INT 21 wait or sleep delay will cause variations in the sampling periodicity. These differences are never recovered and so the data sampling will be skewed in time.

It seems on this basis that using a PC as an embedded system is a pretty hopeless task. However there is a solution. Ideally what is required is an independent timing source that can instruct the data sampling to take place on a regular basis. This can be constructed by using one of the PCs hardware timers and its associated interrupt mechanism. The timer is programmed to generate a periodic interrupt. The interrupt starts an interrupt service routine that collects the data and finishes. It then remains dormant until the next interrupt is generated. The timer now controls the data sample timing and this is now independent of the software execution. Again, the standard PC design can be exploited as there is an 18.2 Hz tick generated to provide a stimulus for the system clock. This again is not as straightforward as might be thought. The system clock on a PC is maintained by software counting these ticks even though a real-time clock is present. The real-time clock is used to set up the system time when the PC is booted and then is not used. This is why PCs that are left on for a long time will loose their time accuracy. The tick is not that accurate and the system time will quickly gain or lose time compared to the time maintained by the real-time clock. Each time the PC is rebooted, the system time is reset to match the real-time clock and the accumulated error removed.

There is an already defined interrupt routine for this in the PC. By programming the timer to generate a different time period tick and then intercepting the interrupt handler and passing control to the data sampler software, accurate timing for the data sampling can be achieved. This is the basic method used in the design.

Sampling the data

The data is requested and received via a serial port that connects the PC to the engine management unit (EMU) in the car. The EMU collates the wheel speeds, current engine revs and other information and outputs this as a series of bytes with each byte representing the current value for the parameter. This value is subsequently multiplied by a factor to derive the actual value that is shown by the display software. These factors are provided within the header of the file that contains the data. This topic will be returned to later in this chapter.

When a sample is needed, a request character is sent and by return, the data sample is sent back. The EMU does not use any handshaking and will send the data sample as fast as the serial port will allow it – essentially as fast as the baud rate will let it. This poses yet another interesting challenge as the PC system must be configured so that none of this data is lost or mixed up. If this happens, the values can go out of sequence and because the position signifies which parameter the byte represents, the data can easily be corrupted. Again this is a topic that will be returned to later on as this problem was to appear in field trials for a different reason.

The first design decision is over whether to use an interrupt-based approach or poll the serial port to get the data. The data arrives in a regularly defined and constant packet size and so there is no need to continually poll the port. Once the request has been sent, the software need only poll or process an interrupt six times to receive the six bytes of data. On initial inspection, there is little to choose between the two approaches: the interrupt method requires some low level code to intercept the associated interrupt and redirect it to the new service routine. Polling relies on being able to process each character fast enough that it is always ready for the next character. For a fast processor this is not a problem but for a slower one it could well be and this takes the design back to being system dependent. The modern serial ports are designed to run at very fast baud rates e.g. 115000 kbaud and to do this have FIFO buffers built into them. These FIFOs can be exploited by a polling design in that they take away some of the tight timing. The buffers enable the serial port to accept and store the incoming data as fast as it is sent but do not rely on the polling software to run at the same speed. Providing the polling software can empty the FIFO so that it never overflows, there is no problem. By designing the software so that the polling software always retrieves its six bytes of data before issuing a new request, the system can ensure that this overflow does not take place. With the large 8 to 16 byte FIFOs being used today, they can actually buffer multiple samples before a problem is encountered. With the polling software easier to write and debug, this was the method chosen.

Controlling from an external switch

While the keyboard is fine for entering data and commands, it can be a little tricky to use in a race car. For a start, the driver wears flame proof gloves and this can make pressing a key a little difficult. Add to that the fact that the PC must be securely mounted inside the car and not placed on the passenger seat or driver's lap and it makes this option a little difficult. One solution would be to set the logging running and leave it like that and then simply ignore or filter out the initial data so that it doesn't appear. This is a simple approach and indeed was used as a temporary measure but it is not ideal. One of the problems is how to define and apply the filter. In practice, all the interesting events occurred after the car left the start line. Looking for wheel movement, followed by a traction control intervention signal could identify the start line. By looking for these two signals, the start of the lap or run could be identified and all data after this event kept. What is really needed is to be able to have a simple switch to start and stop the data logging.

With a desktop PC, this is not a problem as the joystick port can provide this directly. IT uses four switches to indicate which way the stick is moved and has additional switches for the firing triggers. Unfortunately, laptops tend to be a little more conservative in their design and usually do not have a joystick port. They do however have a parallel port and this can be used to read and write data to the outside world.

Normally the port is configured to drive a printer and sends bytes of data to the device using eight pins. Other pins are used to implement a handshake protocol to control the data flow and to provide additional status information. As a result, these pins can provide both input and output functions.

With the joystick port, the procedure is simple: connect the switch between ground and the allocated pin and then read the respective bit to see whether it is high or low (switch closed to ground). With the parallel port, it is assumed that there is some hardware (typically in the printer) that will provide the right signal levels. The usual method is to have a TTL level voltage supply (5 volts) and apply this to the input pin to give a high or to ground (0 volts) to indicate a low. Not difficult except that you need a separate voltage supply.

The trick used in this system was to allocate two pins to the input. One pin is used as an output and is initialised by software to be set high. This will provide a TTL high voltage on the output. This is then connected to the input pin via a 10 kΩ resistor to limit the current and protect the parallel port hardware. The switch is then connected to the input pin and ground. When closed the pin is grounded and a logic 0 will be read. When the switch is open, the pin is pulled high to the same voltage as the output pin and a logic 1 will be read[2]. By reading the port and then looking at the value of the bit associated with the input pin, the software can determine whether the switch is open or closed. This can be used as a test condition within control loops to define when the system should perform certain functions e.g. start and stop logging and so on.

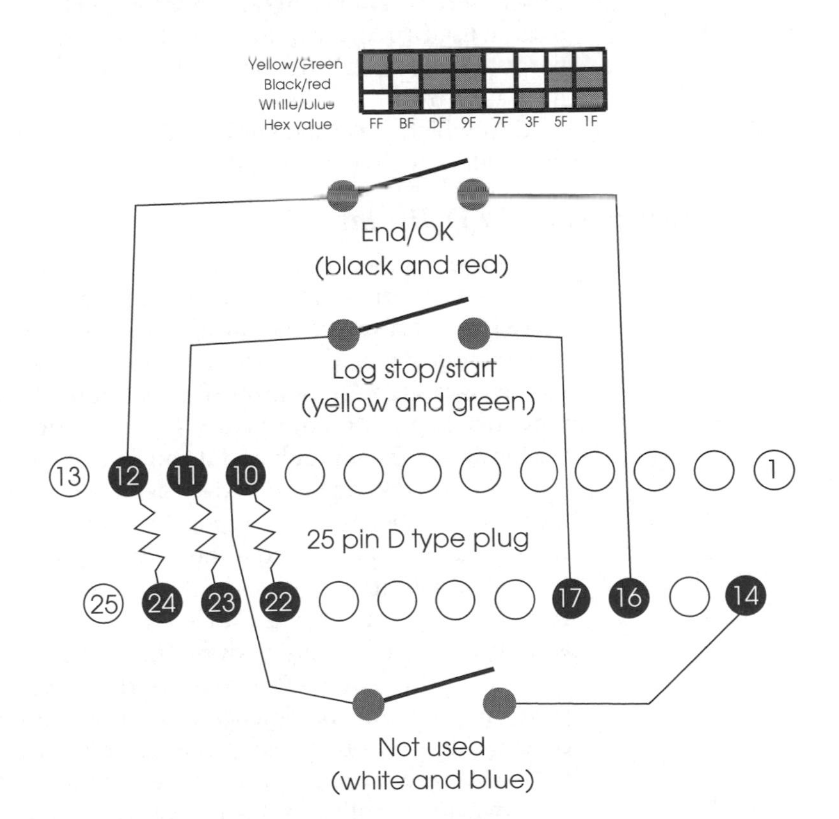

The control switch leadwiring diagram and truth table. The switches are wired to the 25 pin D-type plug and this is plugged into the laptop PC parallel port. The switch settings generate a hexadecimal value depending on their value.

There is no reason why this cannot be repeated to provide several switch inputs. The limit is really the number of output and input pins available. It is important to prevent damage to the port hardware and that the current taken from the output pin is minimised – hence the use of the resistor. This means that it is best to allocate a separate output pin to provide the logical high voltage per input pin. It is possible with some PCs to use this signal to work on several input pins but is dependent on which of the parallel port silicon chips have been used and whether they have been buffered. If high current ports are assumed, then the system will be hardware dependent and may even damage PCs with low current spec ports.

The implementation is not quite as simple due to some specific modifications in the PC hardware design. While most pins are configured to be active high i.e. when the associated bit in the port is set to a logic 1, the pin voltage rises to a logic high, this is not the case for all. Some of the pins are inverted to do the opposite for both the input and output. This means that the values written to the ports to set the output pins high, are not simply created by setting the bits high. Each bit needs to be checked to see if it needs to be inverted or not. You can work this out but I took the easy way out and used a freeware parallel port utility (http://www.lvr.com/parport.htm) that used a nice GUI to program the output pins and displayed their status on screen. This gave the correct binary bit pattern which was then converted to hexadecimal to derive the final value to program into the port.

Driving an external LED display

The parallel port output pins can drive a LED display to provide a confirmation/status signal. While it is possible to connect a LED with a current limiting resistor directly between the pin and ground signal, the power limitations described in the previous paragraph again come into play. It should be assumed that the pin can only provide about 4 mA. This is enough to light a LED but the luminance is not high and it can be difficult to see if the LED is on or not. This can be solved by using a high intensity LED but their power consumption is typically about 20 mA and exceeds the current safely available from the parallel port.

The solution is to use a buffer pack that can supply higher current. These are cheap and easy to use but do require an additional power supply to drive the buffer.

With the ability to drive status LEDs, the software can perform other functions such as indicating the amount of traction control intervention by using this sample to drive a number of LEDs arranged in a bar graph. No intervention and no LEDs are lit. Low level intervention and one LED is lit and so on. This function is incorporated into the software but has not been implemented in the final system.

Testing

Testing was done in two stages: in both stages, it would be necessary to have the system connected to the data source to check that it was being logged and stored correctly, both in workshop and real life conditions. For the first stage, it is a little inconvenient to have a race car up and running generating data to test the logging software. Instead, a simple data generator program was written that behaved like the race car itself and generated dummy test patterns which replicated the car's behaviour. This meant that by using a second PC, the data logging software could be plugged in and tested. Virtually all the debugging was done this way.

This meant that the full functionality could be tested, including the remote switches, without the car. This was fortuitous, as the car was not always available due to it being worked on itself. This meant that the weekend it took to develop the simulator software more than paid for itself by allowing the development to continue independently of the car's availability. This approach parallels many used in other developments, including more complex ones where instruction set simulators and similar tools are used to allow the software development to continue in advance of true hardware availability.

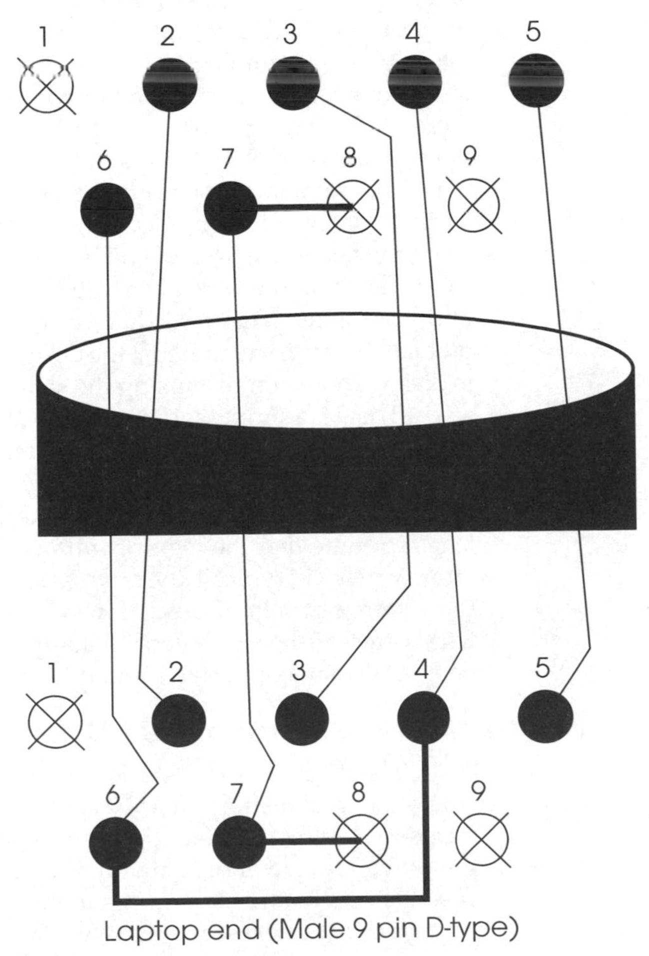

The laptop-data logger RS232 link

With the PC-based testing completed, the system could be installed and tested in the car itself. This was initially done in the garage while stationary and then with a passenger using the system while the car was driven. Once all this was completed, the system and car were taken to a race track and the system used in anger during the day's testing. This final testing revealed several problems with the system.

Problems

Saving to hard disk

The logged data is saved to disk so that it can be read back at a later date. The data is sent to disk every time a sample set is received. This is typically completed before the next sample is required and does not interfere with the timing. This at least was the original thought. In practice, this was the case for most of the time as the data is buffered before it is sent out to disk. When the buffer is emptied, the time taken to complete this is becomes quite long, especially if the disk needs to be woken up and brought up to speed. While this is happening, no samples are taken and the integrity of the sampling rate is broken. Testing during development did not show a problem but when the unit was used in real life, a 2 to 3% timing error was seen. If the logs were kept small, the timing was accurate. If they extended to several minutes, they lost time. The problem turned out to be caused by two design problems: accuracy of the timer counter programming and the buffer being flushed to the hard disk. The timer counter problem will be covered later and was really caused by an underlying behaviour in the compiler. The hard disk issue was a more fundamental problem because it meant that the design could not store data logs to disk without compromising the sampling rate integrity.

There is a simple solution to this. Why use a slow disk when there is plenty of faster memory available? Use a large data array to hold the data samples and then copy the data to disk when the logging is stopped. This has the disadvantage of restricting the length of time that the logging will work to the size of the data array which will typically be far less than the storage and thus the time offered by a hard disk. However with just a few Mbytes of RAM offering the equivalent of many hours logging, this is not necessarily a major obstacle and this solution was looked at.

Data size restrictions and the use of a RAM disk

The PC architecture started out as a segmented architecture where the total memory map is split into 64 kbytes segments of memory. Addressing data in these segments is straightforward providing the size of the data does not go beyond 64 kbytes. If it does the segment register needs to be programmed to place the address into the next segment. While later PC processors adopted larger linear address spaces, many of the compilers were slow to exploit this. This leads to a dilemma: the hardware supports it but the C compiler does not. In this case, the Borland TurboC v2.0 compiler was restricted to data arrays of no larger than 64 kbytes. It could have been possible to use several of these and then implement some housekeeping code that controls when to switch from a filled array to an empty one but initial attempts at that were not successful either. This is a case of a need that looks simple on face value but gets complex when all the potential conditions and scenarios are considered. This led to a fairly major decision: either change to a different or later version of the compiler that supported larger data structures or find some other memory-based solution.

The decision to change a compiler is not to be taken lightly, especially as the application code had been written and was working well. It used the Borland specific functions to access the timer hardware and set up the interrupt routines. Changing the compiler would mean that this code would probably need to be rewritten and tested and this process may introduce other bugs and problems.

A search through the Borland developer archive provided a solution. It came up with the suggestion to use the existing disk storage method but create a RAM disk to store the file. When the logging is complete the RAM disk file can be copied to the hard disk. This copying operation is done outside of the logging and thus the timing problem goes away. There is no change to the fundamental code and as the PC was a laptop with its own battery supply, no real risk of losing the data. One added benefit was that the hard disk was not used and thus powered down during the logging operation and so reduced the power consumption.

Timer calculations and the compiler

The software was designed to calculate the timer values using the parameter passed to the timer set-up routine. All well and good except that the routine was not accurate and the wrong value would be programmed in. The reason was due to rounding and conversion errors in the arithmetic. Although the basic code looked fine and was syntactically correct, the compiler performed several behind-the-scenes approximations that led to a significant error. The problem is further compounded by the need for a final hexadecimal value to be programmed into the timer register. In the end, calculating the exact figure using the Microsoft Windows NT calculator accessory, and then converting the final value to hexadecimal solved this problem. This pre-calculated value is used if the passed parameter matches the required value.

Data corruption and the need for buffer flushing

While the system worked as expected in the lab using the test harness, occasional problems where noticed where the data order unexpectedly changed. The engine RPM data would move to a different location and appear as a wheel speed and the front wheel speeds would register as the rear wheel speeds and so on. It was initially thought that this was to do with an error in the way the data ordering was done. Indeed, a quick workaround was made which changed the sample data ordering but while it appeared to cure the problem, it too suffered from the wrong data order at some point.

Further investigation indicated that this problem occurred when the engine was switched off and re-started while logging or if the logging was started and the engine then switched on. It was also noticed that when the EMU was powered up, it sent out a "welcome" message including the version number. It was this that provided the clue.

The welcome message was stored in the serial port FIFO and when a sample was requested, the real sample data would be sent and added to the FIFO queue. The logger would read the first six bytes from the queue that would be the first six characters of the welcome message. This would then repeat. If the welcome message had a length that was a multiple of six characters, then the samples would be correctly picked up, albeit slightly delayed. The periodicity would still be fine, it's just that sample 1 would be received when sample 2 or 3 would have been. If the message was not a multiple of six, then the remaining characters would form part of the sample and the rest of the sample would be appended to this. If there were three characters left, then the first three characters of the sample would be appended to this and become the last three characters. The last three characters of the data sample are still in the FIFO and would become the first three of the next sample and so on. This would give the impression that the data order had changed but in reality, the samples were corrupted with each logged sample consisting of the end of the previous sample and the beginning of the next.

If the EMU was started before logging was enabled, the characters were lost and the problem did not arise. If the engine stalled during the logging or started during this period, the welcome message would be sent, stored and thus corrupt the sampling. The test chassis did not simulate this, and this was why the problem did not appear in the lab. Another problem was also identified and that was associated with turning the engine off. If this happened while the data sample was being sent, it would not complete the transfer and therefore there is a potential for the system to stall or leave partial data in the queue that could corrupt the data sampling.

Having identified the problem, the solution required several fixes: the first was to clear the buffers prior to enabling the logging so that any welcome message or other erroneous data was removed. This approach was extended to occur whenever a sample is requested to ensure that a mid-logging engine restart did not insert the "welcome" message data into the FIFO. During normal operation, this check has very little overhead and basically adds a read to one of the serial port registers and tests a bit to the existing instruction flow. If characters are detected these are easily removed before operation but as this should only happen during an engine shutdown and restart where timing accuracy is not absolutely essential. Indeed, the data logger marks the data log to indicate that an error has occurred. The sample data collection will also time out if no data is received. This is detected and again the sample aborted and a marker inserted to indicate that a timeout has occurred.

Program listing

The rest of this chapter contains the logging program, complete with comments that describe its operation. These comments are easy to spot as they are in bold text.

```
#include        <stdio.h>
#include        <stdlib.h>
#include        <string.h>
#include        <dos.h>
#include        <dir.h>
#include        <alloc.h>
#include        <conio.h>
#include        <time.h>

#define TRUE     0x01
#define FALSE    0x00
#define DISPLAY  0x00
#define DATALOG  0x01
#define FILECOPY 0x02

/*
 *  VERSION numbers....
 */

#define MAJOR    6
#define MINOR    1

/*
 *  I/O addresses
 */

#define PORT1    0x3F8    /* COM1 port address */
#define PORT2    0x2F8    /* COM2 port address */
#define PORT3    0x3E8    /* COM3 port address */
#define PORT4    0x2E8    /* COM4 port address */

#define LPT1     0x378    /* LPT1 port address */
#define LPT2     0x268    /* LPT2 port address */

/*
 *  Now define the required BAUD rate
 *  38400=0x03    115200=0x01    57600=0x02    19200=0x06
 *  9600=0x0C      4800=0x18     2400=0x30
 */
#define BAUD     0x06     /* 19200 baud rate */
```

```
/*
 *      Data logging utility
 *      Usage: TCLOG
 *
 *      Compile with PACC -R or "_getargs() wildcard expansion"
 *      if you want it to handle wildcards, e.g. CHK *.EXE
 */

/*
 *  This version reprograms the system TICK (18th sec) and daisy chains
 *  onto the 0x1c interrupt where it changes the status of the variable
 *  Waitflag instead of using the int 21 routine which seems to be a little
 *  BIOS unfriendly!
 *  When Waitflag changes, the sampling routine resets and runs again.
 *  In this way the sampling is synchronised to the interrupt rate e.g. 33 Hz
 */

void set_rate(int Hz); /* Programs the tick timer to support Hz sample rate */
void setup_int();                 /* Sets up out vector 1C interrupt routine */
void restore();                      /* Restores the orig vector 1C routine */
void create_header();
void interrupt timer_tick(); /* Changes the value of waitflag from 0 to 128 */
void interrupt (far *old_1C_vect)();          /* Stores the old vector handler */
void tc_display(int tc);                              /* Displays the TC level */

/*
 *  Let's declare the global data...
 */

unsigned int    h,i,j,k;          /* loop counters */
unsigned int    byte_count;       /* Counts six bytes - goes 0 to 6 */
unsigned int    time_out;         /* flag to indicate time out problem */
unsigned int    get_com;          /* bioscom status */
unsigned char   byte;             /* bioscom parameter */
unsigned char     lpt;            /* LPT data */

unsigned char   c[ 512] ;         /* character read from file */
unsigned char   fn[ 20] ;         /* string for target filename */
unsigned int    buf[ 7] ;         /* buffer for incoming data */
FILE *          infile;           /* input file handle */
FILE *          outfile;          /* output file handle */

unsigned char log[ 256] ;         /* Buffer for incoming serial data */

unsigned int waitflag;  /* stores the flag for timer tick status */

main()
{
/*
 *  The main loop may look a little cumbersome in that apart from setup and
 *  restore and the interrupt routine, it does not use any subroutines.
 *  The reason is that timing is quite tight on a slow machine and the
 *  potential overhead of passing large amounts of data can make a difference
 *  at high data rates.
 */

/*
 *  Let's initialise the data
 */

        time_out = 0;
```

```
        waitflag = 0;
        j =0; k =0;
/*
 *      Clear the two spare data channels. These are stored but not used
 *  currently. They can be filled in by reading external data such as
 *  throttle and brake position.
 */
        log[ 6] = 0; log[ 7] =0;

/*
 *  Set up the COM port for 19.200 kbaud, 8B, No P, 1 STOP, FIFO on
 */
        outportb(PORT1 + 1, 0x00);      /* Turn off COM port interrupts */

        outportb(PORT1 + 3, 0x80);      /* Set DLAB on           */
        outportb(PORT1 + 0, BAUD);      /* Set BAUD LO byte      */
        outportb(PORT1 + 1, 0x00);      /* Set BAUD HI byte      */
        outportb(PORT1 + 3, 0x03);      /* 8 bits, No P,1 STOP */
        outportb(PORT1 + 2, 0xC7);      /* Turn on FIFOS         */
        outportb(PORT1 + 4, 0x0B);      /* Set DTR,RTS,OUT2      */

/*
 * Set up the LPT control port to enable the switches...
   bits 3,1 and 0 are used to generate active high signals
   for the status port inputs.

      Bit   7 6 5 4 3 2 1 0
      Data  1 1 0 0 1 0 1 1    0xCB
 */

 outportb(LPT1 + 2, 0xCB);     /* Set up the control port */

/*
 *  Set up the interval timer params.......
 *  This is setup to generate a .033 second timing rate.
 *  This is compatible with the data log software.
 *  ( c[07] = 0x3D;     Set up internal timing for 30 Hz)
 */

      set_rate(30); /* 30 Hz for data logging */
      setup_int();

      printf("Setup complete.  ch = %x\n",check_port());

/*
 *  Print the intro so we know the version and what it does...
 */

        printf("TCLOG version %d.%d  FIFOs enabled 19.2 kbps\n",MAJOR,MINOR);
        printf("This collates data from the traction control.\n");


do  /* Start of main control loop */
 {
        /* Look at external controls */
        /* If logging set then go into log loop */
        /* If not go into display loop */
      /* Currently we go into log loop */

      if (inportb(LPT1+1) < 0x80)
           { /* we are not data logging just displaying the tc values */
             printf("Going into display mode...\n");
```

```
        /* Clear the serial port FIFO */
        printf("Clearing serial port FIFO: ");
        while (inportb(PORT1 + 5) & 1)
        {
        log[10] = inportb(PORT1);
        printf("=X");
        }
        printf("   FIFO Clear\n");

        printf(" A MODE         FL FR RL RR RPM TC  Samples\n");
        j = 0;
        while (inportb(LPT1+1) < 0x80 & bioskey(1) == 0)  /* Check the
switch value and the keyboard */
            {
         /* First send out the M to the data logger to get it */
         /* to send the next 6 bytes of data                  */
         outportb(PORT1,0x4D);
         /* Now lets get the data in  */
         byte_count = 0; /* clear the byte counter */
         do
              {get_com = inportb(PORT1 + 5);
get_com = inportb(PORT1 + 5);
                    if (get_com & 1)
                    {
                    log[byte_count++] = inportb(PORT1);
                    }
                    } while (waitflag != 128 & byte_count != 6);
         /* Now we have six data samples, check if we timed out... */
         j++;
         if (byte_count == 6)
              { /* Display the traction control status */
              tc_display(log[5]);
              /* All done */
                    }
         else tc_display(0); /* RESET the TC status */
         /* all done so wait for timer to expire... */
         printf("Display mode:  %x  %x  %x  %x  %x   %x   %d\r",
         log[0],log[1],log[2],log[3],log[4],log[5],j);
         for(;waitflag != 128;);
         waitflag = 0; /* Now clear the wait flag to repeat */
         }  /* end of data display WHILE loop */
         fcloseall();
         /* restore(); */ /* now performed on program exit */
         printf("\nEnd of data display routine\n");
     } /* end of IF data display = TRUE loop */
   else
          {       /*
                   *  WE ARE DATA LOGGING!
                   */
       /* Set up the data log file  */
       create_header();
       j =0;
       time_out = 0;

       /* Clear the serial port FIFO */
       printf("Clearing serial port FIFO: ");
       while (inportb(PORT1 + 5) & 1)
       {
       log[10] = inportb(PORT1);
       printf("=X");
       }
       printf("   FIFO Clear\n");
```

```
/* Now synchronise with the clock */
printf("Synchronising\n");
printf("A MODE Data logging starting \a\a\a\n");
printf("FL FR RL RR RPM TC  OK   FAIL\n");
for(;waitflag != 128;);
waitflag = 0;


while(inportb(LPT1+1) > 0x80 & bioskey(1) == 0)
/* this would test the switch */
{
/* First send out the M to the data logger to get it */
/* to send the next 6 bytes of data                  */
outportb(PORT1,0x4D);
/* Now lets get the data in  */
byte_count = 0; /* clear the byte counter */
do
{  get_com = inportb(PORT1 + 5);
      if (get_com & 1)
            {  log[ byte_count++]  = inportb(PORT1);
            }
}  while (waitflag != 128 & byte_count != 6);
/* Now we have six data samples, check if we timed out... */
if (byte_count == 6)
    {   /* Success! */
    j++;
    /* Reorg the data correctly
       This has been commented out as it was not needed
       when the real problem was found out. */
    /* Input order: 0 = FL, 1 = FR, 2=RPM,3=TC,4=RL, 5=RR  */
    /* Output order: FL, FR, RL, RR, RPM, TC */
    /* log[9] = log[2];*/  /* Copy RMP to [9] */
    /* log[8] = log[3];*/  /* Copy TC to [8] */
    /* log[2] = log[4];*/  /* Copy RL to [2] */
    /* log[3] = log[5];*/  /* Copy RR to [3] */
    /* log[4] = log[9];*/  /* Copy RPM to [4] */
    /* log[5] = log[8];*/  /* Copy TC to [5] */

    /* Store the data in the file */
    fwrite(log,1,8,outfile);
    /* Display the traction control status */
    tc_display(log[ 5] );

    /* Display the LPT port status */
    /* printf("LPT=%x\n",inportb(LPT1+1)); */

    }
 else /* we timed out.. */
    {
    /* Mark the file so we know we timed out */
    strcpy(log,"********");
    fwrite(log,1,8,outfile);
    time_out++;
    tc_display(0);
    }
 printf("%x %x %x %x %x  %x  %d %d\r", log[ 0] , log[ 1] , log[ 2] ,
 log[ 3] , log[ 4] ,log[ 5] ,j,time_out);
 /* all done so wait for timer to expire... */
 for(;waitflag != 128;);
 waitflag = 0; /* Now clear the wait flag to repeat */
 }  /* end of data logging WHILE loop */
fcloseall();
```

```
        printf("\nTime out = %d",time_out);
          printf(" Transfers attempted: %d\n",j);
        } /* end of data log ELSE */

 } while ( (inportb(LPT1+1) & 0x20) == 0x00);
/*
 * We are quitting the program so restore and close...
*/
restore();
printf(" Quitting TCLOG now...\n");

} /* end of MAIN() */

/*
 *  This is where the sub-routines live that control the tick ...
 */


void set_rate(int Hz /* This is the sample rate we need */)
{
        int rate_hi, rate_lo;   /* low and high bytes to program the timer */
        int num;                                 /* temp value   */

printf("Changing the BIOS 1/18th tick rate to 1/%d\n",Hz);
/*
 *  First calculate the values we need to program...
 *  If the rate is 18... re program to normal BIOS value to restore normality
 */

/*
 * The main clock is 14.31818 MHz divided by 12 to give 1.1931816 MHz
 * Divide by 65536 to give the 18.2 Hz tick.
 * To reprogram it, divide 1,193,181.6 by the Hz value.
 * For 30 Hz this is 39773 or 0x965C
 *
 */
      if (Hz == 18) { rate_hi = 0xFF; rate_lo = 0xFF; }
      else if (Hz == 30) {rate_hi = 0x96; rate_lo = 0x5C;}
        else    {
                        num=65536/(Hz/18.2);
                        rate_hi = num&0xFF00;
                        rate_hi = rate_hi/256;
                  rate_lo = num&0x00FF;
              }
        outportb(0x43,0x36);    /*  Set up 8253 timer for Sq Wave */
      outportb(0x40,rate_lo);  /* program divisor low byte */
      outportb(0x40,rate_hi);  /* program divisor high byte */

/*
 *  The tick has now been set up for the required sampling frequency.
 */
} /* End of set_rate()  */

void interrupt timer_tick()
{
/*
 *  The timer has expired so change the value of waitflag and return..
 */
      waitflag = 128;
} /* end of timer tick interrupt routine */
```

```
void setup_int()
{
        disable();
        old_1C_vect=getvect(0x1C);
        setvect(0x1c,timer_tick);
        enable();
}       /* end of setup_int() */
void restore()
{
        disable();
        set_rate(18);
        setvect(0x1C,old_1C_vect);
        enable();
}       /* end of restore()  */


void create_header()
{
char file_name[ 28] ;       /* Stores the file name */
char temp_str[ 28] ;        /* Holds the file name */
struct ffblk f;             /* File block structure */
int done;                   /* Stores the result of the file find */
int count;                  /* Stores the number of files to work out the
                               next file name for storage               */
/*
 *  This creates the enhanced Racelogic format file header from a
 *  header file called BLANKHDR.BIN
 */

/*
 *  First work out the next file name in the sequence
 */
 count = 0;
 done = findfirst("data_*.dat",&f,0);
 while(!done)      {
            count++;
            printf("Count: %d %s\n",count,f.ff_name);
            done = findnext(&f);
            }
            strcpy(file_name,"data_");
            itoa(count,temp_str,10);
            strcat(file_name,temp_str);
            strcat(file_name,".dat");

            strcpy(fn,file_name);
                printf("file name is %s\n",fn);
            if (!(outfile = fopen(fn,"wb")))
                        {
                        printf("Can't open file %s.  Exitting.\n",fn);
exit(1);
                        exit(1);
                        }
/*
 *  Let's do the file copying now that we know that the
 *  input and output files are open...
 */

               if (!(infile = fopen("BLANKHDR.BIN", "r")))
                        {
                        printf("Can't open BLANKHDR.BIN file for logging\n");
                        exit(1);
                        }
```

```
            fread(c, 260, 1, infile); /* Read the header */
/*
 *  Now update the information
 */
                    c[00] =8;       /* Set up the extra channel 06 to 07 */
                /*    3C = 125Hz, 3E = 7.5Hz, 3f = 202.89/256, 3D = 33Hz  */
                /*    NOTE: <3a,3B,2D, and >3F DON'T WORK! */
                    c[07] = 0x3D;         /* Set up internal timing for 33 Hz */
                /* Name the 1 s time stamp channel */
                c[156]='T'; c[157]='C'; c[158]='6'; c[159]='1';
                c[160]='H'; c[161]='z'; c[162]='3'; c[163]='0';
                /* Store the display factor for the channel */
                    c[164] = 0x00; c[165] = 0x00; c[166] = 0x4C; c[167] = 0x3E;

                /* Name the 20 second time stamp channel */
                    c[180]='T'; c[181]='i'; c[182]='m'; c[183]='e';
                    c[184]=' '; c[185]='s'; c[186]='e'; c[187]='c';
                /* Store the display factor for the channel */
                    c[188] = 0x00; c[189] = 0x00; c[190] = 0xA0; c[191] = 0x41;

/*
 *  Now let's write the channel names.....
 */

                /* Name the first channel */
                    c[12]='L'; c[13]='e'; c[14]='f'; c[15]='t';
                    c[16]=' '; c[17]='F'; c[18]='r'; c[19]=' ';
                /* Store the display factor */
                    c[20] = 0x00; c[21] = 0x00; c[22] = 0x26; c[23] = 0x3F;

                /* Name the second channel   */
                    c[36]='R';c[37]='i';c[38]='g';c[39]='h';
                    c[40]='t';c[41]=' ';c[42]='F';c[43]='r';
                /* Store the display factor */
                    c[44] = 0x00;c[45] = 0x00;c[46] = 0x26;c[47] = 0x3F;

                /* Name the third channel */
                    c[60]='L';c[61]='e';c[62]='f';c[63]='t';
                    c[64]=' ';c[65]='B';c[66]='a';c[67]='k';
                /* Store the display factor */
                    c[68] = 0x00;c[69] = 0x00;c[70] = 0x26;c[71] = 0x3F;

                /* Name the fouth channel */
                    c[84]='R';c[85]='i';c[86]='g';c[87]='h';
                    c[88]='t';c[89]=' ';c[90]='B';c[91]='k';
                /* Store the display factor */
                    c[92] = 0x00;c[93] = 0x00;c[94] = 0x26;c[95] = 0x3F;

                /* Was 259, now 260 for 8th channel */
                    fwrite(c, 260, 1, outfile);  /* write the header */

                /* HEADER completed! */
}       /* end of create_header */

void tc_display(int tc)
{
/*
 * This displays the TC level either on the screen or
 * on an external LED array using the parallel port.
 * Replace with printf to write to the screen
 */
```

```
if (tc == 0)      {outportb(LPT1,0x00);}
else if (tc == 1) {outportb(LPT1,0x01);}
else if (tc == 2) {outportb(LPT1,0x03);}
else if (tc == 3) {outportb(LPT1,0x07);}
else if (tc == 4) {outportb(LPT1,0x0f);}

}  /* end of tc_display() */
```

Index

Symbols

A

B

H

I

J

K

L

M

N

O

P

Q

R

S

X

Z

MIDNIGHT NEVER COME

Also by Marie Brennan and available from Titan Books

THE MEMOIRS OF LADY TRENT

A Natural History of Dragons

The Tropic of Serpents

Voyage of the Basilisk

In the Labyrinth of Drakes (April 2016)

ONYX COURT

In Ashes Lie (June 2016)

A Star Shall Fall (September 2016)

With Fate Conspire (January 2017)

MIDNIGHT NEVER COME

THE ONYX COURT BOOK I

MARIE BRENNAN

TITAN BOOKS

Midnight Never Come

Print edition ISBN: 9781785650734
E-book edition ISBN: 9781785650741

Published by Titan Books
A division of Titan Publishing Group Ltd
144 Southwark Street, London SE1 0UP

First edition: November 2015
2 4 6 8 10 9 7 5 3 1

A CIP catalogue record for this title is available from the British Library.

Printed a[illegible], Croydon, CR0 4YY

This book is dedicated to two groups of people.

To the players of Memento:

Jennie Kaye, Avery Liell-Kok, Ryan Conner, and Heather Goodman.

And to their characters, whose ghosts still haunt this story:

Rowan Scott, Sabbeth, Erasmus Fleet, and Wessamina Hammercrank.

PROLOGUE

The Tower of London: March 1554

Fitful drafts of chill air blew in through the cruciform windows of the Bell Tower, and the fire did little to combat them. The chamber was ill lit, just wan sunlight filtering in from the alcoves and flickering light from the hearth, giving a dreary, despairing cast to the stone walls and meagre furnishings. A cheerless place—but the Tower of London was not a place intended for cheer.

The young woman who sat on the floor by the fire, knees drawn up to her chin, was pale with winter and recent illness. The blanket over her shoulders was too thin to keep her warm, but she seemed not to notice; her dark eyes were fixed on the dancing flames, morbidly entranced, as if imagining their touch. She would not be burned, of course; burning was for common heretics. Decapitation, most likely. Perhaps, like her mother, she would be permitted a French executioner, whose sword would do the work cleanly.

Presuming the Queen's mercy permitted her that

consideration. Presuming the Queen had mercy for her at all.

The few servants she kept were not there; in a rage she had sent them away, arguing with the guards until she won these private moments for herself. As much as solitude oppressed her, she could not bear the thought of companionship in this dark moment, the risk of showing her weakness to others. And so when she waked from her reverie to sense another in the room, her anger rose again. Shedding the blanket, the young woman whirled to her feet, ready to confront the intruder.

Her words died, unspoken, and behind her the fire dipped low.

The woman she saw was no serving-maid, no lady attendant. No one she had ever seen before. A mere silhouette, barely visible in the shadows—but she stood in one of the alcoves, where a blanket had been tacked up to cover the arrow-slit window.

Not by the door. And she had entered without a sound.

"You are the Princess Elizabeth," the woman said. Her voice was a cool ghost, melodious, soft, and dark.

Tall, she was, taller than Elizabeth herself, and more slender. She wore a sleek black gown, close-fitting through the body but flaring outward into a full skirt and a high standing collar that gave her presence weight. Jewels glimmered with dark colour here and there, touching the fabric with elegance.

"I am," Elizabeth said, drawing herself up to the dignity of her full height. "I have given no orders to accept visitors." Nor was she permitted any, but in prison as in court, bravado could be all.

The stranger's voice answered levelly. "I am not a visitor. Do you think this solitude your own doing? The guards allowed it because I arranged that they should. My words are for your ears alone."

Elizabeth stiffened. "And who are you, that you presume to order my life in such fashion?"

"A friend." The word carried no warmth. "Your sister means to execute you. She cannot risk your survival; you are a focal point for every Protestant rebellion, every disaffected nobleman who hates her Spanish husband. She must dispose of you, and soon."

No more than Elizabeth herself had already calculated. To be here, in the stark confines of the Bell Tower, was an insult to her rank. Prisoner though she was, she should have received more comfortable lodgings. "No doubt you come to offer me some escape from this. I do not, however, converse with strangers who intrude on me without warning, let alone make alliances with them. Your purpose might be to lure me into some indiscretion my enemies could exploit."

"You do not believe that." The stranger came forward one step, into a patch of thin, grey light. A cruciform arrow-slit haloed her as if in painful mimicry of Heaven's blessing. "Your sister and her Catholic allies would not treat with one such as I."

Slender as a breath, she should have been skeletal,

grotesque, but far from it; her face and body bore the stamp of unearthly perfection, a flawless symmetry and grace that unnerved as much as it entranced. Elizabeth had spent her childhood with scholars for her tutors, reading classical authors, but she knew the stories of her own land, too: the beautiful ones, the Fair Folk, the Good People, whose many epithets were chosen to mollify their capricious natures.

The faerie was a sight to send grown women to their knees, and Elizabeth was only twenty-one. Since childhood, though, the princess had survived the tempests of political unrest, riding from her mother's inglorious downfall to her own elevation at her brother's hands, only to plummet again when their Catholic sister took the throne. She was intelligent enough to be afraid, but stubborn enough to defy that fear, to cling to pride when nothing else remained.

"Do you think me easier to cozen than my sister? Some say your kind are fallen angels, or in league with the devil himself."

The woman's laugh echoed from the chamber walls like shattering crystal. "I do not serve the devil. I offer you a bond of mutual aid. With my help, you may be freed from the Tower and raised to your sister's throne. Your father's throne. Without it, your life will surely end soon."

Elizabeth knew too much of politics to even consider an offer without hearing it in full. "And in return? What gift—no doubt a minor, insignificant trifle—would you require from me?"

"Oh, 'tis not minor." The faintest of smiles touched

the stranger's lips. "As I will raise you to your throne, you will raise me to mine. And when we both achieve power, perhaps we will be of use to each other again."

Every shrewd instinct and fibre of caution in Elizabeth warned her against this pact. Yet over her hovered the spectre of death, the growing certainty of her sister's bitterness and hatred. She had her allies, surely enough, but they were not here. Could they be relied upon to save her from the headsman?

To cover her thoughts, she said, "You have not yet told me your name."

The fae paused. At last, her tone considering, she said, "Invidiana."

When Elizabeth's servants returned soon after, they found their mistress seated in a chair by the fire, staring into its glowing heart. The air in the chamber was freezing cold, but Elizabeth sat without cloak or blanket, her long, elegant hands resting on the arms of the chair. She was quiet that day, and for many days after, and her gentlewomen worried for her, but when word came that she was to be permitted to walk at times upon the battlements and to take the air, they brightened. Surely, they hoped, their futures—and that of their mistress—were looking up at last.

ACT ONE

Time stands still with gazing on her face,
stand still and gaze for minutes,
houres and yeares, to her giue place:
All other things shall change,
but shee remains the same,
till heauens changed haue their course &
time hath lost his name.

JOHN DOWLAND

THE THIRD AND LAST BOOKE OF SONGS OR AIRES

No footfalls disturb the hush as the man—not nearly so young as he appears—passes down the corridor, floating as if he walks on the shadows that surround him.

His whisper drifts through the air, echoing from the damp stone of the walls.

"She loves me... she loves me not."

His clothes are rich, thick velvet and shining satin, black and silver against pale skin that has not seen sunlight for decades. His dark hair hangs loose, not disciplined into curls, and his face is smooth. As she prefers it to be.

"She loves me... she loves me not."

The slender fingers pluck at something invisible in his hands, as if pulling petals from a flower, one by one, and letting them fall, forgotten.

"She loves me... she loves me not."

He stops abruptly, peering into the shadows, then reaches up with one shaking hand to touch his eyes. "She wants to take them from me, you know," he confides to whatever he sees—or thinks he sees. Years in this place have made reality a malleable thing to him, a volatile one, shifting without warning. "She spoke of it again today. Taking my eyes... Tiresias was blind. He was also a woman betimes; did you know that? He had a daughter. I have no daughter." Breath catches in his throat. "I had a family once. Brothers, sisters, a mother and father... I was in love. I might have had a daughter. But they are all gone now. I have only her, in all the world. She has made certain of that."

He sinks back against the wall, heedless of the grime that mars his fine clothing, and slides down to sit on the floor. This is one of the back tunnels of the Onyx Hall, far from the cold, glittering beauty of the court. She lets him wander, though never far. But whom does she hurt by keeping him close—him, or herself? He is the only one who remembers what this court was, in its earliest days. Even she has chosen to forget. Why, then, does she keep him?

He knows the answer. It never changes, no matter the question. Power, and occasional amusement. These are the only reasons she needs.

"That which is above is like that which is below," he whispers to his unseen companion, a product of his fevered mind. "And that which is below is like that which is above." His sapphire gaze drifts upward, as if to penetrate the stones and wards that keep the Onyx Hall hidden.

Above lies the world he has lost, the world he sometimes thinks no more than a dream. Another symptom of his madness. The crowded, filthy streets of London, seething with merchants and labourers and nobles and thieves, foreigners and country folk, wooden houses and narrow alleys and docks and the great river Thames. Human life, in all its tawdry glory. And the brilliance of the court above, the Tudor magnificence of Elizabetha Regina, Queen of England, France, and Ireland. Gloriana, and her glorious court.

A great light, that casts a great shadow.

Far below, in the darkness, he curls up against the wall. His gaze falls to his hands, and he lifts them once more, as if recalling the flower he held a moment ago.

"She loves me...

"...she loves me not."

RICHMOND PALACE, RICHMOND

17 September 1588

"Step forward, boy, and let me see you."

The wood-panelled chamber was full of people, some hovering nearby, others off to the side, playing cards or engaging in muted conversation. A musician, seated near a window, played a simple melody on his lute. Michael Deven could not shake the feeling they were all looking at him, openly or covertly, and the scrutiny made him unwontedly awkward.

He had prepared for this audience with more than customary care for appearances. The tailor had assured him the popinjay satin of his doublet complemented the blue of his eyes, and the sleeves were slashed with insets of white silk. His dark hair, carefully styled, had not a strand out of place, and he wore every jewel he owned that did not clash with the rest. Yet in this company, his appearance was little more than serviceable, and sidelong glances weighed him down to the last ounce.

But those gazes would hardly matter if he did not impress the woman in front of him.

Deven stepped forward, bold as if there were no one

else there, and made his best leg, sweeping aside the edge of his half-cloak for effect. "Your Majesty."

Standing thus, he could see no higher than the intricately worked hem of her gown, with its motif of ships and winds. A commemoration of the Armada's recent defeat, and worth more than his entire wardrobe. He kept his eyes on a brave English ship and waited.

"Look at me."

He straightened and faced the woman sitting beneath the canopy of estate.

He had seen her from afar, of course, at the Accession Day tilts and other grand occasions: a radiant, glittering figure, with beautiful auburn hair and perfect white skin. Up close, the artifice showed. Cosmetics could not entirely cover the smallpox scars, and the fine bones of her face pressed against her aging flesh. But her dark-eyed gaze made up for it; where beauty failed, charisma would more than suffice.

"Hmmm." Elizabeth studied him frankly, from the polished buckles of his shoes to the dyed feather in his cap, with particular attention to his legs in their hose. He might have been a horse she was contemplating buying. "So you are Michael Deven. Hunsdon has told me something of you—but I would hear it from your own lips. What is it you want?"

The answer was ready on his tongue. "Your Majesty's most gracious leave to serve in your presence, and safeguard your throne and your person against those impious foes who would threaten it."

"And if I say no?"

The freshly starched ruff scratched at his chin and throat as he swallowed. Catering to the Queen's taste in clothes was less than comfortable. "Then I would be the most fortunate and most wretched of men. Fortunate in that I have achieved that which most men hardly dream of—to stand, however briefly, in your Grace's radiant presence—and wretched in that I must go from it and not return. But I would yet serve from afar, and pray that one day my service to the realm and its glorious sovereign might earn me even one more moment of such blessing."

He had rehearsed the florid words until he could say them without feeling a fool, and hoped all the while that this was not some trick Hunsdon had played on him, that the courtiers would not burst into laughter at his overblown praise. No one laughed, and the tight spot between his shoulder blades eased.

A faint smile hovered at the edges of the Queen's lips. Meeting her eyes for the briefest of instants, Deven thought, *She knows exactly what our praise is worth.* Elizabeth was no longer a young woman, whose head might be turned by pretty words; she recognised the ridiculous heights to which her courtiers' compliments flew. Her pride enjoyed the flattery, and her political mind exploited it. *By our words, we make her larger than life. And that serves her purposes very well.*

This understanding did not make her any easier to face. "And family? Your father is a member of the Stationers' Company, I believe."

"And a gentleman, madam, with lands in Kent. He is an alderman of Farringdon Ward within, and has been pleased to serve the Crown in printing certain religious texts. For my own part, I do not follow in his trade; I am of Gray's Inn."

"Though your studies there are incomplete, as I understand. You went to the Netherlands, did you not?"

"Indeed, madam." A touchy subject, given the failures there, and the Queen's reluctance to send soldiers in the first place. Yet his military conduct in the Low Countries was part of what distinguished him enough to be here today. "I served with your gentleman William Russell at Zutphen two years ago."

The Queen fiddled idly with a silk fan, eyes still fixed on him. "What languages have you?"

"Latin and French, madam." What Dutch he had learned was not worth claiming.

She immediately switched to French. "Have you travelled to France?"

"I have not, madam." He prayed his accent was adequate, and thanked God she had not chosen Latin. "My studies kept me occupied, and then the troubles made it quite impossible."

"Good. Too many of our young men go there and come back Catholic." This seemed to be a joke, as several of the courtiers chuckled dutifully. "What of poetry? Do you write any?"

At least Hunsdon had warned him of this, that she would ask questions having nothing to do with his ostensible purpose for being there. "She has standards," the Lord Chamberlain had said, "for anyone she keeps around her. Beauty, and an appreciation for beauty; whatever your duties at court, you must also be an ornament to her glory."

"I do not write my own, madam, but I have attempted some works of translation."

Elizabeth nodded, as if it were a given. "Tell me, which poets have you read? Have you translated Virgil?"

Deven parried this and other questions, striving to keep up with the Queen's agile mind as it leapt from topic to topic, and all in French. She might be old, but her wits showed no sign of slowing, and from time to time she would make a jest to the surrounding courtiers, in English or in Italian. He fancied they laughed louder at the Italian sallies, which he could not understand. Clearly, if he were accepted at court, he would need to learn it. For self-protection.

Elizabeth broke off the interrogation without warning and looked past Deven. "Lord Hunsdon," she said, and the nobleman stepped forward to bow. "Tell me. Would my life be safe in this gentleman's hands?"

"As safe as it rests with any of your Grace's gentlemen," the grey-haired baron replied.

"Very encouraging," Elizabeth said dryly, "given that we executed Tylney for conspiracy not long ago." She turned her forceful attention to Deven once more, who fought the urge to hold his breath and prayed he did not look like a pro-Catholic conspirator.

At last she nodded her head decisively. "He has your

recommendation, Hunsdon? Then let it be so. Welcome to my Gentlemen Pensioners, Master Deven. Hunsdon will instruct you in your duties." She held out one fine, long-fingered hand, the hands featured in many of her portraits, because she was so proud of them. Kissing one felt deeply strange, like kissing a statue, or one of the icons the papists revered. Deven backed away with as much speed as was polite.

"My humblest thanks, your Grace. I pray God my service never disappoint."

She nodded absently, her attention already on the next courtier, and Deven straightened from his bow with an inward sigh of relief.

Hunsdon beckoned him away. "Well spoken," the Lord Chamberlain and Captain of the Gentlemen Pensioners said, "though defence will be the least of your duties. Her Majesty never goes to war in person, of course, so you will not find military action unless you seek it out."

"Or Spain mounts a more successful invasion," Deven said. The baron's face darkened. "Pray God it never come."

The two of them made their way through the gathered courtiers in the presence chamber and out through magnificently carved doors into the watching chamber beyond. "The new quarter begins at Michaelmas," Hunsdon said. "We shall swear you in then; that should give you time to set your affairs in order. A duty period lasts for a quarter, and the regulations require you to serve two each year. In practice, of course, many of our band have others stand in for them, so that some are at court near constantly, others hardly at all. But for your first year, I will require you to serve both assigned periods."

"I understand, my lord." Deven had every intention of spending the requisite time at court, and more if he could manage it. One did not gain advancement without gaining the favour of those who granted it, and one did not do that from a distance. Not without family connections, at any rate, and with his father so new to the gentry, he was sorely lacking in those.

As for the connections he did have... Deven had kept his eyes open, both in the presence chamber and this outer room, populated by less favoured courtiers, but nowhere had he seen the one man he truly hoped to find. The man to whom he owed his good fortune this day. Hunsdon had recommended him to the Queen, as was his privilege as captain, but the notion did not originate with him.

Unaware of Deven's thoughts, Hunsdon went on talking. "Have better clothes made, before you begin. Borrow money if you must; no one will remark upon it. Hardly a man in this court is not in debt to one person or another. The Queen takes great delight in fashion, both for herself and those around her. She will not be pleased if you look plain."

One visit to the elite realm of the presence chamber had convinced him of that. Deven was already in debt; preferment did not come cheaply, requiring gifts to smooth his path every step of the way. It seemed he would have to borrow more, though. This, his father had warned him, would be his lot: spending all he had and more in the hopes of *having* more in the future.

Not everyone won at that game. But Deven's grandfather had been all but illiterate; his father, working as a printer, had earned enough wealth to join the ranks of the gentry; Deven himself intended to rise yet higher.

He even had a notion for how to do it—if he could only find the man he needed. Descending a staircase two steps behind Hunsdon, Deven said, "My lord, could you advise me on how to find the Principal Secretary?"

"Eh?" The baron shook his head. "Walsingham is not at court today."

Damnation. Deven schooled himself to an outward semblance of pleasantry. "I see. In that case, I believe I should—"

His words cut off, for faces he recognised were waiting in the gallery below. William Russell was there, along with Thomas Vavasour and William Knollys, two others he knew from the fighting in the Low Countries. At Hunsdon's confirming nod, they loosed glad cries and surged forward, clapping him on the back.

The suggestion he had been about to make, that he return to London that afternoon, was trampled before he could even speak it. Deven struggled with his conscience for a minute at most before giving in. He was a courtier now; he should enjoy the pleasures of a courtier's life.

* * *

THE ONYX HALL, LONDON

17 September 1588

The polished stone walls reflected the quiet murmurs, the occasional burst of cold, sharp laughter, echoing up among the sheets of crystal and silver filigree that filled the space between the vaulting arches. Chill lights shone down on a sea of bodies, tall and short, twisted and fair. Court was not often so well attended, but something was expected to happen today. No one knew what—there were rumours; there were always rumours—but no one would be absent who could possibly attend.

And so the fae of London gathered in the Onyx Hall, circulating across the black-and-white *pietre dura* marble of the great presence chamber. One did not have to be a courtier to gain entry to this room; among the lords and gentlewomen were visitors from outlying areas, most of them dressed in the same ordinary clothing they wore every day. They formed a plain, sturdy backdrop against which the finery of the courtiers shone all the more vividly. Gowns of cobwebs and mist, doublets of rose petals like armour, jewels of moonlight and starlight and other intangible riches: the fae who called the Onyx Hall home had dressed for a grand court occasion.

They had dressed, and they had come; now they waited. The one empty space lay at the far end of the presence chamber, a high dais upon which a throne sat empty. Its intricate network of silver and gems might have been the web of a spider, waiting for its spinner to return. No one looked at it openly, but each fae present glanced at it from time to time out of the corners of their eyes.

Lune looked at it more often than most. The rest of the time she drifted through the hall, silent and alone. Whispers spread fast; even those from outside London seemed to have heard of her fall from favour. Or perhaps not; country fae often kept their distance from courtiers, out of fears ranging from the well-founded to the ludicrous. Whatever the cause, the hems of her sapphire skirts rarely brushed anyone else's. She moved in an invisible sphere of her own disgrace.

From the far end of the hall, a voice boomed out like the crash of waves on rocky shores. "She comes! From the white cliffs of Dover to the stones of the ancient wall, she rules all the fae of England. Make way for the Queen of the Onyx Court!"

The sea of bodies rippled in a sudden ebb tide, every fae present sinking to the floor. The more modest—the more fearful—prostrated themselves on the black-and-white marble, faces averted, eyes tightly shut. Lune listened as heavy steps thudded past, measured and sure, and then behind them the ghostly whisper of skirts. A chill breeze wafted through the room, more imagined than felt.

A moment later, the doors to the presence chamber boomed shut. "By command of your mistress, rise, and attend to her court," the voice again thundered, and with a shiver the courtiers returned to their feet and faced the throne.

Invidiana might have been a portrait of herself, so still

did she sit. The crystal and jet embroidered onto her gown formed bold shapes that complemented those of the throne, with the canopy of estate providing a counterpoint above. Her high collar, edged with diamonds, framed a flawless face that showed no overt expression—but Lune fancied she could read a hint of secret amusement in the cold black eyes.

She hoped so. When Invidiana was not amused, she was often angry.

Lune avoided meeting the gaze of the creature that waited at Invidiana's side. Dame Halgresta Nellt stood like a pillar of rock, boots widely planted, hands clasped behind her broad back. The weight of her gaze was palpable. No one knew where Invidiana had found Halgresta and her two brothers—somewhere in the North, though some said they had once been fae of the alfar lands across the sea, before facing exile for unknown crimes—but the three giants had fought a pitched combat before Invidiana's throne for the right to command her personal guard, and Halgresta had won. Not through size or strength, but through viciousness. Lune knew all too well what the giant would like to do to her.

A sinuous fae clad in an emerald-green doublet that fit like a second skin ascended two steps up the dais and bowed to the Queen, then faced the chamber. "Good people," Valentin Aspell said, his oily voice pitched to carry, "today, we play host to kinsmen who have suffered a tragic loss."

At the Lord Herald's words, the doors to the presence chamber swung open. Laying one hand on the sharp, fluted edge of a column, Lune turned, like everyone else, to look.

The fae who entered were a pathetic sight. Muddy and haggard, their simple clothes hanging in rags, they shuffled in with all the terror and awe of rural folk encountering for the first time the cold splendour of the Onyx Hall. The watching courtiers eddied back to let them pass, but there was none of the respect that had immediately opened a path for Invidiana; Lune saw more than a few looks of malicious pity. Behind the strangers walked Halgresta's brother Sir Prigurd, who shepherded them along with patient determination, nudging them forward until they came to a halt at the foot of the dais. There was a pause. Then a sound rumbled through the hall: a low growl from Halgresta. The peasants jerked and threw themselves to the floor, trembling.

"You kneel before the Queen of the Onyx Court," Aspell said, with only moderate inaccuracy; two of the strangers were indeed kneeling, instead of lying on the floor. "Tell her, and the gathered dignitaries of her realm, what has befallen you."

One of the two kneeling fae, a stout hob who looked in danger of losing his cheerful girth, obeyed the order. He had the good sense not to rise.

"Nobble Queen," he said, "we hev lost ev'ry thing."

The account that followed was delivered in nearly impenetrable country dialect. Lune soon gave up on understanding every detail; the tenor was clear enough. The hob had served a certain family since time out of mind, but the mortals were recently thrown off their land, and their house burnt to the ground. Nor was he the only one to suffer such misfortune: a nearby marsh had been drained and the entire area, former house and all, given over to a new kind of farming, while a road being laid in to connect some insignificant town to some slightly less insignificant town had resulted in the death of an oak man and the levelling of a minor faerie mound.

When the last of the tale had spilled out, another pause ensued, and then the hob nudged a battered and sorry-looking puck still trembling on the floor at his side. The puck yelped, a sharp and nervous sound, and produced from somewhere a burlap sack.

"Nobble Queen," the hob said again, "we hev browt yew sum gifts."

Aspell stepped forward and accepted the sack. One by one, he lifted its contents free and presented them to Invidiana: a rose with ruby petals, a spindle that spun on its own, a cup carved from a giant acorn. Last of all was a small box, which he opened facing the Queen. A rustle shivered across the hall as half the courtiers craned to see, but the contents were hidden.

Whatever they were, they must have satisfied Invidiana. She waved Aspell off with one white hand and spoke for the first time.

"We have heard your tale of loss, and your gifts are pleasing to our eyes. New homes will be found for you, never fear."

Her cool, unemotional words set off a flurry of bowing

and scraping from the country fae; the hob, still on his knees, pressed his face to the floor again and again. Finally Prigurd got them to their feet, and they skittered out of the chamber, looking relieved at both their good fortune and their departure from the Queen's presence.

Lune pitied them. The poor fools had no doubt given Invidiana every treasure they possessed, and much good would it do them. She could easily guess the means by which those rural improvements had begun; the only true question was what the fae of that area had done to so anger the Queen, that she retaliated with the destruction of their homes.

Or perhaps they were no more than a means to an end. Invidiana looked out over her courtiers, and spoke again. The faint hint of kindness an optimistic soul might have read into her tone before was gone. "When word reached us of this destruction, we sent our loyal vassal Ifarren Vidar to investigate." From a conspicuous spot at the foot of the dais, the skeletally thin Vidar smirked. "He uncovered a shameful tale, one our grieving country cousins dreamed not of."

The measured courtesy of her words was more chilling than rage would have been. Lune shivered, and pressed her back against the sharp edges of the pillar. *Sun and Moon,* she thought, *let it not touch me.* She had played no part in these unknown events, but that meant nothing; Invidiana and Vidar were well practiced in the art of fabricating guilt as needed. Had the Queen preserved her from Halgresta Nellt only to lay this trap for her instead?

If so, it was a deeper trap than Lune could perceive.

The tale Invidiana laid out was undoubtedly false—some trumped-up story of one fae seeking revenge against another through the destruction of the other fellow's homeland—but the person it implicated was no one Lune knew well, a minor knight called Sir Tormi Cadogant.

The accused fae did the only thing anyone could, in the circumstances. Had he not been at court, he might have run; it was treason to seek refuge among the fae of France or Scotland or Ireland, but it might also be safety, if he made it that far. But he was present, and so he shoved his way through the crowd and threw himself prostrate before the throne, hands outstretched in supplication.

"Forgive me, your Majesty," he begged, his voice trembling with very real fear. "I should not have done so. I have trespassed against your royal rights; I confess it. But I did so only out of—"

"Silence," Invidiana hissed, and his words cut off.

So perhaps Cadogant was the target of this affair. Or perhaps not. He was certainly not guilty, but that told Lune nothing.

"Come before me, and kneel," the Queen said, and shaking like an aspen leaf, Cadogant ascended the stairs until he came before the throne.

One long-fingered white hand went to the bodice of Invidiana's gown. The jewel that lay at the centre of her low neckline came away, leaving behind a stark patch of black in the intricate embroidery. Invidiana rose from her throne, and everyone knelt again, but this time they looked up; all of them, from Aspell and Vidar down to the lowliest brainless sprite, knew they were required to witness what came next. Lune watched from her station by the pillar, transfixed with her own fear.

The jewel was a masterwork even among the fae, a perfectly symmetrical tracery of silver drawn down from the moon itself, housing in its centre a true black diamond: not the painted gems humans wore, but a stone that held dark fire in its depths. Pearls formed from mermaid's tears surrounded it, and razor-edged slivers of obsidian ringed the gem's edges, but the diamond was the focal point, and the source of power.

Looming above the kneeling Cadogant, Invidiana was a pitiless figure. She reached out her hand and laid the jewel against the fae's brow, between his eyes.

"Please," Cadogant whispered. The word was audible to the farthest corners of the utterly silent hall. Brave as he was, to face the Queen's wrath and hope for what passed for mercy in her, he still begged.

A quiet clicking was his answer, as six spidery claws extended from the jewel and laid needle-sharp tips against his skin.

"Tormi Cadogant," Invidiana said, her voice cold with formality, "this ban I lay upon thee. Nevermore wilt thou bear title or honour within the borders of England. Nor wilt thou flee to foreign lands. Instead, thou wilt wander, never staying more than three nights in one place, neither speaking nor writing any word to another; thou wilt be as one mute, an exile within thine own land."

Lune closed her eyes as she felt power flare outward from the jewel. She had seen it used before, and knew some of how it worked. There was only one consequence for breaking such a ban.

Death.

Not just an exile, but one forbidden to communicate. Cadogant must have been plotting some treason. And this was a message to his co-conspirators, subtle enough to be understood, without telling the ignorant that a conspiracy had ever existed in the first place.

Her skin shuddered all over. Such a fate might have been hers, had Invidiana been any more enraged by her failure.

"Go," Invidiana snapped. Lune did not open her eyes until the hesitant, stumbling footsteps passed out of hearing.

When Cadogant was gone, Invidiana did not seat herself again. "This work is concluded for now," she said, and her words bore the terrible implication that Cadogant might not be the last victim. But whatever would happen next, it would not happen now. Everyone cast their gaze down again as the Queen swept from the room, and when the doors shut at last behind her, everyone let out a collective breath.

In the wake of her departure, music began to thread a plaintive note through the air. Glancing back toward the dais, Lune saw a fair-haired young man lounging on the steps, a recorder balanced in his nimble fingers. Like all of Invidiana's mortal pets, his name was taken from the stories of the ancient Greeks, and for good reason; Orpheus's simple melody did more than simply evoke the loss and sorrow of the peasant fae, and Cadogant's downfall. Some of those who had shown cruel amusement before now frowned, regret haunting their eyes. One dark-haired fae woman began to dance, her slender body flowing like water, giving form to the sound. Lune pressed her lips together and hurried to the door, before she, too, could be drawn into Orpheus's snare.

Vidar was lounging against one doorpost, bony silk-clad arms crossed over his chest. "Did you enjoy the show?" he asked, that same smirk hovering again on his lips.

Lune longed for a response to that, some perfect, cutting reply to check his surety that he stood in the Queen's favour and she did not. After all, fae had been known to suffer apparent disgrace, only for it later to be revealed as part of some scheme. But no such scheme sheltered her, and her wit failed. She felt Vidar's smirk widen as she shouldered past him and out of the presence chamber.

His words had unsettled her more than she realised. Or perhaps it was Cadogant, or those poor, helpless country pawns. Lune could not bear to stay out in the public eye, where she imagined every whisper spoke of her downfall. Instead she made her way, with as much haste as she could afford, through the tunnels to her own quarters.

The closing of the door gave the illusion of sanctuary. These two rooms were richly decorated, with a softer touch than in the public areas of the Hall; thick mats of woven rushes covered her floor, and tapestries of the great fae myths adorned her walls. The marble fireplace flared into life at her arrival, casting a warmer glow over the interior, throwing long shadows from the chairs that stood before it. Empty chairs; she had not entertained many guests lately. A doorway on the far side led to her bedchamber.

At least she still had this, her sanctum. She had lost the Queen's favour, but not so terribly that she had been forced from the Onyx Hall, to wander like those poor bastards in search of a new home. Not so terribly as Cadogant had.

The very thought made her shiver. Straightening, Lune crossed the room to a table that stood by her bedchamber door, and the crystalline coffer atop it.

She hesitated before opening it, knowing the dreary sight that would meet her eyes. Three morsels sat inside: three bites of coarse bread, who knew how old, but as fresh now as when some country housewife laid them out on the doorstep as a gift to the fae. Three bites to sustain her, if the worst should happen and she should be sent away from the Onyx Hall—sent out into the mortal world.

They would not protect her for long.

Lune closed the coffer and shut her eyes. It would not happen. She would find a way back into Invidiana's favour. It might take years, but in the meantime, all she had to do was avoid angering the Queen again.

Or giving Halgresta any excuse to come after her.

Lune's fingers trembled on the delicate surface of the coffer; whether from fear or fury, she could not have said.

No, she could not simply wait for her chance. That was not how one survived the Onyx Court. She would have to seek out an opportunity, or better yet, create one.

But how to do that, with so few resources available to her? Three bites of bread would not help her much. And Invidiana would hardly grant more to someone out of favour.

The Queen was not, however, the only source of mortal food.

Again Lune hesitated. To do this, she would have to go out of the Onyx Hall—which meant using one of her remaining pieces. That, or send a message, which would be even more dangerous. No, she couldn't risk that; she would have to go in person.

Praying the sisters would be as generous as she hoped, Lune took a piece of bread from the coffer and went out before she could change her mind.

RICHMOND AND LONDON

18 September 1588

So this, Deven thought blearily as he fumbled the lid back onto the close stool, *is the life of a courtier.*

His right shoulder was competing with his head for which ached worse. His new brothers in the Gentlemen Pensioners had taught him to play tennis the previous night, in the high-walled chamber built for that purpose out in the gardens. He'd flinched inwardly at having to pay for entry, but once inside, he took to it with perhaps more enthusiasm than was wise. Then there was drinking and card games, late into the night, until Deven had little memory of how he had arrived here, sharing Vavasour's bed, with their servants stretched out on the floor.

An urgent need to relieve himself had woken him; in the bed, Vavasour slept on. Scrubbing at his eyes, Deven contemplated following his fellow's example, but told himself with resignation that he might as well put the time to use. Otherwise he would sleep until noon and then get caught up once more in the social dance; then it would be too late to leave, so he would stay another night, and so on and so forth until he found himself crawling away from court one day, bleary-eyed and bankrupt.

Checking his purse, he corrected that last thought. Perhaps not bankrupt, judging by his apparent luck at cards the previous night. But such winnings would not finance this life. Hunsdon was right: he needed to borrow money.

Deven suppressed the desire to groan and shook Peter Colsey awake. His manservant was in little better shape than he, having found other servants with whom to entertain himself, but fortunately he was also taciturn of a morning. He rolled off the mattress and confined himself to dire looks at their boots, his master's doublet, and anything else that had the effrontery to require work from him at such an early hour.

The palace wore a different face at this time of day. The previous morning, Deven had been too much focused on his own purpose to take note of it, but now he looked around, trying to wake himself up gently. Servants hurried through the corridors, wearing the Queen's livery or that of various nobles. Outside, Deven heard chickens squawking as two voices argued over who should get how many. Hooves thudded in the courtyard, moving fast and stopping abruptly: a messenger, perhaps. He bet his winnings from the previous night that Hunsdon and the other men who dominated the privy council were up already, hard at work on the business of her Majesty's government.

Colsey brought him food to break his fast, and departed again to have their horses saddled. Soon they were riding out in morning sunlight far too bright.

They did not talk for the first few miles. Only when they stopped to water their horses at a stream did Deven say, "Well, Colsey, we have until Michaelmas. Then I am due to return to court, and under orders to be better dressed when I do."

Colsey grunted. "Best I learn how to brush up velvet, then."

"Best you do." Deven stroked the neck of his black stallion, calming the animal. It was a stupid beast for casual riding—the horse was trained for war—but a part of the fiction that the Gentlemen Pensioners were still a military force, rather than a force that happened to include some military men. Three horses and two servants; he'd had to acquire another man to assist Colsey. That still earned him more than a few glares.

By afternoon the houses they passed were growing closer together, clustering along the south bank of the Thames and stringing out along the road that led to the bridge. Deven stopped to refresh himself with ale in a Southwark tavern, then cocked his gaze at the sky. "Ludgate first, Colsey. We shall see how quickly I can get out, eh?"

Colsey had the sense not to make any predictions, at least not out loud.

Their pace slowed considerably as they crossed London Bridge, Deven's stallion having to shoulder his way through the crowds that packed it. He kept a careful hand on the reins. Travellers like him wended their way one step at a time, mingling with those shopping in the establishments built along the bridge's length; he didn't put it past the warhorse to bite someone.

Nor did matters improve much on the other side. Resigned by now to the slower pace, his horse drifted westward along Thames Street, taking openings where he found them. Colsey spat less-than-muffled curses as his own cob struggled to keep up, until at last they arrived at their destination in the rebuilt precinct of Blackfriars: John Deven's shop and house.

Whatever private estimate Colsey had made about the length of their visit, Deven suspected it was not short. His father was delighted to learn of his success, but of course it wasn't enough simply to hear the result; he wanted to know every detail, from the clothing of the courtiers to the decorations in the presence chamber. He had visited court a few times, but not often, and had never entered such an august realm.

"Perhaps I'll see it myself someday, eh?" he said, beaming with unsubtle optimism.

And then of course his mother Susanna had to hear, and his cousin Henry, whom Deven's parents had taken in after the death of John's younger brother. It worked out well for all involved; Henry had filled the place that might otherwise have been Michael's, apprenticing to John under the aegis of the Stationers' and freeing him to pursue more ambitious paths. The conversation went to business news, and then of course it was late enough that he had to stay for supper.

A small voice in the back of Deven's mind reflected that it was just as well; if he ate here, it was no coin out of his own purse. Why he should dwell on pennies when he was in debt for pounds made no sense, but there it was.

After supper, when Susanna and Henry had been sent off, Deven sat with his father by the fire, a cup of fine malmsey dangling from his fingers. The light flickered beautifully through the Venetian glass and the red wine within, and he watched it, pleasantly relaxed.

"Your place is assured, my son," John Deven said, stretching his feet toward the fire with a happy sigh.

Elizabeth's ominous words about Tylney had stayed in Deven's mind, but his father was right. There were greybeards in the Pensioners, some of them hardly fit for any kind of action. Unless he did something deeply foolish—like conspiring to kill the Queen—he might stay there until he wished to leave.

Some men did leave. Family concerns called them away, or a disenchantment with life at court; some broke their fortunes instead of making them. Seventy marks yearly, a Pensioner's salary, was not much in that world, and not everyone succeeded at gaining the kinds of preferment that brought more.

But then his father drove all money concerns from his mind, with one simple phrase. "Now," John Deven said, "to find you a wife."

It startled a laugh from him. "I have scarcely earned my place, Father. Give me time to get my feet under me, at least."

"'Tis not me you should be asking for time. You have just secured a favourable position, one close to her Majesty; there will be gentlewomen seeking after you like hawks. Perhaps even ladies."

There certainly had been women watching the tennis matches the previous day. A twinge in Deven's shoulder made him wonder how bad a fool he had made of himself. "No doubt. But I know better than to rush into anything, particularly when I *am* serving the Queen. They say she's very jealous of those around her, and dislikes scandalous behaviour in her courtiers." The last thing he needed was to end up in the Tower because he got some maid of honour pregnant.

The best eye to catch, of course, was that of the Queen

herself. But though Deven was ambitious, and her affection was a quick path to reward, he was not at all certain he wanted to compete with the likes of the young Earl of Essex. That would rapidly bring him into situations he could not survive.

"Marriage is no scandal," his father said. "Have a care for how you comport yourself, but do not stand too aloof. A match at court might be very beneficial indeed."

His father seemed likely to keep pressing the matter. Deven dodged it with a distraction. "If all goes as planned, my time will be very thoroughly employed elsewhere."

John Deven's face settled into graver lines. "You have spoken to Walsingham, then?"

"No. He was not at court. But I will do so at the first opportunity."

"Be wary of rushing into such things," his father said. Much of the relaxed atmosphere had gone out of the air. "He serves an honourable cause, but not always by honourable means."

Deven knew this very well; he had done some of that work in the Low Countries. Though not the most sordid parts of it, to be sure. "He is my most likely prospect for preferment, Father. But I'll keep my wits about me, I promise."

With that, his father had to be satisfied.

* * *

LONDON AND ISLINGTON

18 September 1588

Leaving the Onyx Hall was not so simple as Lune might have hoped. In the labyrinthine politics of court, someone would find a way to read her departure as suspicious, should she go out too soon after Invidiana's sentencing of Cadogant. Vidar, if no one else.

So she wandered for a time through the reaches of the Onyx Hall, watching fae shy away from her company. It was an easy way to fill time; though the subterranean faerie palace was not so large as the city above, it was far larger than any surface building, with passages playing the role of streets, and complexes of chambers given over to different purposes.

In one open-columned hall she found Orpheus again, this time playing dance music; fae clapped as one of their number whirled around with a partner in a frenzied display. Lune placed herself along the wall and watched as a grinning lubberkin dragged a poor, stumbling human girl on, faster and faster. The mortal looked healthy enough, though exhausted; she was probably some maidservant lured down into the Onyx Hall for brief entertainment, and would be returned to the surface in the end, disoriented and drained. Those who had been there for a long time, like Orpheus, acquired a fey look this girl did not yet have.

Their attention was on the dance. Unobserved, Lune slipped across to the other side of the hall and out through another door.

She took a circuitous route, misleading to anyone who might see her passing by, but also necessary; one could not simply go straight to one's destination. The Onyx Hall connected to the world above in a variety of places, but those places did not match up; two entrances might lie half the city apart on the surface, but side-by-side down below. It was one of the reasons visitors feared the place. Once inside, they might never find their way out again.

But Lune knew her path. Soon enough she entered a small, deserted chamber, where the stone walls of the palace gave way to a descending lacework of roots.

Standing beneath their canopy, she took a deep breath and concentrated.

The rippling, night-sky sapphire of her gown steadied and became plainer blue broadcloth. The gems that decorated it vanished, and the neckline closed up, ending in a modest ruff, with a cap to cover her hair. More difficult was Lune's own body; she had to focus carefully, weathering her skin, turning her hair from silver to a dull blond, and her shining eyes to a cheerful blue. Fae who were good at this knew attention to detail was what mattered. Leave nothing unchanged, and add those few touches—a mole here, smallpox scars there—that would speak convincingly of ordinary humanity.

But building the illusion was not enough, on its own. Lune reached into the purse that hung from her girdle and brought forth the bread from her coffer.

The coarsely ground barley caught in her teeth; she

was careful to swallow it all. As food, she disdained it, but it served its own purpose, and for that it was more precious than gold. When the last bit had been consumed, she reached up and stroked the nearest root.

With a quiet rustle, the tendrils closed around and lifted her up.

She emerged from the trunk of an alder tree that stood along St. Martin's Lane, no more than a stone's throw from the structures that had grown like burls from the great arch and surrounding walls of Aldersgate. The time, she was surprised to discover, was early morning. The Onyx Hall did not stand outside human time the way more distant realms did—that would make Invidiana's favourite games too difficult—but it was easy to lose track of the hour.

Straightening her cap, Lune stepped away from the tree. No one had noticed her coming out of the trunk. It was the final boundary of the Onyx Hall, the last edge of the enchantments that protected the subterranean palace lying unseen below mortal feet; just as the place itself remained undiscovered, so would people not be seen coming and going. But once away from its entrances, the protections ended.

As if to hammer the point home, the bells of St. Paul's Cathedral rang out the hour from within the tightly packed mass of London. Lune could not repress the tiniest flinch, even as she felt the sound wash over her harmlessly. She had done this countless times before, yet the first test of her own protections always made her nervous.

But she was safe. Fortified by mortal food against the

power of mortal faith, she could walk among them, and never fear her true face would be revealed.

Settling into her illusion, Lune set out, walking briskly through the gate and out of London.

The morning was bright, with a crisp breeze that kept her cool as she walked. The houses crowding the lane soon spaced themselves more generously, but there was traffic aplenty, an endless flow of food, travellers, and goods into and out of the city. London was a voracious thing, chewing up more than it spat back out, and in recent years it had begun to swallow the countryside. Lune marvelled at the thronging masses who flooded the city until it overflowed, spilling out of its ancient walls and taking root in the formerly green fields that lay without. They lived like ants, building up great hills in which they lived by the hundreds and thousands, and then dying in the blink of an eye.

A mile or so farther out, it was a different matter. The clamour of London faded behind her; ahead, beyond the shooting fields, lay the neighbouring village of Islington, with its manor houses and ancient, shading trees. And along the Great North Road, the friendly, welcoming structure of the Angel Inn.

The place was moderately busy, with travellers and servants alike crossing the courtyard that lay between the inn and the stables, but that made Lune's goal easier; with so many people about, no one took particular notice of one more. She passed by the front entrance and went toward the back, where the hillside was dominated by an enormous rosebush, a tangled, brambly mass even the bravest soul would be afraid to trim back.

This, too, had its own protections. No one was there to watch as Lune cupped a late-blooming rose in her hand and spoke her name into the petals.

Like the roots of the alder tree in London, the thorny branches rustled and moved, forming a braided archway starred with yellow blossoms. Inside the archway were steps, leading down through the earth, their wood worn smooth by countless passing feet. Charmed lights cast a warm glow over the interior. Lune began her descent, and the rosebush closed behind her.

The announcement of her name did not open the bush; it only told the inhabitants someone had come. But visitors were rarely kept waiting, outside or in. By the time Lune reached the bottom of the steps, someone was there.

"Welcome to the Angel, my lady," Gertrude Goodemeade said, a sunny smile on her round-cheeked face as she bobbed a curtsy. "'Tis always a pleasure to see you here. Come in, please, please!"

No doubt the Goodemeade sisters gave the same friendly greeting to anyone who crossed their threshold—just as, no doubt, more courtiers came here than would admit it—and yet Lune did not doubt the words were sincere. It was in the sisters' nature. They came from the North originally—brownies were Border hobs, and Gertrude's voice retained traces of the accent—but they had served the Angel Inn since its construction, and supposedly another inn before that, and on back past what anyone could remember. Many hobs were insular folk, attached to a particular mortal family and unconcerned with anyone else, but these two understood giving hospitality to strangers.

The edges of the tension that had frozen Lune's back for days melted away in the warmth of the brownies' comfortable home. Lune suffered Gertrude to lead her into the cosy little chamber and settle her onto a padded bench at one of the small tables. "We haven't seen you here in some time," Gertrude said. She was already bustling about, embroidered skirts swishing with her quick movements, fetching Lune a cup of mead without asking. It was, of course, exactly what Lune craved at that moment. The talents of brownies were homely things, but appreciated all the same.

One brownie, at any rate. Lune opened her mouth to ask where Gertrude's sister was, then paused at sounds on the staircase. A moment later her question was answered, for Rosamund entered, wearing a russet dress that was the twin of Gertrude's save for the embroidery on its apron—roses instead of daisies—just as her cheerful face mirrored that of her sister.

Behind her came others who were less cheerful. Lune recognised the haggard male hob immediately; the others were less familiar, having mostly pressed their faces into the floor of the Onyx Hall when she last saw them.

Gertrude made a sympathetic sound and hurried forward. For a short time the room seemed overfull, wall-to-wall with hobs and pucks and a slender, mournful-faced river nymph Lune had missed among them the first time. But no brownie would suffer there to be confusion or standing guests for long; soon enough a few of the strangers were ensconced at the tables with bread fresh out of the oven and sharp, crumbly cheese, while the more tired among them were bundled off through another door and put to bed.

Lune wrapped her fingers around her mead and felt uncomfortable. She had dismissed her illusion of mortality—she would have felt odd maintaining it inside, as if she had kept a traveling cloak on—but the bite of human bread she had eaten still made her proof against church bells, iron horseshoes, and other anti-faerie charms. How the refugees had gotten to the Angel from the Onyx Hall, she did not know, but she doubted it had been so easy. Rosamund must have been present at court, though. Lune chided herself for not studying the crowd more closely.

Gertrude had not forgotten her. Moments later, the smell of roasted coney filled the room, and Lune was served along with the others. The food was simple, prosaic, and good; one could easily imagine mortals eating the same thing, and it made the elaborate banquets of the court seem fussy and excessive.

Perhaps, Lune thought, *this is why I come here. For perspective.*

Would it be so bad, to leave the court? To find a simpler life, somewhere outside of London?

It would be easier, certainly. In the countryside, there was less need to protect oneself against mortal tricks.

Peasant folk saw fae from time to time, and told stories of their encounters with black dogs or goblins, but no one made trouble of it. Or rarely, at least. They generally only tried to lay creatures who made too much a nuisance of themselves. And out there, one was well away from the intrigues of the Onyx Court.

Next to Lune on the bench, a tuft-headed sprite began to sniffle into his bread.

Wherever these rural fae had come from, it was not far enough to save them from Invidiana.

No, she could not leave London. To be subject to the tides of the court, but unable to affect them...

There was another choice, of course. Across the boundary of twilight, down the pleasant paths that led neither to Heaven nor Hell, and into the deeper reaches of Faerie, where Invidiana's authority and influence did not reach. But few mortals ever wandered so far, and for all the dangers they posed to fae, Lune would not leave them behind. Mortals were endlessly fascinating, with their brief, bright lives, and all the passion that fuelled them.

Rosamund began to shepherd the others off, murmuring about baths and nice soft beds. Gertrude came by as the sprite vacated Lune's bench. "Now then, my lady—forgive me for that. Poor things, they were starved to the bone. Was it just a bite to eat you were looking for, and a breath of good country air?"

Her apple-cheeked face radiated such friendly helpfulness that Lune shook her head before she could stop herself. On the instant, Gertrude's cheerful demeanour transformed to concern. "Oh, dearie. Tell us about it."

Lune had not meant to share the story, but perhaps it was appropriate; she could hardly ask for aid without explaining at least some of why she needed it, and the Goodemeades were generally ignorant of politics. They might be the nearest fae who had *not* already heard.

"I am disgraced at court," she admitted.

She tried to speak as if it were of small moment. Indeed, sometimes it was; if everyone who angered Invidiana suffered Cadogant's fate, there would soon be no court left. But she stood upon the edge of a knife, and that was never a comfortable place to be.

Gertrude made a sympathetic face. "Queen's taken a set against you, has she?"

"With cause," Lune said. "You listen to the talk in the mortal inn, do you not?"

The brownie dimpled innocently. "From time to time."

"Then you know they fear invasion by Spain, and that a great Armada was only recently defeated."

"Oh, we heard! Great battles at sea, or some such."

Lune nodded, looking down at the remnants of her coney. "Great battles. But before them and after, great storms as well. Storms for which we paid too high a price." She had confessed the details only to Invidiana, and would not repeat them; that would only deepen the Queen's wrath. But she could tell Gertrude the shape of it. "I was Invidiana's ambassador to the folk of the sea, and did not bargain well enough. She is displeased with the concession I promised."

"Oh dear." Gertrude paused to assimilate this. "What was so dreadful, then? I cannot imagine she wants us to be invaded; surely it was worth the price."

Lune pushed her trencher away, painting a smile over her ever present knot of worry. "Come, you do not want to talk of such things. This is a haven away from court and its nets—and long may it remain so."

"True enough," the brownie said complacently, patting her apron with plump but work-worn fingers. "Well, all's well that ends well; we don't have any nasty Spanish soldiers trampling through the Angel, and I'm sure you'll find your way back into her Majesty's good graces soon enough. You have a talent for such things, my lady."

The words returned Lune to her original purpose. "I hesitate to ask you this," she admitted, looking at the doorway through which the last of the refugees had vanished. "You have so many to take care of now—at least until they can be settled elsewhere. And I wonder Rosamund could even bring them here safely."

Her reluctance had exactly the desired effect. "Oh, is that all?" Gertrude exclaimed dismissively, springing to her feet. The next Lune knew, the brownie was pressing an entire heel of bread into her hands. It was not much different from what the Goodemeades had served, but any fae could tell one from the other at a touch. Mortality had a distinctive weight.

Looking down at the bread, Lune felt obscurely guilty. The maidservants of the Angel put out bread and milk faithfully; everyone knew that. And Invidiana taxed the Goodemeades accordingly, just as she taxed many country fae. Many more rural humans than city folk put out food for the fae, yet it was in the city that they needed it most. The Onyx Hall shut out the sounds of the bells and other such threats, but to venture into the streets unfed was an assurance of trouble.

She needed this. But so did the Goodemeades, with their guests to take care of.

"Go on, take it," Gertrude said in a soft voice, folding her hands around the bread. "I'm sure you'll find a good use for it."

Lune put her guilt aside. "Thank you. I will not forget your generosity."

MEMORY

May–August 1588

In villages and towns all along the coast of England, piles of wood awaited the torch, and men awaited the first sight of the doom that was coming to devour them.

In the crowded harbour of Lisbon, the ships of the *Grande y Felicícisma Armada* awaited the order that would send them forth, for God and King Philip, to bring down the heretic queen.

In the waters that separated them, storms brewed, sending rain and heavy winds to lash the lands on both sides of the English Channel.

The Armada was a greater thing in story than it was in reality. The five hundred mighty ships that would bear an unstoppable army to England's shores, their holds crammed with implements of torture and thousands of Catholic wet nurses for the English babies who would be orphaned by the wholesale slaughter of their parents, were in truth a hundred and thirty ships of varying degrees of seaworthiness, crewed by the dregs of Lisbon, some of whom had never been to sea before, and commanded by a landsman given his posting only a few months gone. Disease and the depredations of the English scourge Sir Francis Drake had taken their toll on God's weapon against the heretics.

But the worst was yet to come.

In this, the quietest month of the year, when all the experienced seamen had assured the Duke of Medina-Sidonia that the waters would be calm and the winds fair for England, the storms did not subside; instead, they grew in strength. Gales drove the ships back when they tried to progress, and scattered the weaker, less seaworthy vessels. Fat-bottomed merchantmen, Mediterranean galleys unsuited to the blasts of the open sea, lumbering supply ships that slowed the pace of the entire fleet: the Great and Most Fortunate Armada was a sorry sight indeed.

Delays had slain what remained of May; June rotted

away in the harbour of La Coruña, while sailors sickened and starved, their victuals fouled by the green wood of the barrels they were kept in. The commanders of the fleet found new terms by which to damn Drake, who had burned the seasoned barrel-staves the previous year.

In July they sailed again, obedient to God's mission.

Red crosses waved on white flags. The banner of Medina Sidonia's ship carried the Virgin and a crucified Christ, and the motto *Exsurge, Domine, et judica causam tuam!* Monks prayed daily, and even sailors were forbidden to take the Lord's name in vain.

Yet none of it availed.

Beacon fires flared along the coast of England: the Spanish had been sighted. The wind favoured the English, and so did the guns; the trim English ships refused boarding engagements, dancing around their ungainly enemy, battering away with their longer guns while staying out of Spanish range. Like dogs tearing at a chained bear, they harried the Spanish up the coast to Scotland, while the storms kept up their merciless assault.

Storms, always storms, every step of the way.

Storms struck them in the Orkneys, and again off the Irish coast, as the Armada fought to crawl home. From Lisbon into the Channel, around all the islands of England, Scotland, and Ireland—everywhere the fleet went, the wrath of sky and sea pursued.

Sick unto death with scurvy and typhus, maddened by starvation and thirst, the sailors screamed of faces in the water, voices in the sky. God was on their side, but the sea was not. Ever fickle, she had turned an implacable face to them, and all the prayers of the monks could not win her goodwill.

For a deal had been struck, in underwater palaces spoken of only in sailors' drunken tales. The sea answered to powers other than man's, and those powers—ever callous to human suffering—had been persuaded to act in favour of the English cause, against their usual disinterested neutrality.

So it was that the skies raged on command and alien figures slipped through the water, dancing effortlessly around the foundering vessels, luring men overboard and dragging them under, discarding many to wash up, bloated and rotting, on the Irish shore, but keeping a few for future amusement. It was difficult to say who had the more unfortunate fate: those who died, or those who lived.

In Spain, bells rang out in premature celebration, while his most Catholic Majesty awaited news of his most holy mission.

In England, the heretic queen rallied her people, while reports trickled in from Drake and the Lord Admiral, speaking of English heroism.

In the turbulent waters of the Atlantic, the remnants of the Armada, half their number lost, captured, or sunk, limped homeward, and took with them the hopes of a Spanish conquest of England.

* * *

THE ONYX HALL, LONDON

18 September 1588

The mortal guise fell away from Lune like a discarded cloak the moment the alder tree grew shut around her, and she concealed the bread within the deep folds of her skirts. Those who wished to, would find out soon enough that she had it, and where she had obtained it, but she would hide it as best she could. Plenty of lesser courtiers would come begging for a crumb if they knew.

Some of them might smell it on her; certain fae had a nose for mortality. Lune hurried through the Onyx Hall to her chambers, and tucked the heel of bread into her coffer as the door closed behind her.

With it safely stowed, she rested her hands upon the inlaid surface of the table, tracing with one fingertip the outline of its design. A mortal man knelt at the foot of a tower; the artisan had chosen to show only the base of the structure, leaving to the imagination which faerie lady had caught his heart, and whether she returned his love.

It happened, sometimes. Not everyone played with mortals as toys. Some, like hobs, served them faithfully. Others gave inspiration to poets and musicians. A few loved them, with the deathless passion of a faerie heart, all the stronger for being given so rarely.

But mortals were not Lune's concern, except insofar as they might provide her with a route to Invidiana's favour.

She lowered herself onto the embroidered cushion of a

stool. With deliberate, thoughtful motions, Lune began to remove the jewelled pins from her hair, and laid each one on the table to represent her thoughts.

The first she laid down glimmered with fragments of starlight, pushing the boundaries of what she, as a courtier in disgrace, might be permitted to adorn herself with. *A gift,* Lune thought. A rare faerie treasure, or a mortal pet, or information. Something Invidiana would value. It was the commonest path to favour, not just for fae but for humans as well. The difficulty was, with so many gifts being showered at the Queen's feet, few stood out enough to attract her attention.

A second pin. The knob at the end of this one held the indigo gems known as the sea's heart. Lune's fingers clenched around it; she had dressed for court in a rush, and had not attended to which pins she chose. Had Vidar seen it? She prayed not. Bad enough to have lost the Queen's goodwill by that disastrous bargain with the folk of the sea; worse yet to wear in her hair their gift to the ambassador of the Onyx Court.

Dame Halgresta certainly had not seen it; of that, Lune could be sure, because she was not bleeding, or dead.

She set it down on the table, forcing her thoughts back to their task. If not a gift, then what? A removal of an obstacle, perhaps. The downfall of an enemy. But who? The ambassador from the Courts of the North had quit the Onyx Hall in rage after the execution of the mortal Queen of Scots, accusing Invidiana of having engineered her death. There were enemies aplenty in that coalition of Seely and Unseely monarchs, the courts of Thistle and Heather and Gorse. To move against them, however, Lune would have to go there herself: a tedious journey, with no assets or allies waiting for her at the end.

As for other enemies, she was not fool enough to think she could take action against the Wild Hunt and live.

Lune sighed and pulled a third pin from her hair.

Silver locks spilled free as she did so, sending the remaining pins to the floor. Lune left them where they fell, fingering the snowflake finial of the one in her hand. Give the Queen something she wanted, or remove something that stood between her and what she wanted. What else was there?

Amusement. The Queen was a cold woman, heartless and cruel, but she could be entertained. Her favourite jests were those that accomplished some other goal at the same time. Even without that, though, to amuse the Queen…

It was a slim enough thread, but the last thing she could grasp for.

Lune held the snowflake pin, pressing her lips together in frustration. The outlines of her options were simple enough; the difficulty lay in moving from concept to action. Everything she thought of was weak, too weak to do her much good, and she was not positioned to do more. The trap of courtly life: those in favour were the best positioned to gain favour, while those who fell out of it were often caught in a spiral of worsening luck.

She would not accept it. Running her thumb over the sharp, polished points of the snowflake, Lune disciplined her mind. How could she better her position in the Onyx Court?

"Find Francis Merriman."

Lune was on her feet in an instant, the snowflake pin reversed and formed into a slender dagger in her hand. Her private chambers were charmed against intruders, a basic precaution in the Onyx Hall, and no one would break those protections unless they had come to do her harm.

No one, save the slender figure in the shadows.

Lune let out her breath slowly and relaxed her grip on the dagger, though she did not put it aside. "Tiresias."

He was often where he should not be, even where he *could* not be. Now he crouched in the corner, his slender arms wrapped around his knees, his pale, ethereal face floating in the darkness.

Lune avoided Invidiana's mortal pets for a varied host of reasons: Orpheus for fear of the effect his music might have on her; Eurydice for her ghost-haunted eyes; Achilles for the barely contained violence that only the Queen's will held in check. Tiresias was different. She did not fear the gift for which he was named. Sometimes Lune doubted even Invidiana could tell which visions were true, which mere constructs of his maddened brain.

No, it was the madness itself that gave her pause.

He was older than the other pets, they said, and had survived longer than any. Achilles died so often that one of the quickest routes to Invidiana's favour, if only briefly, was to find another mortal with a gift for battle fury and bring him to court; she was forever pitting the current bearer of that name against some foe or another, just for an evening's entertainment. They fought well, all of them, and sooner or later died bloodily.

They rarely survived long enough to suffer the effects of the Onyx Hall.

Tiresias survived, and paid the price.

He had flinched at her sudden movement, fear twisting his face. Now he looked up at her, searchingly. "Are you real?" he whispered.

He asked the question incessantly, no longer able to distinguish reality from his own delusions. It made for great sport among the crueller fae. Lune sighed and let the dagger revert to a pin, then laid it on the table. "Yes. Tiresias, you should not be here."

He shrank farther back into the shadows, as if he would meld into and through the wall. Perhaps he could, and that was how he arrived in such unexpected places. "Here? 'Tis nothing more than a shadow. We are not here. We are in Hell."

Lune moved away from the table, and saw his eyes linger on the coffer behind her. A few bites of mortal bread could not lift the faerie stain from his soul; after untold years in the Onyx Hall, she doubted anything could. If he set foot outside, would he crumble to dust? But he hungered for mortality, sometimes, and she did not want him thinking of the bread she had. "Go back to your mistress. I have no patience for your fancies."

Tiresias rose, and for a moment Lune thought he might obey. He wandered in the wrong direction, though—neither toward the door, nor the coffer. The back of his sable doublet was torn, a thin banner of fabric fluttering behind him like a tiny ghost of a wing. Lune opened her mouth to order him away again, but stopped. He had said something, which she had overlooked in her fright.

Moving slowly, so as not to startle him, Lune approached Tiresias's back. He would always have been a slender man, even had he lived as a normal human, but life among the fae had made him insubstantial, wraithlike. She wondered how much longer he would last. Mortals could survive a hundred years and more among the fae—but not in the Onyx Hall. Not under Invidiana.

He was fingering the edge of a tapestry, peering at it as if he saw something other than the flooded shores of lost Lyonesse. Lune said, "You spoke a name, bade me find someone."

One pale finger traced a line of stitchery, moonlight shining down upon a submerged tower. "Someone erred, and thus it sank. Is that not what you believe? But no—the errors came after. Because they misunderstood."

"Lyonesse is ages gone," Lune replied, with tired patience. She might not have even been there, for all the attention he paid to her. "The name, Tiresias. Who was it you bade me find? Francis Merriman?"

He turned and fixed his sapphire gaze on her. The pupils of his eyes were tiny, as if he stared into a bright light; then they expanded, until the blue all but vanished. "Who is he?"

The innocence of the question infuriated her, and in her distraction, she let him slip past. But he did not go far, halting in the centre of the room, reaching for some imagined shape in the air before him. Lune let her breath out slowly. Francis Merriman: a mortal name. A courtier? A likely chance, given the political games Invidiana played. No one Lune knew of, but they came and went so quickly.

"Where can I find him?" she asked, trying to keep her voice gentle. "Where did you see him? In a dream?"

Tiresias shook his head violently, hands scrabbling through his black hair, disarranging it. "I do not dream. I do not dream. Please, do not ask me to dream."

Lune could imagine the nightmares Invidiana sent him for her own entertainment. "I will do nothing to you. But why should I seek him?"

"He knows." The words came out in a hoarse whisper. "What she did."

Her heart picked up its pace. Secrets—they were worth more than gold. Lune tried to think who Tiresias might mean. "She. One of the ladies? Or—" Her breath caught. "Invidiana?"

Bitter, mocking laughter greeted the suggestion. "No. Not Invidiana; that is not the point. Have you not been listening?"

Lune swallowed the desire to tell him she would start listening when he said something of comprehensible substance. Staring at the seer's tense face, she tried a different tack. "I will search for this Francis Merriman. But if I should find him, what then?"

Slowly, one muscle at a time, his body eased, until his hands hung limp at his sides. When he spoke at last, his voice was so clear she thought for a heartbeat that he was in one of his rare lucid periods—before she listened to his words. "Stand still, you ever-moving spheres of Heaven..." A painful smile curved his lips. "Time has stopped. Frozen, cold, no heart's blood to quicken it to life once more. I told you, we are all in Hell."

Perhaps there had never been any substance in it to begin with. Lune might be chasing an illusion, pinning too much hope on the ramblings of a madman. Not everything he said came from a vision.

But it was the one possibility anyone had offered her, and the only one she was likely to receive. Her best hope otherwise was to bargain her bread for information that might be of aid. There were plenty of courtiers who would have use for it, playing their games in the world above.

When she made her bargains, she would ask after this Francis Merriman. But secretly, so she did not betray her hand to the Queen. Surprise might count for a great deal.

"You should go," she murmured, and the seer nodded absently, as if he had forgotten where he was, and why. He turned away, and when the door closed behind him, Lune returned to her table and collected the scattered pins that had fallen from her hair.

There were possibilities. She simply had to bring one to fruition.

And quickly, before the whirlpools of the court dragged her down.

RICHMOND PALACE, RICHMOND

29 September 1588

"…You shall be retained to no person nor persons of what degree or condition, by oath, livery, badge, promise, or otherwise, but only to her Grace, without her special license…"

Deven suppressed a grimace at those words. How strictly were they enforced? It might hamstring his plans for advancement at court, if the Queen were jealous with that license; he would be bound to her service only, without any other patron. Certainly some men served other masters, but how long had they petitioned to be allowed to do so?

Hunsdon was still talking. The oath for joining the Gentlemen Pensioners was abominably long, but at least he did not have to repeat every word of it after the band's captain; Deven only affirmed the different points that Hunsdon outlined. He recognised Elizabeth as the supreme head of the Church; he would not conceal matters prejudicial to her person; he would keep his required quota of three horses and two manservants, all equipped as necessary for war; he would report any fellow remiss in such matters to the captain; he would keep the articles of the band, obey its officers, keep secrets secret, muster with his servants when required, and not depart from court without leave. All enumerated in elaborately legalistic language, of course, so that it took twice as long to say as the content warranted.

Deven confirmed his dedication to each point, kneeling on the rush matting before Hunsdon. As ordered, he had dressed himself more finely, driving his Mincing Lane tailor to distraction with his insistence that the clothing be finished in time for today's Michaelmas ceremony. The doublet was taffeta of a changeable deep green, slashed with cloth-of-silver that blithely violated the sumptuary laws, but one visit to court had been enough to show Deven how few people attended to those restrictions. The aglets on his points were enamelled, as was the belt that clasped his waist, and he was now a further fifty pounds in debt to a goldsmith on Cheapside. Listening to Hunsdon recite the last words, he prayed the expense would prove worthwhile.

"Rise, Master Deven," the baron said at last, "and be welcome to her Majesty's Gentleman Pensioners."

The Lord Chamberlain settled a gold chain about his shoulders when he stood, the ceremonial adornment for members of the band. Edward Fitzgerald, lieutenant of the Gentlemen Pensioners, handed him the gilded poleax he would bear while on duty, guarding the door from the presence chamber to the privy chamber, or escorting her Majesty to and from chapel in the morning. Deven was surprised by the heft of the thing. Ceremonial it might be, and elaborately decorated, but not decorative. The Gentlemen Pensioners were the elite bodyguard of the monarch, since Elizabeth's father Henry, eighth of that name, decided his dignity deserved better escort than it had previously possessed.

Of course, before Deven found himself using the gilded polearm, any attacker would have to win through the Yeomen of the Guard in the watching chamber, not to mention the rest of the soldiers and guardsmen stationed at any palace where the sovereign was in residence. Still, it was reassuring to know that he would have the means to defend the Queen's person, should it become necessary.

It meant that *he* was not purely decorative, either.

His companions toasted their newest member with wine, and a feast was set to follow. In theory, the entire band assembled at court for Michaelmas and three other holidays; in practice, somewhat less than the full fifty were present. Some were assigned to duties elsewhere, in more distant corners of England or even overseas; others, Deven suspected, were at liberty for the time being, and simply had not bothered to come. A man might be docked pay for failure to attend as ordered—that was in the articles he had sworn to obey—but a rich enough man hardly need worry being fined a few days' wages.

Despite the revelry, Deven's mind kept returning to the question of patronage. His eyes sought out Hunsdon, across the laughing, boisterous mass of men that filled the chamber where they dined. The officers of the band sat at a higher table—Hunsdon and Fitzgerald, plus three others who were the company's standard bearer, clerk of the check, and harbinger.

He could ask Hunsdon. But that would be tantamount to telling the baron that he intended to seek another master.

Surely, though, that would come as no surprise. Hunsdon knew who had secured Deven's position in the Gentlemen Pensioners.

Deven reached reflexively for his wine, grimaced, then grinned at himself. He did not know how he was going to handle his patronage, but one thing he *did* know: making any plans about such things while this drunk was not wise. Attempting to ask delicate questions of his captain would be even less wise. Therefore, the only course for a wise man to follow was to go on drinking, enjoy the night, and worry about such matters on the morrow.

RICHMOND PALACE, RICHMOND
30 September 1588

Deven had been among military men; he should have expected what the morrow would bring. William Russell, who either possessed the constitution of an ox or had not drunk nearly as much as he appeared to the previous night, arrived in his chamber at an hour that would have been reasonable had Deven gone to bed before dawn, and rolled him forcibly out of bed. "On your feet, man; we can't keep the Queen waiting!"

"Nnnnnngh," Deven said, and tried to remember if

there was anything in the articles that forbade him to punch one of his fellows.

Between the two of them, Colsey and Ranwell, his new manservant, got him on his feet and stuffed him into his clothes. Deven thought muzzily that someone had arranged for a Michaelmas miracle; he didn't have a hangover. Round about the time he formed up with the others for the Queen's morning procession to chapel, he realised it was because he was still drunk. And, of course, Fitzgerald had assigned him to duty that day, so he was on display in the presence chamber when the inevitable hangover came calling. He clung grimly to his poleax, tried to keep it steady, and prayed he would not vomit in front of his fellow courtiers.

He survived, though not happily, and passed the test to which he had been put. Moreover, he had his reward; the Queen emerged from her privy chamber just as he was handing off his position to Edward Greville, and she gifted him with a nod. "God give you good day, Master Deven."

"And to you, your Majesty," he answered, bowing reflexively; the world lurched a little when he did, but he kept his feet, and then she was gone.

The Queen remembered his name. It shouldn't have pleased him so much, but of course it did, and that was why she did it; Elizabeth had a way of greeting a man that made him feel special for that instant in which her attention lighted upon him. Even his headache did not seem so bad in the aftermath.

It came back full force as he left, though. Handing

off his poleax to Colsey, he suffered Ranwell to feed him some concoction the man swore would cure even the worst hangover; less than a minute later his stomach rebelled and he vomited it all back up. "Feed me that again," Deven told his new servant, "and you'll find yourself sent to fight in Ireland."

Colsey, who still did not appreciate having to share his master with an interloper, smirked.

Deven cleaned his mouth out and took a deep breath to fortify himself. He wanted little more than to collapse back into bed, but that would never do, so instead he addressed himself to the business at hand.

It made no sense to ask Hunsdon about the permissibility of acquiring another patron, if he did not have such a man already friendly to him. Deven hoped he did, but until that was confirmed, best not to broach the subject with Hunsdon at all.

Squaring his shoulders, Deven gritted his teeth and went in search, hangover and all.

But luck, which had preserved him through the morning's ordeal, was not on his side in this matter. The Principal Secretary, he learned, was ill and thus absent from court. His inquiries led him to another man, ink-stained and bearing a thick sheaf of papers, who was attending the meetings of the privy council in Walsingham's absence.

Deven made bold enough to snatch a moment of Robert Beale's time. After introducing himself and explaining his business, at least in broad outline, he asked,

"When might the Principal Secretary return to court?"

Beale's lips pressed together, but not, Deven thought, in irritation or offense. "I could not say," the Secretary said. "He requires rest, of course, and her Majesty is most solicitous of his health. I would not expect him back soon—for some days at least, and possibly longer."

Damnation, again. Deven forced a smile onto his face. "I thank you for your time," he said, and got out of Beale's way.

He could hardly go asking favours of a man on his sickbed. He could send a letter—but no. Better not to press the matter. As much as it galled him, he would have to wait, and hope the Principal Secretary recovered soon.

THE ONYX HALL, LONDON
20 October 1588

Tens of thousands of mortals lived in London, and more in the towns and villages that surrounded it. In the entirety of England, Lune could not begin to guess how many there were.

Except to say there were too many, when she was trying to find a particular one.

She had to be discreet with her inquiries. If Tiresias was to be trusted—if he truly had a vision, or overheard something while lurking about—then this Francis Merriman knew something of use. It followed, then, that she did not want to share him with others. But so far discretion had availed her nothing; the mortal was not easily found.

When a spindly little spriteling came to summon her before Vidar, her first thought was that it had to do with her search. There was no reason to think that, but the alternatives were not much more appealing. Concealing these thoughts, Lune acknowledged the messenger with a nod. "Tell the Lord Keeper I will come when I may."

The messenger smiled, revealing sharp, goblinish teeth. "He demanded your immediate attendance."

Of course he did. "Then I will be pleased to come," Lune said, rising as she mouthed the politic lie.

In better times, she might have made him wait. Vidar's exalted status was a new thing, and Lune had until recently been a lady of Invidiana's privy chamber, one of the Queen's intimates—inasmuch as she was intimate with anyone. That freedom was gone now; if Vidar said to leap, then leap she must.

And of course he kept her dangling. Vidar's rise to Lord Keeper had made him a most desirable patron, rich in both wealth and enchantment, and now his outer chamber thronged with hopeful courtiers and rural fae begging some favour or another. He might have demanded her immediate attendance, but he granted audience to a twisted bogle, two Devonshire pisgies, and a travel-stained faun in Italian dress before summoning Lune into the inner chamber.

He lounged in a chair at the far end of the room, and did not rise when she came in. Some fae held to older fashions of clothing, but he closely followed current styles; the crystals and jet embroidered onto his doublet winked in the light, an obvious mimicry of Invidiana's clothing. Rumour had it the black leather of his tall, close-fitting boots was the skin of some unfortunate fae he had captured, tortured, and executed on the Queen's behalf, but Lune knew the rumours came from Vidar himself. It was ordinary doeskin, nothing more. But the desire for that belief was telling enough.

She gave him the curtsy rank demanded, and not a hair more. "Lord Ifarren."

"Lady Lune." Vidar twiddled a crystal goblet in his bony fingers. "How good of you to come."

She waited, but he did not offer her a seat.

After a leisurely study of her, Vidar set aside the goblet and rose. "We have known one another for a long time, have we not, my lady? And we have worked together in days past—to mutual benefit, as I weigh it. It pains me to see you thus fallen."

As a stag in season was pained to see a rival fall to a hunter's arrow. Lune cast her gaze modestly downward and said, "'Tis kind of you to say it, my lord."

"Oh, I have a mind to offer you more kindness than just a sympathetic word."

She instantly went on guard. Lune could think of nothing Vidar might gain by offering her true help, but that did not mean he *would* not. As cut off as she had been from the inner circles of court gossip, he might have some gambit in play she did not see. But what would she have to offer him?

No way to find out, save to walk farther into his trap.

"I would be most glad to hear anything your lordship might extend to me."

Vidar snapped his fingers, and a pair of minor goblins hurried to his side. At his gesture, they began unlacing the points of his sleeves, drawing them off to reveal the black silk of his shirt underneath. Ignoring them, Vidar asked, "You once lived for an extended period of time among mortals, yes?"

"Indeed, my lord." He raised one needle-thin eyebrow, and she elaborated. "I was a waiting-gentlewoman to Lady Hereford—as Lettice Knollys was known, then. Her Majesty bid me thence to keep a daily eye upon the mortal court, and report to her its doings."

The skeletal fae shuddered, a twitchy, insect-like motion. "Quite a sacrifice to make on the Queen's behalf. To live, day and night, under a mask of mortality, cut off from all the glory of our own court... Ash and Thorn. I would not do it again."

It might be the first sincere statement he had made since Lune entered. Vidar's own mortal masquerade, the one that had earned him his new position, had been more sporadic than sustained, and he had not enjoyed it. She said temperately, "I was pleased to serve her Majesty in such a capacity."

"Of course you were." He let the cynical note hang in the air, then offered, "Wine?"

Lune nodded, and took the cup a goblin brought to her. The wine was a fine red, tasting of the smoky, fading light of autumn, the flamboyant splendour of the leaves and their dry rustle underfoot, the growing bite of winter's chill. She recognised it from the first sip: surely one of the last remaining bottles brought as a gift to Invidiana when Madame Malline le Sainfoin de Veilée replaced the old ambassador from France. Some years hence, that had been. Madame Malline had remained at the Onyx Court when the ambassador from the Courts of the North departed, but relations were strained. There would be no more such gifts, not for a long time.

"You might," Vidar said, breaking her reverie, "have a chance to serve her Majesty again."

She failed to hide entirely the sharp edge that put on her interest. "Say on."

"Return to the mortal court."

The blunt suggestion made her breath catch. To live among mortals again... it was exhausting, dangerous, and exhilarating. Few fae had the knack for it, or even a liking. No wonder Vidar had sent for her.

But what purpose did he have in mind? Surely not her former assignment, Lettice Knollys. If the fragments of gossip Lune had heard were correct, she was no longer at court; she was in mourning for the death of her second husband, the Earl of Leicester.

She took another sip of wine. This one burned more than the first. "Return, my lord? To what end?"

"Why, to gather information, as you did before." Vidar paused. "And, perhaps, to gain access to—even leverage over—a certain individual."

She had concerned herself too much of late with fae

politics: the bargain with the folk of the sea, the raids of alfar ships, the never-ending tensions with the Courts of the North. Lune cursed herself for not keeping a closer eye on the doings of mortals: she did not know who was prominent now, whom she might be dispatched to trouble. She might not even recognise the name Vidar gave. "And who might that be, my lord?"

"Sir Francis Walsingham."

Cut crystal dug into her fingers.

Lune said carefully, "I believe I recognise that name."

"You should. He has lasted quite a long time, for a mortal, and risen high. Principal Secretary to Queen Elizabeth, he is now." Vidar gestured, and a goblin brought him his wine cup again. "Have you ever met him?"

"He did not come to court until after I had ceased my masquerade." Though she knew who he was, enough to be afraid.

"You will find him easily enough. The mortal court is at Richmond now, but they will shift to Hampton Court before long. You can join them there."

Lune handed off her goblet to a servant. The wine tasted too much of regret, and impending loss. "My lord, I have not yet said I would undertake this task."

A thin, predatory smile spread across Vidar's face. In a purring voice, he said, "I do not think you have a choice, Lady Lune."

As she had feared. But which was the greater risk: refusal or acceptance? Whatever honey Vidar used to coat it, she was not being offered this assignment out of a desire to see her redeem her past mistakes. Walsingham was not merely Principal Secretary; he was one of the foremost spymasters of Elizabeth's court. And his Protestantism was of a puritan sort, that assumed all fae to be devils in disguise. Any attempt to approach him, much less keep watch over him, might result in him catching her out, and if he caught her out...

Only mortal food given in tithe to the fae protected glamours and other magics. A short period of imprisonment could have disastrous results.

Food. Lune said, "Such masquerades are costly, Lord Ifarren. To maintain a plausible presence at court, one must be there every day. Mortal bread—"

"You will have it," Vidar said dismissively. "A spriteling will bring it to you each morning—or evening, if you prefer."

He had capitulated far too easily. "No. Such a plan leaves no margin of safety. Were I to be bidden to some duty elsewhere and missed the messenger, we would risk exposure. A whole loaf at a time, or more than one." A whole loaf, eaten only when needed, could cover quite a long journey in mortal disguise. Long enough, perhaps, to reach safety in another land.

If worse should truly come to worst.

Vidar's cynical eye seemed to see her thoughts. "You overestimate her Majesty's trust in you. But a week could be arranged. On Fridays, perhaps. Mortals assume we favour that day; we might as well oblige their fancies. And then you need not fear their holy day. I take it by this hard bargaining that you have agreed?"

Had she? Lune met Vidar's gaze, searching the flat blackness of his eyes for some hint of—something. Anything. Any crumb of information that might guide her.

She could not even be certain these orders came from Invidiana. Vidar might have concocted them, as a means of removing her permanently.

No. Even he would not endanger the Onyx Court in such fashion, to risk her true nature being revealed.

...Or would he? His desires were no secret in the higher circles of court. Even Invidiana knew her councillor coveted her throne. Where the Wild Hunt would destroy the Queen and tear the Onyx Hall down stone by stone, scattering her court to the four winds, Ifarren Vidar was more subtle; he would leave all as it was, but claim the Crown for himself. If he could but find a way.

Was this it? Was Lune to become a pawn in some hidden scheme of his?

If so—if she could discover the pattern of it, and inform Invidiana—

There was more than one route to favour.

Lune spread her skirts, and gave him no more humble a curtsy than she had before. Humility would be more suspicious to him than pride. "I am most grateful for the chance to be of service to her Majesty."

"Of course." Vidar eyed her with satisfaction. "Would it please you to be seated, Lady Lune? I have prepared a description of the role you are to take—"

"Lord Ifarren." She took pleasure in interrupting him. "My task is to be as you said? To gather information, and gain access to Sir Francis Walsingham?"

"And leverage, of whatever sort may offer itself."

Nothing would *offer* itself, but she might create something. But that was neither here nor there. "Then I will create my own role, as her Majesty trusted me to do in the past."

Displeasure marred the line of his mouth. "Her Majesty likewise trusted you to bargain sharply against the sea people." Lune damned the day she had ever been sent beneath the waves. Vidar had not been there, with the task of convincing the inhabitants of an alien land that the doings of mortal nations were their concern. Fae they might be, but unlike their landbound brethren, the mermaids and roanes and other denizens of the sea had not adopted current customs of courtly rule. And their idea of interaction with humans involved shipwrecks and the occasional lover, not politics. She had been lucky to find *anything* they wanted.

But to say so would sound peevish and weak. Instead she said, "You disdain mortal life, Lord Ifarren. Would you ride a horse raised by one who detested animals?"

"I know Walsingham," Vidar said.

"And I will most humbly hear your advice where he is concerned. But you asked for me because there is none in the Onyx Court more talented at this art than I. When I approach the mortal court, I will do so on my own terms."

The challenge hung in the air between them. Then Vidar waved one hand, as if it did not matter. "So be it. I will inform you of the court's movements. And *you* will inform *me* of your chosen role, before you go to join them."

"I will need some bread before then."

"Why?"

Now he was the one sounding peevish. Lune said calmly, "To familiarise myself with the situation, my lord. I have not been among that court in many years."

"Oh, very well. Now get out of my sight; I have other things to attend to."

Lune made her curtsy and withdrew. If Vidar had meant to position her where she would fail, she had at least escaped one trap. And with the allotment she would be given, she could afford to trade her own bread to other fae for information.

Once upon a time, she had clawed her way up from insignificance to favoured status, by shrewd trading and well-timed service. If she had done it once, she could—and would—do so again.

HAMPTON COURT PALACE, RICHMOND
14 October 1588

Deven rode into the spacious Base Court of the palace and dismounted almost before his bay gelding came to a halt. The October air had picked up a distinct chill since sunset, nipping at his cheeks, and his fingers were cold inside his gloves. There was a storm building, following in the aftermath of the day's gentle autumn warmth. He tossed his reins to a servant and, chafing his hands together, headed for the archway that led deeper into the palace.

Stairs on his left inside the arch led upward to the old-style Great Hall. No longer the central gathering place of the monarch and nobles, at Hampton Court the archaic space was more given over to servants of the household, except on occasions that called for great pageantry. Deven passed through without pausing and headed for the chambers beyond, where he could find someone that might know the answer to his question.

The Queen was not using that set of rooms as her personal quarters, having removed to a different part of the sprawling palace, but despite the late hour, a number of minor courtiers were still congregated in what was sometimes used as the Queen's watching chamber. From them, he learned that Elizabeth was having a wakeful night, as she often did since the recent death of her favourite, the Earl of Leicester. To distract herself, she had gone to a set of rooms on the southern side of the Fountain Court to listen to one of her ladies play the virginals.

The door was guarded, of course, and Deven was not in that elite rank of courtiers who could intrude on the Queen uninvited. He bowed to his two fellows from the Gentlemen Pensioners, then turned to the weary-eyed usher who was trying unsuccessfully to stifle a yawn.

"My most sincere apologies for disturbing her Majesty,

but I have been sent hither to bring her a message of some importance." Deven brought the sealed parchment out and passed it over with another bow. "It was Sir James Croft's most express wish that it be given to her Grace as soon as may be."

The usher took it with a sigh. "What does the message concern?"

Deven bit back the acid response that was his first reflex, and said with ill-concealed irritation, "I do not know. 'Tis sealed, and I did not inquire."

"Very well. Did Sir James wish a reply?"

"He did not say."

"Wait here, then." The usher opened the door and slipped inside. A desultory phrase from the virginals floated out, and a feminine laugh. Not the Queen's.

When the usher re-emerged, he had something in his hand. "No response to Sir James," he said, "but her Majesty bids you carry these back to the Paradise Chamber." He held out a pair of ivory flutes.

Deven took them hesitantly, trying to think of a way around embarrassing himself. He failed; the usher gave him a pitying smile and asked, "Do you know the way?"

"I do not," he was forced to confess. Hampton Court had grown by stages; now it was a sprawling accretion of courtyards and galleries, surpassed in England only by Whitehall itself, which his fellows reassured him was even more confusing to explore.

"The quickest path would be through these chambers

to the Long Gallery," the usher said. "But as they are in use, go back to the Great Hall..."

It wasn't as bad as he feared. A pair of galleries ran north to south through the back part of the palace, connecting to the Long Gallery of the south side, with the chambers where Elizabeth had chosen to reside for this visit. At the most south-easterly corner of the palace, and the far end of the Long Gallery, lay the Paradise Chamber.

Deven unlocked the door and nearly dropped the flutes. The candle he bore threw back a thousand glittering points of light; raising it, he saw that the dark chamber beyond was crammed to the walls with riches beyond words. Countless gems and trifles of gold or silver; tapestries sumptuously embroidered in coloured silks; pearl-studded cushions; and, dominating one wall, an unused throne beneath a canopy of estate. The royal arms of England decorated the canopy, encircled by the Garter, and the diamond that hung from the end of the Garter could have set Deven up in style for the rest of his life.

He realised he had stopped breathing, and made himself start again. No, not the rest of his life. Ten years, maybe. And ten years' fortune would not do him much good if he were executed for stealing it.

The entire contents of the room, though...

No wonder they called it Paradise.

He set the flutes on a table inlaid with mother-of-pearl and backed out again, locking the door on the blinding wealth within, before it could tempt him more. *They would hardly miss one small piece, in all that clutter...*

Perhaps it was his own guilty thoughts that made him so edgy. When Deven heard a sound, he whirled like an animal brought to bay, and saw someone standing not far from him.

After a moment, he relaxed a trifle. Rain had begun to deluge the world outside, obscuring the moon, and so the Long Gallery was lit only by his one candle, not enough to show him the figure clearly, but the silhouette lacked the robe or puffed clothing that would mark an old courtier or a young one. Nor, he reminded himself, did he have anything to feel guilty about; he had done nothing more than what he was ordered to, and no one, servant or otherwise, could hear the covetous thoughts in Deven's mind.

But that recalled him to his duty. Though the Queen was not present, surely he also had a duty to defend that which was hers. "Stand fast," he said, raising the candle, "and identify yourself."

The stranger bolted.

Deven gave chase without thinking. The candle snuffed out before he had gone two strides; he abandoned it, letting taper and holder fall so he could lunge for the door through which the stranger had vanished. It stood just a short distance from the Paradise Chamber, and when he flung himself through it, he found himself on a staircase, with footsteps echoing above him.

The stranger was gone by the time he reached the third floor, but the steps continued upward in a secondary staircase, cramped and ending in a half-height door that was obviously used for maintenance. Deven yanked the door open and wedged himself through, into the cold, drenching rain.

He was on the roof. To his right, low crenellations guarded the drop-off to the lower Paradise Chamber. He looked left, across the pitched sheets of lead, and just made out the figure of the stranger, running along the roof.

Madness, to give chase on a rooftop, with his footing made uncertain by rain-slicked lead. But Deven had only an instant to decide his course of action, and his blood was up.

He pursued.

The rooftop was an alien land, all steep angles and crenellated edges, with turrets rising here and there like masts without sails. The path the stranger took was straight and level, though, unbroken by chambers, and that was what oriented Deven in his fragmentary map of Hampton Court: they were running along the roof of the Long Gallery, back the direction he had come.

In his head, he heard the usher say, *The quickest path would be through these chambers...*

The gallery led straight toward the room where Elizabeth sat with her ladies, whiling away her sleeplessness with music.

Deven redoubled his efforts, flinging caution to the wind, keeping to his feet mostly because his momentum carried him forward before he could fall. He was gaining on the stranger, not yet close enough to grab him, but nearly—

Lightning split the sky, half-blinding him, and as thunder followed hard on its heels Deven tried too late to stop.

Brick cracked him across the knees, halting his stride instantaneously. But his weight carried him forward, and he pitched over the top of the crenellations, hands flying out in desperation, until his left fingers seized on something and brought him around in a shoulder-wrenching arc. His right hand found brick just in time to keep him from losing his grip and falling a full story to the lower rooftop below.

He hung from the crenellations, gasping for air, with the rain sending rivers of water through his hair and clothes to puddle in his boots.

His left shoulder and hand ached from the force of stopping his fall, but Deven dragged himself upward, grunting with effort, until at last he could hook one foot over the bricks and get his body past the edge. Then he collapsed in the narrow wedge where the pitch of the roof met the low wall of the crenellations and let himself realise he wasn't about to fall to his death.

The stranger.

Deven twisted to look over the wall, onto the roof of the chambers where Elizabeth listened to the virginals. He saw no sign of the intruder anywhere on the rain-streaked lead, and no hatches hung open in the turrets that studded the corners of the extension; through the grumble of the storm, he heard a faint strain of music. But that meant nothing save that no one had been hurt yet.

Even if Deven could have made the jump down, he

could not burst in on the Queen, soaked to the bone and with his doublet torn, its stuffing leaking out like white cotton entrails. He hauled himself to his feet, wincing as his bruised knees flared, and began his limping progress back along the Long Gallery, to the door that had led him up there to begin with.

His news, predictably, caused a terrible uproar, and soon a great many people were roused out of bed, but the intruder had vanished without a trace. Some time later, no longer dripping but still considerably damp, Deven found himself having to relate the story to Lord Hunsdon, from his arrival at Hampton Court that night up to the present moment.

"You saw nothing of his face?" Hunsdon asked, fingers tapping a worried beat on the desk before him.

Deven was forced to shake his head. "He wore a cap low on his head, and we stood some ways apart, with only one candle for light. He seemed a smallish fellow, and dressed more like a labourer than a gentleman, but beyond that I cannot say."

"Where do you think he went, after you lost him?"

The chambers there connected at their corner to the courtiers' lodgings that ringed the Base Court; from there, the man might have run nearly anywhere, though the soaring height of the Great Hall would have forced him to circumnavigate the courtyard if he wished to go somewhere else. There was no good access to the ground; everything was at least two storeys. With rope, he might have gone through a second-floor window, but they found no such rope, nor sign of a very wet man coming in anywhere.

The last Deven had seen of the man was when they reached the end of the gallery, and the stranger... leapt over the edge.

No, not quite. The man had leapt, yes, but upward, into the air—not as a man would jump if he intended a landing on a pitched roof below.

After that, his memory only offered him the flapping of wings.

He shook his head again, shivering in his damp, uncomfortable clothes. "I do not know, my lord. Out into the gardens, perhaps, and from thence into the Thames. Or perhaps there was a boat waiting for him." How he would have gotten from a second- or third-floor roof to the gardens, Deven could not say, but he had no better explanation to offer.

Nor, it seemed, did Hunsdon. The baron's mouth was set in a grim line. "It seems the Queen is safe for now. But we shall stay alert for future trouble. If you see the fellow again..."

Deven nodded. "I understand, my lord." He might walk past the man in the street and not know him. But Deven believed now, as he had not truly before, that the Queen's enemies might stage threats against her life. His duty was more than simply to stand at her door with a gold-covered axe.

He prayed such a threat would not come again. But if it did, then next time, he would be more effective in stopping it.

MEMORY

12 July 1574

The sleeping man lay in an untidy sprawl on his bed. The covers, kicked aside some time earlier, disclosed an aging body, a sagging belly usually hidden by the peasecod front of his doublets, and his dark hair was thinning. He was still fit enough—not half so far gone as some other courtiers—but the years were beginning to tell on him.

In his mind, though, in his dreams, he was still the young man he had been a decade or two before.

Which suited very well the purposes of the being that came to visit him that night.

How it slipped in, no observer could have said. Under the edge of the door, perhaps, or out of the very stuff of shadows. It showed first as a stirring in the air, that coalesced into an indistinct shape, which drifted gently through the chamber until it reached the bed.

Hovering over the sleeping man, the figure took more distinct shape and colour. A fluttering linen chemise, freed from the constraints of bodice and kirtle and the usual court finery. Auburn hair, flowing loose, its tips not quite brushing the man below. A high forehead, and carmined lips that parted in an inviting smile.

The man sighed and relaxed deeper into his dream.

Robert Dudley was hunting, riding at a swift canter through open fields, pursuing hounds that gave the belling cry of prey sighted. At his side rode a woman, a red-haired woman. He thought, faintly, that she had been someone else a moment ago—surely it was so—but now she was younger, her hair a darker shade of red.

And they were not riding, they were walking, and the hounds had vanished. A pleasant stream laughed to itself, hidden somewhere in the reeds to one side. The sunlight was warm, casting green-gold light down through the trees; up ahead the landscape opened into a grassy meadow, with something in it. A structure. A bower.

Curtains fluttered invitingly around the bed that stood within.

Clothing vanished at a thought, leaving skin upon skin, and together they tumbled into bed. Auburn hair cascaded around him, a second curtain, and Robert Dudley gazed adoringly into the face of Lettice Knollys, all logic and reason crumbling before the onslaught of passion that overwhelmed him.

Easy enough, to fan the flames of an early flirtation into a conflagration. He would not remember this upon waking, not as anything more than an indistinct dream, but it would serve its purpose nonetheless. And if Lettice Knollys were in truth Lettice Devereux, Lady Hereford, and wed elsewhere, it did not matter. Dudley did not have to marry her. He had only to give his heart, turning it from the target at which it had ever been fixed: his beloved Queen Elizabeth.

Robert Dudley, Earl of Leicester, moaned deep in his throat as he writhed on the bed, aware of nothing but the dream that suffused his mind. Above him hovered the ghostly form of Lettice Knollys, perfect as she had never been, even in the blossom of her youth.

The scholars of Europe spoke of demons they called succubi. But more than one kind of creature in the world wielded such power, and not all served the devil.

Some served a faerie Queen, and did her bidding with pleasure, dividing from the mortal Queen her most loyal and steadfast admirer.

A man might die of such surfeit. The ghostly figure lost its definition, fading once more into indistinct mist, and with an unfulfilled sigh the Earl of Leicester subsided into dreamless sleep.

There would be other nights. The creature that visited him considered itself an artist. It would work upon him by slow degrees, building his desire until he thought of no one else. And when his heart turned away from the mortal Queen, and the creature's work here was done...

There would be other mortals. Invidiana always had use for this creature's talents.

WHITEHALL PALACE, WESTMINSTER

3 November 1588

Deven stood in front of the polished mirror and ran one hand over his jaw, checking for stubble. Colsey had shaved him that morning, and his hair was newly trimmed into one of the more subdued styles currently fashionable; he wore a rose-red doublet with a falling collar, collected from the tailor only yesterday when the court completed its move into Whitehall, and even his low shoes were laced with silk ribbons. He looked better than he had when he was first presented to the Queen, but felt very nearly as inadequate.

From behind him, Colsey said, "Best you get moving, master."

The reminder was appreciated, though a little presumptuous—Colsey occasionally forgot he was not Deven's father, to order him about. It made Deven take a deep breath and turn away from his blurred reflection in the mirror, setting himself toward the door like a man at the tilt.

Tilting. He had thought about entering the upcoming Accession Day jousts, but knew it would be a waste of his time and coin; certainly one could catch the Queen's eye by performing well, but he was at best indifferent with the lance. He would have to content himself with the usual pageantry of the Gentlemen Pensioners, who would make a brave show around Elizabeth during the celebrations.

He had a hard time focusing on pageantry, though, when his feet were leading him toward a real chance for success at court.

He fingered the tabs at the bottom of his new doublet and wondered if it looked too frivolous. A useless thought—he had not the time to go change—but he was second-guessing himself at every turn today.

Deven gritted his teeth and tried to banish his nerves.

Several men were in the chamber when he arrived, and

a number more came and went. Such was the inevitable consequence of absence from court, even with someone like Beale to cover one's duties. But Deven was expected, and so he waited very little before being ushered into the chamber beyond, where the Principal Secretary sat behind a small mountain of paper.

Deven advanced halfway across the floor and then knelt on the matting. "Master Secretary."

Sir Francis Walsingham looked tired in the thin November sunlight that filtered through the palace's narrow windows. They had not been lying, when they said he was ill; the marks of it showed clearly. Deven had met him twice before—the rest of their dealings had been through intermediaries—and so he had sufficient basis for comparison. Walsingham was dark complected for an Englishman, but his skin had a pale, unhealthy cast to it, and there were circles under his eyes.

"I am glad," Deven said, "that God has seen fit to restore you to health."

Walsingham gestured for him to rise. "My illness was unfortunate, but 'tis past. Beale tells me you have some matter you would beg of me."

"Indeed." He had expected more small talk beforehand, but given the pile of work facing Walsingham, perhaps he should not be surprised the man wished to cut directly to what was relevant. That encouraged Deven to speak plainly, as he preferred, rather than larding his words with decoration, which seemed to be a substantial art form at court.

He clasped his hands behind his back and began. "I

wished to thank you in person for your good office in securing for me the position I now hold in the Gentlemen Pensioners."

"'Tis no great matter," Walsingham said. "You did me good service among the Protestants in the Low Countries, and your father has much aided her Majesty in the suppression of seditious pamphlets."

"I am glad to have been of service," Deven answered. "But I hope my use might not end there."

The dark eyes betrayed nothing more than mild curiosity. "Say on."

"Master Secretary, the work I did on your behalf while on the continent made it clear to me that the defence of her Majesty—the defence of England—depends on many types of action. Some, like armies and navies, are public. Others are not. And you are clearly a general in the secret sort of war."

The Principal Secretary's lips twitched behind their concealing beard. "You speak of it in poetic terms. There is little of poetry in it, I fear."

"I do not seek poetry," Deven said. "Only a chance to make my mark in the world. I have no interest in following my father in the Stationers' Company, nor does Gray's Inn hold me. To be utterly frank, my desire is to be of use to men such as yourself, who have the power and the influence to see me rewarded. My father earned the rank of gentleman; I hope to earn more."

And that, he hoped, would strike a sympathetic chord. Walsingham had been born to a family with far greater connections than Deven's own, but he had earned his knighthood and his position on the privy council. Whether Deven could strike a target so high, he doubted—but he would aim as high as he could.

Or perhaps his words would turn, like a knife in his hands, and cut him. Walsingham said, "So you serve, not out of love for England and her Queen, but out of ambition."

Deven quelled the urge to flinch and salvaged what he could. "The two are not in conflict with one another, sir."

"For some, they are."

"I am no dissident Catholic, Master Secretary, nor a traitor tied to the purse strings of a foreign power, but a good and true-hearted Englishman."

Walsingham studied him, as if weighing his every virtue and vice, weakness and use, with his eyes alone. He was, in his way, as hard to face as Elizabeth.

Under the sharp edge of that gaze, Deven felt compelled to speak on, to lay on the table one of the few cards he possessed that might persuade the Principal Secretary and undo the damage of his own previous words. "Have you heard of the incident at Hampton Court?" Walsingham nodded. Of course he had. "Then you know 'twas I who came across the intruder."

"And pursued him over the rooftops."

"Even so." Deven's fingers had locked tight around each other, behind his back. "You have no reason to believe me, Master Secretary—but ambition was the farthest thing from my mind that night. I pursued that man without concern for my own safety. I do not tell you this out of pride; I wish you to understand that, when I had only an instant to think, I thought of the Queen's safety. And when the man was gone—vanished into the night—I blamed myself for my failure to catch him.

"I have no wish to run across rooftops again. But you, Master Secretary, are dedicated to making such things unnecessary, by removing threats before they can approach so near to her Grace. That is a task to which I will gladly commit myself. I had rather be of more use to the Queen and her safety than simply standing at her door with a gold axe in my hands."

He hadn't meant to speak for so long, but Walsingham had let him babble without interruption. A shrewd move; the more Deven spoke, the less planned his words became, and the more inclined he was to speak from the heart. He just hoped his heart sounded more like a fervent patriot than a callow, idealistic boy.

Into the silence that followed his conclusion, the Principal Secretary said, "Then you would do what? Fight Catholics? Convert their faithful? Spy?"

"I am sworn to her Majesty's service here at court," Deven said. "But surely you have need of men here, not to find the information, but to piece together what it means." He offered up an apologetic smile. "I—I have always liked puzzles."

"Have you." The door creaked behind Deven; Walsingham waved away whoever it was, and then they were alone again. "So the short of it is, you would like to solve puzzles in my service."

And to benefit thereby—but Deven was not fool enough to say that again, even if they both heard those words still hanging in the air. He hesitated, then said, "I would like the chance to prove my worth in such matters to you."

It was the right answer, or at least a good one. Walsingham said, "Inform Beale of your wishes. You shall have your chance, Michael Deven; see you do not squander it."

He was kneeling again almost before the words were finished. "I thank you, Master Secretary. You will not regret this."

ACT TWO

There is no treasure that doeth so vniuersallie profit, as doeth a good Prince, nor anie mischeef so vniuersallie hurt, as an yll Prince.

BALDESAR CASTIGLIONE
THE COURTYER

The chamber is a small one, and unfurnished; whatever lay here once, no one claims it now. There are rooms such as these in the Onyx Hall, forgotten corners, left vacant when their owners died or fled or fell from favour.

For him, they feel like home: they are neglected, just as he is.

He came in by the door, but now he cannot find it again. Instead he wanders to and fro, from one wall to another, feeling the stone blindly, as if the black marble will tell him which way to go, to be free.

One hand touches the wall and flinches back. He peers at the surface, leaning this way and that, as a man might study himself in a mirror. He stares for a long moment, then blanches and turns away. "No. I will not look."

But he will *look; there is no escape from his own thoughts. The far wall now draws his eye. He crosses to it, hesitant in his steps, and reaches out until his fingers brush the stone, tracing the image he sees.*

A face, like and unlike his own. A second figure, like

and unlike her. He spins about, but she is not there with him. Only her likeness. Only in his mind.

Against his will, he turns back, wanting and not wanting to see.

Then he is tearing at the stone, pulling at the mirror he imagines until it comes crashing down, but that brings him no respite. All about him he sees mirrors, covering every wall, standing free on the floor, each one showing a different reflection.

A world in which he is happy. A world in which he is dead. A world in which he never came among the fae, never renounced his mortal life to dwell with immortal beings.

A world in which...

He screams and lashes out. Blood flowers from his fist; the silvered glass is imaginary, but it hurts him just the same. Beyond that one lies another, and soon he is lurching across the room, breaking the mirrors, casting them down, pounding at them until they fall in crimson fragments to the floor. His hands strike stone, again and again, lacerating his flesh, cracking his delicate bones.

Until he no longer has the will to fight, and sinks into a crouch in the centre of the chamber, mangled fingers buried deep in his hair.

All around him, the pieces of his mind reflect a thousand broken other lives.

He could see much, if he looked into them. But he no longer has the will for that, either.

"There are no other lives," he whispers, trying to make

himself believe it, against all the evidence of his eyes. "What is over and done cannot be redone. 'Tis writ in stone, and will not fade."

His bleeding hands drift downward and begin to write strange, illegible hieroglyphs upon the floor. He must record it. The truth of how it went. Else those who come after will be lost in the maze of mirrors and reflections, never knowing reality from lies.

It will not matter to them. But it matters to him, who tried for so long to tell the truth of the futures he saw. That gift has turned traitor to him, bringing nothing more than pain and despair, and so he takes refuge in the past, writing it out amidst the shattered pieces of a hundred might-have-beens.

HAMPTON COURT PALACE, RICHMOND

6 January 1590

The winter air carried a crisp edge the sunlight did little to blunt, but for once there was hardly more than a breeze off the Thames, scarcely enough to stir the edge of Deven's cloak as he hurried through the Privy Garden. He passed bare flowerbeds protected beneath layers of straw, squinting at the brightness. He had been assigned to serve the Queen at supper for the Twelfth Night feast, but that duty hardly precluded one from participating in the merriment. Deven had no idea how many cups of hippocras he had downed, but it felt like a dozen too many.

Nor was he the only one who had overindulged, but that was to his advantage. Deven had not risen at this ungodly hour without reason. With so many courtiers and the Queen herself still abed, he could snatch a few moments for himself, away from prying eyes—and so could the one he was hurrying to meet.

She was waiting for him in the Mount Garden, standing in the lee of the banqueting house, well-muffled in a fur-trimmed cloak and gloves. The hood fell back as Deven reached for her face, and lips met cold lips in a kiss that quickly warmed them both.

When they broke apart, Anne Montrose said, a trifle breathlessly, "I have been waiting for some time."

"I hope you are not too cold," Deven said, chafing one slender hand between his own. "Too much hippocras, I fear."

"Of course, blame the wine," she said archly, but smiled as she did so.

"'Tis a thief of men's wits, and of their ability to wake." Frost glittered on the ground and the bare branches of trees like ten thousand minuscule diamonds, forming a brilliant setting for the gem that was Anne Montrose. With her hood fallen back, her unbound hair shone palest gold in the sun, and her wide eyes, a changeable grey, would not have looked out of place on the Queen of Winter that featured prominently in last night's masque. She was not the greatest beauty at court, but that mattered little to him. Deven offered her his arm. "Shall we walk?"

They strolled sedately through the hibernating gardens, warming themselves with the exercise. It was not forbidden for them to be seen together; Anne was the daughter of a gentleman, and fit company for him. There were, however, difficulties. "Have you spoken to your mistress?" Deven asked.

He was hesitant to broach the topic, which might ruin the glittering peace of this morning. It had weighed heavy upon him, though, since he first voiced it to Anne, some months prior. The increased duties of winter court and the never-ending ceremonies of the Christmas season had prevented them from doing more than exchange brief greetings whenever they passed, and now he fretted with impatience, wanting an answer.

Anne sighed, her breath pluming out in a cloud. "I have, and she has promised to do what she may. 'Tis difficult, though. The Queen does not like for her courtiers to marry."

"I know." Deven grimaced. "When Scudamore's wife asked permission, the Queen beat her so badly she broke Lady Scudamore's finger."

"I am glad I do not serve *her*," Anne said darkly. "The stories I hear of her temper are dreadful. But I am not the one who will bear the brunt of her wrath; she cares little what a gentlewoman in service to the Countess of Warwick does. You, on the other hand..."

Marriage is no scandal, his father had said, when he went into service at court, over a year ago. *Get thee a wife,* his fellows in the band had said. It was the way of the world, for men and women to marry—but not the way of the Queen. She remained virginal and alone, and so would she prefer her courtiers to be.

"She is envious," Anne said, as if she had heard that thought. "There is no love in her life, and so there should be none in the lives of those who surround her—save love for her, of course."

It was true as far as it went, but also unfair. "She has had love. I do not credit the more sordid rumours about her and the late Earl of Leicester, but of a certainty she was fond of him. As they say she was of Alençon."

"Her froggish French prince. That was politics, nothing more."

"What would you know of it?" Deven said, amused. "You could not have been more than ten when he came to England."

"Do you think the ladies of court have ceased to gossip about it? Some say it was genuine affection, but my lady of Warwick says not. Or rather, she says that any affection the Queen may have felt was held in check by her awareness of politics. He was, after all, Catholic." Anne reflected on this. "I think it was desperation. Mary was old when she married; Elizabeth would have been older, in her forties. It was her last chance. And, having lost it, she now vents her frustration on those around her who might find happiness with another."

The breeze off the Thames was picking up, forging a sharper edge. Anne shivered and pulled up the hood of her cloak. Deven said, "Enough of the Queen. I am one of her Gentlemen Pensioners; she calls me fair, gives me minor gifts, and finds me amusing at times, but I'll never be one of her favourites. She cannot take much offense at the prospect of my marriage." It had been Mary Shelton, chamberer to her Majesty, not John Scudamore of the Pensioners, who suffered the broken finger.

Anne laughed unexpectedly from within the depths of her hood. "So long as you do not get me with child, and end up in the Tower for it, like the Earl of Oxford."

"We would run away, first." It was a romantic and stupid thing to say. Where would they go? The only places he knew were London and Kent, and the Netherlands. The former were too near the Queen's grasp, and the latter, no refuge at all. But Anne favoured him with an amused smile, one he could not help returning.

All too soon, though, frustration returned to plague him, as it so often did. They walked a little way in silence; then Anne, sensing his mood, asked, "What troubles you?"

"Practicalities," he confessed. "A growing awareness that my ambition and I dwell in separate spheres, and I may well never ascend to meet it."

Her gloved hand rose and tucked itself into the crook of his elbow. "Tell me."

This was why he loved her. At court, a man must always watch what he said; words were both currency and weapons, used to coax favour from allies and strike down enemies. And the ladies were little better; Elizabeth might forbid her women to engage heavily in politics, but they kept a weather eye on the Queen's moods, and could advance the causes of petitioners when they judged the moment right—or hinder them. Even those without the Queen's ear could carry tales to those who had it, and a man might find his reputation poisoned before he knew it, from a few careless words.

He never felt the need for such caution with Anne, and she had never given him cause, not in the year he had known her. She had said once, last autumn, that when in his company she could be at ease, and he felt the same. She was not the greatest beauty at court, nor the richest catch, but he would gladly trade those for the ability to speak his mind.

"I look at Lord Burghley," he said, approaching the subject from a tangent. "Much of what Walsingham does is built on foundations laid by Burghley, and in fact the old baron still maintains his own links with agents and informants. When Burghley dies, or retires from her Majesty's service—which won't happen until after the Second Coming—his son Robert will inherit his barony, his offices, and his agents."

When he paused, Anne said, "But you are not Robert Cecil."

"Sidney might have been—he was married to Walsingham's daughter, before either of us came to court—but he's dead. And I am not sufficiently in Walsingham's affections to take his place, nor ever likely to be."

Anne squeezed his arm reassuringly. They were walking too close together, her farthingale shoving at his leg with every stride, but neither of them moved to separate. "Do you need to be?"

"To do what Walsingham does? Yes. I haven't the wealth to support such an enterprise, nor the connections. Beale and I are forever passing letters and petitions up and down the chain, obtaining licenses for foreign travel, pardons for prisoners who might be of use, requests for gifts or pensions to reward those who have been of service. They do not often receive payment, but the important thing is that they believe they might. I cannot promise that and be believed. And even if I could... I am not in the Queen's councils." Deven's mouth twisted briefly in inarticulate frustration. "I am the son of an unimportant gentleman, distinguished enough by my conduct in the Netherlands to be rewarded with a position at court, pleasing enough to be granted the occasional preferment—but nothing more. Nor ever likely to be."

That speech, delivered in a low monotone from which familiarity had leached all the passion, carried them back to the centre of the garden where the banqueting house stood. The morning was upon them in full; the Queen would be waking soon, and he had to be there for the honour guard when she processed to chapel for the service of Epiphany. But the chambers of the palace were close and stuffy, too full of people flocking to the winter court; out here the air was clean and simple, and he did not want to leave.

Anne turned to face him and took his gloved hands in her own, buff-coloured leather against brown. "You are twenty-seven," she pointed out. "The men you speak of are *old* men. They achieved their positions over time. How old was Walsingham, when Elizabeth made him her Secretary?"

"Forty-one. But he had connections at court—"

"Also built over time."

"Not all of them. Much of it is a matter of family: fathers and sons, brothers and cousins, links by marriage—"

Her fingers tightened fractionally on his, and Deven caught himself. "I'll not lay you aside for political advantage," he promised.

The words brought a smile to her face that warmed her grey eyes. "I did not think you would."

"The true problem is the Queen. I do not speak against her," he added hastily, and could not restrain a quick glance around, to reassure himself they were alone in the garden. "I am her loyal servant. But her preference is for those of families she knows—often those bound to her already by ties of blood. Of which I am not one."

Anne relinquished his hands so she could straighten her hood. "Then what will you do?"

He shrugged. "Be of use to Walsingham, as much as I can be. Hope that he will reward me for my service."

"Then I have something for you."

Deven cast a startled glance at her, then frowned. "Anne, I have told you before—'tis neither meet nor safe for you to carry tales."

"Gossip is one of the great engines of this court, as you well know. I am not listening at keyholes, I promise you." She was a tallish woman, the top of her hood at eye level for him, and so she did not have to tilt her head back much to look at him; instead she tilted it to the side, eyes twinkling. "Are you not the least bit curious?"

He was and she knew it. "You will find a way to tell me, regardless."

"I could be more subtle, but this is so much easier." Anne folded her hands demurely across the front of her cloak. "'Tis a minor thing, to my eyes, but I never know when some minor thing fits into the greater patterns you and your master see. You are aware of Doctor Dee?"

"The astrologer? He had an audience with the Queen a month gone, at Richmond."

"Do you know the substance of it?"

Deven shook his head. "He was at court only a day or two, and I did not speak to him."

"My lady of Warwick tells me 'tis some difficulty with his house and books. Someone despoiled them while he was abroad; he seeks redress. You may expect to see more of him, I should think—or at least to hear people arguing on his behalf."

"People such as your countess?"

"I thought you did not want me carrying tales." She laughed as he mock-scowled at her. "I imagine your master knows of his situation—they are friends, are they not?—but I can learn more if you would like."

This, he was unpleasantly aware, was often how espionage worked. Few of those who fed Walsingham information did so in an organised and directed fashion, deliberately infiltrating places where they did not belong, or masquerading as that which they were not. Most of the intelligence that reached the Principal Secretary came from men who simply kept their eyes and ears open, and wrote to him when they saw or heard something of interest.

Men, and the very rare woman.

As if she had heard that thought—he must be as transparent as glass to her—Anne said, "'Tis not as if I were offering to return information from the court of the Holy

Roman Emperor, or the Pope's privy closet. I will simply tell you if Doctor Dee calls on the countess again."

"I cannot ask a woman to spy," Deven said. "It would be infamous."

"'Tis listening, not spying, and you are not asking me. I do it of my own free will. Consider it a dowry of an intangible sort, paid in advance." Anne took his hand again, and tugged him a step forward, so they stood in the shadow of the banqueting house. There she cupped his jaw in her gloved fingers and kissed him again. "Now I must return; my lady will be rising."

"As will mine," Deven murmured, over the rapid beating of his heart. "You will tell me what the countess says—whether the Queen would be angry at the thought of our marriage?"

"I will," Anne promised. "As soon as I may."

MEMORY

21 December 1581

Many parts of the subterranean palace consisted of adjoining chambers, one opening into the next with never a break. Some were arranged around cloistered courtyards of sculpture or night-blooming plants; others connected via long galleries, hung with tapestries and paintings of rich hue.

But there were other passages, secret ones. Few fae ever saw them, and almost no mortals.

The man being escorted through the tunnel was a rare exception.

Of the other mortals who had been brought that way, most were attractive; those who were not held influential positions at court or in trade, and compensated for their lack of handsomeness by their use. This one was different. His cowl taken from him, his clipped, mutilated ears were bared for all to see, and though he was not old, cunning and suspicion—and at the moment, fear—robbed his face of any beauty. Nor was he a powerful man.

He was no one. But he knew a little of faeries, and now his investigations had brought him here, to a world whose existence he had never so much as suspected.

A door barred the way at the end of the passage, bronze-bound and painted black. One of the escorting fae, a hunched, goblinish thing, raised his bony-knuckled hand and knocked. No response came through the door, but after a moment it swung open on oiled hinges, as if of its own accord.

The chamber into which the mortal man stepped was as sumptuous as the corridor outside was bleak. The floor was bare of either rushes or carpets, but it was a fine mosaic in marble, strange figures that he would have liked to study more closely. Cool silver lights gleamed along the walls; out of the corner of his eye, he thought he saw wings moving within their depths. The walls were likewise marble, adorned at regular intervals by tapestries of coloured silk studded here and there with jewels. The ceiling was a masterwork display of astrological notation, reflecting the current alignment of the stars far above.

But all this richness was dominated into insignificance by the curtain before him.

Was it black velvet, worked elaborately in silver? Or cloth-of-silver, painstakingly embroidered with black silk? His escort, his guards, stood between him and it, as if he would have approached to examine it. Some of the gems encrusting the fabric appeared to be diamonds, while others were more brilliant and alive than any diamonds he had ever seen. Pearls as large as hummingbird eggs weighted its bottom edge. The curtain alone displayed wealth only the crowned heads of Europe could hope to equal, and not even all of them.

He was not surprised when one of his escorts kicked him in the back of the knee, forcing him to the floor.

The stone pressed hard and cold against his knees as he waited.

And then a voice spoke from behind the curtain.

"You seek after magic, Edward Kelley."

"I do." The words came out rusty and faint on the first try; he wet his lips and said it again. "I do. And I have found it." Found more than he had ever dreamed of.

A soft sound came from behind the curtain, a cool laugh. The voice was melodious and controlled, and if the face that accompanied it was anything to match, she must be the most beautiful faerie lady to ever call England home.

Lady—or queen? Even among fae, he doubted such riches were common.

The lady spoke again from her concealment. "You have found only the meanest scraps from the table of magic. There is more, far more. You wish to know the secrets of creation? We have them bound in books. You wish to transform base metal into gold? 'Tis child's play, for such as us."

Faerie gold. It turned to leaves or stones before long—but a man could do a great deal with it, while it still shone. And though it was a poor substitute for true transformation, the Philosopher's Stone, learning of it might advance his alchemical work.

Yes, there was a feast here for him.

"I would be your ladyship's most humble student," he said, and bowed his head.

"I am sure," the lady said. "But you must know, Edward Kelley—all gifts carry a price. Especially those from fae."

He was a learned man. Some believed fae to be devils in different guise. Others placed them midway between Heaven and Hell: above men in the hierarchy of creation, but below the celestial forces that served God.

Regardless of the explanation, all agreed that to strike a bargain with their kind was a dangerous business. But having seen this much, no man who laid claim to intellectual curiosity could be expected to turn back.

He had to swallow before his voice would work. "What price would you demand?"

"Demand?" The lady seemed offended. "I will not ask for your soul, or your firstborn child. I merely have a request of you, that I think you will find it easy enough to grant."

That was more ominous than a straightforward demand would have been. He waited, eyes on the hanging pendants of pearls, to hear more. They did not quite touch the floor, and in the shadows beyond he thought he could see just the hem of a glittering skirt.

At length the lady said, “There is a mortal scholar known as Doctor John Dee.”

Kelley nodded, then remembered the lady could not see him. “I know of him.”

“He seeks to speak with angels. For this purpose he has contracted the services of a man named Barnabas Saul. My request is that you take Saul’s place. The man is nothing more than a charlatan, a cozener who seeks to take advantage of Doctor Dee. We will arrange for him to be discredited, and you will replace him as scryer.”

“And then?” Kelley knew it would not end there. “Once I am in Dee’s confidences—assuming I can make it there—”

“’Tis easily arranged.”

“Then what would you have me do?”

“Nothing damaging,” the hidden lady assured him. “He will never speak to angels, whatever scryer he contracts to assist him. But ’tis in our interests that he should think he has done so. You will describe visions to Dee, when he asks you to gaze into the crystal. You may invent some if you wish. From time to time, one of my people will visit you in that glass, and tell you what to say. And in exchange, we will teach you the secrets you wish to learn.”

Kelley had never met the man; what did he care if Dee

was led astray by faeries? Yet it made him nervous all the same. "Can you promise me the things I say will not harm him? Can you give me your word?"

All around him, the silent fae of his escort stiffened.

Silence from behind the curtain. Kelley wondered how badly he had offended. But if the lady fed him visions that would incite Dee to treason, or something else harmful…

"I give you my word," the lady said in a clipped, hard tone unlike her previous voice, "that I will give no orders for visions that will harm Doctor John Dee. If you lead him astray with your own invention, that is no fault of ours. Will that suffice, Edward Kelley?"

That should be enough to bind her. He hoped. He dared not press for more. Yet he had one further request, unrelated to the first. "I am most grateful," he said, and bowed his head again. "You have already given me more than I ever dreamed of, bringing me here. But though it be presumptuous of me, I do have one more thing to ask. Your voice, lady, is beauty itself; might I have the privilege of gazing upon your face?"

Another silence, though this time his escort did not take it as strongly amiss.

"No," the lady said. "You shall not see me tonight. But on some future day—if your service pleases—then perhaps, Edward Kelley, you shall know who I am."

He wanted to see her beauty, but she had surmised correctly; he also wanted to see whom he was serving. But it was not to be.

Was he willing to accept that, in exchange for what the fae might teach him?

He had answered that question before he ever agreed to accompany them beneath the streets of London.

Edward Kelley bowed until his forehead touched the cold marble and said, “I will serve you, Lady, and go to Doctor Dee.”

THE ONYX HALL, LONDON

5 February 1590

The night garden of the Onyx Hall had no day garden with which to contrast, but still it bore that name. It was enormous, comparable in size to the great presence chamber, but very different in character; in place of cold, geometric stone, there was instead the softness of earth, the gentle arch of branches. The quiet waters of the Walbrook, the buried river of London, bisected the garden’s heart. Paths meandered through carefully arranged beds of moonflower, cereus, and evening primrose; angel’s trumpet wound its way up pillars and around fountains. Here and there stood urns filled with lilies from the deeper reaches of Faerie. Night lasted eternally here, and the air was perfumed with gentle scents.

Lune breathed in deeply and felt something inside her relax. As much as she enjoyed living among mortals, it was exhausting beyond anything else she knew. Easy enough to don human guise for a trip to Islington and back; living among them was a different matter. Being in the Onyx Hall again was like drinking cool, pure water after a long day in the sun and wind.

The ceiling above was cloaked in shadow, and spangled with brilliant faerie lights: tiny, near-mindless creatures below even a will-o'-the-wisp—barely aware enough to be called fae. The constellations they formed changed from time to time, as much a part of the garden's design as the flowerbeds and the delicate streams that rippled through them. Their shape now suggested a hunter, thrown through the air from the antlers of a stag.

That was troubling. There must have been some recent clash with the Wild Hunt.

Lune was not the only one in the garden. A small clutch of four fae had gathered a little distance away, under a sculpted holly tree. A black-feathered fellow perched in the branches, while two ladies gathered around a third, who sat on a bench with a book in her hands. Whatever she was reading aloud to them was too quiet for Lune to hear, but it sparked much mirth from her audience.

Footsteps on the flagstones made her turn. Lady Nianna Chrysanthe hurried to her side, saying breathlessly, "You must have come early."

"I finished my business sooner than expected." Vidar had not questioned her nearly as closely as he might have. Lune was not sure whether to find that worrisome, or merely a sign that he was not as competent as he liked to believe. "What do you have for me?"

The honey-haired elfin lady cast a glance around, then beckoned Lune to follow. They went deeper into the garden, finding their own bench on which to sit. They were still within sight of the group beneath the holly tree, and now another pair at the edge of a fountain, but the important thing was that no one could overhear them.

"Tell me," Nianna whispered, more out of excitement than caution, "what does—"

"Give me your news first," Lune said, cutting her off. "Then I will tell you."

Pushing Nianna was dangerous. Lune's work among mortals had gone some way toward restoring her status, but not her former position in the privy chamber, and Nianna alone of her former companions there deigned to speak with her much at all. Lune did not want to lose her most reliable source of information. But she knew Nianna, and knew how far the lady could be pushed. Nianna pouted, but gave in. "Very well. What do you wish me to begin with?"

Lune pointed at the faerie lights in their constellation. "How stand matters with the Wild Hunt?"

"Not well." Nianna deflated a little. Her slender fingers plucked at the enamelled chain that hung from her girdle. "There are rumours they will ally at last with the Courts of the North—"

"There have always been such rumours."

"Yes, but this time they seem more serious. Her Majesty has formed a provisional agreement with Temair. A regiment of Red Branch knights, to fight the Wild Hunt—or the Scots, or both—if she uses her influence on the mortal court to affect events in Ireland on their behalf. They're willing to consider it, at least; the unanswered question is how much aid she would have to give them in return."

Lune let her breath out slowly. Red Branch knights; they would be quite an asset, if Invidiana could get them. The English fae could hold their own against Scotland, even if the redcaps along the Border took the other side, but the Wild Hunt was, and always had been, a very different kind of threat. The only thing that kept them at bay so far was their absolute refusal to fight this war on any mortal front. And Invidiana was not foolish enough to leave the safety of the Onyx Hall, buried in the midst of mortal London, and meet them on their own terms.

"Has there been an active threat from the Hunt?"

"An assassin," Nianna said dismissively. "A Catholic priest. I believe he is still strung up in the watching chamber, if you wish to see him."

She did not. Would-be assassins, and their punishments, were a common occurrence. "But from the *Hunt*?"

"That is why everyone thinks they may have a true alliance with the Courts of the North."

If true, it was worrisome. The Scottish fae had no compunctions about using mortals in the fight; they had been forming pacts with witches for years and sending them south to cause trouble. The leaders of the Hunt all claimed kingship over one corner of England or another;

they might have decided they wanted lands and sovereignty badly enough to look the other way while their allies soiled their hands with mortal tools.

Lune doubted it, but she wasn't in a position to judge. "What else?"

Nianna put a painted fingernail to her lip, considering. "Madame Malline has been asking around, attempting to discover how the bargain with the folk of the sea was struck. I do not know what the Cour du Lys would want with such knowledge, but there must be something. You should stay away from her."

That, or barter with her. If she could do it in a way that wouldn't anger Invidiana. "Continue."

"That is all I know about her aims. Let me see… there have been a few more fae in from country areas, complaining that their homes have been destroyed." Nianna dismissed this with a wave. She had been a lady of the privy chamber to Invidiana for a long time; the travails of country folk were insignificant to her. "Oh yes, and a delegation of muryans from Cornwall—they just arrived; no one knows yet what they want. What else? Lady Carline has a new mortal she's stringing along. She might get to keep him for a time; Invidiana has taken Lewan Erle back to her bed, so her attention is elsewhere. Or not."

Bedroom politics did not interest Lune. They might be of consuming fascination to some, but they rarely affected the matters she attended to. What had she not yet asked about? "The Spanish ambassador?"

"Don Eyague is still here, and has hosted a few visitors. The rumour is that he is being courted by a growing faction in Spain that are dissatisfied with living in a Catholic nation. But I do not know what they intend; I doubt they have the numbers or resources to make any kind of substantial change to the mortal government, even if they decided to mimic her Majesty. For all I know, they may be considering emigration—here, or to the Low Countries. Who can say?"

Fae rarely emigrated, but no matter. Lune had only one question left for Nianna. She glanced casually toward the pair of fae at the nearby fountain, to make sure they were not listening. The lady sat on the stone coping of the edge, framing herself and her red silk gown prettily against the elaborate grotesque that spouted water in five directions, while the gentleman stood and read to her from a book. The sight momentarily distracted her; it looked like the same book the others had been reading beneath the holly tree.

Upon that thought, the book burst into flame.

The gentleman cried out in shock and dropped the burning pages. Lune caught a fleeting glimpse of something streaking across the garden; she turned to track it and saw the book the others held also catch fire. Silhouetted briefly atop the pages was a tiny, glowing, lizard-shaped creature. Then the salamander darted down again, and vanished amidst the flowers of the garden. A trail of smoke showed briefly where it went; then it was gone.

Lune asked, very carefully, "What was that?"

"I did not see, but I believe it was a sal—"

"I am not asking about the creature. What did it just burn?"

"Oh." Nianna tittered, then hid it guiltily behind her hand. "A mortal book. It has been ever so popular at court, but it seems her Grace does not find it amusing."

Lune rolled her eyes and seated herself once more on the bench. "Ah. Would this be that poem called *The Faerie Queene*?"

"You know of it?"

"The Queen—the mortal Queen—likes it greatly, so of course every courtier who wishes to curry favour must be seen with a copy, or heard quoting it at every suitable occasion." And a few that were not so suitable; Lune was growing heartily tired of the work. "I am not surprised Invidiana dislikes it."

"Well, it *is* very inaccurate. But would she not like that? After all, if the mortals knew the truth, they would come down here with crosses and priests and drive us out."

Nianna was a brainless fool. The insult was not that it was inaccurate; no one expected accuracy out of poets. Peasants might know the truth about fae—to an extent—but poets took that truth, dried it out, ground it to powder, mixed it with strange chemicals, and used it to dye threads from which to weave a tapestry that bore only the most passing of resemblances to real fae and their lives.

No, the insult was that the Faerie Queen of the title was a transparent symbol for Elizabeth. That would be what Invidiana could not stand.

The gentleman at the fountain was now extemporaneously composing a poem, in a very loud voice, about the swift and merciless wrath of the Queen. He was not very good. Lune ignored him and asked Nianna, "What of Francis Merriman?"

"What?" Nianna had been attending to the poem. "Oh, yes. A few people have heard that name."

"They have?" Lune restrained her eagerness. She had not forgotten her conversation with Tiresias, over a year ago. There was no Francis Merriman in Elizabeth's court, at least not yet. She suspected she would find him there—perhaps Elizabeth herself was the one whose secret Merriman held?—but she had to consider other possibilities. "Who knows of him?"

Nianna began to count the names off on her jewelled fingers. "That wretched Bobbin fellow... Lady Amadea Shirrell... one or two others, I think. It does not matter. They all said the same thing. Tiresias mentioned the name to them, at one time or another. Some of them looked for him, but no one turned anything up."

Hope vanished like a pricked soap bubble. Lune quelled a frustrated sigh; she did not want Nianna to think the matter too important to her. She already had a time of it, impressing upon the lady the merit of keeping her inquiries discreet, lest Invidiana catch wind of them before Lune had this mortal in hand.

Nianna's words reinforced her growing conviction that she was simply too early. Tiresias was a seer. When he was not raving in confusion about things that never existed, he spoke of the future. Francis Merriman might not even have been born yet.

But when he did appear, Lune intended to be the first to find him.

"Lady Lune," Nianna breathed, plucking at her sleeve, "I have gathered information for you, as you asked—told you all the news of the court, and answered all your questions. Will you not tell me now of my love?"

Ah yes. Lune said, "He will meet you Friday next, at the tavern called The Hound, that lies near Newgate."

Many fae at court had mortal friends or lovers. Some knew their companions were fae; a great many more did not. John Awdeley, a clerk of the chaundry at Whitehall Palace, did not know, and so Nianna required assistance in arranging assignations with him, in such a time and place as she could keep up her pretence of being a maidservant in a London mercer's house.

She did not love him, of course, but she was infatuated with him for now, and being fae, now was all that mattered.

Lune answered Nianna's subsequent questions with rapidly diminishing patience; she had no interest in describing the state of Awdeley's beard, nor recounting how many times he had asked after his maidservant love. "Enough," she said abruptly, when Nianna asked for the seventh time whether he had shown any favour to other women, and if so, their names and stations, so she might curse them and teach them the foolishness of being rival to a fae. "I do not follow the man his every waking moment; I have little to do with him. And our bargain was not that I would recount his doings to you, down to the contents of his supper. You have your meeting with him, and I the latest news of the court. Now begone with you."

Nianna drew herself up in a graceful, offended ripple. "How dare you speak so dismissively to me? Who do you think you are?"

"Who am I?" Lune smiled, giving it a malicious edge. "While you trail after Invidiana, carrying her gloves and her fans and enduring the brunt of her wrathful moods, I eat mortal bread every day and report to her the secret doings of Elizabeth's court. I am, moreover, your pander to this mortal you have set your sights upon. What aid I give you there, I can revoke. Turning his thoughts to another woman would be easy."

The lady hissed, all warmth of manner instantly gone. "Your body would be floating in Queenhithe by the next morning for fish and gulls to peck at."

"Would it?" Lune met her gaze unblinking. "Are you sure?"

The spectre of the Queen was so easy to invoke. And while Invidiana certainly did not come to the aid of every fae who claimed the possibility, she took no offense at being so named; it served her purposes to be a figure of terror to her courtiers. She even made good on the threats occasionally, simply to keep everyone guessing. The threats would lose their meaning if they never bore teeth.

Nianna backed down, but not graciously. Lune might have to find a new informant to keep her abreast of matters. She would not have antagonised Nianna so, but the conference with Vidar had put her back up, and left her with no patience for the lady's passing mortal infatuation.

They parted on coldly courteous terms, and Lune wandered through the garden alone. Two faerie lights drifted loose from the constellation above and floated about her shoulders. Lune brushed them away. Since losing her position in the privy chamber, she had not the rank or favour to merit such decoration, and did not want anyone carrying tales of her presumption. By the time they reached the ears of those in power, the casual wanderings of two faerie lights would be a halo of glory she had shaped and placed on her own head.

The conversation with Nianna left her weary. She had intended to spend more time in the Onyx Hall, to see for herself how the patterns of alliance and power had shifted, but all that talk of Awdeley had turned her thoughts back to the mortal court. Nianna's infatuation was simple to understand. Lune herself spent a great deal of time feigning just such an attachment.

Her feet sought out the chamber of the alder roots. She had a brief leave from the duties of her masquerade; she would go to Islington and rest herself at the Angel, under the friendly care of the Goodemeades, before returning to the life and duties of Anne Montrose.

RICHMOND PALACE, RICHMOND

2 February 1590

The Christmas season had gone, and with it went a great deal of gaiety and celebration. While some privileged courtiers continued to dwell at Hampton Court nearby—the Countess of Warwick among them—the core of Elizabeth's court removed to Richmond, a much smaller palace, and more used for business than pleasure.

Deven would not have minded, were it not that he no longer saw Anne even in passing. The entire corps of Gentlemen Pensioners was obliged to attend the Queen at Christmas, and the increased numbers did not lighten anyone's load; indeed, the extra effort required to organise the full band during the elaborate ceremonies of the season was draining. Life at Richmond was simpler, if more austere, and free time easier to come by.

Easier to come by, and easier to spend: Deven found himself often closeted with Robert Beale, Walsingham's secretary. As much as the common folk might like to believe that the defeat of the Armada ended the threat from Spain, they were not so lucky, as the reports pouring in from agents abroad showed. Philip of Spain still had his eye fixed on heretic England and its heretic Queen.

"Did you hear," Beale said, laying down a paper and rubbing his eyes, "that Essex wants command of the forces being sent to Brittany?"

The room was small enough that the fire made it stuffy;

Deven took advantage of Beale's pause to set down his own reading and unbutton the front of his doublet so he could shrug out of it. The green damask was pulling apart at the shoulder, but the garment was comfortable, and Fitzgerald had not assigned him to duty today. He had not even bothered with a collar or cuffs that morning; he had slipped from his quarters to here in a state of half-dress, intending to spend the entire day cloistered away from court ritual.

He laid the doublet over an unused chair and dragged his thoughts back to what Beale had said. "You're ahead of me, as usual," Deven admitted, returning to his seat. "I did not even know the Brittany expedition had been agreed to."

"It hasn't, but it will be."

Deven shook his head. "The Queen will never let him go. She's overfond of that one."

"Which means he will be insufferable with frustration. The man should be allowed to go abroad and kill things; it might cool his hot head."

"If these reports are accurate, he will have his chance soon enough." Deven scowled at the note in his hands. Someone had got hold of a message in cipher, and Walsingham had passed it to his stenographer Thomas Phelippes. The report Phelippes had returned to them was written in a clear hand, and Deven's fledgling Spanish was sufficient to interpret it; the problem must be in the message itself. "I doubt the accuracy of this one, though, unless Philip plans to arm every man, woman, child, and cow in Spain."

"Forget Spain." The voice came from behind Deven; he

twisted in his chair in time to see Walsingham closing the door behind himself.

The Principal Secretary's face looked pinched, and his words were startling. Forget Spain? They were the Great Enemy; Deven would no more expect Walsingham to forget Spain than for Philip to forget Elizabeth.

"Not you, Robin," Walsingham said, as the Secretary moved to set down the papers in his hand. Beale sighed and kept them. "I have a task for you, Deven."

"Sir." Deven rose and bowed, wishing he had at least kept his doublet on. His breeches, only loosely laced on in the absence of a doublet to be tied to, threatened to flap at the waist.

Walsingham ignored his state of undress. "Ireland."

"Ireland?" He sounded foolish, repeating the Principal Secretary's statement, but it was entirely unexpected. "What of it, sir?"

"Fitzwilliam has accused Perrot of treason—of conspiring with Philip to overthrow her Majesty."

Sir John Perrot was a name Deven had only recently become familiar with; one of Walsingham's men, he had returned a year and a half before from a stint as Lord Deputy in Ireland. Fitzwilliam, then, must be Sir William Fitzwilliam, his successor.

Beale had been listening, not reading; now he said, "Impossible."

Walsingham nodded. "Indeed. And this is why, Deven, you will turn your thoughts from Spain to Ireland.

Fitzwilliam has a grievance with Perrot; he resents that Perrot sits on the privy council and advises her Majesty on Irish affairs, and resents more that the lords in Ireland have taken to writing him directly, bypassing Fitzwilliam's own authority. I am not surprised by his antagonism. What surprises me is the form it has taken. Why this accusation, and why now?"

Deven didn't want to voice the thought that had come into his head, but no doubt Walsingham had already thought the same. "The answers to that, sir, most likely lie in Ireland."

He did not want to go. Aside from the general unpleasantness of traveling to Ireland, it would take him away from court and Anne, neither of which he wanted to leave for long. But he could not pledge his service to Walsingham, and then balk when asked to serve him elsewhere. It would be a mark of the Secretary's trust; Beale himself had been sent on diplomatic missions before.

But Walsingham was shaking his head. "It may come to that, but not yet. I suspect some cause here at court. Fitzwilliam is Burghley's man; I doubt Burghley has goaded him to this, but there may be factional forces I am not seeing. Before I send you anywhere, I will have you look about court. Who shows an interest in Ireland? Who is formulating petitions regarding affairs there, that have not yet reached the privy council?"

"It might have nothing to do with Ireland. Perhaps this strike has entirely to do with Perrot, and Fitzwilliam is simply a convenient route to it."

Beale nodded at this, but Walsingham again shook his head. "I do not think so. Keep your eyes open, certainly, for anything regarding Perrot—but Ireland is your focus. Search out anything I may have missed."

"Yes, sir." Deven bowed again and reached for his doublet.

"*Anything,*" Walsingham repeated, as Deven quickly looped his points through the waist of the doublet and started on the buttons. "Even things in the past—years past. Whatever you may find."

"Yes, sir." Dressed enough to go out once more, Deven took his leave. He would have to find Colsey and put himself together properly, starting with a doublet that wasn't coming unsewn. His quiet day in private, it seemed, would have to wait.

HAMPTON COURT PALACE, RICHMOND

13 February 1590

The banked coals in the fireplace cast a dim, sullen glow over the bedchamber, barely enough to highlight its contents: the chests containing clothes and jewels, the bed heaped with blankets, the pallets of sleepers on the floor.

Anne Montrose lay wakeful, eyes on the invisible fretwork of the ceiling above, listening to her companions breathe. A gentle snore began; the countess had drifted off. One of the other gentlewomen made quiet smacking noises and rolled over. A few sparks flared up the chimney as a glowing log end crumbled under its own weight. The snoring ceased as the countess lapsed into deeper sleep.

Silent as a ghost, Anne rolled back her own blankets and stood.

The rushes pricked at her bare feet as she stole across the floor. The hinges did not creak when she opened the door; she took particular care to keep them well oiled at all times. A muted thud was the only sound to betray her when she left.

Midnight had passed already, and the palace lay sleeping. Even the courtiers indulging in illicit trysts had retired by now. The dark cloak she wore was symbolic as well as practical; it served as a useful focal point for the minor charm she called up. Anyone who might be awake would not see her unless she wanted them to.

By the thin bars of light that came in through the courtyard windows, she made her way along the gallery and to the privy stair that led down to the gardens. Snow had dusted the ground during the day. In the moonlight, their shoulders and heads capped with white, the heraldic beasts that marked out the squares of the Privy Garden seemed even stranger than usual, like frosted gargoyles that might leap into motion without warning. She cast sidelong glances at them as she passed, but they remained lifeless stone.

Up ahead, the banqueting house loomed tall and sinister in the Mount Garden, surrounded by trees pruned carefully into grotesques. And on the far side of that, murmuring to itself under a thin shell of ice, the Thames.

A figure waited for her in the shadows of the Water Gallery, just above the river's edge.

"You are late."

Lune kept the illusion of Anne Montrose over her features; she did not want the nuisance of reconstructing it. She did not have to think like a human, though, and so she stood barefoot on the icy ground, the cloak now flapping free in the wind off the river.

"I am not late," she said, as a bell ringer inside the palace clock tower began to toll the second hour after midnight.

Vidar smiled his predatory smile. "My mistake."

Why Vidar? Ordinarily he dispatched a minor goblin to bring her bread. Lune supposed she would learn the answer soon enough, but she would not satisfy him by asking. Instead she held out one hand. "If you please. I am near the end of my ration."

She was unsurprised when Vidar did not move. "What matters that? Unless you expect a priest to leap out of the river and bid you begone, in the name of his divine master, you are in no immediate danger of being revealed."

Months before, Lune had snatched a few days of solitude for herself, pleading an ill kinswoman in London to cover for her absence. Those days spent wearing her true face had allowed her to shift the schedule of her ration; the goblins delivered it on Fridays, but she ate it on Tuesdays. The margin of safety might be important someday. But Vidar did not know that, and so she feigned the apprehension she should have felt, hearing him come so close to naming the mortal God to her face.

"My mistress may wake and find me gone," she said, sidestepping Vidar's jibe. "I should not tarry."

He shrugged his bony shoulders. "Tell her you slipped off for a tryst with that mortal toy of yours. Or some other tale. I care not what lie you give her." He settled his back against the brickwork of the Water Gallery, arms crossed over his narrow chest. "What news have you?"

Lune tucked her reaching hand back inside her cloak. So. Again she was unsurprised; she knew she was hardly the Onyx Court's only source of information regarding the mortals. But it meant something, that Vidar considered the matter pressing enough to seek her out here. Like the mistress he strove to emulate and eventually unseat, he rarely left the sanctuary of the Onyx Hall.

"Sir John Perrot has been accused of treason," she said, allowing the pretence that Vidar did not already know. "He is a political client of Walsingham's, and so the Principal Secretary is moving to defend him. Deven has been assigned to investigate: who is taking an interest in Irish affairs, and to what end." That Deven and Walsingham both were at Richmond, she did not say. The Countess of Warwick had been bidden there the other day to attend the Queen, by which fortunate chance Lune had been able to learn of Deven's assignment; Vidar would be displeased if he knew how rarely she saw him at the moment. He was her link to Walsingham. Without him, she had very little.

Vidar tapped a sharp fingernail against a jewelled clasp that held the sleeve of his doublet closed. "Has your toy asked you to tell him of what you hear?"

"No, but he knows I will do it regardless." Lune's eyes went from the tapping fingernail to Vidar's face, his sunken eyes hidden in shadow. "Is the accusation our doing?"

The fingernail stopped. Vidar said, "You are here to do the bidding of the Queen, not to ask questions."

How the removal of Perrot would advance Invidiana's bargaining with the Irish fae, Lune could not guess, but Vidar's attempt to dodge the question told her it would. "The better I understand the Queen's intentions, the better I may serve her."

She startled a bitter but honest laugh out of Vidar. "What a charming notion—understanding her intentions. Dwelling among mortals has made you an optimistic fool."

Lune pressed her lips together in annoyance, then smoothed her features out. "Have you instructions for me, then? Or am I simply to listen and report?"

Vidar considered it. Which, again, told her something: Invidiana was permitting him some measure of discretion in this matter. He had not come here just as a messenger. And that told her why it was Vidar, and not a goblin, bringing her bread tonight.

She tucked that information away, adding it to her meagre storehouse of knowledge.

"Seek out the interested parties," he said, the words guarded and thoughtful. "Assemble a list of them. What they desire, and why, and what they would be willing to do in exchange."

Then no bargain had yet been settled with the Irish fae. If it had been, Lune would be assigned more specifically to cultivate a particular faction. Had the accusation of Perrot been simply a demonstration of Invidiana's power, to convince the Irish of her ability to deliver on her promises?

She could not tell from here, and Lady Nianna, even when feeling friendly, was not enough to keep her informed. It was pleasant to dwell among mortals, close enough to the centre of the Tudor Court to bask in its glory without being caught in its net, and to enjoy the illusion of freedom from the ever more vicious intrigues of the Onyx Court, but she could never forget that it was an illusion. Nowhere was safe. And if she ever let that slip her mind, she would discover what it meant to truly fall from favour.

"Very well," Lune said to Vidar, allowing a note of boredom to creep into her voice. Let him think her careless and inattentive; it was always better to be underestimated. "Now, my bread, if you will."

He remained motionless for a few breaths, and she wondered if he would try to extort some further service out of her. But then he moved, and drew from inside his cloak a small bundle of velvet.

Holding it just short of her extended hand, Vidar said, "I want to see you eat it now."

"Certainly," Lune replied, easily, with just a minor note of surprise. It made no difference to her; adding a week of protection now would not negate the remnant she still enjoyed. Vidar must suspect her of hoarding the bread, instead of eating it. Which she had done, a little, but only with great care. The last thing she wanted was to see her glamour destroyed by a careless invocation to God.

The bread this time was coarse and insufficiently baked. Whether Vidar had chosen it from among the country tithes, or Invidiana had, or someone else, it was clear the chooser meant to insult her. But it did the job whether it was good bread or bad, so Lune swallowed the seven doughy bites, if not with pleasure.

The instant she was done, Vidar straightened. "There will be a draca in the river from now on. If anything of immediate import develops, inform it at once."

This time Lune failed to completely hide her surprise, but she curtsied deeply. "As my lord commands."

By the time she straightened, he was gone.

RICHMOND PALACE, RICHMOND
3 March 1590

As much as Deven would have liked to present Walsingham with a stunning revelation set in gold and decorated with seed pearls, after a fortnight of investigating the Irish question, he had to admit defeat.

It wasn't that he had learned nothing; on the contrary, he now knew more than he had ever expected to about the peculiar subset of politics that revolved around their neighbour island to the west. Which included a great deal about the Irish Earl of Tyrone, and the subtleties of shiring Ulster; there were disputes there going back ages, involving both Sir John Perrot and the current Lord Deputy Fitzwilliam.

But it all added up to precisely nothing Walsingham would not have known already.

So Deven laid it out before his master, hoping the Principal Secretary would make something of it he could not. "I will keep listening," he said when he was done, and tried to sound both eager and determined. "It may be there is something I have missed."

For this conference he had been permitted into Walsingham's private chambers for the first time. They were not particularly splendid; Deven knew from Beale the financial difficulties the Principal Secretary faced. He had understood from his earliest days at court that many people there were in debt, but the revelation of Walsingham's own finances had disabused him of any lingering notion that the heaviest burdens lay on ambitious young men such as himself. A few hundred pounds owed to a goldsmith paled into insignificance next to tens of thousands of pounds owed to the Crown itself.

Of course, Elizabeth herself was in debt to a variety of people. It was the way of the world, at least at court.

But Walsingham did not live in penury, either. His furnishings were understated, like his clothing, but finely made, and the chamber was well lit, both from candles and the fire burning in the hearth to drive away the damp chill. Deven sat on a stool near that fire, with Walsingham across from him, and waited to see if his master saw something he did not.

Walsingham rose and walked a little distance away, hands clasped behind his back. "You have done well," he said at length, his measured voice giving nothing away. "I did not expect you to discover so much about Tyrone."

Deven bent his head and studied his hands, running his thumb over the rough edge of one fingernail. "You knew about these matters already."

"Yes."

He could not entirely suppress a sigh. "Then what was the purpose? Simply to test me?"

Walsingham did not respond immediately. When he did, his voice was peculiarly heavy. "No. Though you have, as I said, performed admirably." Another pause; Deven looked up and found the Principal Secretary had turned back to face him. The firelight dancing on his face made him look singularly unwell. "No, Michael—I was hoping you might uncover something more. The missing key to a riddle that has been troubling me for some time."

The candour in his voice startled Deven. The admission of personal failure, the use of his given name—the choice of this chamber, rather than an office, to discuss the matter—Beale had said before that Walsingham had an occasional and surprising need to confide in others about sensitive matters. Others that included Beale.

Others that had not, before now, included Deven.

"Fresh eyes may sometimes see things experienced ones cannot," he said, hoping he sounded neither nervous nor intrusive. Walsingham held his gaze, as if weighing something, then turned away. His hand trailed over a chess set laid out on a table; he picked up one piece and held it in his hand, considering. Then he set it down on a smaller table next to Deven. It was a queen, the black queen. "The matter of the Queen of Scots," he said. "Who were the players in that game? And what did they seek?"

The non sequitur threw Deven for a moment—from Ireland to Scotland, with no apparent connection. But he was accustomed by now to the unexpected ways in which Walsingham tested his intelligence and awareness, so he marshalled his thoughts. The entire affair had begun when he was very young—possibly before, depending on how one counted it—and had ended before he came to court, but the Scottish queen had more influence on English policy than most courtiers could aspire to in their lives, and her echoes were still felt.

"Mary Stewart," he said, picking up the chess piece. It was finely carved from some wood he could not identify, and stained dark. "She should be considered a player herself, I suppose. Unless you would call her a pawn?"

"No one who smuggled so many letters out through the French embassy could be called a pawn," Walsingham said dryly. "She had little with which to fill her time but embroidery and scheming, and there must be limits to the number of tapestries and cushions a woman can make."

"Then I'll begin with her." Deven tried to think himself in her place. Forced to abdicate her throne and flee to a neighbouring country for sanctuary—sanctuary that became a trap. "She wanted... well, not to be executed, I imagine. But if we are considering this over a longer span of time, then no doubt she wished her freedom from confinement. She was imprisoned for, what, twenty years?"

"Near enough."

"Freedom, then, and a throne—any throne, from what I hear. English, Scottish, probably French if she could have got it back." Deven rose and crossed to the chessboard. If Walsingham had begun the metaphor, he would continue it. Selecting the white queen, he set her down opposite her dark sister. "Elizabeth, and her government. They—you—wished security for the Protestant throne. Against Mary as a usurper, but there was a time, was there not, when she was considered a possible heir?"

Walsingham's face was unreadable, as it so often was, particularly when he was testing Deven's understanding of politics. "Many people have been so considered."

He hadn't denied it. Mary Stewart had Tudor blood, and Catholics considered Elizabeth a bastard, incapable of inheriting the throne. "But it seems she was a greater threat than a prospect. If I may be so bold as to say so, my lord, I think you were one of the leading voices calling for her removal from the game."

The Principal Secretary did not say anything; Deven had not expected him to. He was already considering his next selection from the chessboard. "The Protestant faction in Scotland, and their sovereign, once he became old enough to rule." The black king went onto the table, but Deven placed him alongside the white queen, rather than the black. "I do not know James of Scotland; I do not know what love he may bear his late mother. But she was deposed by the Protestant faction, and branded a murderess. They, I think, did not love her."

Having mentioned the Protestants, the next components were clear. Deven hesitated only in his choice of piece. "Catholic rebels, both in England and Scotland." These he represented with pawns, one black and one white, both ranged in support of the black queen. It would be a mistake to assume the rebels all unlettered recusant farmers, but ultimately, whatever their birth, they were pawns of the Crowns that backed them. "Their motives are clear enough," he said. "The restoration of the Catholic faith in these countries, under a Catholic queen with a claim to both thrones."

What had he not yet considered? Foreign powers, that backed the rebels he had depicted as pawns. They did not fit the divide he had created, using black pieces for the Scots and white for the English; in the end he picked up the two black bishops. "France and Spain. Both concerned, like the rebels, with the restoration of Catholicism. France has long invested her men and munitions in Scotland, the better to bedevil us, and Spain sent the Armada as retaliation for Mary Stewart's execution."

Walsingham spoke at last. "More a pretext than an underlying cause. But you have them rightly placed."

The pieces were arrayed on the table in a strange, disorganised game of chess. On one side stood the black queen with her two bishops, a pawn, and a second pawn from the white; on the opposing side, the white queen and the black king. Had he missed anyone? The English side was grievously outnumbered. But that was a true enough representation. She had Protestant allies, but none whose involvement in the Scottish matter was visible to Deven.

The only group he still wondered about was the Irish, with whom this entire discussion had begun. But he did not know of any involvement on their part, nor could he imagine any that made sense.

If he had failed the test, then so be it. He faced Walsingham and made a slight bow. "Have I passed your examination, sir?" By way of reply, Walsingham took a white knight and laid it on the English side. "Her Majesty's privy council," he said. Then he moved the white queen out into the centre, the empty space between the two. "Her Majesty."

Dividing Elizabeth from her government. "She did not wish the Queen of Scots to be executed?"

"She was of two minds. As you observed, Mary Stewart was a potential heir, though one who would never be acceptable to those of our Protestant faith. Her Majesty also feared to execute the anointed sovereign of another land."

"For fear of the precedent it would set."

"There were those who sought our Queen's death, of course, regardless of precedent. But if one Queen may be killed, so may another. Moreover, you must not forget they were kinswomen. Her Majesty recognised the threat to her own safety, and that of England, but she was most deeply reluctant."

"Yet she signed the execution order in the end."

Walsingham smiled thinly. "Only when driven to it by overwhelming evidence, and the patient effort of us her privy councillors. Bringing that about was no easy task, and her secretary Davison went to the Tower for it." At Deven's started look, he nodded. "Elizabeth changed her mind in the end, but too late to prevent Mary's execution; Davison bore the weight of her wrath, though little he deserved it."

Deven looked down at the table, with the white queen standing forlornly, indecisively, between the two sides. "So what is the riddle? Her Majesty's true state of mind regarding the Queen of Scots?"

"There is one player you have overlooked."

Deven bit his lip, then shook his head. "As much as I am tempted to suggest the Irish, I do not think they are who you mean."

"They are not," Walsingham confirmed.

Deven studied the chess pieces once more, both those on the table and those unused on the board, then made himself close his eyes. The metaphor was attractive, but easy to get caught in. He mustn't think of knights, castles, and pawns;

he must think of nations and leaders. "The Pope?"

"Ably represented by those Catholic forces you have already named."

"A Protestant country, then." Mustn't think in black and white. "Or someone farther afield? Russia? The Turks?"

Walsingham shook his head. "Closer to home."

A courtier, or a noble not at court. Deven could think of many, but none he had cause to connect to the Scottish Queen. Defeated, he shook his head. "I do not know."

"Nor do I."

The flat words brought his head up sharply. Walsingham met his gaze without blinking. The deep lines that fanned out from his eyes were more visible than ever, and the grey in his hair and beard. The vitality of the Principal Secretary's intellect made it easy to forget his age, but in this admission of defeat he looked old.

Not defeat. Walsingham would not be outplayed. But it seemed he had, for the moment, been stymied. "What do you mean?"

Walsingham gathered his long robe around him and sat once more, gesturing for Deven to do the same. "Her Majesty had little choice but to execute Mary Stewart; the evidence against her was unquestionable. And years in the assembling, I might add; I knew from previous experience that I would need a great deal. Yet for all the efforts of the privy council, and all that evidence, it was a near thing—as her treatment of Davison shows."

"You think someone else persuaded her in the end? Or was

arguing against it, and turned her back after her decision?"

"The former."

Deven's mind was racing, pursuing these new paths Walsingham had opened up. "Not anyone on the Catholic side, then." That ruled out a good portion of Europe, but fewer in England, even with some educated guesses as to who was a closeted papist.

"There is more." Walsingham steepled his ink-stained fingers, casting odd shadows over his weary face. "Some of the evidence against the Queen of Scots fell too easily into my hands.

"There are certain strokes of good fortune that seem too convenient, certain individuals whose assistance was too timely. Not just at the end, but throughout. During the inquiry into her husband's death, she claimed that someone had forged the letters in that casket, imitating her cipher in order to incriminate her. An implausible defence—but it may have been true."

"Someone among the Protestant Scots. Or Burghley."

But Walsingham shook his head before the words were even out. "Burghley has long had his agents, as did Leicester, before his death. But though we have not always been free with the knowledge we gain, I do not think they would, or could, have kept such an enterprise concealed from me. The Scots are a better guess, and I have spent much effort investigating them."

His tone said enough. "You do not think it was them, either."

"It does not end with the Queen of Scots." Walsingham rose again and began to pace, as if his mind would not allow his tired body to remain still. "That was the most obvious incident of interference, and the longest, I think, in the founding and execution. But I have seen other signs. Courtiers presenting unexpected petitions, or changing stances that had seemed firmly set. Or the Queen herself."

Elizabeth, not Mary. "Her Grace has always been of a… mercurial temperament."

Walsingham's dry look said he needed no reminder. "Someone," the Secretary said, "has been exercising a hidden influence over the Queen. Someone not of the privy council. I know my fellows there well enough; I know their positions. These interventions I have seen have, from time to time, matched the agenda of one councillor or another… but never one consistently."

Deven respected Walsingham enough to believe that evaluation, rather than assume someone had successfully misled him for so long. Yet someone must have, had they not? Incredible as it was, someone had found a way to play this game without being uncovered.

The Secretary continued. "It would seem our hidden player has learned, as we all have, that to approach the Queen directly is less than productive; he more often acts through courtiers—or perhaps even her ladies. But there are times when I can think of no explanation save that he had secret conference with her Majesty, and persuaded her thus."

"Within the last two years?"

Walsingham's dark gaze met Deven's again. "Yes."

"Ralegh."

"He is not the first courtier her Majesty has taken to her bosom without naming him to the council; indeed, I sometimes think she delights in confounding us by consulting others. But we know those individuals, and account for them. It is not Ralegh, nor any other we can see."

Deven cast his mind over all those with the right of access to the presence and privy chambers, all those with whom he had seen the Queen walk in the gardens. Every name he suggested, Walsingham eliminated. "But your eye is a good one," the Secretary said, with a wry smile. "There is a reason I took you into my service."

He had been pondering this matter since Deven's appearance at court? Since before then, from the sound of it; Deven should not be surprised to find himself a pawn in this game. Or, to switch metaphors, a hound, used to tease out the scent of prey. Yet a poor hound he was turning out to be. Deven let his breath out slowly. "Then, my lord, by your arguments, there are a number of people it cannot be, but a great many more who it *might* be. Those you can set aside are a few drops against an ocean of possibilities."

"Were it not so," Walsingham said, his voice flat once more, "I had found him out years ago."

Accepting this rebuke, Deven hung his head.

"But," the Secretary went on, "I am not defeated yet. If I cannot find this fellow by logic, I will track him to his lair—by seeing where he moves next."

Now, at last, they were coming to the true reason Walsingham had picked up that first chess piece and asked about the Queen of Scots. Deven did not mind playing the role of the Secretary's hound, when set to such a compelling task. And he even knew his quarry. "Ireland."

"Ireland," Walsingham agreed.

With these recent revelations in mind, Deven tried to see the hand of a hidden player in the events surrounding Perrot, Fitzwilliam, and Tyrone. Yet again the muddle defeated him.

"I do not think our player has chosen a course yet," Walsingham said when he admitted this. "I had suspected him the author of the accusation against Perrot, but what you have uncovered makes me question it. There are oppositions to that accusation I did not expect, that might also be this unknown man's doing."

Deven weighed this. "Then perhaps he is playing a longer game. If, as you say, he manipulated events surrounding the Queen of Scots, he has no aversion to spending years in reaching his goal."

"Indeed." Walsingham passed a hand over his face, pinching the bridge of his nose. "I suspect he is laying the foundations for some future move. Which is cautious of him, and wise, especially if he wishes his hand to remain unseen. But his caution also gives us time in which to track him."

"I will keep listening," Deven said, with more enthusiasm than he had felt when he said it before. Now that he knew what to listen *for,* the task was far more engaging. And he did not want to disappoint the trust

Walsingham had shown him, revealing this unsolved riddle in the first place. "Your hidden player must be good, to have remained unseen for so long, but everyone makes mistakes eventually. And when he does, we will find him."

MEMORY

December 1585

The man had hardly stepped onto the dock at Rye when he found two burly fellows on either side of him and a third in front, smiling broadly and without warmth. "You're to come with us," the smiler said. "By orders of the Principal Secretary."

The two knaves took hold of the traveller's elbows. Their captive seemed unassuming enough: a young man, either clean-shaven or the sort who cannot grow a beard under any circumstances, dressed well but not extravagantly. The ship had come from France, though, and in these perilous times that was almost reason enough on its own to suspect him. These men were not searchers, authorised to ransack incoming ships for contraband or Catholic propaganda; they had come for him.

He was one man against three. The captive shrugged and said, "I am at the Secretary's disposal."

"Too right you are," one of the thugs muttered, and they marched him off the dock into the squalid streets of Rye.

With their captive in custody, the men rode north and

west, under a grey and half-frozen sky. Three cold, miserable days brought them to a private house near the Palace of Placentia in Greenwich, and the next day brought a knock at the door. The leader muttered, "Not before time, either," and went to open it.

Sir Francis Walsingham stepped through, shaking out the folds of his dark cloak. Outside, two men-at-arms took up station on either side of the door. Walsingham did not look back to them, nor at the men he had hired, though he unpinned the cloak and handed it off to one of those men. His eyes were on the captive, who had risen and offered a bow. It was difficult to tell whether the bow was meant to be mocking, or whether the awkwardness of his bound hands led to that impression.

"Master Secretary," the captive said. "I would offer you hospitality, but your men have taken all my possessions—and besides, the house isn't mine."

Walsingham ignored the sarcasm. He gestured for the two thugs to depart but their leader to remain, and when the three of them were alone in the room, he held up a letter taken from the captive, its seal carefully lifted. "Gilbert Gifford. You came here from France, bearing a letter from the Catholic conspirator Thomas Morgan to the dethroned Queen of Scots—a letter that recommends you to her as a trustworthy ally. I trust you recognise what the consequences for this might be."

"I do," Gifford said. "I also recognise that if those consequences were your intent, you would not have come here to speak privately with me. So shall we skip the threats and intimidation, and move on to the true matter at hand?"

The Principal Secretary studied him for a long moment. His dark eyes were unreadable in their nest of crow's-feet. Then he sat in one of the room's few chairs and gestured for Gifford to take the other, while Walsingham's man came forward and unbound his hands.

When this was done, Walsingham said, "You speak like a man who intends to offer something."

"And you speak like one who intends to negotiate for something." Gifford flexed his hands and examined them, then laid them carefully along the arms of the chair. "In plain terms, my position is this: I come bearing that letter of recommendation, yes, and have every intention of putting it to use. What I have not yet determined is the use to which I will put it."

"You offer your services."

Gifford shrugged. "I have taken stock of the other side. No doubt you have a file somewhere detailing it all, Douai, Rome, Rheims—"

"You became a deacon of the Catholic church in April."

"I would be disappointed if you did not know. Yes, I studied at their seminaries, and achieved some status therein. Had I not, Morgan would not now be recommending me to Mary Stewart. But if you know of those things, you also know of my conflicts with my supposed allies."

"I am aware of them." Walsingham sat quietly, with none of the fidgeting that marked lesser men. "You mean to say, then, that these conflicts of yours were a sign of true disaffection, and that your recent status was achieved in order to gain their trust."

Gifford smiled thinly. "Perhaps. I would like to be of use to someone. I have no particular passion for the Catholic faith, my family notwithstanding, and I judge your cause to be in the ascendant. Though perhaps my willingness to switch sides is reason enough for you not to trust me. I am no ideologue for anyone's faith, yours included."

"I deal with men of the world as much as with ideologues."

"I am glad to hear it. So that is my situation: I was sent to find some way to restore secret communications for Mary Stewart, so that her allies here and abroad might be able to plot her release once more. If you wish to block that communication, you can stop me easily enough—but then they will find someone else."

"Whereas if I make use of you, I will know what is being said."

"Assuming, of course, that I am the only courier, and not sent to distract you from the real channel of communication."

The two men sat silently, watched by the third, while the fire crackled and gave out its warmth. There was no illusion of warmth between Walsingham and Gifford—but there was opportunity, and in all likelihood both preferred that to warmth.

"You have been publicly arrested," Walsingham said at last. "What will you tell Morgan?"

"He's a Welshman. I will tell him I told you that I came

here to advance the interests of the Welsh and English factions against the Jesuits."

"Whereupon I, favouring any sort of internal strife among my enemies, released you to proceed about that business."

Gifford smiled mockingly.

Walsingham weighed him for a long moment, his shrewd eyes unblinking. At last he said, "You will keep me informed as you make contact with the Queen of Scots, so that we may devise a way to keep her correspondence under our eye." He took up the letter from Morgan and passed it back to Gifford. "Go to Finch Lane, near Leadenhall Market. There is a man there named Thomas Phelippes. You will give him this letter, and keep in his company until I send you onward. The delay will not be remarked; the Scottish woman's residence is being moved, and until it is settled once more no one will expect you to contact her. Phelippes will return the letter to you when you go."

Gifford accepted the letter and tucked it away. "May this be profitable to us both."

But when they released him, he did not proceed as instructed to Leadenhall. The time for that would come, but he had other business first.

The house he sought out stood hard by the fishy stench of Billingsgate, but despite its location, its windows and doors were boarded up. By the time he arrived there, he had shaken off the Secretary's men who had been following him, and so he entered the tiny courtyard alone.

Dusk was falling as he knelt with fastidious care on

the ground and ran his long fingers around the edge of one flagstone. With a small grimace of effort, he pried it from its rest, revealing not dirt beneath, but a vertical passage, with a ladder propped against one wall. When the flagstone settled back into place above him he reached out, found smooth stone, and laid his lips on it in a grudging kiss.

A whisper of sound as the stone shifted, and a rush of cool light.

He stepped through into a place which both was and was not beneath the courtyard of the house near Billingsgate. As he did so, the last vestiges of the facade that was Gilbert Gifford fell away, and with a disturbingly fluid shrug, a new man revealed himself.

He had cut it very fine; much longer and his protection would have faded. He had underestimated how cursedly slow travel could be, when confined to slogging along ordinary roads on ordinary horses, and food not specifically given in offering did nothing to maintain his facade. But he had reached the Onyx Hall in safety, and he could explain away his delay in getting to Phelippes.

First, he would report to his Queen.

With one last rippling shiver that shook off the lingering stain of humanity, Ifarren Vidar set off deeper into the Onyx Hall, to tell of his work against the Scottish woman, and to prepare himself for more time spent imprisoned in mortal guise.

RICHMOND PALACE, RICHMOND

6 March 1590

The draca was not the only fae around Elizabeth's court. Lune was the only one living as a human, but others came and went, to gather secrets, visit lovers, or simply play tricks. Sometimes she knew of their presence; other times she did not.

But she assumed, even before she left the Onyx Hall to lay the groundwork for Anne Montrose's entrance, that someone had been set to watch her. With the endless peregrinations of Elizabeth's court, it was necessary; they could not spend more than a month or two in one residence before it became fouled by habitation, and the Queen's whim could send everyone packing on a moment's notice. The sprites and goblins assigned to bring Lune bread on Fridays had to know where to find her.

That was hardly the watcher's only purpose, though. She never let herself forget that someone else was reporting back on her actions.

Since Vidar's appearance at Hampton Court, Lune had conducted herself with even more care than usual. Her ostensible purpose there was to monitor Walsingham and gain access to him via Deven, but she might at a moment's notice be asked to take action on the Irish affair. A less subtle fae might charm one or more courtiers into behaving as desired; Lune knew her value lay in her ability to work through human channels. Invidiana did not want her influence over the mortal court betrayed by indiscreet use of faerie magic.

So she gathered secrets, and tallied favours, and waited to see what would happen, one eye ever on the few tiny scraps of information about her own court she was able to glean from her contacts.

The draca in the river was useful. Water spirits were often garrulous, and this one was no exception; it might not have access to the daily life of the Onyx Court—it never went past the submerged entrance in the harbour of Queenhithe—but it spoke to other water-associated fae, and even (it claimed) to the Thames itself, which was the lifeblood of London. More news came its way than one might expect. When surprisingly warm and sunny weather descended on them one afternoon, Anne Montrose persuaded the countess to go out along the river, and Lune spent nearly an hour talking to the draca, learning what it knew.

When Elizabeth summoned the countess to attend her at Richmond, the draca followed them downriver. Lune never used it to send word to Vidar and Invidiana, but one blustery day in March, the draca gave her a warning: Vidar himself would bring her bread that night.

Lune was not surprised—she had expected she might see him again, given the apparent importance of Ireland to both courts at present—but she might have been startled, if she came upon him unawares. She thanked the draca, rewarded it with a gold earring purloined from the countess, and went about her business as if nothing were unusual.

Someone was always watching.

Richmond was smaller, and more difficult to sneak around in. Lune left the countess's chamber well in advance of her appointed rendezvous, and sacrificed her careful illusion of Anne Montrose for the purpose of disguise. It was possible to turn mortal eyes away, but tiring; far easier to appear as someone who had a right to be up and about, even at odd hours.

A servant of the household in this part of the palace; a man-at-arms in that part, though she impersonated men badly and would not have wanted to attempt a conversation as one.

One careful stage at a time, she made her way outside and into the night.

When at Hampton Court, she met her courier along the river; here, the appointed meeting place lay within the shadows of the orchard. She ducked beneath the drooping, winter-stripped branches of a willow and, straightening, discarded her appearance of mortality entirely.

The fae who waited for her there was not Vidar.

Lune swore inwardly, though she kept her face smooth. A change of plans? A deliberate deception on Vidar's part? Or just the draca lying for its own amusement or self-interest? It did not matter. Gresh, one of her more common contacts, was waiting for her.

"Your bread," he grunted, and tossed it at her without ceremony. Like most goblin fae, he was a squat and twisted thing; ceremony would have been a painful mockery on him. Lune sometimes thought that was why so few of them occupied places of importance in the Onyx Court. In addition to their chaotic and unrefined natures, which disrupted the elegance Invidiana prized, they did not look the part. Mostly they operated as minions of the elfin fae, or stayed away from court entirely.

But elegance and beauty were not the only things that mattered; Dame Halgresta and her two brothers proved that. Raw ugliness and power had their places, too.

Lune caught the bread and examined it; she had been shorted on her ration more than once. The lump was large enough to make up the requisite seven bites, though, and so she tucked it away in her purse. "So what you got?" Gresh asked, scratching through the patchy, wiry hairs of his beard. "Make it quick—just the important stuff—got better things to do with my time than sit under some drippy tree listening to gossip." He glared up at the willow's swaying branches as if personally offended by them.

She had spent much of the day planning out what she would say to Vidar, how she would answer the questions she anticipated him asking. Faced with only Gresh, she felt rather deflated. "The Earl of Tyrone is likely to come before the privy council again soon," she said. "If her Majesty wishes to take some action, that would be an opportune time; he is an ambitious man, and a contentious one. He can be bought or provoked, as needed."

Gresh picked something out of his beard, examined it, then threw it away with a disappointed sigh. "What about what's-his-face? The one you supposed to be watching. Not the Irish fellow."

"Walsingham may be my assignment," Lune said evenly, "but he is not the only way to advance our Queen's interests at court. I began with the most significant news."

"Planning to bore me with insignificant news?"

"*Less* significant is not the same as *insignificant*."

"Dunna waste time arguing; just get on with it."

Tedious experience had taught her that Gresh could neither be charmed nor intimidated into better behaviour. It simply wasn't in his nature. Lune swallowed her irritation and went on. "Walsingham continues to defend Sir John Perrot against the accusation of treason. His health has been poor, though. If he has to take another leave of absence, Robert Beale will likely stand in for him with the privy council, as he has done before, but while Beale will follow his master's wishes, he will be less effective of an advocate for Perrot."

Gresh scrunched his brows together in either pain or intense thought, then brightened. "Walserthingy, Wasserwhatsit… oh, right! Something I was supposed to ask you." He feigned a pensive look. "Or should I make you wait for it?"

Lune didn't bother to respond to that; it would only amuse him more.

"Right, so, Water-whoever. Got some mortal fellow dangling that serves him, right?"

"Michael Deven."

"Sure, him. How loyal's he?"

"To me?"

"To his master."

She hadn't expected that question. To buy herself time, Lune said, "He has been in Walsingham's service since approximately a year and a half ago—"

Gresh snorted, a phlegmy sound. "Ain't asking for a history. Would your mortal pup betray him?"

Her nerves hummed like harpstrings brought suddenly into tune. Lune said carefully, "It depends on what you mean by betrayal. Would he act directly against Walsingham's interests?" She didn't even have to ponder it. "No. Deven, like his master, is dedicated to the well-being of England and Elizabeth. At most, his opinion on how to serve that well-being might differ from Walsingham's. I suppose if it differed enough, and he thought the situation critical enough, he might take action on his own. But a direct betrayal? Never. The most he has done so far is indiscreetly share some information he should have kept secret."

"That so?" Gresh greeted this with an eager leer. "Like what?"

Lune kept her shrug deliberately careless. "Matters I have already shared with her Majesty. If you are not privy to them, that is no concern of mine."

"Aw, c'mon." The goblin pouted—a truly hideous sight. "No new scraps you could toss the way of this poor, bored soul?"

Why was he pressing? "No. I have nothing new to report."

It could have been the wind that stirred the branches of the willow. By the time she realised it wasn't, the knife was already at her throat.

"Really," Vidar breathed in her ear, his voice soft with malice. "Would you care to rethink that statement, Lady Lune?"

Gresh cackled and did a little dance.

She closed her eyes before they could betray her. More than they already had. With sight gone, her other senses were sharpened; she heard every quiet tap as the willow's bare branches met and parted, the chill whistling of the damp spring breeze. Frost left a hard crust on the ground and a hard scent in the air.

The edge against her throat rasped imperceptibly across her flesh as she inhaled, its touch light enough to leave the skin unbroken, firm enough to remind her of the blade's presence.

"Have you heard something to the contrary, Lord Ifarren?" she asked, moving her jaw as little as possible.

His left arm was wrapped around her waist, nails digging in hard enough to be felt through the boning of her bodice. Vidar was taller than her, but with his skeletal build, he weighed about the same. What would Gresh do, if she tried to fight Vidar?

There was no point in trying. Even if she got the knife away from him, what would she do? Kill him? Invidiana had been known to turn a blind eye to the occasional murder, but Lune doubted this one would go unremarked. If she could even best the faerie lord.

He laughed silently; she felt it where his body pressed against her back. "How very evasive an answer, Lady Lune." Vidar pronounced her title like a threat. "I have heard something very interesting indeed. I have heard that you spoke with that mortal toy of yours."

"I speak with him often." The words came out perfectly unruffled, as if they stood in ordinary discourse.

"Not so often as you might. You should have told me he was at Richmond without you."

"My apologies, my lord. It was an oversight."

Another silent laugh. "Oh, I am sure. But this recent conversation—that is the one that interests me. A little whisper has said he told you something of import." His grip tightened around her waist, and the knife pressed closer. "Something you have not shared."

She knew the conversation he meant. There was no way the draca could have heard it; they stood in the palace kennel at the time, well away from the river. What manner of fae could have overheard them without being seen?

A black dog, perhaps—some skriker or brash. Hidden in among the hounds. But how much had it heard?

Not everything, or Vidar would not be here now, forcing the information out of her. But she had to be very careful of what she said.

She opened her eyes. Gresh had vanished, his duty done; if he was eavesdropping, that would be Vidar's problem.

"Sir Francis Walsingham," she said, "has begun to suspect."

Vidar went still against her back. Then his arm uncurled; the elfin lord kept the knife against her throat as he circled around to stand in front of her. His black eyes glittered in the near-total darkness.

"What did you say?" he whispered.

She wet her lips before she could suppress the nervous movement. "The Principal Secretary has begun to suspect that someone unknown to him has a hand in English politics."

"What has he seen?"

The question lashed out like a whip. But it was easier for Lune to retain her composure, even with the knife still against her skin, now that her body was not pressed to Vidar's in violently intimate embrace. "Seen? Nothing. He suspects only." She had to give him more than that. "The recent events concerning Ireland have caught his attention. He is beginning to look back at past matters, such as the Queen of Scots."

Vidar's shoulders rose fractionally with tension. Lune knew that one would worry him.

"And what," Vidar said, his voice now hard with control, "will he do with his suspicions?"

Lune shook her head, then froze as she felt the knife scrape her throat once more. "I do not know. Deven does not know. Walsingham spoke of it only briefly, and that in a confused fashion. He has not been well; Deven thinks this a feverish delusion brought on by overwork."

She stood motionless, briefly forgotten as Vidar considered her words. The black dog—if that was the watcher in question—could not have heard them over the racket the other hounds were making. He had only seen them talking, and surmised from her reaction that whatever Deven spoke of was important. Vidar's sharp reaction to her first declaration had made that plain.

Which meant that she could afford to bend the truth—within limits.

Vidar's gaze sharpened and turned back to her. "So," he said. "You learned of this—a clear and immediate threat to the Queen's grace and the security of our people—and you chose to keep the information to yourself." His lips peeled back from his teeth in mockery of a smile. "Explain why."

Lune sniffed derisively. "Why? I should think it obvious, even to you. This is the kind of situation that makes people stop thinking, sends them into a blind panic wherein they strike out at the perceived threat, thinking only to destroy it. Which might be a terrible waste of opportunity."

"Opportunity." Vidar relaxed his arm; the knife moved away, though it still glimmered in his hand, unsheathed and ready. "Opportunity for Lady Lune, perhaps—at the expense of the Onyx Court, and all the fae who shelter under its power."

She wondered if this rhetoric came from their habit of copying mortals. The greater good of the Onyx Court, and the faerie race as a whole, was occasionally deployed as a justification for certain actions, or an exhortation to loyalty. It might have carried more force had it not been only an occasional device—or if anyone had believed it to be more than empty words. "Not in the slightest," she said, keeping her voice even and unperturbed. "I am no fool; what gain could there possibly be for me, betraying her Majesty in such a manner? But I am better positioned than any to see which direction Walsingham moves, what action he takes. And I tell you that quick action would be inadvisable here. Far better to watch him, and to move subtly, when fortune should offer us a chance." She allowed herself an ironic smile. "Even should he uncover the times and places in which we have intervened, I hardly expect he will imagine fae to be the culprits."

And that was true enough. But Vidar's malicious smile had returned. "I wonder what her Majesty would think of your logic?" Beneath the facade of her composure, Lune's heart skipped several beats.

"I might not tell her." Vidar examined the point of his dagger, scraping some imagined fleck of dirt off it with one talon-like fingernail. "It would be a risk to me, of course—if she found out… but I might be willing to offer you that mercy, Lady Lune."

She had to ask; he was waiting for it. "At what price?"

His eyes glittered at her over the blade in his hands. "Your silence. At some point in the future, I will bid you keep some knowledge to yourself. Something commensurate with what I do for you now. And you will be bound, by your word, to keep that matter from the Queen."

She could translate that well enough. He was binding her to be his accomplice in some future bid to take the Onyx Throne.

Yet what was her alternative? Say to him, *So tell the Queen, and be damned,* and then warn Invidiana of Vidar's ambition? She knew of it already, and he had not said anything specific enough to condemn him. At which point Lune would be dependent on nothing more than the mercy of a merciless Queen.

Lune kept from grinding her teeth by force of will, and said in a voice that sounded only a little strained, "Very well."

Vidar lowered the knife. "Your word upon it."

He was leaving nothing to chance. Lune swallowed down bile and said, "In ancient Mab's name, I swear to repay this favour with favour, of commensurate kind and value, when you should upon a future occasion ask for it, and to let no word of it reach the Queen."

That, or remove him as a threat before he ever had occasion to ask. *Sun and Moon,* Lune thought despairingly, *how did I reach such a state, that I should be swearing myself to Vidar?*

The dagger vanished as if it had never been. "Excellent," Vidar said, and smiled that toothy smile. "I look forward to hearing your future reports, Lady Lune."

OATLANDS PALACE, SURREY

14 March 1590

Standing at attention beside the door that led from the presence chamber to the privy chamber, Deven fixed his eyes on the far wall and let his ears do the work. It was a tedious duty, and a footsore one—shifting one's weight was frowned upon—but it did afford him a good opportunity to eavesdrop. He had come to suspect that Elizabeth's penchant for conversing in a variety of languages was as much an obfuscatory tactic as a demonstration of her learning; her courtiers were a polyglot assortment, following the lead of their Queen, but few could speak every language she did. He himself was often defeated by her rapid-fire speech, but he had enough Italian now to sift out the gist of a sentence, and his French was in fine practice. So he stared off into the distance, poleax held precisely upright, and listened.

He listened particularly for talk of Ireland.

A hidden player, Walsingham had said. Deven had already calculated that any such player must either have the right of entree to the presence chamber—likely the privy chamber as well—or else must have followers with such a right. The former made more sense, as one could not effectively influence the machinery of court at a distance for long, but he couldn't assume it too firmly.

Unfortunately, though a great many people were barred from entry, a great many were not. Peers of the realm, knights, gentlemen—even some wealthy merchants—

ambassadors, too. Could it be one of them? Neither the Spanish nor the French would have reason to urge Mary's execution on Elizabeth, and they had little enough reason to care what happened in Ireland, though part of the accusations against both Perrot and the Earl of Tyrone were that they had conspired with the Spanish.

But ambassadors came and went. If Walsingham was correct, this player had been active for decades. They made poor suspects, unless the true players were their more distant sovereigns—but those were already on the board, so to speak.

Round and round Deven's thoughts went, while out of the corners of his eyes he watched courtiers come and go, and he eavesdropped on every scrap of conversation he could.

A tap on his brocade shoulder roused him from his reverie. Focusing so much on the edges of the room, he hadn't paid any attention to what was in front of him.

William Tighe stood before him, ceremonial polearm in hand. On the other side of the doorway, John Darrington was changing places with Arthur Capell. Deven relaxed his stance and nodded thanks to Tighe.

As he stepped aside, he saw something that distracted him from the endless riddle in his mind. Across the chamber, the Countess of Warwick laid aside her embroidery hoop and rose from her cushion. She made a deep curtsy to Elizabeth, then backed away. Transferring the poleaxe to his left hand, Deven moved quickly to the outer door, where he bowed and opened it for the countess.

He followed her out into the watching chamber, past the Yeomen of the Guard and usher at that door, and as soon as they passed out of earshot he said, "Lady Warwick. If I could beg a moment of your time?"

She looked mildly surprised, but nodded and gestured for him to walk at her side. Together they passed out of the watching chamber, filled with those courtiers hoping for an opportunity to gain entrance to the more restricted and privileged domain beyond. Deven waited until they had escaped those rapacious ears before he said, "I humbly beg your pardon for troubling you with this matter; I am certain there are many other, more pressing cares that demand your ladyship's time. But I am sure you can understand how affection drives a man's heart to impatience. Have you any sense yet how her Majesty's disposition lies, with respect to my desire?"

It was far from the most elaborate speech he had ever delivered at court, yet he seemed to have puzzled the countess. "Your desire?"

"Mistress Montrose, your waiting-gentlewoman," Deven said. "She tells me she has asked your ladyship to discern which way the wind blows with the Queen—whether her Grace would be angered by the notion of our marriage."

Her step slowed marginally. Working in Walsingham's service, Deven had questioned a variety of dubious men; he had learned to read body language very well. What he read in her hesitation chilled him. "Master Deven... she has made no such request of me."

They walked on a few more strides, Deven's legs carrying him obediently onward, because he had not yet told them to do otherwise.

"No such request," he repeated, dumbly.

The look she gave him was guarded, but compassionate. "If you wish it, I can discover her Majesty's inclination on the matter. I am sure she would not object."

Deven shook his head, slowly. "No... no. That is... I thank you, my lady." The words came out by rote. "I may ask for your good office in this matter later. But I... I should speak to Anne."

"Yes," the countess said softly. "I imagine you should. God give you good day, Master Deven."

OATLANDS PALACE, SURREY
15 March 1590

He did not seek out Anne until the next day. He spent the evening alone in his chamber, sending Ranwell off on a spurious errand. Colsey waited on him alone that night, and was permitted to stay because he would keep his mouth shut.

The place his thoughts led him was not pleasant, but he could not avoid it. And delay would not improve matters. When he had leisure the following afternoon, he went in search of Anne.

She was not with the countess; she had been assigned other tasks that day. Oatlands was a small palace compared with Hampton Court or Whitehall, yet it seemed the proverbial haystack that day, and Anne the needle that kept eluding his search. Not until nearly dusk did he find her, when he went again to check the countess's own chambers, and found her making note of a delivery of books.

He stopped on the threshold, his movement suddenly arrested, and she looked up from her paper. The smile that lit her face made him hope it was all a simple misunderstanding—but he did not believe it.

"We must speak," he said without preamble.

Anne put down the pen and bit her lip. "The countess may return soon; I should—"

"She will forgive you this absence."

A thin line formed between her pale brows, but she rose from her seat. "Very well."

He would not have this conversation inside; there were always ears to overhear, whether they belonged to courtiers or servants of the household. Anne fetched a cloak. Deven had not thought to bring one for himself. Together they went out into the orchard, where the trees only intermittently protected against the spring wind.

Anne walked with him in silence, granting him the time he needed. The words were prepared in his mind, yet they did not come out easily. Not with her at his side.

"I spoke with the countess today."

"Oh?" She seemed guardedly curious, no more.

"About the Queen. About—you, and me, and the matter of our marriage." His cheeks and lips were going cold already. "That was the first she heard of it."

Anne's step slowed, as the countess's had before her.

Deven made himself turn to face her. His gut felt tight, like he was holding himself together by muscle alone. "If you do not wish to marry me, all you need do is say so."

The words were spoken, and she did not immediately dispute them. Instead she dropped her chin, so that her hood half-concealed her face. That gesture, too, spoke clearly to him. He waited, trying not to shiver, and almost missed it when she whispered, "'Tis not that I do not wish to. I cannot."

"Cannot?" He had resigned himself to her cooled affections, or tried to; now he seized on this word with mingled hope and confusion. "Why?"

She shook her head, not meeting his gaze.

"Does your father not approve?"

Another shake of her head. "I—I have no father."

"Are you promised already to another? *Wed* to another, God forbid?" Again she denied it. Deven groped for other possible reasons. "Are you Catholic?"

A wild, inappropriate laugh escaped her, then cut off abruptly. "No."

"Then in God's name, why not?"

He said it louder than he meant to. Anne flinched and turned away, presenting her cloaked back to him. "I—"

Her voice was ragged, like his, but determined. He knew well how strong her will was, but it had never been turned against him before. "I am sorry, Michael. You deserve an explanation, and I have none. But I cannot marry you."

The strained beats of his heart marked the time as he stared at her, waiting for further words, that were not forthcoming. "Now, or ever?"

Another painfully long pause. "Ever."

That flat declaration drained the warmth out of him faster than the bitter air ever could. Deven swallowed down the first three responses that came to his tongue; even now, in his bafflement and pain, he did not want to hurt her, though the urge flared within him. Finally he said, hearing the roughness in his own voice, "Then why did you let me believe you would?"

She turned back at last, and the tears that should have been in her eyes were absent. She had a distant look about her, and though it might simply be how she showed pain, it angered him. Had this meant nothing to her?

"I feared you would leave me, when you knew," she said. "You are your father's heir, and must marry. I did not wish to lose you to another."

The words were too manipulative. He was expected to protest, to tell her there was no other in his heart, and though it was true he would not say it. "If you wished me to stay, then you should not have kept me like a fish on a hook. I believed you trusted me more than that—as I trusted you."

Now tears sparkled at the corners of her eyes. "Forgive me."

He shook his head, slowly. There was some riddle here he could not solve, but he had not the will to untangle it. If he stayed any longer, he would say something he would regret.

Turning, he left her in the dead wilderness of the orchard, with her cloak rippling in the cold wind.

OATLANDS PALACE, SURREY
19 March 1590

The countess was a kind woman, as ladies of the court went. She kept a weather eye not just on the Queen she served, but on the women who served her in turn. It did not escape her that a problem had arisen between her waiting-gentlewoman Anne Montrose and Michael Deven of the Gentlemen Pensioners, and following the revelation of that problem, the two of them had fallen out.

Lune would have preferred Lady Warwick to be less concerned for her well-being. As it was, she almost resorted to faerie magic to convince her human mistress to leave her be. Anne Montrose needed to be upset, but not *too* upset, lest the countess pry too closely; hidden behind that mask, Lune had to shake off the practiced habits of her masquerade, and figure out what to do next.

She cursed herself for the misstep. Originally she had fostered his worry about Elizabeth's possible jealousy because it provided a convenient delaying tactic; their romance was useful, but she could not possibly afford to go through with an actual marriage. And he was not, unfortunately, the sort of courtier to indulge in an illicit affair for years on end without worrying about scandal. It would have been easier if he were. But he soon made it clear he wished to wed her, and so she had to find ways of putting him off.

She should have expected he would speak to the countess directly. She should have known, the first time she lied and said she had asked her mistress to look into the matter, that a time would come when she must produce an answer.

It was a problem she could not solve as Anne Montrose, because Anne loved Deven; the mortal woman she pretended to be would marry him and be done with it. As Lune, her one bitter consolation was that Deven was unlikely to spot fae manipulation at court, now that the nearest fae manipulator had become estranged from him.

But what now? She had no answer to that all-important question. Walsingham had other confidants—Robert Beale, Nicholas Faunt—but if she approached them in her current guise, all she would do was rouse suspicion. The more effective course of action, in the long term, would be to retreat and return under a different glamour and persona, but with the Principal Secretary searching for evidence of Invidiana's hand, Lune could not afford the months it would take to reintegrate herself to any useful extent.

She might have no choice but to resort to more direct

methods: concealment, eavesdropping, theft of papers, and other covert activities. To do so would require extensive use of charms, and so as far as she was concerned they were a last resort—but she might be at that point. Vidar expected her to provide information, and soon.

Deven's absence left a palpable hole in her life. Their duties often kept them apart, but it had become habit to seek out occasions to meet, even if they saw each other only in passing, exchanging a smile while going opposite ways down a gallery or through a chamber. Now she avoided him, and he her. Being near each other was too uncomfortable.

What would she do without him? Not until he was gone did she realise how much she had depended on him. She saw Walsingham twice, at a distance, and fretted over what the Principal Secretary might be doing.

What Deven might be doing. He was, as he had said, Walsingham's hound.

She lay awake late into the night the following Thursday, staring into the darkness as if it would provide an answer. And so she was awake when the countess rose from her bed and reached for a dressing gown.

Anne Montrose whispered, "My lady?"

"I cannot sleep," the countess murmured back, pulling on the padded, fur-trimmed gown. "I need air. Will you walk with me?"

Anne shed her blankets and helped her mistress, fetching a coif to keep her head and ears warm outside. They both slipped on overshoes, then exited the chamber, leaving the other gentlewomen undisturbed.

Deep in the recesses of her mind, where Anne Montrose gave way to Lune, the faerie thought: *Something is wrong.*

The countess did not walk quickly, but she moved with purpose, through the palace and toward the nearest exterior door. Anne followed her, squinting to see in the near-total darkness, and then they were outside, where the air rested unnaturally still.

In the silence, she thought she heard a sound.

Music.

Music intended only for the countess's ears.

Anne Montrose's face took on a wary, alert expression her mistress would have been surprised to see—had she eyes for anything other than the miniature stone tower of the herber up ahead.

Who would summon her? Who would play a faerie song, to lure the Countess of Warwick from her bed and into the shadows of night?

They rounded the herber, and found someone waiting for them.

Orpheus's rangy body was wrapped tenderly around the lyre, his fingers coaxing forth a melody that was still all but inaudible, to all but its intended target. The countess sank to the ground before him, heedless of the damp that immediately soaked into and through her dressing gown; her mouth hung slack as she gazed adoringly up at the mortal musician and listened to his immortal song.

Heavy footsteps squelched in the wet soil behind Lune,

and again, as with Vidar, she realised the truth too late.

The countess was not the target. She was merely the lure, to draw Lune outside, away from mortal eyes.

Lune flung herself to the left, hoping to evade the one behind her, but hands the size of serving platters were waiting for her. She dropped to the ground—the fingers clamped shut above her shoulders, just missing their grip—but then a boot swung forward and struck her squarely in the back, sending her face-first into the dirt.

Two paces away, the countess sat serenely, oblivious to the violence, held by the power of Orpheus's gift.

A knee planted itself in Lune's back, threatening to snap her spine with its sheer weight. She cried out despite herself, and heard a nasty chuckle in response. Her arms were twisted up and back, bound together with brutal efficiency; then her captor hauled her up by her hair and flung her bodily against the stone wall of the herber.

Coughing, stumbling, eyes watering with pain, Lune could still make out the immense and hated bulk of Dame Halgresta Nellt.

"You fucked up, slut," the low, rocky voice growled. Even through the venom, the pleasure was unmistakable. "The Queen forgave you once—Mab knows why. But she won't forgive you this time."

Lune forced her lungs to draw in air. "I haven't," she managed, then tried again. "Vidar knows. Everything I know. What goes on here. I report to him."

Halgresta hadn't come alone. Six goblins materialised

out of the shadows, armed and armoured and ready to catch Lune if she tried to run—as if she could outrun a giant. She had to talk her way out of this.

Talk her way out, with Halgresta. It would be like using a pin to dismantle an iron-bound door.

The giantess grinned, showing teeth like sharpened boulders. "Vidar knows everything, eh? Well, he does now. But not from you."

What? What had Vidar learned? How had he gotten someone closer to Walsingham than she had?

"You lost your toy, bitch." Halgresta's voice struck her like another blow, knocking all the wind out. "You lost that mortal of yours."

Deven.

The stabbing pain in her ribs was subsiding; Lune didn't think anything was broken. She made herself stand straighter, despite her awkwardly bound arms. "I am not finished," she said, with as much confidence as she could muster. "Deven is only one route to Walsingham. There are other ways to deal with the problem—"

Halgresta spat. The wad of spittle hit the countess's shoulder and slid down, unnoticed; Orpheus's melody was still ghosting through the air, plaintive and soft. "Right. Other ways. And other people to take care of them. You? You're coming back to the Onyx Hall."

"Let me talk to Vidar," Lune said. Had she reached such a nadir that he seemed like a thread of hope? Yes. "I am sure he and I can reach an accord."

The giantess leaned forward, until her ugly, stony face was the only thing Lune could see, almost invisible in the darkness. "Maybe you and Vidar could," Halgresta growled. "Who knows what plots you and that spider have hatched. But he's not the one who told me to bring you in.

"The Queen is."

ST. PAUL'S CATHEDRAL, LONDON

7 April 1590

Sir Philip Sidney, late husband of Sir Francis Walsingham's daughter, had been buried in a fine tomb in St. Paul's Cathedral when he died in 1586.

Now the tomb was opened again, to receive the body of Sir Francis Walsingham.

The ceremony was simple. The Principal Secretary had died in debt; his will, found in a secret cabinet in his house on Seething Lane, had requested that no great expenditure be made for his funeral. They buried him at night, to avoid attracting the attention of his creditors.

And so there was no great procession, no men-at-arms wearing matching livery—not even the Queen. She had quarrelled often with Walsingham, but in the end, the two respected one another. She would have come if she could.

Deven stood alongside Beale and others he knew more distantly: Edward Carey, William Dodington, Nicholas Faunt. Some small distance away stood the pale, grieving figures of Ursula Walsingham and her daughter Frances. The gathering was not large.

The priest's voice rolled sonorously on, his words washing over Deven and vanishing up into the high Gothic reaches of the cathedral. The body was placed in the tomb, and the tomb closed over it.

The body. Deven had seen death, but never had he so much difficulty connecting a living man to the lifeless flesh he left behind.

He could not believe Walsingham was dead.

The priest pronounced a benediction. The gathered mourners began to depart.

Standing rooted to his spot, eyes fixed on the carved stone of the tomb, Deven thought bleakly, *Master Secretary—what do I do now?*

ACT THREE

O heauens, why made you night, to couer sinne?
By day this deed of darknes had not beene.

THOMAS KYD
THE SPANISH TRAGEDIE

They dance in intricate patterns, coming together and parting again, skirts and long sleeves swaying a counterpoint to their rhythm. But his ears cannot hear the music, or the sound of their laughter. His world has wrapped him in silence. To his eyes, those around him are ghosts: they dance beneath the earth, which is the realm of the dead, and the dead have no voices with which to speak. Aeneas fed his ghosts blood, and Odysseus, too, but no such heroes exist here. There is no blood that might quicken their voices to life once more.

He hovers against a pillar, entranced and afraid, and the other ghosts stare at him. No—not ghosts. He remembers now. They are alive. They speak, but he cannot hear them. Only whispers, ghost sounds, unreal.

They wonder why he does not speak to them. That is what the living do; they talk, they converse, they prove their existence with words. But where Tiresias was blind, the man who bears his name is mute. He cannot—dares not—speak.

His jaw aches from being clenched tightly shut. Words

beat within him like caged birds, terrified, desperate, fighting to break free, and when he keeps them trapped within, they stab at him with talons and beaks, until he bleeds from a thousand unseen wounds. He cannot speak. If he makes a sound, the slightest sound—

Are you real? *He is desperate to know. If they are real… if he could be sure, then perhaps he would have the courage.*

No. *No courage. It died, broken on the rack of this place. He sees too much of what will come—or what came, what might come, what could never be. He no longer believes in a difference. A difference would mean his choices matter. His choices, and his mistakes. Everyone's mistakes.*

Fire. Fire and ash and blood fill his vision. The dance vanishes. The walls are broken open, stones shattered, the sky brought down to fight the earth. He presses his hands against his head, his eyes, harder, harder, slams himself against the pillar—did he cry out? Fear grips him by the throat. No sound. No sound. Certain words are the wrong words; the only safe words are no words.

They stare at him and laugh, but he hears nothing.

The silence chokes him. Perhaps he should speak, and be done with it.

But no. He cannot do it; he lacks the strength. Too much has been lost. The man he needs is gone, gone beyond recall. Alone, mute, he has no will to act.

She has seen to that.

He curls up on the stone, not knowing where he lies,

not caring, and wraps his trembling hands around his throat. The birds want to fly. But he must keep them safe, keep them within, where they will harm no one but him.

None of this is real. But dreams have the power to kill.

THE TOWER OF LONDON

9 April 1590

The light hurt Lune's eyes, but she refused to let it show. "Tiresias. The Queen's seer. Will you bid—" No, not bid. She had no right to demand such things. "Will you beg him to visit me?"

A harsh laugh answered her. Sir Kentigern Nellt's voice rumbled an octave below his sister's, and was twice as ugly. It matched the rest of him, from his rough-hewn face to the cruelty of his spirit. Whether he even bothered to pass along her requests, Lune did not know, but she had to ask.

Vidar first; she was already in debt to him, but she would have promised more to get out of this cell. He did not come, though. Nor did Lady Nianna, which was no surprise. Lune had been on good terms with the previous Welsh envoy, the bwganod Drys Amsern, but the Tylwyth Teg changed their ambassadors regularly; they did not like anyone to remain for too long under the corrupting influence of the Onyx Court. Amsern was gone. And the Goodemeades had no political influence with which to aid her.

The seer was the last person she could think to ask for.

At a gesture from Sir Kentigern, her goblin jailers

heaved on the heavy bronze door of her cell and swung it shut once more. The resulting blackness was absolute. Her protection against human faith had long since worn off; beyond that, she could not tell how long she had been there. Nor when, if ever, she would get out. The Onyx Hall did not extend beyond the walls of London, but the Tower lay within those walls, and it, too, had its reflection below. These cells were used for people Invidiana was very displeased with. And while a mortal died quickly if you deprived him of food—even more quickly, without water—it was not so with fae. Wasting away might take years.

Sitting in the darkness, Lune thought, *Sun and Moon. When did I become so alone?*

She missed... *everything*. The entire false life she had constructed for herself, torn away in an instant. She missed Anne, which made no sense; Anne had never been real.

But it reminded her of memories long buried. Not just her recent time at Elizabeth's court; that distant, mist-shrouded age—how long ago?—before she came to the Onyx Hall. Lune could no longer recall where she lived then, nor who was around her, but she knew that life had been different. Gentler. Not this endless, lethal intrigue.

She was so very tired of intrigue. Tired of having no one she could honestly call "friend."

"Too much mortal bread," Lune whispered to herself, just to break the silence. A year of it had changed her, softened her. Made her regret the loss of such mortal things as warmth and companionship. She was inventing memories now, losing herself in delusion like Tiresias, pining for a world that could never be.

It wasn't true, though. There *were* fae like that, fae who could be friends. The Goodemeades were living proof of its possibility.

Not in the Onyx Hall, though.

And if Lune wanted to survive, she could not afford to indulge such fancies. She gritted her teeth. To escape the intrigue, first she had to scheme her way free. The only way out was through.

Keys rattled outside her cell. The lock clanked and thunked, and then the familiar protest of the hinges as the door swung open. Light flooded through. Lune stood, using the wall to steady herself, and looked flinchingly toward the opening, raising her gaze a degree at a time as her eyes could bear the light.

A figure came through. Not twisted enough for a goblin. Not tall enough for Kentigern. But not Tiresias, either. The silhouette had a broad, triangular base: a woman, in court dress.

"Leave the door open," an accented and melodious voice said. "She will not flee."

The goblin outside bowed, and stepped back.

Lune's eyes were adjusting at last. Her visitor moved to one side, so she was no longer backlit, and with a confused shock Lune recognised her. "Madame," she said, and sank into a curtsy.

The ambassador from the Cour du Lys seemed all the

more immaculate for her dirty surroundings, wearing a crystalline gown in the latest fashion, her lovely copper hair curled and swept up under a pert little hat. Malline le Sainfoin de Veilée eyed Lune's filthy skin with distaste, but inclined her head in greeting. "The *chevalier* has given me permission to speak with you, and a promise of discretion."

Kentigern would probably keep that promise, if only because he was not subtle enough to seek out a buyer for his information. "Is discretion needed, *madame ambassadrice*?"

"If you choose to accept the bargain I offer you."

Lune's mind felt as rusty as the door hinges. That the French envoy was offering her help, she could understand, but why? And what did she want in return?

Madame Malline did not explain immediately. Instead she snapped her jewelled fingers, and when a head appeared in the doorway—a sprite belonging to the embassy, not the goblin jailer—she spoke imperiously in French, demanding two stools. A moment later these were brought in, and Madame Malline arrayed herself on one, gesturing for Lune to take the other.

When they were both seated, and Madame Malline had arranged her glittering skirts to her satisfaction, she said, "You may know I am in negotiations with your Queen, regarding the conflict with the Courts of the North, and which side my king will take. You, Lady Lune, are not valuable enough as a bargaining piece to be worth much in that debate, but you are worth a little. I am prepared to offer Invidiana certain concessions of neutrality—minor ones, nothing more—in exchange for your freedom from this cell."

"I would be in your debt, madame," Lune said reflexively. Sitting in the light, on a cushioned stool, had warmed up the stiff muscles of her mind. She remembered now how she might repay that debt.

She hoped the ambassador meant something else.

"Indeed," the French fae murmured. "I know of you, Lady Lune, though we have not spoken often. You have more between your ears than fluff; you would not have survived for so long were it not so. You know already what I will ask."

Lune wished desperately for a bath. It seemed a trivial thing, her unwashed state, when laid against her political predicament, but the two were not unconnected; grimy, with her hair straggling around her face in strands dulled from silver to grey, she felt inferior to the French elf. It would undermine her in the bargaining that must come.

But she would do her best. "My Queen," Lune said, "also knows what you will ask. I would be foolish indeed to betray her."

Madame Malline dismissed this with a wave of one delicate hand. "*Certainement*, she knows. But she, knowing, permitted me to come here. We may therefore conclude that she does not see it as a betrayal."

"What we may conclude, *madame ambassadrice*, is that it amuses her to grant me enough rope with which to hang myself." Lune gestured at the walls of her cell.

"That I have been kept here means she has not made up her mind to destroy me. This might be her way of making her decision: if I tell you more than she wishes me to, then she will declare it treason and execute me."

"But then the bird would already have flown, *non*? I would have the information she does not wish me to have. Unless you suggest she would strike you dead even as you speak."

Invidiana could do it, with that black diamond jewel. But she had not used it on Lune, and the envoy had a point; by the time Lune was dead, the information would already have been passed on. And despite everything, Lune did not think Invidiana would breach protocol so inexcusably as to kill a foreign ambassador on English soil. She often bent the rules of politics and diplomacy until they wept blood, but to break them outright—especially during such negotiations—would ensure an alliance against her that even the Onyx Court could not survive.

It was a slim enough thread on which to hang her life. But what was her alternative? Invidiana might yet decide to kill her anyway—or worse, forget her. One day the door would cease to open, and then Lune would dwindle to nothingness, alone in the dark, screaming away her final years inside a stone box.

"I am willing to negotiate," she said.

"Bon!" Madame Malline seemed genuinely pleased. "Let us speak, then. I will have wine brought, and you will tell me—"

"No." Lune cut her off as the French elf raised her hand to summon a servant again. She stood and straightened her skirts, resisting the urge to brush dirt off them. It wouldn't help, and it would make her look weak. "You secure my release, and *then* I tell you what I know."

The warmth in the ambassador's smile dwindled sharply at the insinuation. She could not be surprised, though; distrust and suspicion were the daily bread of the Onyx Court. And indeed, she played the same card in return. "But once you are free of this cell, what is to reassure me *I* will have what *I* seek?"

Lune had not expected so obvious a trick to succeed, but it had been worth a try. "I will tell you some things now, and more once I am free."

Madame Malline pursed her full lips, considering it. "Tell me, and I will see to it you are moved to a better cell, and so on from there."

The alternative was to give her word, and Lune's half of the bargain was necessarily too vague for that to work. "Very well."

Servants appeared again. One sprite poured the wine, while another bowed deeply and presented Lune with a platter of fresh grapes. She made herself eat these slowly, as if she did not really need them. Negotiations were not over. She still could not afford to look weak.

"The folk of the sea," she said when the sprites had bowed and retreated to the edge of the room. "They take offense if you call them fae, and in truth I do not know if they are. 'Tis a question for philosophers to debate. I went among them for politics."

Madame Malline nodded. "The mortal Armada, yes."

"They are a secretive people; they do not welcome commerce with outsiders, and reckon themselves to have little care what goes on above the surface of the water. Indeed, in some cases they bear hostility toward those who live on land." Such as, for example, the Cour du Lys, the strongest faerie court in the north of France. Lune did not know what offense had been committed there, but she knew there had been one. She would have to be careful not to offer the ambassador any information that might be useful in healing that breach.

But her explanations had to seem natural and unaffected. "They would not speak directly to anyone who lives on the surface," Lune said, "but they will talk to our river nymphs, sometimes. We had occasional contact through the estuary at Gravesend. It was through this that Invidiana arranged for my embassy. They agreed to let me come among them; I do not know what she promised them for that concession."

"Did you go alone?" the envoy asked.

"Two of the estuary nymphs accompanied me, their tolerance for saltwater being higher than their river-bound sisters. Beyond that, I was served by the folk of the sea."

"And how did you go among them?"

She could still feel the air whistling past her cheeks, the gut clench of fear that this had all been some cruel jest of

Invidiana's. Lune closed her eyes, then made herself open them and meet Madame Malline's gaze. "I leapt from the cliffs of Dover. And that, *madame ambassadrice,* is all I will say for now." She rose, stepped clear of her stool, and spread her soiled skirts in a curtsy. "If you would know more, then show me what you can do on my behalf."

Madame Malline studied her, then nodded thoughtfully. "*Oui,* Lady Lune. I will do so. And I look forward to hearing the continuation of your tale."

A moment later she was gone, and the door closed again, blocking out all light. But a stool stayed behind, a promise of assistance to come.

ST. JAMES' PALACE, WESTMINSTER
10 April 1590

In the end, his urine came forth at his mouth and nose, with so odious a stench that none could endure to come near him.

The report crumpled abruptly in Deven's hand; he made his fingers unclench. Laying the paper on the table, he smoothed it out, and suppressed the urge to fling it in the fire.

Walsingham was barely in his tomb, and already the Catholics were rejoicing, and spreading damnable rumours in their glee. They made of the Principal Secretary's death something so utterly vile—

The paper was creasing again. Deven snarled and turned his back on it.

He did so in time to see Beale enter the room. The older man looked as if he had not slept well the previous night, but he was composed. Beale's gaze flicked past Deven to the battered report.

"They saw him as their chief persecutor," he said quietly, brushing a strand of greying hair out of his eyes. "So terrible a figure cannot die like an ordinary man, and so they invent stories, which confirm their belief that he was an atheist and font of rank corruption."

Deven's jaw ached as he moved it, from having been clenched so tight. "No doubt there will be a festival in Spain, when the news reaches Philip."

"No doubt." Beale came farther into the room, sought out a chair and sank into it. "While here we mourn him. The English Crown has lost a great supporter. A great man."

At the moment when we need him the most.

The thought was casual, reflexive—and then the implications struck him.

His jerk of movement drew Beale's eye. "Indeed," Deven said, half to himself. "The Catholics are very glad of it. But he said he did not think the guilty party was Catholic."

Beale frowned. "'Guilty party'?"

Deven turned to face him, driven by a sudden energy. "He must have spoken of it to you—he held you in great trust. A hidden player, he told me, scarcely a month gone. Someone with a hand in our court, who operates in secret."

"Ah," Beale said, and his frown deepened. "Yes."

"Do you doubt him?"

"Not entirely." Beale's hands moved to straighten the papers scattered over the desk, as if they needed something to do while his brain and mouth were otherwise occupied. "He told you of the Queen of Scots, I presume? In that matter, I agree with him. I was closely involved with certain parts of that affair, and I do believe someone was influencing the Queen. Regarding the recent events with Perrot... I am not so sure."

Disregarding this latter part, Deven said, "But you do believe there is such a player."

"Or was. He may be gone now."

"Walsingham set me to hunt this man. He hoped fresh eyes might see what his could not. And now he's dead."

The paper shuffling stopped, as Beale saw the mark he aimed at. "Deven," he said, clearly choosing his words with care, "Sir Francis has—had been sick for a long time. Think of his absence last year. This is not a new-sprung development, risen out of nowhere in the last month."

"But if the hidden player *is* still around, and is involved with the Irish matter—"

"If, if," Beale said impatiently. "I am not convinced of either. And even were it so, why not eliminate *you*? After all, you are the one up to your eyebrows in the trouble surrounding Perrot. If anyone was about to uncover the secret, it would be you."

Deven snorted. "I do not have so high an opinion of

myself as to think I pose a greater threat than Walsingham. If I did not uncover it, someone else would, and pass it along to him."

Beale rose and came around the corner of the table to take him by the shoulders. "Michael," the older secretary said, soft but firm. "I know it would be easier to believe that someone poisoned or cursed Sir Francis, and brought about his untimely death. But he was a sick man, one who had shaken off illness often before in his determination to continue his work. He could not do so forever. God willed it that his time should end. That is all the explanation there is."

The grip on his shoulders threatened his self-control. Just a short month before, Deven had seen before himself a bright and intriguing future, with both a patron and a wife to lend it purpose. Now he had no prospect of either.

All he had was the duty the Principal Secretary had laid upon him.

Deven stepped back, out of Beale's hands. His voice came out steadier than he expected as he said, "No doubt you are right. But it does not answer the matter of this hidden player. You do not know if he is still around, but you also do not know that he is gone. I intend to find out. Will you help me?"

Beale grimaced. "As I may. Sir Francis's death has put matters into disarray. If anything is to be preserved of the work he has done, the agents and informers he acquired, I'll have to find someone else to take them on."

This broke through the desolate fog that had gripped

Deven's mind. He had not thought of that, but of course Beale was right; only someone well placed on the privy council could make good use of Walsingham's people. "Did you have someone in mind?"

"Burghley has made overtures, which I expected. But Essex also expressed an interest."

"Essex?" Deven knew it was disrespectful, but he could not repress a snort. "He hasn't the patience for intelligence work." Or the mind.

"No, he hasn't. But he married Sir Francis's daughter."

"*What?*"

Beale sighed heavily, sitting once more. "In secret. I don't know when, and I don't know if Sir Francis knew. But Essex told me, as a means of strengthening his position." His tired eyes shifted back up to Deven. "Do not tell the Queen."

"And risk her throwing a shoe at me? I think not." Essex had been her favourite since his stepfather Leicester's death, though God alone knew why. The Queen's affection was easy enough to understand; she was in her late fifties, and Essex not yet twenty-five. But Deven did not believe the man held much affection for his sovereign. Elizabeth might still be admired for her wit and political acumen, but not for her beauty, and Essex did not seem the type to love her mind. His affection would last precisely as long as the tangible rewards of her favour.

"Unfortunately," Beale went on, "when all is said and done I cannot pass on everything intact, even if Burghley or Essex concedes the ground to the other. Too much of it was in Sir Francis's head, and never committed to writing. Even I do not know who all his informants were."

With that, Deven could not help. Walsingham had never shared all his secrets with anyone, and now, without a proper patron, Deven lacked the influence to be of use politically. He would have to scrabble hard for favour and preferment.

Unless...

If Walsingham was right, and the hidden player had occasional direct access to the Queen, then Elizabeth certainly knew who it was. But was she pleased with that situation? Knowing her distaste for being managed by her councillors, no. If Deven could uncover the man's identity, and use the knowledge to break his influence...

He hadn't Essex's beauty. But he did not want the burden of being Elizabeth's favourite; all he wanted was her favour.

This might earn it for him.

Deven settled himself back into his chair, and shoved the report of Catholic rumours aside without looking at it. "Tell me," he said, "what you know of the hidden player."

THE ONYX HALL, LONDON

9–12 April 1590

The improvements to her circumstances came one tantalising step at a time. First it was the stool, left in her cell, followed shortly by a torch and a pallet on which to sleep. Then removal to a better cell, one that did not lie at the roots of the White Tower. To earn that one, Lune had to tell Madame Malline of her leap from the cliffs of Dover, plummeting three hundred feet into the choppy waters of the English Channel. It was no jest: the strangely shimmering pearl she'd been given to swallow permitted her to survive underwater, though not to move with the grace of her nymph escorts, or the merfolk who waited for her below.

The merfolk. The roanes. The evanescent sprites born from the spray of the crashing waves. Stranger things, in deeper waters. She did not see the Leviathan itself, but lesser sea serpents still occasionally haunted the Channel between England and France. Were the folk of the sea fae? What defined fae nature? They were alien, enchanting, disturbing, even to one such as Lune. No wonder mortals told such strange stories of them.

But they were little touched by human society. That, she told Madame Malline, was the most difficult thing about them. Those fae who dwelt in the cracks and shadows of the mortal world did so because of their fascination with humans and human life. The Onyx Court was only the most vivid proof of that fascination, the most intensive mimicry of mortal habits. The folk of the sea were more like the inhabitants of deeper Faerie, less touched by the currents of change. But at least those who dwelt in Faerie breathed air and walked on the earth; beneath the waves lay a world where up and down were little different from north or east, where events flowed according to inscrutable rhythms.

Even speaking of them, she fell back into the metaphors of speech she had acquired there, likening everything to the subtle behaviour of water.

That information got Lune into a more comfortable cell. A primer on the diplomacy of underwater society took her back to her own chambers, where she lived under house arrest, with Sir Prigurd Nellt instead of Sir Kentigern commanding the guards that bracketed her door.

Then came the final negotiation, the one she had been anticipating for some time.

"Now," Madame Malline said when they had dispensed with the pleasantries, "you know what I wish to hear. Stories of how you went to the sea, what you found there—these are interesting, and I thank you for them. But I have shown you my goodwill in helping you thus far, and the time has come for you to repay it."

They were seated by the fire in Lune's outer chamber, with glasses of wine at hand. Not the fine French vintage Vidar had offered the day he set Lune on Walsingham's trail, but a good wine nevertheless. Lune could almost ignore the way her chambers had been ransacked after her downfall, her charms breached, her jewels and her little store of mortal bread stolen away by unknown hands.

"Au contraire, madame ambassadrice," Lune said, dropping briefly into the envoy's own tongue to soften the rudeness she was about to offer. "Secure my freedom; have the guards removed from my door. Then I will give

you the information you seek."

Madame Malline's smile was beautiful and utterly without warmth. "I do not think so, Lady Lune. Should I do so, there would be nothing save gratitude that binds you to help me further. And though grateful you may be, when weighed against your fear of angering Invidiana..." She lifted her wine goblet in one graceful, ringed hand, and her smile turned just the faintest bit malicious. "*Non.* You will tell me, and take your chances with your Queen."

All as Lune had expected. And, in a way, as she had needed.

"Very well," she said, letting the words out reluctantly. "You wish to know, then, what I agreed to. What price I offered them, in exchange for their assistance against the Spanish Armada."

"Oui."

"Peace," Lune said.

One delicately plucked eyebrow arched upward. "I do not understand."

"The folk of the sea do not ignore everything that goes on in the air. I do not know who spread the rumours; perhaps a draca or other water spirit eavesdropped on someone's indiscreet conference, then spoke to another, and so on until the news flowed downriver and reached them. Invidiana intended to make war against the folk of the sea. And the concession I offered them was an agreement to abandon that course."

Madame Malline studied her, eyes narrowed and full lips

pursed. At last she said, meditatively, "I do not believe you."

Lune met her gaze without flinching. "It is true."

"Your Queen has an obsession with mortal ways, mortal power. Even her wars against the Courts of the North have their origin in mortal affairs, the accusation that she sabotaged the Queen of Scots. There is no human court out on the water. Why should Invidiana desire power over the folk of the sea? What cares she for what they do beneath the surface?"

"She cares nothing for it," Lune said. "But she cares a great deal for what the folk of the sea can do about mortals *on* its surface. Had her Majesty's plans moved more quickly, we would not have had to negotiate for their assistance against the Armada, but she was not yet fully prepared to assert sovereignty over the undersea. If she had done so... imagine what she could do, were they bound to obey her." Lune paused, to let Madame Malline consider it. Break the back of Spanish shipping. Give fair weather to English vessels, and foul to their enemies. Strike coastal areas with crippling storms.

The ambassador's mind quickly moved ahead to the next complication. "But you have told me yourself that they are not organised, they have no *Grand Roi*—that you had to bargain with a dozen nobles of one sort or another to reach any agreement. Even the Courts of the North have unified themselves more than that. Your Queen could not hope to control the oceans."

"She would not need to. A small force would do. They

are highly mobile, the folk of the sea, and adapt with speed; a few dedicated, obedient groups would be able to wreak quite enough havoc to suit her purposes."

Lune took advantage of the pause to reach for her own wine and conceal her face behind the rim. Madame Malline was staring into the fire, clearly working through the ramifications of this. Searching for a way to turn it to the benefit of the Cour du Lys. They had their own conflicts with Spain, with Italy, with heathen fae across the Mediterranean Sea.

Surely Madame Malline could work out why Lune would fear Invidiana's retaliation, once she had spoken.

The French elf's eyes finally moved back to Lune's face. "I see," the ambassador said, her voice slightly breathless. "I thank you, Lady Lune. Your Queen has listeners on this room, of course, but I have paid them off. For your honesty, I will do more than have you freed; I will also protect you from her retaliation. She will hear from her spies that you told me a persuasive lie. I cannot promise it will be enough, but it is all I may do."

Lune smoothed the lines of worry from her own face. Rising from her seat, she curtsied to the envoy. "You have my most humble thanks, *madame ambassadrice*."

* * *

THE STRAND, OUTSIDE LONDON

13 April 1590

The list Beale gave Deven was depressingly short.

Gilbert Gifford had been granted a handsome pension of a hundred pounds a year for his work in passing along the letters of the Queen of Scots, but Thomas Phelippes had reported more than two years ago that he'd been arrested by French authorities and slung in prison. So far as Beale knew, Gifford was still there. By all accounts, he was as untrustworthy and mercenary a man as Walsingham had ever hired; rumours said he'd later tried to arrange Elizabeth's murder with Mendoza, the former Spanish ambassador to England. He might well have been serving another master. But Deven could not very well question him when he was in a French jail. And his cousin among the Gentlemen Pensioners, though a dubious character in his own right, was not useful to Deven.

Henry Fagot was another informer Walsingham had suspected of coming too easily to hand, but he was even less accessible than Gifford; no one knew who he had been. He had passed information out of the French embassy some six or seven years before, but hid behind a false name. The potential suspects, of course, were long gone from England.

And those were his two strongest prospects. From there, the list degenerated even more. Some individuals were dead; others were gone; others weren't individuals at all, but rather suspicions of "someone in the service of Lord and Lady Hereford," or leads even less concrete than that.

This was the information Walsingham had not given him, for fear of prejudicing his mind and leading his thoughts down paths others had already explored. Having considered it, Deven had to agree; the past would not give him the answer. He had to look at the present. If Walsingham was right, and the player was still active, with a hand in the Irish situation... a great many ifs, as Beale said. But what other lead could he follow?

Nothing save his suspicions about Walsingham's death. And Beale had argued well against those.

Carrying a message from a council meeting at Somerset House to St. James' Palace, his cloak pulled tight around him in feeble protection against a driving rainstorm, Deven abruptly remembered Beale's words.

"I know it would be easier to believe that someone poisoned or cursed Sir Francis..."

Poison, no. But Deven could think of at least one man who might have the capacity to bring about a man's death through infernal magic.

Doctor John Dee.

He raised his head, heedless of the water that streamed down his face, and stared blindly through the grey curtain of rain. Dee. A necromancer, they said, who trafficked with demons and bound spirits to his will. But also Walsingham's friend; would Dee have betrayed him so foully?

There were other problems. Dee had been on the continent for six years—six crucial years, in the tale of the

Queen of Scots. But Fagot's work in the embassy had begun around the time that Dee departed. And Gifford, too, had conveniently shown up in that time.

Could they have been working for the astrologer, while he was abroad?

Someone had persuaded Elizabeth, possibly by meeting with her in person. Dee could not have done that, unless someone had gone to a great deal of effort to fabricate rumours about his travels with Edward Kelley. It was a stretch to imagine the man working so effectively through intermediaries. And what would Dee care about events in Ireland?

Deven shook his head, sending water flying. Beneath him, his bay gelding kept stolidly putting one foot down after another, ignoring both the rain and the preoccupation of his rider. Too many questions without answers—but it was the strongest possibility yet. Before his departure for the continent, Dee had spun out grand visions of England's destiny in the world, with Elizabeth upon the throne. The Queen of Scots would have been an obstacle to those visions, one he might take steps to remove.

And perhaps his difficulties now stemmed, at least in part, from Elizabeth's disillusionment over how she'd been managed into killing her Scottish cousin.

What did Deven know about Dee's activities now, the positions and benefits for which he was petitioning the Crown?

The answers came obediently to mind—and with them, something else. The reason why he knew those answers.

Anne.

"'Tis listening, not spying, and you are not asking me. I do it of my own free will."

Yes, she had volunteered information on Doctor Dee quite eagerly. Deven knew all about the man's penury, the theft of books and priceless instruments from his house at Mortlake, the dispute with his wife's brother over the ownership of that house. Even Burghley's attempts to get Dee's confederate Edward Kelley back to England, so he could put his Philosopher's Stone to work producing gold for Elizabeth. Information Deven had taken in and set to one side, because he could not see what to do with it.

The thought of Anne twisted like a knife in him. They hadn't spoken since that confrontation in the orchard; shortly thereafter, according to the Countess of Warwick, Anne had begged and received permission to leave her service. Deven did not know why, nor had he asked; the subject was too painful, the unresolved questions between them too sharp. These thoughts, however, cast the entire situation in a new and unpleasant light.

What she could possibly be doing in Dee's service, he did not know. But if Dee were the player...

More ifs. He had so few names to chase, though. And going after Dee directly would not be wise.

Was he thinking of this because he truly suspected Anne, and thought finding her would accomplish something? Or did he just wish to see her again?

"A bit of both," he admitted out loud, to no one in

particular. The gelding flicked his ears, scattering droplets of rain.

By the time he arrived at St. James' Palace, drenched and shivering, he had made up his mind. He stopped to change clothes only because it would not do to drip on the floor of a peer.

The Countess of Warwick frowned when Deven asked what reason Anne had given for leaving. "She did not speak of your argument, though I suspect that played a part. No, she named some other cause..."

Deven stood in his wet hair and dry clothes, and tried not to chafe with impatience.

"'Tis hard to recall," the countess admitted at last, looking embarrassed. "I am sorry, Master Deven. An ailing family member, perhaps. Yes, I remember, that was it—her father, I believe."

"I have no father," Anne had said, when he asked her why she could not marry.

So either she had lied to the countess, or to him. And she had lied to him before.

He put on a look of solicitous concern. "I am very sorry to hear it. Perhaps it was concern for her father that led to our troubles. Do you know where her family lives? I have been given a leave of absence from my duties; I might call upon her, to offer my sympathies if nothing else."

The countess's confusion melted away, and she smiled indulgently at him, no doubt thinking of young love. "That would be very kind of you. She is London-born, from the parish of St. Dunstan in the East."

Little more than a stone's throw from Walsingham's house, south and west along Tower Street. Deven would have ridden to Yorkshire, but he need not go far at all.

"I thank you, my lady," Deven said, and left with all the haste decency would allow.

THE ONYX HALL, LONDON

14 April 1590

Though almost everything of value had been stripped from Lune's chambers following her disgrace, her gowns remained. No one, apparently, wanted to be seen wearing the clothing of a traitor imprisoned beneath the White Tower.

She dressed herself in raven's feathers, simple but elegant, with an open-fronted collar and cuffs that swept back from her hands in delicate lacework. Now, of all times, she wanted to show her loyalty to Invidiana by wearing the Queen's colours. The plain pins holding up her silver hair were her only adornment; humility, alongside loyalty, would be her watchword tonight.

When she was ready, she took a steadying breath, then opened the door to her chamber and stepped outside.

"Are you ready?" Sir Prigurd asked in his resonant bass voice, and waited for her brief curtsy. "Come along, then."

Two guards accompanied them through the palace. Lune was not taken to the presence chamber. A good sign, or a bad one? She could only speculate. Prigurd led her onward, and soon Lune knew where they were going.

The Hall of Figures was a long gallery, sunken below the usual level of the rooms by the depth of a half-flight of stairs. Statues lined it on both sides, ranging from simple busts to full figures to a few massive works large enough to fill a small chamber on their own. Some were made by mortal artisans, others by fae; some had not been crafted at all, unless the basilisk could be called a crafter.

Lune prayed the stories were not true, that Invidiana kept a basilisk in some hidden confine of the Onyx Hall.

Prigurd and the guards stayed on the landing at the top of the stairs. Lune went down alone. As her slipper touched the floor, she saw movement out of the corner of her eye; she flinched despite herself, thinking of basilisks.

No monster. In her distraction, she had simply taken the man for a statue. The mortal called Achilles had more to recommend him to Invidiana than just his battle furies; his nearly naked body might have been a sculpted model for the perfection of the human form.

He took her by the arm, his hard fingers communicating the violence that always trembled just below the surface. Lune knew better than to think it directed at her, but she also knew better than to think herself safe from it. She offered no resistance as Achilles led her down the gallery, past the watching statues.

A chair had been placed partway down the Hall of Figures, and a canopy of estate erected above it. Before Lune came anywhere near it, she sank gracefully to her knees—as gracefully as she could, with Achilles still holding one arm in an iron grip.

"Bring her closer."

The mortal hauled Lune to her feet before she could stand on her own, towed her forward a few steps, and shoved her down again.

The moments passed by in silence, broken only by breathing, and a scuff at the entrance to the gallery as Sir Prigurd shifted his weight.

"I am given to understand," Invidiana said, "that you have been telling Madame Malline lies."

"I have," Lune said, still kneeling in a sea of raven feathers. "More than she realises."

A few more heartbeats passed; then, on some unspoken signal from the Queen, Achilles released Lune's arm. She remained kneeling, her eyes on the floor.

Invidiana said, "Explain yourself."

There was no point in repeating the early steps of it; Invidiana knew those already. She might even know what Lune had said at the end. But that was the part she wished to hear, and so Lune related, in brief, honest outline, the lie she had told the ambassador. "She believed me, I think," Lune said when she was done. "But if she does not, 'tis no matter; the lie tells her nothing she can use."

"And so you gained your freedom," Invidiana said. Her voice was as silken and cold as a dagger of ice, that could kill and then melt away as if it had never been. "By slandering your own sovereign."

Lune's heart thudded painfully. "Your Majesty—"

"You have spread a lie that will damage my reputation in other lands. You have given the *ambassadrice du Lys* information about the undersea that might be turned against England. You have sold details of a royal mission, for the sake of your own skin." The whip crack of her words halted. Invidiana murmured the next part softly, almost intimately. "Tell me why I should not kill you."

Feathers crumpled in her fingers, their broken shafts stabbing at her skin. Lune's heart was beating hard enough to make her body tremble. But she forced herself to focus. Invidiana was angry, yes, but the anger was calculated, not heartfelt. A sufficiently good reply might please the Queen, and then the rage would vanish as if it had never been.

"Your Majesty," she whispered, then made her voice stronger. "When those in other lands hear that you dream of extending your control over the folk of the sea, they will fear you, and this is no bad thing. As for Madame Malline, indeed, I *hope* she tells her king what I have said, and he attempts to pursue it; if he threatens war undersea, thinking to win himself some concession thereby, then we will have the pleasure of watching those proud and powerful folk destroy him. Moreover, by satisfying her with this lie, I have ended her prying questions, that might otherwise have uncovered the truth of my embassy, and the secrets I have kept on your Grace's behalf."

Having offered her political reasons, Lune risked a

glance upward. A flash of white caught her eye, and she found herself meeting an unfocused sapphire gaze. Tiresias knelt now at Invidiana's feet, leaning against her skirts as a hound might, with her spidery fingers tangled in his black hair. He wore no doublet, and the white of his cambric shirt blazed in the darkness of the hall.

She swallowed and lifted her chin higher, fixing her attention just below Invidiana's face. "And if I may be so bold as to say it, your Majesty—no fae who cannot find a way to benefit herself while also serving the Onyx Throne belongs in your court."

Invidiana considered this, one hand idly stroking Tiresias's hair. He leaned into the touch, as if there were no one else present.

"Pretty words," the Queen said at last, musingly. She tightened her grip on Tiresias, dragging his head back until he gazed up at her, mouth slackened, throat exposed and vulnerable. The Queen gazed down into her seer's eyes, as if she could see his visions there. "But what lies behind them?"

"Your Grace." Lune risked the interruption; silence might kill her just as surely. "I will gladly return to the service I left. I told Dame Halgresta I had other options available to me; give me my freedom, and I will discover all you wish to know about Walsingham."

Tiresias laughed breathlessly, still trapped by Invidiana's hand. "A body in revolt, the laws of nature gone awry. It cannot happen. Yet the stories say it did, and are not stories true?" One hand rose, as if seeking something; it faltered mid-air, came to rest below the unlaced collar of his shirt. "Not those that are lies."

His words hardened Invidiana's black eyes. She trailed one fingernail down the seer's face; then her hand moved to hover near the jewel in the centre of her bodice, the black diamond edged by obsidian and mermaid's tears. The sight transfixed Lune with fear. But when the Queen scowled and returned her attention to Lune, she left the jewel where it was pinned. "Walsingham is no longer a problem. You may be. But I am loathe to cast aside a tool that may yet have use in it, and so you will live."

Lune immediately bent her head again. "I am most grateful for—"

"You will live," Invidiana repeated in honeyed, venomous tones, "as a warning to those who might fail me in the future. Your chambers are no longer your own. You may remain in the Onyx Hall, but for hospitality you will be dependent upon others. Anyone giving you mortal food will be punished. If hands turn against you, I will turn a blind eye. Henceforth you are no lady of my court."

The words struck like hammer blows on stone. Lune's hands lay slack and nerveless in her lap. She might have wept—perhaps Invidiana wanted tears, begging, a humble prostration on the floor, a display of sycophantic fear. But she could not bring herself to move. She stared, dry eyed, at her Queen's icy, contemptuous face, and tried to comprehend how she had failed.

"Take her," the Queen said, her voice now indifferent,

and this time Achilles truly did have to drag Lune to her feet and out of the hall.

MEMORY
6 April 1580

It began as a trembling, a rattling of cups and plates on sideboards, a clacking of shutters against walls.

Then the walls themselves began to shake.

People fled into the streets of London, fearing their houses would fall on them. Some were killed out there, as stones tumbled loose and plummeted to the streets. Nothing was exempt: a masonry spire on Westminster Abbey cracked and fell; the Queen felt it in her great chamber at Whitehall; across all of southern England, bells tolled in church steeples, without any hand to ring them.

God's judgment, the credulous believed, was come to them at last.

The judgment, though, did not come from God—nor was it intended for them.

Out in the Channel, the seabed heaved and the waves rose to terrifying heights. The waters swamped all under, with no respect for country; English, French, and Flemish, all drowned alike as their ships foundered and sank.

Some few were close enough to see the cause of the tremor, in the short moments before their death.

The bodies struck the waves with titanic force. Those

few, hapless sailors saw colossal heads, hands the size of cart horses, legs thicker than ancient trees. Then the waters rose up, and they saw nothing more.

At Dover, a raw white scar showed where a segment of the cliff had cracked and fallen in the struggle.

In the days to come, mortals on both sides of the Channel would feel the aftershocks of the earthquake, little suspecting that beneath the still unsteady waves, terrible sea beasts were tearing at the corpses of Gog and Magog, the great giants of London, who paraded in effigy through the streets of the city every Midsummer at the head of the Lord Mayor's procession.

Rarely did the conflicts of fae become so publicly felt. But the giants, proud and ancient brothers, had long refused to recognise any Queen above them, and Invidiana did not take kindly to rebellion. Some said she had once been on friendly terms with them, but others scoffed; she had no friends. At most, they might have once been useful to her.

Now their use had ended.

Giants could not be disposed of quietly. She sent a legion of minions against them, elf knights and hobyahs, barguests and redcaps from the north of England, and the brutal Sir Kentigern Nellt to lead them. On the cliffs of Dover the battle had raged, until first one brother and then the other fell to their opponents. In a final gesture of contempt, Nellt hurled their bodies into the sea, and shook the earth for miles around.

While the mortals cowered and prayed, the warriors

laughed at their fallen enemies. And when the waves had subsided and there was no more to see, they retired to celebrate their bloody triumph.

TOWER WARD AND FARRINGDON WITHOUT, LONDON

15 April 1590

A monumental, stone Elizabeth gazed down on Deven as he rode up Ludgate Hill toward the city wall, making him feel like a small boy that had been caught shirking his duties. He had leave from the lieutenant of the Gentlemen Pensioners to be absent that day, but still, he breathed more easily when he and Colsey passed through the gate, with its image of the Queen, and into London.

The rains that had deluged the city of late had washed it moderately clean for once. The smaller streets were still a treacherous sludge of mud, but Deven kept to wider lanes, where cobbled or paved surfaces glistened after their dousing. Only when he turned north onto St. Dunstan's Hill did he have to be careful of his horse's footing.

In the churchyard, he halted and tossed his reins to Colsey. He cleared the steps leading to the church door in two bounds, passing a puzzled labourer who was scrubbing them clean, and went inside.

The interior of the church was murky, after the rain-washed brilliance outside. Deven's eyes had not yet adjusted when he heard a voice say, "How may I be of service, young master?"

The words came from up ahead, on his left. Deven turned his head that way and said, "I seek a parishioner of yours, but I do not know where the house lies. Can you direct me?"

"I would be glad to. The name?"

His vision had cleared enough to make out a balding priest. Deven said, "The Montrose family."

The priest's brow furrowed along well-worn lines. "Montrose... of this parish, you said?"

"Yes. I am searching for Anne Montrose, a young woman of gentle birth, who was until recently in service to the Countess of Warwick."

But the priest shook his head after a moment of further thought and said, "I am sorry, young master. I have no parishioners by that name. Perhaps you seek the church of St. Dunstan in the West, outside the city walls, near to Temple Bar?"

"I will ask there," Deven said mechanically, then thanked the priest for his assistance and left. The countess would not have confused the two parishes. Yet some vain hope made him ride a circuit around St. Dunstan's, asking at all the churches that stood near it, then cross the breadth of the city again to visit the other St. Dunstan's, which he had passed on his way in from Westminster that morning.

Only one church, St. Margaret Pattens, had any parishioners by the surname of Montrose: a destitute family with no children above the age of six.

Colsey stayed remarkably silent through this entire enterprise, given how Deven had told him nothing of the day's purpose. When his master emerged from St. Dunstan in the West, though, the servant said tentatively, "Is there aught I can do?"

The very hesitance in Colsey's voice told Deven something of his own expression; in the normal way of things the man never hesitated to speak up. Deven made an effort to banish the blackness he felt to somewhere less public, but his tone was still brusque when he snapped, "No, Colsey. There is not."

Riding back along the Strand, he wrestled with that blackness, struggling to shape it into something he could master. Anne Montrose was false as Hell. She had lied to her mistress about her home and her family. Doubtless she was not the only one at court to have hidden inconvenient truths behind a falsehood or two, but in light of the suspicions Deven had formed, he could not let the trail die there.

The ghost of Walsingham haunted his mind, asking questions, prodding his thoughts. So Anne was false. What should be his next step?

Trace her by other means.

* * *

ST. JAMES' PALACE, WESTMINSTER
16 April 1590

Hunsdon looked dubious when he heard Deven's request. "I do not know... Easter will be upon us in a week. 'Tis the duty of her Majesty's Gentlemen Pensioners to be attendant upon her during the holiday. *All* of them."

Deven bowed. "I understand, my lord. But never in my time here has every single member of the corps been present at once, even at last month's muster. I have served continually since gaining my position, taking on the duty periods of others. This is the first time I have asked leave to be absent for more than a day. I would not do so were it not important."

Hunsdon's searching eye had not half the force of Walsingham's, but Deven imagined it saw enough. He had not been sleeping well since the Principal Secretary's death—since his rift with Anne, in truth—and only the joint efforts of Colsey and Ranwell were keeping him from looking entirely unkempt. No one could fault him in his performance of his duties, but his mind was elsewhere, and surely Hunsdon could see that.

The baron said, "How long would you be absent?"

Deven shook his head. "If I could predict that for you, I would. But I do not know how long I will need to sort this matter out."

"Very well," Hunsdon said, sighing. "You will be fined for your absence on Easter, but nothing more.

With everyone—or at least most of the corps—coming to court, finding someone to replace you until the end of the quarter should not be difficult. You have earned a rest, 'tis true. Notify Fitzgerald if you intend to return for the new quarter."

If this matter occupied him until late June, it was even worse than he feared. "Thank you, my lord," Deven said, bowing again.

Once free of Hunsdon, he went straightaway to the Countess of Warwick again.

She had taken Anne on as a favour to Lettice Knollys, the widowed Countess of Leicester, who had last year married for the third time, to Sir Christopher Blount. A question to her new husband confirmed that his wife, out of favour with Elizabeth, was also out of easy reach; she had retired in disgrace to an estate in Staffordshire. Blount himself knew nothing of Anne Montrose.

Deven ground his teeth in frustration, then forced himself to stop. Had he expected the answer to offer itself up freely? No. So he would persist.

Inferior as Ranwell's personal services were to Colsey's, the newer servant could not be trusted with this. Deven sent Colsey north with a letter for the countess, and made plans himself to visit Doctor John Dee.

* * *

THE ONYX HALL, LONDON
18 April 1590

Lune's own words mocked her, until she thought she heard them echoing from the unforgiving walls of the palace: *No fae who cannot find a way to benefit herself while also serving the Onyx Throne belongs in your court.*

It was true, but not sufficient. Lune did not believe for an instant that Invidiana was angry at the lie she had given Madame Malline; that was simply an excuse. But the Queen had set her mind against Lune before that audience ever happened—before Lune ever went to the Tower. Would anything have changed that?

Ever since she went undersea, her fortunes had deteriorated. The assignment to Walsingham had seemed like an improvement, but only a temporary one; in the end, what had it gained her?

Time among mortals. A stolen year, hovering like a moth near the flame of the human court. A lie far preferable to the truth she lived now.

Living as an exile in her own home, hiding in shadows, trying to keep away from those who would hurt her for political advancement or simple pleasure, Lune missed her life as Anne with a fierce and inescapable ache. Try as she did to discipline her mind, she could not help thinking of other places, other people. Another Queen.

Elizabeth had her jealousies, her rages, and she had thrown her ladies and her courtiers in the Tower for a variety of offenses. But for all that her ringing tones echoed from the walls of her chambers, threatening to chop off the heads of those who vexed her, she rarely did so for anything short of genuine, incontrovertible treason.

And despite those rages, people flocked to her court.

They went for money, for prestige, for connections and marriages and Elizabeth's reflected splendour. But there was more to it than that. Old as she was, contrary and capricious as she was, they loved their Gloriana. She charmed them, flattered them, wooed them, bound them to her with charisma more than fear.

What would it be like, to love one's Queen? To enjoy her company for more than just the advantage it might bring, without concern for the pit beneath one's feet?

Lune felt the eyes on her as she moved through the palace, never staying long in one place. A red-haired faerie woman, resplendent in a jewelled-black gown that spoke of a rapid climb within the court, watched her with a sharp and calculating eye. Two maliciously leering bogles followed Lune until to escape them she had to dodge through a cramped passageway few knew about and emerge filthy on the other side.

She kept moving. If she stayed in one place, Vidar would find her. Or Halgresta Nellt.

Without mortal bread, going into the city was impossible. But when she heard a familiar, heavy tread, she ran without thinking; the nearest escape lay in the Threadneedle Street well, one of the exits from the Onyx Hall.

Luck afforded her this one sign of favour; with no sense of what hour it was in the mortal city, Lune found herself above ground in the dead of night. She wasted no time in flinging a glamour over herself and dodging into the shadows of a tiny lane, where she waited until she was certain the giantess had not followed.

It was a dangerous place to be. One of the nearest things to an inviolable rule in the Onyx Hall forbade drawing too much attention among mortals. Night allowed more freedom of movement than day, but without bread or milk, she would be limited to a goblin's skulking mischief.

Or she could flee.

Like a needle pointing to the north star, her head swivelled unerringly to look up Threadneedle, as if she could see through the houses to Bishopsgate and the road beyond. Out of London.

Invidiana wanted her to stay and suffer. But did she have to?

Wherever Lune had been before she came here, London was her home now. Some few fae migrated, even to foreign lands, but she could no more leave her city to live in Scotland than she could dwell among the folk of the sea.

She looked back at the well. Dame Halgresta lacked the patience to lie in wait; whether she had been chasing Lune, or simply passing by, she would be gone now.

Lune stepped back out into Threadneedle Street, laid her hand on the rope, and descended down the well, back into the darkness of the Onyx Hall.

MORTLAKE, SURREY

April 25, 1590

Deven rode inattentively, his eyes fixed on the letter in his hand, though he knew its contents by heart already.

I arranged a position for Mistress Montrose with Lady Warwick at the request of her cousin, a former waiting gentlewoman in my own service, Margaret Rolford.

Colsey was no fool. He knew why his master had searched London from one end to the other; he asked the next logical question before he left Staffordshire, knowing that otherwise he would have to turn around and go back. The answer was waiting in the letter.

Margaret Rolford lives now in the parish of St. Dunstan in the East.

The manservant had that answer waiting, too. "No Rolfords, either. Not there, nor in Fleet Street. I checked already."

No Margaret Rolford. No Anne Montrose. Deven wondered how Margaret had come into Lettice Knollys's service, but it wasn't worth sending again to Staffordshire to ask; he no longer believed he would uncover anything useful by that route. Anne seemed to have come from nowhere, and to have vanished back to the same place.

He scowled and tucked the letter into his purse.

Cottages dotted the land up ahead, placid and pastoral, with a modest church spire rising above them. Had he reached the right village? Deven had given both his servants a day's liberty and ridden out alone; Colsey would not approve of him coming here. So he himself had to flag down a fellow trudging along the riverside towpath with a basket on his back and ask, "Is this the village of Mortlake?"

The man took in his taffeta doublet, the velvet cap on his head, and bowed as much as the weight of the basket would allow. "Even so, sir. Can I direct you?"

"I seek the astrologer Dee."

He half-expected his words to wipe the pleasant look from the man's face, but no such thing; the fellow nodded, as if the scholar were an ordinary citizen, not a man suspected of black magic. "Keep along this road, sir, and you'll find him. There's a cluster of houses, but the one you want is the largest, with the extra bits built on."

The villager caught the penny Deven tossed, then quickly sidestepped to regain control of his burden as it slipped.

Deven soon saw what the man had meant. The "extra bits" were extensions easily as large as the house to which they had been added, making for a lopsided, rambling structure that encroached on the cottages around it. Flagstone paths connected that building to several nearby ones, as if they were all part of the same complex. And none of it was what Deven expected; nothing about the exterior suggested necromancy and devilish conjurations.

He dismounted, looped his horse's reins around a fence post, and knocked at the door. It was opened a moment later by a maidservant, who promptly curtsied when she found a gentleman on the step.

A twinkling later he was in the parlour, surreptitiously eyeing the unremarkable furnishings. But he did not have long to look; soon an older man with a pointed, snow-white beard entered.

"Doctor Dee?" Deven offered him a polite bow. "I am Michael Deven, of the Queen's Gentlemen Pensioners, and formerly in service to Master Secretary Walsingham. I beg your pardon for the imposition—I should have sent a letter in advance—but I have heard much of you from my master, and I hoped I might beg assistance from such a learned man."

His nerves hummed as he spoke. If his suspicions were correct, he was foolish to come here, to expose himself thus to his quarry. But he had not been able to talk himself out of this journey; the best he could do was to deliberately omit to send a letter, so that Dee would have no warning of his coming.

But what did he expect to find? There were no mystic circles on the floor, no effigies of courtiers awaiting burial at a crossroads or beneath a tree. And Dee did not flinch at Walsingham's name. The man might be the hidden player, but it was increasingly difficult for Deven to believe he might have killed Walsingham by foul magic.

"Assistance?" Dee said, gesturing for Deven to take a seat.

Deven contrived to look embarrassed; he might as well

put his flush to use. "I—I have heard, sir, that you are as able an astrologer as dwells in England. I am sure your time is much occupied by working on behalf of the Queen's grace, but if you might spare a moment to help a young man in need..."

Dee's alert, focused eyes narrowed slightly at this. "You wish me to draw up a horoscope? To what end?"

Glancing away, Deven permitted himself a nervous, self-deprecating laugh. "I—well, that is—you see, there's a young woman."

"Master Deven," the astrologer said in unpromising tones, "I do occasionally calculate on behalf of some of her Majesty's court, but not often. I am no street corner prophet, predicting marriage, prosperity, and the weather for any who pass by."

"Certainly not!" Deven hastened to reassure the man. "I would not even ask, were it simply a matter of 'will she or won't she.' But I have run into difficulty, and having tried everything at my disposal, I am at a loss as to how to proceed." He had to skirt that part carefully; he did not want to give Dee any more information than necessary. Assuming the man had not already heard his name from Anne. "I am sure you have many more important researches to occupy your time—I would be more than happy to fund them in some small part."

The words were perfectly chosen. Dee would have taken offense at the suggestion of being paid for his work; no doubt the man wanted to distinguish himself as no common magician. But an offer of patronage, no matter how fleeting and minor, did not go amiss, especially given the astrologer's financial difficulties.

Dee's consideration did not take long. "A horary chart is simple enough to draw up. I imagine, by your flushed complexion, that the matter is of some urgency to you?"

"Indeed, sir."

"Then come with me; we can answer your question directly." Deven followed his host through the cottage and into one of the extensions, where he stopped dead on the threshold, awed into silence by the sight that greeted him. The room was lined with shelves, a great library that dwarfed those held by even the most learned of Deven's own acquaintances. Yet it had an air of recent abuse, that called to mind what Anne had said about Dee's troubles; there were blank stretches of shelving, scars on the woodwork, and a conspicuous lack of reading podiums or other accoutrements he expected of a library.

Dee invited him over to the one table the room still held, with a stool on either side of it and a slew of paper on top. The papers were swept away before Deven could attempt to read them, and fresh sheets brought out, with an inkwell and a battered quill.

"First," Dee said, "we pray."

Startled, Deven nodded. The two men knelt on the floor, and Dee began to speak. His words were English, but they did not come from the Book of Common Prayer; Deven listened with sharp interest. Not Catholic, but perhaps not entirely Church of England either. Yet the man apparently considered prayer a requisite precursor to any kind of mystical work.

None of it was what he had expected.

When the prayer was done, they sat, and Dee sharpened his quill with a penknife. "Now. What is the question you wish answered?"

Deven had not formulated its precise wording in his mind. He said, choosing his words with care, "As I said, there's a young gentlewoman. She and I have had difficulties, that I wish to smooth over, but she has gone away, and despite my best efforts I cannot locate her. What..." He reconsidered the question before it even came out of his mouth. "How may I find her again?"

Dee sat with his eyes closed, listening to this, then nodded briskly and began marking out a square on the paper that lay before him.

After watching the astrologer work for a few minutes, Deven said hesitantly, "Do you not wish to know my date of birth?"

"'Tis not necessary." Dee did not even look up. "For a horary chart, what matters is the moment at which the question was formulated." He selected a book from a stack on the floor behind him and consulted it; Deven glimpsed orderly charts of numbers and strange symbols, some of them marked in red ink.

He waited, and tried not to show his relief. That had worried him the most, the prospect of giving Dee such information about himself. A magician might do a great deal with that knowledge. As it stood now, he might be any ordinary gentleman, asking after any ordinary woman; he had not even mentioned Anne's name.

But had she mentioned his?

Dee worked in silence for several minutes, examining the chart in the book, making calculations, then noting the results on the square horoscope he sketched out. It did not take long. Soon Dee leaned back on his stool and studied the paper, one hand idly stroking his pointed white beard.

"Be of good cheer, Master Deven," Dee said at last in absent, thoughtful tones at odds with his words. "You will see your young woman soon. I cannot say when, but look you here—the Moon is in the Twelfth House, and the Stellium of Mars, Mercury, and Venus—her influence has not yet passed out of your life."

Deven did not look where the ink-stained finger pointed; instead he watched Dee. The chart meant nothing to him, while the astrologer's pensive expression meant a great deal. "Is there more?"

The sharp eyes flicked up to meet his. "Yes. Enemies threaten—her enemies, I think, but they may pose a danger to you as well. The gentlewoman's disposition is obscure to me, I fear. Conflict surrounds her, complicating the matter. Death will send her into your path again."

Death? A chill touched Deven's spine. Was that a threat? He did his best to feign the concern of the lovestruck man he pretended to be, while searching for any hint of malice in the other's gaze. Perhaps the chart really did say that. He wished he knew something of astrology.

Deven bent over the paper, lest Dee read too much out of his own expression. "What should I do?"

"Be wary," the philosopher said succinctly. "I do not think the woman means you harm, but she may bring harm your way. Saturn's presence in the Eighth House indicates authority is set against this matter, but the Trine with Jupiter..." He shook his head. "There are influences I cannot read. Allies, perhaps, where you do not expect them."

It might be nothing more than a trick, something to send him running in fear. But at the very least, it did not sound like the kind of horoscope an impatient man might invent to placate a lovelorn stranger. Either it was a coded warning, or it was genuine.

Or both.

"I thank you, Doctor Dee," he said, covering his thoughts with courtesy. "They say knowledge of the stars helps prepare a man for that which will come; I only hope it shall be so with me."

Dee nodded, still grave. "I am sorry to have given you such ill tidings. But God guides us all; perhaps 'twill be for the best."

Recalling himself, Deven removed his purse and laid it on the table. It was more than he had meant to pay, but he could not bring himself to fish through it for coins. "For your researches.

I pray they lead you to knowledge and good fortune."

THE ONYX HALL, LONDON
25 April 1590

A clutch of chattering hobs and pucks passed through the room, laughing and carefree. All the fae of England were abuzz with the preparations for May Day, and the courtiers were no exception. Every year they took over Moor Fields north of the wall, enacting charms and enchantments that would keep mortals away. And if a few strayed into their midst, well, May Day and Midsummer were the two occasions when humans might hope for kindlier treatment at fae hands. Even the cruelty of the Onyx Court subsided for a short while, at those great festivals.

Lune watched them go from her perch high above. The chamber had a great latticework of arches supporting its ceiling, and it was upon one of these that she rested, her skirts tucked up around her feet so they would not trail and attract notice. It was an imperfect hiding place; plenty of creatures in the palace had wings. But it gave her a brief respite both from malicious whispers, and from those who sought to harm her.

When all around her was silent, she lowered herself slowly to the floor. Her gown of raven feathers was suitable for hiding, and she had long since discarded her velvet slippers; the pale skin of her bare feet might betray her, but it was much quieter when she moved. She lived like a rat in the Onyx Hall, hiding in crevices, stealing crumbs when no one was looking.

She hated every heartbeat of it.

But hatred was good; anger was good. They gave her the energy to keep fighting, when otherwise she would have given up.

She would not let her enemies defeat her like this.

Lune slipped barefoot out of the chamber, down a passageway that looked all but disused, lifting the ragged hem of her skirts so they would not leave traces in the dust. Until she began her rat's life, she had never realised how many forgotten corners the palace held. It was enormous, far larger than any mortal residence, and if it served the function of both hall and city to the fae that dwelt therein, still it was more than large enough for their needs.

Up a narrow staircase and through a door formed of interwoven hazel branches, and she was safe—as safe as she could get. No one seemed to know of this neglected chamber, which meant she had already bypassed one part of Invidiana's sentence upon her, that she be dependent on others for a place to lay her head. This place was hers alone.

But someone else had found it.

Lune's body froze, torn between fight and flight, assuming on the instant that it was Vidar. Or Dame Halgresta. Or one of their servants. Her hands flexed into claws, as if that would be of any use, and her bare feet set themselves against the dusty floor, ready to leap in any direction.

She saw no one. But someone was there.

Lune knew she should run. That was life these days; that was how she survived. But the chamber's scant furnishings, some of them scavenged from elsewhere in the palace, could not possibly be concealing the tall, heavy form of the Captain of the Onyx Guard, and if it were just some goblin minion…

She should still run. Lune was no warrior.

Instead she moved forward, one noiseless step at a time.

No one crouched behind the narrow bed, with its mattress stuffed with straw. No one stood in the shadow of a tall mirror that had been there when Lune found the room, its crystalline surface so cracked and mazed that nothing could be seen in its depths. No one waited between the cobwebbed, faded tapestries and the stone walls.

She paused, listening, and heard nothing. And yet…

Guided by instinct, Lune knelt and looked into the space beneath the bed.

Tiresias's face stared back at her, pale and streaked with tears.

Lune sighed in disgust. Her tension did not vanish entirely, but a good deal of it evaporated; she had never once seen the madman attack anyone. And he did not look like he was spying; he looked like he was hiding.

"Come out from under there," she growled. How had he fit? Small as he was, she never would have expected the seer could curl up in that narrow space. He shook his head at her words, but the violation of even this tenuous sanctuary angered Lune; she reached under the bed and dragged him out bodily. Invidiana was unlikely to execute her simply for manhandling one of her pets.

Emerging into the dim light, Tiresias gave her a twisted

smile that might have been meant to be bright. "Not everything is found so easily," he said gravely. "But if one's cause is good... you might do it."

"Get out," Lune spat. She barely restrained herself from striking him, venting the anger she dared not release on anyone else in the Onyx Hall. "You are one of *her* pets, *her* tools. For all I know, she sent you to me—and anything you say might be a trap she has laid. Everything is a trap, with her."

He nodded, as if she had said something deeply wise. "One trap begets another." Hiding under the bed had sent his hair into disarray, strands tangling with the tips of his eyelashes, twitching when he blinked. "Would you like to break the traps? All of them?"

Lune laughed bitterly, retreating from him. "Oh, no. I will not hear you. One deranged, pointless quest is enough—or would this be the same one? Will you tell me again to seek Francis Merriman?"

Tiresias had begun to turn toward the door, as if to wander off mid-conversation, but his motion arrested when she said that, and he pivoted back to face her. "Have you found him?"

"Have I found him," she repeated, flat and unamused. "No. I have not. He is no one at the mortal court—no gentleman or lord, no wealthy merchant, no officer serving in any capacity. He is not a poet or playwright or painter in the city, nor a prisoner in the Tower. If he lives in some future time that you have foreseen, then I doubt me I will be here to see him come, unless my fortune changes a great deal for the better. If he lives now, then he is no one of any note, and I have no reason to seek him." She glared at him, full of fury, as if all her fall in station were his fault. It was not, but she could and did blame him for how long she had spent chasing a vain, false hope. "I believe you invented Francis Merriman, out of your own mad fancies."

"Perhaps I did." It came out unutterably weary, heavy with resignation. He glanced down, his delicate shoulders slumping under a familiar weight of pain. "Perhaps only Tiresias is real."

The words stole the breath from her body. Anger died without warning, as his meaning became clear. "You," Lune whispered, staring at him. "*You* are Francis Merriman."

His eyes held lifetimes of wistful sadness. "Long ago. I think."

Invidiana's pets, with their classical names, each one collected for a special talent. Lune had given little thought to where they came from, who they were before they fell into the shadows of the Onyx Court. And how long had Tiresias been there? After so many years, who would bother to recall Francis Merriman?

Except him. And not always then. "Why?" Lune asked, hands lifting in wordless confusion. "You scarcely even remember who you were. What changing tide brought you to speak that name again?"

He shook his head, hair falling forward like a curtain too short for him to hide behind. "I do not know."

"'Twas in my chamber," Lune said, remembering. "I was considering my situation. I asked myself how I might better my standing in the Onyx Court—and then you spoke. Do you remember?"

"No." A tear glimmered at the edge of his sapphire eye.

A swift step brought her close; she took him by the arms and shook him once, restraining the urge to violence. Could she have avoided her downfall, had she seen what lay under her very eyes? "Yes, you *do*. Madman you may be, but 'twas no accident you said those words. You said you knew what she did. Who?"

"I cannot." His breath caught raggedly in his throat, and he twisted in her grip. "I cannot. If I—" He shook his head, convulsively. "Do not ask me. Do not make me do this!"

He tore himself free and stumbled away, catching himself against the wall. Lune studied his back for a moment, noting in pitiless detail the trembling of his slender shoulders, the whiteness of his fingers where they pressed against the stone. He feared something, yes. But her *life* hung in the balance; she could not stay ahead of her enemies forever.

If the price of her survival was forcing him to speak, then she would not hesitate.

"Francis Merriman," she said, enunciating the name with soft precision. "Tell me."

The name stiffened his whole body. He might have done anything in that moment; Lune tensed, wondering if he would strike her. Instead he whispered, almost too faint to hear, "Forgive me, Suspiria. Forgive me. 'Tis all I can do for you now. Forgive me..."

His voice trailed off. Francis Merriman lifted his head and turned back to face her, and Lune saw the transcendent effort of his will push back the fogs and shadows of untold years among the fae, leaving his eyes drawn and strained, but clear. The resulting lucidity, the determination, frightened her more than his madness ever had.

With a deliberate motion, he reached out and gripped Lune's arms, fingertips digging into the thin tissue of her sleeves.

"Someone must do it," he said. "I have known that for years. You have asked, and you have little left to lose; therefore I lay it upon you. You must break her power."

Lune wet her lips, willing herself not to look away. "Whose power?"

"Invidiana's."

The instant he spoke the name, a paroxysm snapped his head back, and his hands clenched painfully on Lune's arms. She cried out and reached for him, thinking he would collapse, but he kept his feet and brought his head down again. Six points of red had blossomed in a ring on his brow, flowers of blood, and they poured forth crimson ribbons as he spoke rapidly on, through gritted teeth. "I saw, but did not *understand*—and neither did she. 'Tis my fault she formed that pact, and we have all suffered for it, fae and mortals alike. You must break it. 'Twas not right. She is still c—"

The words rasped out of him, ever wilder and more strained, until the only thing keeping him on his feet was their mutual grip and the splintering remnants of his will. Now his voice died in an agonised cry, and his legs gave way. He slipped free of Lune's hands and crumpled bonelessly to the floor, his face a mask of blood.

The only sound in the room was the pounding of Lune's heart, and the ragged gasping of her own breath as she stared down at him.

I cannot, he had said, when she demanded he speak. *If I—If I do, I will die.*

Lune remembered where she was. In a chamber of the Onyx Hall, with the Queen's mad seer lying bloody and dead at her feet. She ran.

MORTLAKE AND LONDON

25 April 1590

A man might not be thought strange if he took an early supper before riding the eight miles back to London, nor if he spoke cheerfully of his purpose in coming to Mortlake. Deven's observations on his way in were true; though some in the village were suspicious of Dee's conjurations, casual chatter over his food revealed that the astrologer often served as a mediator in local problems, settling disputes and offering advice.

Deven was not sure what to think.

The delay meant a late start back to London, though, and full dark came well before he reached the Southwark end of London Bridge. The bankside town offered many inns, but without a manservant it would be irritating, and Deven was in no mood to stop yet; his mind was too full of thoughts. Though the great bell at Bow had long since rung curfew, he bought his way through the Great Gate House that guarded the bridge, trading on his coin and his status as a gentleman and a Gentleman Pensioner.

Dee could not have murdered Walsingham by black magic. Deven simply did not believe it. But did that mean that Walsingham had died of purely natural causes, as Beale insisted, or merely that Deven had pinned his suspicions on the wrong man? The astrologer might still be the hidden player, without being a murderer. Was he working with Anne, or not? And if so, how much stock—if any—should Deven put in the man's predictions?

He thought he was keeping at least marginally alert for movement around him. Cloak Lane was deserted, empty of others who like him were braving the curfew, but there might be footpads; alone, without a manservant, Deven had no intention of being taken by surprise.

Yet he was, when a figure stumbled abruptly out of the blackness of a narrow alley.

The bay horse reared, as surprised as his rider, and Deven fought to control the beast with one hand while reaching for his sword with the other. Steel leapt free, his gelding's hooves thudded into the unpaved street, and he raised his blade in readiness to strike—

—then the figure lifted its face, and Deven recognised her.

"Anne."

She shied back from him, hands raised as if to defend herself. The sword was still in his hand. Deven scanned Cloak Lane quickly, but saw no one else.

Dee had spoken of enemies and conflict.

He had said that death would send Anne into his path again.

She was backed against the shuttered wall of a shop, like an animal brought to bay. The sight slipped under his defences, sparking sympathy against his will. Deven compromised; he dismounted, so as not to loom over her, but kept the sword out, relaxed at his side. "Anne. 'Tis me—Michael Deven. Is someone chasing you? Are you in trouble?"

She had changed, since last he saw her; the bones of her face stood higher, as if she had lost weight, and her hair looked paler than ever. Her clothing was a sad imitation of a gentlewoman's finery, and—she must be running from someone—she stood barefoot in the dirt.

"Michael," she whispered. The whites of her eyes stood out starkly in her stricken face. She started to say something, then shook her head furiously. "Go. Leave me!"

"No," he said. "You are in trouble; I know it. Let me help you." A foolish offer, yet he had to make it. He extended his left hand, as if toward a wild horse that might bolt.

"You *cannot* help me. I have told you that already!"

"You have told me nothing! Anne, in God's name, what is going on?"

She flinched back at his words, hands flying up to defend her face, and Deven's blood froze as she changed.

Hair—silver. Gown—black feathers, trembling with her. And her face, imperfectly warded by her hands, refined into otherworldly beauty, high-boned and strange, with silver eyes wide in horror and fear.

The creature that had been wearing Anne Montrose's face stood a moment longer, pressed against the wall like she expected to be struck down on the spot.

Then she cried out and fled into the darkness of the city.

THE ANGEL INN, ISLINGTON
25 April 1590

The veil of glamour she threw over herself as she ran covered her imperfectly, a bad attempt at a human seeming, until she was nearly to Aldersgate. Then the bells tolled and it shredded away like mist, leaving her exposed. Lune fled the city as if the Wild Hunt were at her heels.

She fled north, without pausing to consider her course, and arrived panting at the rosebush behind the Angel Inn.

What she would tell them, she did not know. But she cried out until the doorway revealed itself, then threw herself down the steps to the room below.

Both of the Goodemeades were there, Rosamund

catching her as she came through. "My lady," the brownie said in surprise, then looked up at her face. All at once her expression changed; the concern stayed, but steely determination rose up behind it. "Gertrude," she said, and the other brownie moved.

At a gesture, the rushes and strewing herbs covering the floor whisked away into tidy piles, revealing the worn wooden boards beneath. Then these groaned and flexed aside, and where they parted Lune saw more stairs, with lights blooming into life below. She had no chance to ask questions, and no mind to frame them; the hobs hurried her through this secret door, and the boards grew shut behind them.

The room below held two comfortable beds and a hearth now flickering with fire, but no other inhabitants. Rosamund led her to one bed and got her to sit, putting Lune at eye level with the little brownie. Her face still showed concern, and determination, and a sharp-eyed curiosity that was new.

"Now, dear," she said in a gentle voice, holding Lune's hands, "what has happened?"

Lune drew in a ragged, shuddering breath. She hadn't thought about what to say, what story she would offer them to explain her distracted state; too much had happened, Invidiana, the seer, Michael. All her wary instincts failed. "Tiresias is dead."

Soft gasps greeted her statement. "How?" Rosamund whispered. Her plump fingers trembled in Lune's. "Who killed him?"

Lune could not suppress a wild, short laugh. "He did. He knew it would mean his death, yet still he spoke."

The sisters exchanged startled, sorrowful looks. Tears brimmed in Gertrude's eyes, and she pressed one hand to her heart. "Ah, poor Francis."

"What?" Lune snatched her hands from Rosamund's, staring at Gertrude. "You knew who he was?"

"Aye." Gertrude answered her, while Rosamund pressed one kind hand against Lune's shoulder, to keep her from rising. "We knew. Francis Merriman... we remember when he bore that name, though precious few others do. And if he died as you say..."

Rosamund finished her sentence. "Then he has betrayed her at last."

The brownie did not have to try hard to keep Lune in place; her knees felt like water, trembling from her headlong flight, with Deven's oath and the tolling of the bells still reverberating in her bones. Lune dug her fingers into the embroidered coverlet. "How—"

"The jewel," Rosamund said. "The one she wears on her bodice. We've suspected for ages that she laid it on him, not to speak of certain things, on pain of death. 'Twas the only explanation we could find for his silence. And we could not ask him to speak—not when it would carry such a price."

Lune remembered the six points of blood appearing on his brow, where the claws of the jewel had touched. Never before had she seen its power strike home.

She swallowed down the sickness in her throat. *She* had

asked him to speak. Forced him.

"Lass," Gertrude said, coming forward to lay a hand on Lune's other shoulder, so she was hemmed in by both sisters. "I would not question you, so soon after his death, but we must know. What did he say?"

His blazing, lucid eyes swam in her vision. Lune shivered, feeling suddenly closed in; the brownies let her go when she tried to rise, and she went toward the hearth, as if its flames could warm the cold spot in the pit of her stomach. "He told me to break her power. That she… that she had formed some kind of pact. And that it was harming everyone, both mortal and fae."

She did not see the sisters exchange a glance behind her back, but she felt it. Standing in the hidden room beneath their home, Lune's sense finally gathered itself enough for her to wonder. The Goodemeades helped those in need—that was why she had come to them—but otherwise they stayed out of the politics of the Onyx Court. Everyone knew that.

Everyone who had not heard their questions, had not seen the alert curiosity in Rosamund's eyes.

They paid more attention than anyone credited.

"This pact," Rosamund said from behind Lune. "What did he tell you about it?"

Lune shivered again, remembering his hoarse voice, desperately grinding out words through the pain that racked him. "Very little. He… he could barely speak. And it struck him down, the—the jewel did—before he could tell me all. She misinterpreted some vision of his." Hands wrapped tightly around her elbows, she turned and faced the Goodemeade sisters. "What vision?"

Gertrude shook her head. "We do not know. He never spoke of it to us."

"But this pact," Lune said, looking from Gertrude to Rosamund. Their round, friendly faces were unwontedly solemn, but also wise. "You know of that, don't you?" The sisters exchanged glances again, a silent and swift communication. "Tell me."

A flicker of wings burst into the room before they could speak. Lune twitched violently at the motion; her nerves were frayed beyond endurance, and the fear-inspired energy that drove her this far had faded. But the little brown bird settled on Gertrude's hand, flirting its reddish tail, and she saw it was merely a nightingale—not even a fae in changed form.

But it must have been touched by fae magic, for it chirped energetically enough, and the brownies both nodded as if they understood. They asked questions of it, too, questions that stirred more fear in Lune's heart—"Who?" and "How many?" and "How long before they arrive?"

And then, after another burst of birdsong, "Tell us what he looks like."

Finally Gertrude nodded. "Thank you, little friend. Keep watch still, and warn us when they draw near."

The nightingale launched itself into the air, flew to an opening in the wall Lune had not attended to before, and vanished. Rosamund turned once more to Lune. "They are searching for you, my lady. A half-dozen soldiers, and that horrible mountain Halgresta. They cannot know you are here, I think, but they always suspect us when someone's in trouble. Never fear, though; we are good at turning their suspicions aside."

"But it also seems," Gertrude added, "that we have a visitor skulking around our rosebush. Tell me, are they aware of that nice young man you were with at the mortal court?"

"Nice young..." Lune's heart stuttered. "Yes, they are."

Gertrude nodded decisively. "Then we must take care of him, too."

LONDON AND ISLINGTON

25 April 1590

Delay had cost him any hope of keeping the silver-haired creature in sight. But she left a trail: raven feathers, shed from her gown as she fled.

Deven followed them through the cramped and twisted streets of London. The woman eschewed Watling Street, Old Change, Cheapside, instead making her way northwest by back lanes, until he found a feather beneath the arch of Aldersgate itself.

The gate should have been shut for the night, but the heavy doors hung open, the guards there blinking and disoriented.

The trail led north. Mounted now, Deven should have lost sight of the black feathers in the night, but their faint glimmer drew his eye. By the time he reached Islington, he had a fistful of the things, iridescent and strange.

The last feather he found impaled on the thorn of a rosebush behind the Angel Inn.

Light showed here and there along the inn's back wall, and he knew they would still welcome a traveller at the front. But the woman could not have gone that way—

Unless, his mind whispered uneasily, *she put on Anne's face again.*

The feathers rustled in his fist. Despite himself, Deven paced around the rosebush, as if he would find some other sign. The thorned branches stood mute.

The hairs on the back of his neck rose. Deven glanced up at the sky, but it stood clear from one horizon to the next, with not a cloud in sight. Why, then, did he feel a thunderstorm approaching? He drew his blade again, just for the comfort of steel, but it did him little good. Something was coming, and every nerve screamed at him to run.

"Master Deven! This way, quickly!"

He spun and saw a woman beckoning from a doorway that glowed with warm, comforting light. He was on the staircase before he realised the doorway stood in the rosebush, in the comfortable tavern before he considered that he had just passed underground, through the opening in the floor before he asked himself, *Who is this woman? And why did you just follow her?*

"There," a northern accent said with satisfaction, from somewhere in the vicinity of his belt. "I wouldn't normally resort to charms, but we couldn't rightly stand there and argue. My apologies, Master Deven."

The sword trembled in his hand.

The woman who had lured him below was joined by a second, just as short, and alike as only a sister could be. They wore tidy little dresses covered with clean, embroidered aprons, and their apple-cheeked faces spoke of friendliness and trust—but they came only to his belt, and were no more human than the figure silhouetted in front of the fire, her hair shining like silver washed with gold.

"Michael," she breathed.

He retreated a step, risked a glance over his shoulder, saw that the floor had grown shut behind him. Levelling his sword-point at the three of them, Deven said, "Come no closer."

"Truly," one of the little women said, the one with roses embroidered on her apron, "there's no need for that. We brought you below, Master Deven, because there are some rather unpleasant people coming this way, and you will be safer down here. I promise, we mean no harm."

"How in *God's* name am I supposed to believe that?"

All three cringed, and one of the women gave a muffled squeak—the one with the daisies on her apron. "Now, now," the rose woman said, a trifle more severely. "That isn't very gentlemanly of you. Not to mention that we shouldn't like to see our house pop up out of the ground without so much as a by-your-leave, or an apology to the folk above. We are fae, Master Deven; surely you must know what that means."

Ominous thudding answered before he could; all four of them looked up. "They're at the rosebush," the daisy woman said, and then a snarl reverberated through the chamber, deep and hard, like thunder in an ugly storm.

"Open, in the name of the Queen."

The two short ones exchanged glances. "I am the better liar," the daisy woman said, and the rose woman answered, "but they will be suspicious if they do not see us both." She fixed Deven with a stare that was no less effective for coming from a creature so small. "You will put up your sword, good master, and refrain from invoking certain names while in our house. We are protecting you from what's above, which is good for both you and us. Once we have gotten rid of these nuisances, we shall answer any question you have."

"As many as we know the answers to," the daisy woman corrected her. "Come, we must hurry."

Upon which the two of them whisked off their aprons, mussed their hair, yawned theatrically, and hurried up the stairs, looking for all the world as if they had just been roused from bed.

The floor stretched open to let them pass, then shut again, like a cellar without a door.

Deven said, half to himself, "What..."

"Hush," the silver-haired creature hissed. She had not spoken since uttering his name, and now her attention was not more than half on him; she still looked upward, listening as heavy boots clomped across the floor.

"Where is she?" The voice he had heard before. It made Deven feel as if his bones were grinding together.

"I beg your pardon, Dame Halgresta—" The words were punctuated by a yawn. "We had just retired for the night. Would you like some mead?"

A clanking splash, as of a metal tankard being knocked to the floor. "I would not. *Tell me where she is.*"

The other sister: "Who?"

"Lady—" The deep voice cut off in a noise something between a growl and a laugh. "Lady no more. The bitch Lune."

Deven glanced across the hidden room at his involuntary companion. The silver-haired woman shivered unconsciously, her hands rising to cup her elbows. Upstairs, the two sisters parried the stranger's questions with a masterful blend of innocence, confusion, and well-timed misdirection. No, they had not seen the lady—beg pardon, the woman Lune. Aye, of course they would say if they had; were they not the Queen's loyal subjects? No, they had not seen her in some time—very rarely at all, since she went to the mortal court.

At that, finally, the fae woman looked across the room at him. Her eyes shone unmistakably silver, no common grey… but the set of them was familiar, from many a fond study.

Neither of them dared speak, with danger so near above. Instead they stared at each other, until the fae woman—Lune—broke and turned away.

He had not listened to the rest of the conversation

above. More heavy footsteps, lighter voices trying to press the departing visitor to take some sweetmeats, or ale for the ride back to the city. Then silence, and the feeling of oppression lifted.

Deven decided to risk it. Crossing the floor, he approached Lune as closely as he dared, and in a voice pitched to carry no further than her ears, he said, "What has become of Anne Montrose?"

The pointed chin lifted a fraction. Her voice equally soft, Lune said, "She was always thus, beneath the mask."

He turned away, realised the sword was still in his hand, sheathed it. And then they waited for the sisters to return.

"Dame Halgresta's gone," Rosamund said to Lune, when they came downstairs again. "I presume you listened? They know nothing of Francis's death; someone saw you flee the palace, is all. Be careful, my lady. She very much wishes to kill you."

Gertrude nudged her sister in the ribs while tying her daisy-flowered apron back on. "Manners, Rosamund. Now that we haven't got that awful giantess breathing down our necks, we should take care of our guest."

"Oh! Of course!" Rosamund made a proper curtsy to Deven. "Welcome to our house, Master Deven. I am Rosamund Goodemeade, and this is my sister Gertrude. And this is the Lady Lune."

Ever since she and Deven had lapsed into silence, Lune's

attention had been fixed on the fireplace, the safest target she could find. Now she said wearily, "He knows." She turned to find the brownies wide-eyed and a little nervous. Relaxing her arms from their tight positions across her body, she added, "He drove the glamour from me when I was on my way here."

His blue eyes might have been shuttered against a storm, so little could she read out of them. Walsingham's service had taught him well—but he had never used such defences against her before. Well, she could not blame him. "So there you have it, Master Deven," Lune said to him, hearing her own voice as if it belonged to a stranger. "There are faeries at the mortal court. Though most of them come in secret, and do not disguise themselves as I did."

A muscle worked in his jaw. When Deven spoke, it sounded almost nothing like his natural voice, either. "So 'twas you all along. I suspected Dee."

Gertrude said in confusion, "She was what all along?"

"The hidden player," Lune said, still looking at Deven. "The secret influence on English politics that his master Walsingham has begun to suspect."

Bitterness twisted the corner of his mouth. "You were under my eyes, the entire time."

Lune matched him with her own sour laugh. "'Tis a night for such things, it seems. You are both right and wrong, Master Deven. I was a lead to your hidden player—not the player herself. There are two Queens in England. You serve one; you seek the other."

Her words broke through the stoic facade he had constructed while they waited, revealing startlement beneath. "*Two* Queens..."

"Aye," Rosamund said. "And that may be the answer to the question you asked us, Lady Lune, before we were interrupted."

It was enough to distract her from Deven. "What?"

Gertrude had scurried off to the far end of the room while they spoke. Something bumped the back of Lune's farthingale; she looked down to see the brownie pushing a stool almost as tall as she was. "If we're going to have this conversation," Gertrude said, with great firmness, "then we will sit while we do so. I've been on my feet all day, baking and cleaning, and you two look about done in."

"I have not said I will stay," Deven said, with another glance over his shoulder to the sealed top of the staircase.

Lune smiled ironically at him. "But you will. You want answers—you and your master."

"Walsingham is *dead.*" In the time it took him to say that, two strides ate up the distance between them and Deven was in her face, his anger beating at her like the heat from the fire. "I suppose I have you to thank for that."

Her knees gave out; she dropped without grace onto the stool Gertrude had put behind her. "He—what? Dead? When?"

"Do not pretend to be innocent," he spat. "You knew he was looking for you, for evidence of your Queen's hand. He was a threat, and now he's dead. I may be the world's greatest fool—you certainly played me as such—but not so great a fool as that."

Rosamund's hand closed over the silk of his right sleeve, drawing his fingers back from the sword hilt they had unconsciously sought. "Master Deven," the brownie said. The man did not look down at her. The uneven shadows of firelight turned his face monstrous, warping the clean lines of his features. "Lady Lune was imprisoned when your master died. She could not have killed him."

"Then she gave the order for it to be done."

Lune shook her head. She could not hold Deven's gaze; she felt naked, exposed, confronting him while wearing her true face. He would not have glared at Anne with such anger and hate. "I did not. But if he's dead... how?"

"Illness," Deven said. "Or so it was made to seem."

Walsingham had often been sick. He might have died by natural means. Or not. "My task," she said, staring fixedly at the battered feathers of her skirt, "was to watch over Walsingham, to know what he was about. And, if I could, to find a means of influencing him."

Deven met this with flat disgust. "Me."

"He is—was—an astute man," Lune said, dodging Deven's implicit question. She could not explain her choice, not now. "I believe my Queen feared he was coming near the truth. You may be right to blame me, Master Deven, for I told Vidar—a fae lord—what the Principal Secretary was about. After I was taken from Oatlands, he may have taken steps to remove that threat. But I never ordered it."

Gertrude had Deven's other sleeve now, and the gentle but insistent tugging from the brownies got him to back up a step, so that he no longer towered over Lune on the stool. "Why?" Deven asked at last. Some of the anger had gone from his voice, replaced by bewilderment. "Why should a faerie Queen care what happens in Ireland, or what became of Mary Stewart?"

"If you will sit," Gertrude said, returning with patient determination to her point of a moment before, "we may be able to answer that question."

When they were all seated, with mugs of mead in their hands—the brownies' family name, Deven realised, was more than mere words—the rose-flowered woman, Rosamund, began to speak.

"My lady," she said, bobbing her curly head at Lune. "How long have you been at the Onyx Court?"

Lune had straightened the remnants of her feathered gown and smoothed her silver hair, but her bare feet were still an incongruous note, the slender arches freckled with mud. "A long time," she said. "Not so long as Vidar, I suppose, but Lady Nianna and Lady Carline are more recently come than I. Let me see—Y Law Carreg was the ambassador from the Tylwyth Teg then..."

It reminded Deven powerfully of his early days at Elizabeth's court. A flood of names unknown to him, currents of alliance and tension he could not read. Somehow it made the notion more real, that there truly was another court in England.

When Lune's recitation wound down, Rosamund said, "And how long has Invidiana been on the throne?"

The elfin woman blinked in astonishment. "What manner of question is that?" she said. "An age and a day; I do not know. We are not mortals, to come and go in measured time." And indeed, Deven realised, in all her explanation of her tenure at court, she had not once named a date or span of years.

The sisters looked at each other, and Gertrude nodded. Rosamund said, with simple precision, "Invidiana became the Queen of faerie England on the fifteenth day of January, in the mortal year fifteen hundred and fifty-nine."

Lune stared at her, then laughed in disbelief. "Impossible. That is scarcely thirty years! I myself have been at the Onyx Court longer than that."

"Have you?" Gertrude said, over the top of her mead.

The elfin woman's lips parted, at a loss for words. Deven had been quiet since they sat down, but now he spoke. "That is the day Elizabeth was crowned Queen."

"Just so," Rosamund answered.

Now he was included in Lune's disbelieving stare. "That is not possible. I *remember*—"

"Most people do," Gertrude said. "Not specific memories, tied to specific mortal years—no, you're quite right, we do not measure time so closely. Perhaps if we did, more fae would notice the change. The Onyx Court as such has only existed for thirty-one years, perhaps a bit longer, depending on how one considers it. Vidar has been there longer. But all your memories of Invidiana's reign do not go further back than that. You just believe they do, and forget what came before."

Rosamund nodded. "My sister and I are some of the only ones who remember what came before. Francis was another. She let him remember on purpose, I believe, and we were with him when it happened; he kept us from forgetting. Of the others who know, every last one now rides with the Wild Hunt."

Lune's silver eyes widened, and she set her mug down with careful hands. "They claim to be kings."

"And they were," Gertrude confirmed. "Kings of faerie England, one corner of it or another. Until Elizabeth became Queen, and Invidiana with her. In one day—one moment—she deposed them all."

Deven had not forgotten where the conversation began. "But why? This cannot be usual for your kind." It was not usual for *his* kind, to be sitting in a hidden cellar of a faerie house, speaking with two brownies and an elf. His mead sat untouched on the table before him; he knew better than to drink it. "Why the connection?"

"We are creatures of magic," Rosamund said, as casually as if she were reminding him they were English. "And in its own way, a coronation ceremony *is* magic; it makes a king—or a queen—out of an ordinary mortal. Gertrude and I have always assumed Invidiana took advantage of that ritual to establish her own power."

Lune's voice came from his right, unsteady and faint. "But she did more than that, didn't she? Because there was a pact."

"'Pact?'" The word chilled Deven. "What do you mean?"

For a moment, he thought he perceived both sorrow and horror in her expression. "Do you recall me asking after a mortal named Francis Merriman?" Deven nodded warily. "He was under my eyes, as I was under yours. He... died tonight. He told me of a pact formed by Invidiana, my Queen, that he said was harming mortals and fae alike. And he begged me to break it."

Deven said, "But a pact..."

"Must be known to both parties," Rosamund finished for him. "Any fae with an ounce of political sense knows that Invidiana regularly interferes with the mortal court, and uses that court to control her own people. And from time to time a mortal learns that he or she has dealings with fae—usually someone enough in thrall that they will not betray it. But if what Francis said is correct... then someone on the other side knows precisely what is going on."

The words were trembling in Deven's throat. He let them out one by one, fearing what they meant. "The Principal Secretary... he told me of a hidden player. And he believed that player did—not often, but at times—have direct access to her Majesty."

He missed their reactions; he could not bring himself to look up from his clenched fists. The suggestion was incredible, even coming from his own mouth. That Elizabeth might know of faeries—not simply know of them, but traffic with them…

"I believe it," Lune whispered. "Indeed, it makes more sense than I like."

"But *why*?" Frustrated fear and confusion boiled out of Deven. "Why should such a pact be formed? What would Elizabeth stand to gain?"

An ironic smile touched Lune's thin, sculpted lips. "The keeping of her throne. We have worked hard to ensure it, at Invidiana's command. The Queen of Scots you have already named; Invidiana took great care to remove her as a threat. Likewise with other political complications. And the Armada…"

Her sentence trailed off, but Gertrude finished it, quite cheerfully. "You have Lady Lune to thank for those storms that kept the Spanish from our shores."

The bottom dropped out of Deven's stomach. Lune said, "I negotiated the treaty only. I have no power to summon storms myself."

He desperately floundered his way back to politics, away from magic. "And your Queen gained her own throne in return."

The black feathers he'd collected along the way had fallen from his hand at some point after he came downstairs. Lune had the broken tip of one in her fingers, and with it was tracing invisible patterns on the tabletop, her gaze unfocused. "More than that," she said, distant with thought. "Elizabeth is a Protestant."

Rosamund nodded. "Whereas Mary Tudor and Mary Stewart were both Catholics."

"What means that to you? Surely you cannot be Christian."

"Indeed, we are not," Lune said. "But Christianity can be a weapon against us—as you yourself have seen." Nor had Deven forgotten; he would use it again, if necessary. "Catholics have rites against us—prayers, exorcisms, and the like."

"As does the Church of England. And many puritan-minded folk call your kind all devils; surely that cannot be to your advantage."

"But the puritans are few in number, and the Church of England is a new-formed thing, which few follow with any ardour. 'Tis a compromise, designed to offend as few as possible as little as possible, and it has not existed long enough for its rites to acquire true power. The Book of Common Prayer is an empty litany to most people, form without the passionate substance of faith." Lune laid the feather tip down on the table and turned her attention to him. "This might change, in years to come. But for now, the ascendancy of your Protestant Queen is a boon to us."

He could taste his pulse, so hard was his heart beating. The chessboard in his mind rearranged itself, pieces of new colours adding themselves to the fray. Walsingham had surely never dreamed of this. And when Beale heard...

If Beale heard.

In personal beliefs, Walsingham had been a Protestant reformer, a "puritan" as their opponents called them; he would have loved to see the Church of England stripped of its many remaining papist trappings. But Walsingham was also a political realist, who knew well that any attempt at sweeping reformation would provoke rebellion Elizabeth could not survive. Beale, on the other hand, was outspoken about his beliefs, and often agitated for puritan causes at court.

Should Beale ever hear that Elizabeth, the great compromiser of religion, had formed a pact with a *faerie queen*—

England was already at war with Catholic powers. She could not fight another one within her own borders.

Deven looked from Rosamund, to Gertrude, to Lune. "You said this Francis Merriman of yours begged you to break the pact."

Lune nodded. "He said it was a mistake, that both sides had suffered for it." Her hesitation was difficult to read; the silver eyes were alien to him. "I do not know the effects of this pact, but I know Invidiana. I can imagine why he wanted me to break it."

"Do you intend to do so?"

The question hung in the air. This deep underground, there was no sound except their breathing, and the soft crackling of the fire. The Goodemeade sisters had their lips pressed together in matching expressions; both of them were watching Lune, whose gaze lay on the broken feather tip before her.

Deven had known Anne Montrose—or thought he had. This silver-haired faerie woman, he did not know at all. He would have given a great deal to hear her thoughts just then.

"I do not know how to," Lune said, very controlled.

"That is not what I asked. I do not know the arrangements of your court, but two things I can presume: first, that your Queen would not want you to interfere with this matter, and second, that you are out of favour with her. Else you would not be here, barefoot and in hiding, with her soldiers hunting you out of the city. So will you defy her? Will you try to break this pact?"

Lune looked to the Goodemeades. The brownies' faces showed identical resolution; it was not hard to guess what they thought should be done.

But what he was asking of her was treason.

Deven wondered if Walsingham had ever felt such compunctions, asking his agents to betray those they professed to serve.

Lune closed her eyes and said, "I will."

* * *

MEMORY

14–15 January 1559

Despite the cold, people packed the streets of London. In the southwestern portions of the city, in the northeast—in all those areas removed from the centre—men wandered drunkenly and women sang songs, while bonfires burned on street corners, creating islands of light and heat in the frozen air, banners and the clothing of the wealthy providing points of rich colour. Everywhere in the city was music and celebration, and if underneath it all many worried or schemed, no such matters were permitted to stain the appearance of universal rejoicing.

The press was greatest in the heart of the city, the great artery that ran from west to east. Crowds packed so tightly along the route that hardly anyone could move, save a few lithe child thieves who took advantage of the bounty. Petty Wales, Tower Street, Mark Lane, Fenchurch, and up Gracechurch Street; then the course straightened westward, running down Cornhill, past Leadenhall, and into the broad thoroughfare of Cheapside. The cathedral of St. Paul awaited its moment, and then the great portal of Ludgate, all bedecked with finery. From there, Fleet Street, the Strand, and so down into Westminster, and every step of the way, the citizens of London thronged to see their Queen.

A roar went up as the first members of the procession exited the Tower, temporarily in use once more as a royal residence. By the time the slender figure in cloth of gold and silver came into view, riding in an open-sided litter and waving to her people, the noise was deafening.

The procession made its slow way along the designated route, stopping at predetermined points for pageants that demonstrated for all the glory and virtue of the new sovereign. No passive spectator she, nor afraid of the chill; when she could not hear over the noise of the crowd, she bid the pageant be performed again. She called responses to her loyal subjects, touching strangers for a moment with the honour and privilege of royal attention. And they loved her for that, for the promise of change she brought, for the evanescent beauty that would all too soon fade back to show an architecture of steel beneath.

She reached Westminster late in the day, exhausted but radiant from her ordeal. The night passed: in drunkenness for the people of London, in busy preparation for the great officials in Westminster.

Come the following morning, when she set forth again, a shadow mirrored her elsewhere.

In crimson robes, treading upon a path of blue cloth, one uncrowned woman passed from Westminster Hall to the Abbey.

In deepest black, moving through subterranean halls, a second uncrowned woman passed from the Tower of London to a chamber that stood beneath Candlewick Street.

Westminster Abbey rang with the sonorous speeches and ceremony of coronation. Step by step, a woman was transformed into a Queen. And a few miles away, the passages and chambers of the Onyx Hall, emptied for this day, echoed back the ghostlike voice of a fae, as she stripped herself of one name and donned another.

A sword glimmered in her hand.

The presiding bishop spoke traditional words as the

emblems of sovereignty were bestowed upon the red-haired woman. The sound should not have reached the Onyx Hall, any more than the shouts of the crowd should have, but it was not a matter of loudness. For today, the two spaces resounded as one.

Then the fanfares began, as one by one, a succession of three crowns were placed upon an auburn head.

As the Onyx Hall rang with the trumpet's blast, the sword flashed through the air and struck a stone that descended from the ceiling of the chamber.

Drunken revellers in London heard the sound, and thought it a part of the celebrations: the tolling of a terrible, triumphant bell, marking the coronation of their Queen. And soon enough the bells would come, ringing out in Westminster and spreading east to the city, but this sound reached them first, and resonated the most deeply. Sovereignty was in that sound.

Those citizens who were on Candlewick Street at the time fell silent, and dropped to their knees in reverence, not caring that the object they bowed to was a half-buried stone along the street's south edge, its limestone surface weathered and scarred, unremarkable to any who did not know its tale.

Three times the stone tolled its note, as three times the sword struck it from below, as three times the crowns were placed. And on the third, the sword plunged into the heart of the stone.

All mortal England hailed the coronation of Elizabeth,

first of her name, by the Grace of God, Queen of England, France, and Ireland, Defender of the Faith, et cetera; and all faerie England trembled at the coronation of Invidiana, Queen of the Onyx Court, Mistress of the Glens and Hollow Hills.

And a dozen faerie kings and queens cried out in rage as their sovereignty was stripped from them.

Half-buried in the soil of Candlewick Street, the London Stone, the ancient marker said to have been placed there by the Trojan Brutus, the mythical founder of Britain; the stone upon which sacred oaths were sworn; the half-forgotten symbol of authority, against which the rebel Jack Cade had struck his sword a century before, in validation of his claim to London, made fast the bargain between two women.

Elizabeth, and Invidiana.

A great light and her great shadow.

ACT FOUR

O no! O no! tryall onely shewes
The bitter iuice of forsaken woes;
Where former blisse present euils do staine;
Nay, former blisse addes to present paine,
While remembrance doth both states containe.

SIR PHILIP SIDNEY
"THE SMOKES OF MELANCHOLY"

Sunlight caresses his face with warmth, and grass pricks through the linen of his shirt to tickle the skin inside. He smiles, eyes closed, and lets his thoughts drift on the breeze. Insects sing a gentle chorus, with birds supplying the melody. He can hear leaves rustling, and over the crest of the hill, her laughter, light and sweet as bells.

The damp soil yields softly beneath his bare feet as he runs through the wood. She is not far ahead—he can almost glimpse her through the shifting, dappled emerald of the shadows—but branches keep hindering him. A silly game. She must have asked the trees to help her. But they play too roughly, twigs snagging, even tearing his shirt, leaf edges turning sharp and scoring his face, while acorns and rocks batter the soles of his feet. He leaves a trail of footprints that fill with blood. He does not like this game anymore.

And then he teeters on the edge of a pit, almost falling in. Below, so far below…

She might be sleeping. Her face is peaceful, almost smiling.

But then the rot comes, and her skin decays, turning

mushroom-coloured, wrinkling, swelling, bloating, sinking in at the hollows of her face, and he cries out but he cannot go to her—the serpent has him fast in its coils, and as he fights to free himself it rears back and strikes, sinking its fangs into his brow, six stabbing wounds that paralyze him, steal his voice, and she is lost to him.

The fae gathered around laugh, taking malicious pleasure in his blind struggles, but it loses all savour when he slumps into the vines they have bound around him. His dreams are so easy to play with, and the Queen never objects. Bored now by his silent shudders, they let the vines fall away as they depart.

He is left in the night garden, where the plants have never felt neither sun nor breeze. High above, cold lights twinkle, spelling out indecipherable messages. There might be a warning in them, if he could but read it.

What good would it do him? He had warnings before, and misunderstood them.

Water rushes along at his side. Like him it is buried, forgotten by the world above, disregarded by the world below, chained to serve at her pleasure.

It has no sympathy for him.

He weeps for his loss, there on the bank of the brook—weeps bloody tears that stain the water for only an instant before dissolving into nothingness.

He has lost the sunlit fields, lost the laughter, lost her. *He shares her grave, here in these stone halls. It only remains for him to die.*

But he knows the truth.
Even death cannot bring him to her again.

THE ANGEL INN, ISLINGTON

25 April 1590

"We must get you back into the Onyx Hall," Rosamund said to Lune.

Gertrude was in the corner, murmuring to a sleek grey mouse that nodded its understanding from within her cupped hands. Lune was watching her, not really thinking; her thoughts seemed to have collapsed in fatigue and shock after she committed herself to treason. It was a reckless decision, suicidal even; tomorrow morning she would regret having said it.

Or would she? Her gaze slid once more to Deven, like iron to a lodestone. His stony face showed no regrets. She had never expected him to become caught in this net, and could not see a way to free him. However lost he might be right now, he would not back away. Though this pact might benefit Elizabeth, it was also harming her; so Tiresias had said—no, Francis Merriman. The seer had fought so hard to reclaim that self. Having killed him, the least Lune could do was grant him his proper name.

Francis Merriman had believed this pact was wrong. The Goodemeades obviously agreed with him. And Deven's master might well have been murdered at Invidiana's command. She knew him too well to think he would let that pass.

Lune herself had nothing left to lose save her life, and even that hung in the balance. But was that sufficient reason to betray her Queen?

Faint memories stirred in the depths of her mind. The thought, so fleetingly felt, that once things had been different. That once the fae of England had lived warmer lives—occasionally scheming against one another, yes, occasionally cruel to mortals, but not always. Not this unrelenting life of fear, and the ever-present threat of downfall.

Even those who lived far from the Onyx Hall dwelt in its shadow.

The Onyx Hall. Rosamund's words finally penetrated. Lune sat bolt upright and said, "Impossible. I would be executed the moment I set foot below."

"Not necessarily," Gertrude said. The mouse had vanished; now the brownie was prodding the fire, laying an additional log so that bright flames leapt upward and illuminated the room. "I've sent Cheepkin to see if anyone has found Francis's body. So far as we know, that jewel doesn't tell Invidiana when someone dies, so she may not yet know."

Lune's stomach twisted at the mere thought of being in the same room as the Queen when she learned of it. "She will know *how* he died, though. And she will wonder to whom he betrayed her."

Rosamund's nod was not quite complacent, but it didn't show half the alarm Lune felt it should. "Which is why we shall give her another target to suspect. And do you some good in the bargain, I think, as you will be the one to tell her." The brownie's soft lips pursed in thought. "She will be angry regardless, and afraid; how much, she will wonder, did Francis manage to say before he died? But that cannot be helped; we cannot pretend he died by other means. What we must do is make certain she does not suspect *you*."

"Who did you have in mind?" Gertrude asked her sister.

"Sir Derwood Corr. We can warn him to leave tonight, so he'll be well clear of the palace before she tries to arrest him."

Deven was looking at Lune, but she had no more idea than he what the Goodemeades meant. "Who is Sir Derwood Corr?"

"A new elf knight in the Onyx Guard. Also an agent of the Wild Hunt."

Gertrude nodded her approval. "She fears them anyway; it cannot do much harm."

They seemed to be serious. An agent of the Wild Hunt, infiltrating the Onyx Guard itself—and somehow the Goodemeades knew about it, and were eager to get the knight out of harm's way. "Are you working with the Wild Hunt?"

"Not *exactly*," Gertrude said, hedging. "That is, they would like us to be. We choose not to help them, at least most of the time; someone else brought Sir Derwood in. But we do keep an eye on their doings."

Lune had no response to this extraordinary statement. Deven, slouched on his stool as much as his stiff doublet would allow, snorted. "The Principal Secretary said 'twas infamous to use women agents, but I vow he would have made an exception for you."

They are not spies, Lune thought. *They are spymasters. With the very birds and beasts of the field their informants.*

"So," Rosamund said briskly. "As soon as Cheepkin reports in, Lady Lune, we shall smuggle you back into the Onyx Hall. You can tell Invidiana that Sir Derwood is an ally of the Wild Hunt; she will discover that he has fled; she will assume Francis spoke to him, and not to you. With any luck, that will sweeten her mind toward you, at least a bit."

Lune did not hold out much hope for that. Was she truly about to return to her rat's life, hiding from Vidar and Dame Halgresta and everyone else who might think to curry favour by harming or eliminating her?

The low, smouldering fire that had lived in her gut since her imprisonment—no, since her inglorious return from the sea—had an answer for that.

Yes, she would. She would go back, and tear every bit of it down.

Then I am a traitor indeed. May all the power of Faerie help me.

"Very well," she murmured.

Deven took a deep breath and sat up. "What may I do?"

"No time for that now," Gertrude said. "We must

return Lady Lune, before someone finds Francis. Might I ask a favour of you, Master Deven?"

He looked wary. "What is it?"

"Nothing dangerous, dearie; just a bit of dodging around Invidiana. Come with me, I'll show you." Gertrude took him by the hand and led him upstairs.

Lune watched them go, leaving her behind with Rosamund. "Is this safe?" she asked quietly. "I did not think of it before I came, but Invidiana has spies everywhere. She may learn of what we have said here."

"I do not think so," Rosamund said, and now she *did* sound complacent. "We're beneath the rosebush, here—very truly *sub rosa*. Nothing that happens here will spread outside this room."

For the first time, Lune looked upward, to the ceiling of the hidden chamber. Old, gnarled roots spread finger-like across the ceiling, and tiny roses sprang improbably from their bark, like a constellation of bright yellow stars. The ancient emblem of secrecy gave her a touch of comfort. For the first time in ages, she had friends she could trust.

She should have come to the Goodemeades sooner. She should have asked them about Francis Merriman.

They lied too well, convincing everyone that they stayed out of such matters. But if they did not, they would never have survived for so long.

Lune realised there was something she had not said. The words came awkwardly; she spoke them so often, but so rarely with sincerity. "I thank you for your kindness," she whispered, unable to face Rosamund. "I will be forever in your debt."

The brownie came over and took her hands, smiling into her eyes. "Help us set this place right," she said, "and the debt will be more than repaid."

A lantern glowed by the door of the inn, and light still showed inside. Lying as it did along the Great North Road, the Angel was a major stopping point for travellers who did not gain the city before the gates closed at dusk, and so there was always someone awake, even at such a late hour.

Deven led his horse toward the road in something of a daze. The part of him that was accustomed to following orders had for some reason decided to obey the little brownie Gertrude, but his mind still reeled. Faeries at court. How many of them? He remembered the rooftop chase, and the stranger that had vanished. Perhaps he had not imagined the flapping of wings.

He mounted up, rode into the courtyard of the inn, and dismounted again, so that anyone inside would hear his arrival. Looping his reins over a post, he stepped through the door, startling a sleepy-eyed young man draped across a table. The fellow sat up with a jerk, dropping the damp rag he held.

"Sir," he said, stumbling to his feet. "Needing a room, then?"

"No, indeed," Deven said. "I have some ways to ride

before I stop. But I am famished, and need something to keep me going. Do you have a loaf of bread left?"

"Uh—we should—" The young man looked deeply confused. "You're riding on, sir? At this hour of the night? The city gates are closed, you know."

"I am not going into the city, and the message I bear cannot wait. Bread, please."

The fellow sketched a bad bow and hastened through a door at the far end of the room. He emerged again a moment later with a round, crusty loaf in his hand. "This is all I could find, sir, and 'tis a day old."

"That will do." At least he hoped it would. Deven paid the young man and left before he would have to answer any more questions.

He rode away, circled around, came back to the rosebush. Gertrude had provided him with a bowl; now he set it down by the door of one of the inn's outbuildings, with the loaf of bread inside, and feeling a great fool, he said, "Food for the Good People; take it and be content."

The little woman popped up so abruptly he almost snatched out his blade and stabbed her. The night had not been good on his nerves. "Thank you, dearie," Gertrude said with a cheerful curtsy. "Now if you could pick it up again? We have some of our own, of course, a nice little supply—we so often have to help out others—but if Invidiana finds we've been giving Lady Lune mortal bread… well, we aren't giving it to her, are we? You are. So that's all right and proper. Never said anything about *mortals* giving her bread or milk, and not as if she has any right to tell you what to do. Not that it would stop her, mind you."

Bemused, Deven picked up the bowl and followed the still chattering brownie back to the rosebush, which opened up and let them pass below.

Lune was still in the hidden room, washing her feet in a basin of clear water. She glanced up as he entered, and the sight made his throat hurt; the motion was so familiar, though the body and face had changed. He thrust the bowl at her more roughly than he meant to, and tried to ignore the relieved pleasure on her face as she took the bread. "I shall have to think where to hide this," she said. "You are clever, Gertrude, but Invidiana will still be angry if she learns."

"Well, eat a bite of it now, my lady," the brownie said, retrieving the bowl from Deven. "You could use a good night's sleep here, but we can't risk it; you need to go back as soon as possible. Has Cheepkin returned?"

"While you were out," Rosamund said. "No one has found Francis yet. I've made sure Sir Derwood knows to leave."

"Good, good. Then 'tis time you went back, Lady Lune. Are you ready?"

Deven, watching her, thought that she was not. Nonetheless, Lune nodded her agreement. Holding the small loaf in her hands as if it were a precious jewel, she pinched off a bite, put it in her mouth, chewed, and swallowed. He watched in fascination, despite himself; he had never seen anyone eat bread with such attentive care.

Rosamund said to him, "It strengthens our magic against those things that would destroy it. Traveling through mortal places is dangerous without it."

As he had seen, earlier that very night. No wonder Lune treated it as precious.

"Now," Gertrude said briskly. "Master Deven, would you escort her back to London? 'Twould go faster riding, and unless Lady Lune makes herself look like a man, she should not be traveling alone."

The comment about disguise brought him back to unpleasant matters with a jolt. Lune was towelling her feet dry with great concentration. He very much wanted to say no—but he made the mistake of looking at Gertrude and Rosamund. Their soft-cheeked faces smiled up at him in innocent appeal. His mouth said, "I would be glad to," without consulting his mind, and thus he was committed.

Lune stood, dropped the towel on her stool, and walked past him. "Let us go, then."

By the time he followed, she was gone from the main room upstairs. He found her outside, waiting with her back to him. Words stuck in his throat; he managed nothing more than a stiff, "My horse is this way." His bay stopped lipping at the grass when Deven took hold of the reins. No footsteps sounded behind him, but when he turned, he found her just a pace away.

Except it wasn't her. She wore a different face, a human one. Not, he was desperately relieved to see, the face of Anne Montrose.

"Who is that?" he said, and could not keep the bitterness out of it.

"Margaret Rolford," Lune said, coolly.

Deven's mouth twisted. "Once a waiting-gentlewoman to Lettice Knollys, as I understand it."

Margaret Rolford's eyes were probably brown in sunlight; at night, they looked black. "I congratulate you, Master Deven. You followed me farther than I realised."

There was nothing he could say to that. Steeling himself, Deven put his hands around Margaret's waist—thicker than Lune's, and Anne's—and lifted her into the saddle; then he swung himself up behind her.

He had not realised, when he agreed to Gertrude's request, that it would mean riding the distance to London with his arms around the faerie woman.

Deven set his jaw, and touched his heels to the flanks of his gelding.

The tiny sliver of a moon had set even before he returned from Mortlake; they rode in complete darkness toward the few glimmering lights of London. Margaret Rolford's body was not shaped like Anne Montrose's—she had a sturdier frame, and was shorter—but still it triggered memories. A crisp, sun-washed autumn day, with just enough wind to lift a maiden's unbound hair. Both of them released from their duties, and diverting themselves with other courtiers. The young ladies all rode tame little palfreys, but Anne wanted more, and so he put her up on the saddle of his bay and galloped as fast as he dared the length of a meadow, her slender body held safely against his.

Silence was unbearable. "Doctor Dee," he said, without preamble. "He has nothing to do with it, then?"

She rode stiffly, her head turned away from him even though she sat sideways in the saddle. "He claims to speak with angels. I doubt he would speak with us."

Us. She might look human when she chose to, but she was not. *Us* did not include him.

"But you have agents among—among mortals."

"Of course."

"Who? Gilbert Gifford?"

A considering pause. "It depends on which one you mean."

"Which *one*?"

"The Gifford who went to seminary in France was exactly who he claimed to be. The Gifford that now rots in a French jail is someone else—a mortal, enchanted to think himself that man." She sniffed in derision. "A poor imitation; he let himself be arrested so foolishly."

Deven absorbed this, then said, "And the one who carried letters to the Queen of Scots?"

She paused again. Was she doubting her decision to array herself against her sovereign? Deven knew what Walsingham did with double agents who then crossed him in turn. Could he do that to Lune?

"Lord Ifarren Vidar," she said at last. "When he was done, a mortal was put in his place, in case Gifford might be of use again."

Not so long as he was imprisoned in France. Deven asked, "Henry Fagot?"

"I do not know who that is."

How much of this could he trust? She had lied to him for over a year, lied with every particle of her being. He trusted the Goodemeades, but why? What reason had he to trust *any* faerie?

They were nearing the Barbican crossroads. "Where am I going?"

She roused, as if she had not noticed where they were. "We should go in by Cripplegate. I'll use the entrance near to it."

Entrance? Deven turned his horse east at the crossroads, taking them through the sleeping parish of St. Giles. At the gate, he bribed the guards to let them pass, and endured the sly expressions on their faces when they saw he rode with a lady. Whatever the faerie had done to the men at Aldersgate, he did not want to see it happen here.

Then they were back inside the city, the close-packed buildings looming dark and faceless, with only the occasional candle showing through a window. The hour was extremely late. Deven followed Wood Street until she said, "Left here," and then a moment later, "Stop."

He halted his gelding in the middle of Ketton. The narrow houses around them looked unexceptional. What entrance had she meant?

She slipped down before he could help her and made for a narrow, shadowed close. No doubt she would have left him without a word, but Deven said, "'Tis dangerous, is it not? What you go to do."

She stopped just inside the close. When she turned about, Margaret Rolford was gone; the strange, inhuman face had returned.

"Yes," Lune said.

They stared at one another. He should have let her go without saying anything. Now it was even more awkward.

The words leapt free before he could stop them.

"Did you enchant me? Lay some faerie charm upon me, to make me love you?"

Lune's eyes glimmered, even in the near total darkness. "I did not have to."

A moment later she was gone, and he could not even see how. Some door opened—but he could see no door in the wall—and then he was alone on Ketton Street, with only his tense muscles and the rapidly fading warmth along his chest to show there had ever been a woman at all.

THE ONYX HALL, LONDON
26 April 1590

A faerie queen did not process to chapel in the mornings, as a mortal queen might, but other occasion was found for the ceremony that attended Elizabeth's devotions. Invidiana left her bedchamber with an entourage of chosen ladies, acquired an escort of lords in her privy chamber, then passed through a long, columned gallery to the chamber of estate, where a feast was laid for her each day. It was an occasion for spectacle, a demonstration of her power, wealth, and importance; any fae aspiring to favour attended, in hopes of catching her eye.

Lune hovered behind a pillar, her pulse beating so loudly she thought everyone must hear it. This was the moment at which she trusted the Goodemeades, or she did not; she put her life in their hands, or she ran once more, and this time did not return.

A rustling told her that the fae in the gallery were withdrawing to the sides, out of the way of the procession that was about to enter. Hunting horns spoke a brief, imperious fanfare. She risked a glance around the pillar, and saw the Queen. Vidar was not with her, but Dame Halgresta was, and Lord Valentin Aspell, Lady Nianna, Lady Carline… did she want to do this so very publicly?

The moment was upon her. She must decide.

Lune dashed out into the centre of the gallery and threw herself to the floor. She calculated it precisely; her outstretched hands fell far enough short of Invidiana's skirts that the Queen did not risk tripping over her, but close enough that she could not be ignored. Once there, she lay very still, and felt three trickles of blood run down her sides where the silver blades of Invidiana's knights pricked through her gown and into her skin.

"Your Grace," Lune said to the floor, "I bring you a warning of treachery."

No one had run her through—yet. She dared not breathe. One nod from Invidiana...

The cool, measured voice said, "Would this be your own treachery, false one?"

Obedient laughter greeted the question.

"The Wild Hunt," Lune said, "has placed a traitor in your midst."

The hated, growling voice of Dame Halgresta spoke from behind Invidiana. "Lies, your Majesty. Let me dispose of this vermin."

"Lies hold a certain interest," the Queen said. "Entertain me, worm. Who am I to believe a traitor?"

Lune swallowed. "Sir Derwood Corr."

No voices responded to her accusation. She had the name right, did she not?

One of the blades piercing her back vanished, and then Lune cried out as the other two dug in deeper; someone grabbed her by the tattered remnants of her high collar and wrenched her to her feet. Standing, Lune found herself under the blazing regard of a handsome elf knight, black-haired, green-eyed, and transfigured with fury.

"Lying slut," he spat, twisting his left hand in her battered collar. A sword still hovered in his right. "Do you think to rise from where you have been thrown by accusing me, a faithful knight in her Majesty's service?"

Sun and Moon. He did not leave.

Lune dared not look at Invidiana. Even the slightest hint of hesitation... "A faithful knight?" she asked, heavy with derision. “How long have you served the Queen, Sir Derwood? An eyeblink, in the life of a fae. What tests have proved your loyalty to her? Has it been so very strenuous, parading about in your fine black armour, keeping a pleasant smile on your face?” She wished she dared spit, but trapped as she was, it could only go into his face. “Your service is words only. Your heart belongs with the Hunt.”

Corr snarled. “Easy enough for a worm to make a baseless accusation. My service may be new, but it is honest. Where is your proof of my guilt?”

“You received a message last night,” Lune said. “From outside the Onyx Hall.”

For the first time, she saw his confidence falter. “’Tis common enough.”

“Ah, but with whom did you communicate? And what answer did you send back?” She saw a crack, and hammered it. “They say the Hunt is in the north right now. If we send that way, will we find your messenger seeking them? What news does he bear?”

Riders of the Wild Hunt were deadly foes in combat, but they had not the subtlety and nerve to survive in the Onyx Court.

Lune’s collar ripped free as she flung herself backward. Not fast enough: the tip of Corr’s sword raked across the skin above her breast. One of his fellow guardsmen reached for his arm, meaning to stop him; Invidiana did not tolerate murders in front of her that she had not commanded herself. But Corr was too new, and did not understand that. Metal shrieked as his blade skidded uselessly off the other knight's armour.

Curled up tight to protect herself from the feet suddenly thundering around her, Lune did not see exactly what happened to Corr. The press of bodies was too great regardless, with the fae of the Onyx Guard flocking to protect their Queen, and Sir Prigurd wading in with his giant's fists, his normally placid face showing betrayed anger at the failure of his newest protégé.

Corr did his best to sell his life dearly, but in the end, his was the only body that fell.

You should have left, Lune thought, when she heard the rattle of his armour crashing to the floor. *Your true loyalty was too strong. This is no place for faithful knights such as you.*

She did not resist when she was hauled to her feet once more. The guardsman who held her said nothing; he just kept her upright as she lifted her face to Invidiana.

Lune did not see the Queen at first, just the muscled bulk of Dame Halgresta. Then, at an unspoken signal, the Captain of the Onyx Guard stepped aside, abandoning her protective pose, but keeping her wide-bladed sword in hand.

Invidiana's cold black eyes took in the sorry remnants of Lune's gown, the blood that now coated her breast. "Well, worm," she said. "It seems you spoke true—this time."

Lune could not curtsy, with the guardsman holding her. She settled for inclining her head. "I would not have inflicted my presence upon your Grace without great reason." And that was true enough.

Around the two of them, the array of lords and ladies, guardsmen and attendants waited, every last one of them ready to smile or turn away in disdain, following their Queen's lead in how Lune was to be treated now.

"Release her," Invidiana said to the guardsman, and the hands on Lune's shoulders vanished.

Lune immediately knelt.

"You are filthy," Invidiana said in bored tones, as if the very sight of Lune tasted bad. "Truly like a worm. I do not tolerate filth in my court. Have your wounds dressed, and clean yourself before you show your face here again."

"I will most humbly obey your Majesty's command."

The instant Lune rose to a crouch and backed the requisite three steps away, off to one side, the procession reassembled itself and swept onward down the gallery. Only a few goblins remained behind, to collect and dispose of the corpse of Sir Derwood Corr.

Lune permitted herself one glance down at his slack, blood-spattered face. No one would investigate the message he received last night; they would assume it came from the Hunt. But it seemed he *had* sent a reply, and not to the Goodemeades. What had he told the Hunt? That the Goodemeades were interfering?

She needed to warn them. And to apologize for having brought about Corr's death. Lune did not mourn him, but they would.

The stinging cut across her breast, the smaller wounds along her back, gave her all the cause she needed. Some fae at court practiced healing arts, but no one would think it strange if she went to the Goodemeades.

Corr's body, dragged by the heels, scraped along the floor and out of the gallery, leaving a smear of blood behind. Lune lifted her gaze from it and saw those fae still in the chamber staring at her and whispering amongst themselves.

Invidiana had given her leave to wash and be healed. It was a tiny sign of acceptance, but a sign nonetheless. She was no longer to be hunted.

Bearing her head high, Lune exited the gallery, with all the dignity and poise of the favoured lady she no longer aspired to be.

LONDON AND ISLINGTON

26 April 1590

In the morning, it all seemed so terribly unreal.

Colsey's silently disapproving glances chastised Deven for his late return the previous night; the manservant affected to have been asleep when he came in, but Deven doubted it. He had gone to bed straightaway, and suffered uneasy dreams of everyone he knew removing masks and revealing themselves to be fae; now he awoke in brilliant sunlight, with nothing to show for his strange night except a feeling of insufficient sleep.

Had any of it happened?

Deven rose and dressed, then suffered Colsey to shave him, scraping away the stubble Ranwell had left behind. With his face now peeled—Colsey had attended to his task with perhaps a little too much care, as if to show up his upstart fellow—Deven wondered, blankly, what to do with himself.

Whereupon he saw the letter on the windowsill.

Staring at the folded paper as if it were a viper, he did not approach immediately. But the letter stayed where it was, and moreover stayed a letter; at last he drew near and, extending one cautious hand, picked it up.

The top read "Master Michael Deven" in a round, untutored secretary hand. Pressed into the sealing wax was a fragment of dried rose petal.

Deven held his breath and broke the seal with his thumb.

To Master Michael Deven, Castle Baynard Ward, London, from the sisters Gertrude and Rosamund Goodemeade of the Angel in Islington, sub rosa, greetings.

The paper trembled in his hand. Not a dream, then.

We hope this letter finds you well rested and in good health, and we beg your presence at the Angel Inn when occasion shall serve, for there are matters we neglected to discuss with you before, some of them of great importance. Speak your name at the rosebush when you arrive.

Deven exhaled slowly and refolded the paper. Brownies. He was receiving letters from brownies now.

Colsey leapt to his feet when his master came clattering downstairs. "My sword and cloak," Deven said, and the servant fetched them with alacrity. But when Colsey would have donned his own cloak, Deven stopped him with an outstretched hand. "No. You may have another day of leisure, Colsey. Surely after so many days in the saddle, you could do with some time out of it, eh?"

The servant's eyes narrowed. "You're most gracious, master—but no thank you. I'm fit enough to ride some more."

Deven let out an exasperated breath. "All right—I shall be more blunt. You're staying here."

"Why, sir?" Colsey's jaw was set in a determined line. "You know you can trust my discretion."

"Always. But 'tis not a matter of discretion. I simply must go alone."

"You riding to see that necromancer again?"

"There's no evidence of Dee practicing necromancy, and no, I am not going to Mortlake." Deven gave his servant a quelling look. "And you are not to follow me, either."

The disappointed expression on Colsey's face made him glad he'd issued the warning.

Deven hit upon something that would stop him—he hoped. "I am about Walsingham's business, Colsey. And though I trust you, there are others who would not. You will stay behind, lest you foul what I am attempting to do here."

Though Colsey kept the rest of his grumbling objections behind his teeth, Deven imagined he could hear them pursuing him as he rode back out through Aldersgate, retracing the path of black feathers he had followed the night before. Knowing his destination, he rode faster, and came soon to the sturdy structure of the Angel.

He rode past it, tethered his horse, and made his way to the spot behind the inn.

The rosebush was there, looking innocuous in daylight. Feeling an utter fool—but who was there to hear him, if he were wrong?—he approached it, cleared his throat, bent to one of the roses, and muttered, "Michael Deven."

Nothing happened for a few moments, and his feeling of foolishness deepened. But just when he would have walked away, the rosebush shivered, and then there was an opening, with a familiar figure emerging from it.

Familiar, but far too tall. Gertrude Goodemeade arranged her skirts and smiled up at him from a vantage point much closer to his collarbone than his navel. "I am sorry to keep you waiting, but we did not expect you so soon, and I had to put the glamour together."

She still looked herself—just larger. Deven supposed a woman less than four feet tall might attract attention in broad daylight. "Aren't you afraid someone will see us standing here, with the rosebush… open?"

Gertrude smiled cheerily. "No. We are not found so

easily, Master Deven." Bold as brass, she reached out and took his arm. "Shall we walk?"

The rosebush closed behind her as she towed him forward. Deven had thought they might go into the woods behind the Angel, but she led him in quite the opposite direction: to the front door.

Deven hung back. "What is in here?"

"Food and drink," Gertrude said. "Since you do not trust our own."

He had slept far later than his usual hour; now it was the noontime meal. Gertrude secured them a spot at the end of one of the long tables, and perhaps some faerie charm gave them privacy, for no one sat near them. "You can drink the mead here," the disguised brownie said. "'Tis our mead anyway—the very same I gave you last night—but perhaps you will trust it when you see others drink it."

A rumbling in Deven's stomach notified him that he was hungry. He ordered sausage, fresh bread, and a mug of ale. Gertrude looked a trifle hurt.

"We have your best interests at heart, Master Deven," she said quietly.

He met her gaze with moderate cynicism formed during his ride up to Islington. "Within reason. You also wish to make use of me."

"To the betterment of her you serve. But we also wish you to be safe, my sister and I; else we should not have brought you in last night, but left you out where Dame Halgresta could find you." Gertrude lowered her voice and leaned in closer. "That is one thing I wished to warn you of. They know Lady Lune had close dealings with you, and that you served Walsingham; they may yet come after you. Be careful."

"How?" The word came out sharp with resentment. "It seems you can make yourselves look however you wish. Some faerie spy could replace Colsey, and how would I ever know?" The thought gave him a jolt.

Gertrude shook her head, curls bouncing free of the cap on her head. "'Tis very hard to feign being a familiar person; you would know. But 'tis also true that we can disguise ourselves. You have a defence, though." She took a deep breath, then whispered, "The name of your God."

She did not shrink upon uttering the word. Deven took a bite of his sausage, and thought of the bread he had given Lune last night.

"They'll be protected against it, of course," Gertrude said in normal tones. "Most of them, anyhow. But most will still flinch if you say that name, or call on your religion in any fashion. 'Tis the flinch that will warn you."

"And then what?"

The brownie shrugged, a little sheepishly. "Whatever seems best. I would rather you run than fight—many of those she might send against you do not deserve to die—but only you can judge how best to keep yourself safe. And we *do* want you safe."

"Who are 'we,' in this matter?"

"My sister and I, certainly. I have no right to speak for

Lady Lune. But I believe in my heart that she, too, wishes you safe."

Deven stuffed a hunk of bread into his mouth, so he would not have to reply.

Glancing around the inn, Gertrude seemed willing to change the subject. "Tell me, Master Deven: what do you think of this place?"

He chewed and swallowed while he considered the room. The day was sunny and warm; open shutters allowed a fresh breeze into the room, while tallow dips augmented the natural light. Dried lavender and other strewing herbs sweetened the rushes on the floor, and the benches and tables were well scrubbed. The ale in his leather jack was good—surprisingly tart—the bread fresh, the sausage free of unpleasant lumps. What reason had she for asking? "'Tis agreeable enough."

"Have you spent any nights here?"

"Once or twice. The beds were refreshingly clear of unwanted company."

"They should be," Gertrude said with a sniff. "We beat them out any night they are not in use."

"You beat..." Deven's voice trailed off, and he set his bread down.

Her smile had a kind of pleased mischief in it. "Rosamund and I *are* brownies, Master Deven. Or had you forgot?"

He had not forgotten, but he had not yet connected their underground home to the inn—and he should have. As he looked around the room with new eyes, Gertrude went on. "We do a spot of cleaning every night—scrubbing, dusting, mending such as needs it—that has been our task since before there was an Angel, since a different inn stood on this site. Even last night, though I don't mind saying we were a bit rushed to get our work done, after you left."

Deven could not resist asking; he had always wondered. "Is it true you leave a house if the owner offers you clothes?" Gertrude nodded. "Why?"

"Mortal clothes are like mortal food," the brownie said. "Or fae clothes and fae food, for that matter. They bring a touch of the other side with them. Wear them, eat them, and they start to change you. Your average brownie, he'll be offended if you try that with him; we're homebodies, and not often keen to change. But some fae crave that which is mortal. It draws them, like a moth to a candle flame."

The solemnity in her voice was not lost on Deven. "Why did you summon me here, Mistress Goodemeade?"

"To eat and drink in the Angel." She held up one hand when he would have said something in retort. "I am quite serious. I wished you to see this place, to see what Rosamund and I make of it."

"Why?"

"To stop you, before you could grow to hate us." Gertrude reached out hesitantly, and took his hands in her own. Her fingers were warm, and somehow both calloused and soft, as if the gentleness of her touch made up for the marks left by lifetimes of sweeping and scrubbing. "Last night you heard of politics and murder, saw Lady Lune as a fugitive, hiding from a heartless Queen and her minions. The Onyx Court hides a great deal of ugliness behind its beautiful face—but that is not all we are.

"Some of us find purpose and life in helping make human homes warm and welcoming. Others show themselves to poets and musicians, giving them a glimpse of something more, adding fire to their art." She met his gaze earnestly, her dark honey eyes beseeching him to listen. "We are not all to be feared and fought."

"Some fae," Deven said in a low voice, so that others would not hear, "play tricks on mortals—even unto their deaths. And others, it seems, play at politics."

"'Tis true. We have pucks aplenty—bogy beasts, portunes, will-o'-the-wisps. And our nobles have their games, as yours do. But the wickedness of some humans does not turn you against them all, does it?"

"You are not human." Yet it was so easy to forget, with her hands gripping his across the table. "Should I judge you by the same standards?"

Sombreness did not sit well on Gertrude; her face was meant for merriment. "We follow your lead," she said. "There is a realm of Faerie, that lies farther out—over the horizon, through twilight's edge. Some travel to it, mortal and fae alike, and some fae dwell there always. That realm rarely concerns itself with mortal doings. But here, in the shadows and cracks of your world... when your leaders took chariots into battle, ours soon went on wheels as well.

When they abandoned chariots for horses, our elf knights took up the lance. We have no guns among us, but no doubt that will change someday. Even those who do not crave contact with mortals still mimic your ways, one way or another."

"Even love?" He had not meant to say it.

A heartbreaking smile touched Gertrude's face. "Especially love. Not often, but it does happen."

Deven pulled his hands free, nearly upsetting his ale jack. "So you wish me to remember that 'tis your Queen I work against, and not the fae people as a whole." Not Lune. "Is that it?"

"Aye." Gertrude folded her hands, as if she had not noticed the vehemence with which he moved.

"As you wish, then. I will remember it." Deven threw a few coins onto the table and stood.

Gertrude caught up with him at the door. "You should come below for a moment before you leave; I have something for you. Will you do that for me?"

He needed time away from fae things, but he couldn't begrudge the request. "Very well."

"Good." She passed by him, out into the bright sunlight, and called back over her shoulder, "By the by? We also brew their ale."

* * *

THE ANGEL INN, ISLINGTON
26 April 1590

"Mistress Goodemeade." Lune nodded her head formally to Rosamund. "At her Majesty's command, I seek healing for these wounds I have suffered. Few if any in the Onyx Court hold any love for me, given my Queen's recent displeasure; therefore I come here, to ask for aid."

"Of course, my lady." Rosamund offered an equally formal curtsy in response. "Please, come with me, and I will tend to you."

They descended the staircase, and then descended again, and the rose-marked floorboards closed behind them.

"My lady!" Formality gave way to distress. "Cheepkin told us some of what passed, but not all. Sit, sit, and let me see to you."

Lune had no energy to disobey, and no desire to. She let Rosamund press her onto the stool Deven had occupied the previous night—it seemed like ages ago. "I am so very sorry. Corr had not left—"

The brownie clicked her tongue unhappily. "We know. Oh, if he had only listened..."

Deft fingers untied those sleeve and waist-points that had not already broken, then unlaced her bodice at the back. Lune winced as the material of her undergown pulled free where dried blood had glued it to her skin. She would need to obtain new clothing somehow, or else resort to glamours to cover up her tattered state. People would know she wore an illusion, but at least in the Onyx Hall she need not fear it being broken.

Naked to the waist, she closed her eyes while Rosamund dabbed at her cuts with a soft, wet cloth. "I fear I have put you in danger. With Corr there, I had to cast suspicion on him somehow, and I said he had received a message the previous night. If they trace it back to you—"

"Never you mind," Rosamund said. "We would not be here, Gertrude and I, if we could not deal with little problems like that."

"He also seems to have sent a message out, to the Hunt. At least, he panicked when I accused him of it. But I do not know what it said."

The ministering cloth paused. A heartbeat later, it resumed its work. "Something touching on my sister and me, I expect. We shall see."

Lune opened her eyes as Rosamund began daubing her wounds with a cool, soothing ointment. "Lord Valentin questioned me before I left. Where I had gotten mortal bread—I told him a simple lie—and how I had found out about Corr. They found Francis's body while I was there. I led Aspell to believe the two were connected."

"Then we must be sure they do not catch the messenger. Does this feel better?"

"Very much so. Thank you." The fire seemed to have the knack of warming the room just enough; the cool, damp chill of an underground chamber was perfectly offset, so Lune did not shiver as Rosamund fetched bandages from a small chest. At least not from cold.

The brownie swathed her ribs and collarbone in clean white linen, with soft pads over the cuts themselves. "They should be well in three days," Rosamund said, "and you may take the bandages off after one."

Before Lune could say anything more to that, footsteps sounded above. She had not heard anyone speak through the rosebush, as she had when Dame Halgresta came the previous night. Gertrude, no doubt, but her entire body tensed.

The floor bent open, and the brownie's feet appeared on the top stair, in stout slippers. But a pair of riding boots followed, belonging to someone much larger.

Lune snatched up the bodice of her gown just as Michael Deven came into view.

"Oh!" Gertrude exclaimed, as Deven flushed scarlet and spun about. The floor had already closed behind him; unable to escape, he kept his back resolutely turned. "My lady, I am so very sorry. I did not know you were here."

Lune did not entirely believe her. Irritation warred with an unfamiliar feeling of embarrassment as Rosamund helped her into the stained remnant of her clothing. Fae were often careless of bodily propriety among themselves, particularly at festival time, but mortals were another matter. Especially *that* mortal.

"I just wanted to give Master Deven a token," Gertrude said, opening a chest that sat along one wall. "So our birds can find him if he isn't at home. They will carry messages for you, Master Deven, should you need to send to us. Lady

Lune, I would give you one as well—"

"But it might be found on me," Lune finished for her. The sleeves of her dress were not yet reattached, but at least she was covered now. "I quite understand."

"Aye, exactly." Gertrude carried something over to where Deven yet stood on the staircase; it looked like a dried rosebud, but seemed much less fragile. "Here you are."

He moved enough to accept the token and examine it. "Roses again, I see."

Gertrude clicked her tongue. "*I* would have planted something other than a rosebush, but my sister was so very fond of the notion. Now everything we do is roses, and everyone always thinks of her. I should have had a flower in my name."

Rosamund answered her with mild asperity, and the two sisters bickered in friendly fashion while they helped Lune finish dressing. It lowered the tension in the room, as no doubt they intended, and after a few moments Deven risked a glance over his shoulder, saw Lune was decent again, and finally turned to face them all.

"I did not mean to burst in thus," he said to her, with a stiff bow. "Forgive me."

The rote apology hit Lune with far more force than it should have. His eyes were a lighter blue than the seer's had been, and his hair brown instead of black—he had none of the fey look brought on by life in the Onyx Hall—but in her memory, a wavering, nearly inaudible voice echoed him, "*Forgive me.*"

"Rosamund," she said, cutting into the amiable chatter of the two sisters. "Gertrude. Last night... I did not think to ask; too much else was happening. But before he died, Tiresias—Francis spoke a name. Begged forgiveness of her. A fae woman, I think. Suspiria."

She expected the brownies would recognise the name. She did not expect it to have such an effect. Both sisters gasped, their faces suddenly stricken, and tears sprang into Gertrude's eyes.

Startled, Lune said, "Who is she?"

Rosamund put one arm around her sister's shoulders, comforting her, and said, "A fae woman, aye. Francis loved her dearly, and she him."

Such romances often ended in tragedy, and more so under Invidiana's rule. "What happened to her?"

The brownie met her gaze gravely. "She sits on a throne in the Onyx Hall."

The notion was so incredible, Lune found herself thinking of the Hall of Figures, trying to recall any enthroned statues there. But Rosamund met her gaze, unblinking, and there was only one throne in all the buried palace, only one who sat upon it.

Invidiana.

The cold, merciless Queen of the faerie court, who could no more love a mortal—love anyone—than winter could engender a rose. Who kept Tiresias as the most tormented of her pets, bound by invisible chains he could only break in death. That Francis Merriman might once have loved her, Lune could almost believe; mortals often loved where it was not wise. But Rosamund said Invidiana loved him in return.

Gertrude said to her sister, through her sniffles, "I told you. He remembered her. Even when his mind was gone, when everything else was lost to him, he did not forget."

Deven was staring at them all, clearly lost. Lune was not certain even she followed it. "If this is true—why did you not speak of it before? Surely this is something we needed to know!"

Rosamund sighed and helped Gertrude onto a stool. "You are right, my lady. But last night, we had no time; we had to get you back to the Onyx Hall, before someone could suspect you of Francis's death. And you were distraught, Lady Lune. I did not wish to add to it."

Lune thought of her confrontation with Invidiana that morning. "You mean, you did not wish me to face Invidiana, try to regain her goodwill, knowing the man who died at my feet had once loved her."

The brownie nodded. "As you say. You are good at dissembling, Lady Lune—but could you have done that, without betraying yourself?"

Claiming a seat for herself, Lune said grimly, "I will have to, now. What have you not said?"

Deven leaned against the wall, arms folded over his chest and that shuttered look on his face; it was a mark of Rosamund's own distress that she did not try to coax him into sitting down, but perched on the edge of one of the beds and sighed. "'Tis a long story; I pray you have patience.

"Gertrude and I once lived in the north, but we came here... oh, ages ago; I don't remember when. Another inn stood on this spot then, not the Angel. The mortals had their wars, and then, when a new king took the throne, the first Henry Tudor, a woman arrived on our doorstep.

"She..." Rosamund searched for words. "We thought she was in a bad way when we saw her. Later, we saw how much worse it could be. Suspiria was cursed, you see, for some ancient offense. Cursed to suffer as if she were mortal. 'Twasn't that she was old; fae can be old, if 'tis in their nature, and yet be very well. She *aged*. She sickened, grew weak—suffered all the infirmity that comes with mortality, in time."

Deven made a small noise, and the brownie looked up at him. "I know what you must think, Master Deven. Oh, how terrible indeed, that one of our kind should suffer a fate every mortal faces. I do not expect you to have much sympathy for that. But imagine, if you can, how it would feel to suffer so, when 'tis a thing *not* natural to you."

Whether he felt sympathy or not, he gave no sign. Rosamund went on. "She told us she was condemned to suffer thus, until she atoned for her crime. Well, for ages she had thought her suffering *was* atonement, like the penance mortals do for their sins. But she had come to realise that she must do more—that her suffering would continue until she made up for what she had done wrong."

"A moment," Deven said, breaking in. "How elderly

was she, if she had suffered 'for ages'? There's a limit to how old one can become. Or was she turning into a cricket, like Tithonus?"

Gertrude answered him, her voice still thick. "No, Master Deven. You are quite right: it cannot go on forever. She grew old, and when the span a mortal might be granted was spent, she... died, in a way. She shed her old, diseased body and came out young and beautiful once more, to enjoy a few years of that life before it all began again."

Lune felt sick to her stomach. It was one thing to don the appearance of mortality, as a shield. To sicken and die like a mortal, though—to crawl out of rotting, degraded, liver-spotted flesh, and know to that she must come again—

"We helped her as best we could," Rosamund said. "But her memory suffered like a mortal's; she could not clearly recall what the cause was for which she had been cursed. She knew, though, that her offense had happened here, in the place that became London, and so she had returned here, to seek out those who might know what she should do." The brownie laughed a little, more as if she remembered amusement than felt it. "We thought her mad when she told us what plan she had formed, to lift her curse."

A hundred possibilities sprang to Lune's mind, each madder than the last. More to stop her own invention than to prod Rosamund onward, she said, "What was it?"

The brownie shook her head, as if she still could not believe it. "She vowed to create a faerie palace, beneath all the city of London."

Lune straightened. "Impossible. The Onyx Hall—she cannot have made it."

"Oh?" Rosamund gave her a small smile. "Think, my lady. Where else in the world do you know of such a place? Where else is faerie magic so proof against the powers of iron and faith? Fae live in forests, glens, hollow hills—not cities. Why is there a palace beneath London?"

Rosamund was right, and yet the thought stunned her. Miles of corridor and gallery, hundreds or even thousands of chambers, the Hall of Figures, the night garden, the hidden entrances... the magnitude of the task dizzied her.

"She had help, of course," Rosamund said, as if that somehow reduced it to a manageable scale. "Oh, tremendous help—but one person especially."

Gertrude whispered, "Francis Merriman."

Her sister nodded. "A young man Suspiria had come to know. She met him after her body had renewed itself, and she was desperate to keep him from ever seeing her old, to lift her curse before it came to that. But I think he knew anyhow. He had the gift of sight—visions of the future, or of present things kept secret." Her expression trembled, holding back tears. "She often called him her Tiresias."

Deven looked on, not comprehending. Of course: he did not understand how that name had been warped. It wasn't just that Francis Merriman had been obliterated; the man had become one of a menagerie of human pets, a term of love become a term of control.

"So she lifted her curse," Lune said. Gertrude was

sniffling again, making the silence uncomfortable.

To her surprise, Rosamund shook her head. "Not then. She created the Hall, but when it was done, Suspiria still aged as she had before. She hid behind glamours, to keep Francis from knowing. And oh, it pained her—seeing him stay young, living as he did in the Onyx Hall with her, while she grew ever older. But he knew, and a good thing, too; 'twas him helped her lift the curse at last, one of his visions. Not long after that Catholic woman took the throne, it was." The look of sorrow was back. "We were all so happy for her."

Deven shifted his weight, and the tip of his scabbard scraped against the plastered wall. "Four or five years later—if my sums are right—you say she formed this pact."

Rosamund sighed. "She did *something*. In one day, not only did she become the only faerie queen in all of England, she erased Suspiria from the world. After her curse was lifted, she had begun to gather a court around her; that, Lady Lune, is when Vidar came to the Onyx Hall. Before she was crowned. But he would no more recognise the name Suspiria than he would remember the court he once belonged to. To him, as to everyone else, there has only ever been Invidiana: the cruel mistress of the Onyx Hall."

Attempting to dry her face with a mostly soaked handkerchief, Gertrude whispered, "But we remember her. And that is why we do not help the Wild Hunt. They would tear down the Onyx Hall, every stone of it, scatter its court to the four corners of England… and they would kill Invidiana. And though she is lost to us, we do not wish to see Suspiria die."

Deven straightened and fished a clean handkerchief out of his cuff, offering it to Gertrude. She took it gratefully and repaid him with a watery, wavering smile.

Lune sat quietly, absorbing this information, trying to fit it alongside the things she had seen during her years in the Onyx Court. Fewer years than she had thought. Even the palace itself was new, by fae standards. "You were fond of Suspiria."

"Aye," Rosamund said, unapologetic. "She was warmer then, and kinder. But all kindness left her that day. You have never known the woman we helped."

Nor the woman Francis had loved.

Deven came forward and placed his hands along the edge of the table, aligning them with studious care. "So how do we break the pact?"

Now Rosamund gave a helpless shrug. "We only just learned of its existence, Master Deven. And I imagine the list of people who know its terms is short, indeed. If Francis knew, he died before he could say."

"Which leaves only two that must know," Lune said. "Invidiana and Elizabeth."

"Assuming we are right to begin with," Deven still had his eyes on his carefully placed hands. "That the pact was formed with her."

"Assuming *you* are right," Lune countered, a little sharply. "You are the one who suggested it last night."

The minuscule slump in his shoulders said he remembered all too well, and regretted it—but his silence told her he had no better explanation to offer in its place.

A muscle rose into relief along his jaw, then subsided. "I do not suppose you could trick your Queen into revealing the terms of the pact?"

The sound Lune made was nothing like a laugh. "You are asking me to trick the most suspicious and politically astute woman I have ever met."

"Elizabeth is the same," he flared, straightening in one fluid motion. "Or do you think my Queen a greater fool than yours?"

Lune met his gaze levelly. "I think your Queen less likely to have one of her courtiers murder you for an afternoon's entertainment."

She watched the contentious pride drain out of his face, one drop at a time. At first he did not believe her; then, as her stare did not waver, he did. And when she saw him understand, an ache gripped her throat, so sudden it brought tears to her eyes. What had life been like, when she lived under a different sovereign? She wished she could remember.

Lune rose to her feet and turned away before he could see her expression break. Behind her back, she heard Deven murmur, "Very well. I will see what I may learn. 'Twill not be easy—" He gave the quiet, rueful laugh she remembered, and had not heard in some time. "Well. Walsingham taught me how to ferret out information that others wish to keep hidden. I never expected to use it against a faerie queen, is all."

"Let us know what you learn," Rosamund said, and Gertrude echoed her after blowing her nose one last time. They went on, but Lune could no longer bear to be trapped in the claustrophobic hidden room with the three of them.

"I should return," she said, to no one in particular. "I have been here too long already." She went up the stairs before remembering the floor was closed above her, but it opened when her head neared its planks, two feminine farewells pursuing her as she went. Lune paused only long enough to restore the glamour she had dropped, and began her journey back to the confines of the Onyx Hall.

MEMORY

12 November 1547

The twisting web of streets, the leaning masses of houses and shops, alehouses and livery halls—it all obscured an underlying simplicity.

In the west, Ludgate Hill. Once home to a temple of Diana, now it was crowned with the Gothic splendour of St. Paul's Cathedral.

In the east, Tower Hill, the White Mount. The structure atop it had once been a royal palace; now it more often served as a prison.

In the north, the medieval wall, curving like the arc of a

bow, pierced by the seven principal gates of the city.

In the south, the string of the bow: the straight course of the Thames, a broad thoroughfare of water.

An east–west axis, stretching from hilltop to hilltop, with temporal power on one end, spiritual power on the other. A north–south axis, barrier in the north, access in the south, with the Walbrook, the *wall-brook,* bisecting the city and connecting the two poles.

The buried waters of the Walbrook ran hard by the London Stone, which lay very near to the centre of the city. Near enough to suffice.

A shadow moved through the cloudless autumn sky.

Two figures stirred within the solid earth and stone of the hills. Unseen, their colossal bodies standing where there was no space for them, they reached out and took hold of the power of the earth, which was theirs to command.

Two more stood at the London Stone, blind to the activity of the city around them. A man and a woman, a mortal and a fae.

They waited, as the light around them began to dim.

Slowly, one person at a time, the bustle of the city's streets began to falter and halt. Faces turned upward; some people fled indoors. And the world grew ever darker, as the shadow of the moon moved across the face of the sun, until only a ring of fire blazed around its edge.

"Now," the woman whispered.

The giants Gog and Magog, standing within the hills of Ludgate and the Tower, called upon the earth to obey. The

Roman well that lay at the foundations of the White Tower shivered, its stones trembling; an ancient pit used in the rites of Diana opened up once more below the cathedral; and at the bottom of each, something began to grow.

Standing at the London Stone, Suspiria and Francis Merriman reached out and linked hands, mortal and fae, to carry out a working the likes of which the land had never dreamed.

The shadowed light of the sun fell upon the city and cast stranger shadows, a penumbral reflection of London, like and yet unlike. It sprang forth from the buildings, the streets, the gardens, the wells, and sank downward into the ground.

In the earth beneath London, the shadows took shape. Streets became corridors; buildings, great chambers. They transformed as they went, twisting, flowing, settling into new configurations, defying the orderly relations of natural geometry. And then, when all was in place, stone sprang forth, black and white marble, crystal, onyx, paving the floors, sheathing the walls, supporting the ceilings in round half-barrels and great vaulting arches.

Together they made this, Suspiria and Francis, drawing on the fae strength of the giants; the mortal symbolism of the wall; the wisdom of Father Thames, who alone of all beings understood the thing that was London, having witnessed its growth from its earliest days. In the sun's shadowed light, they formed a space that bridged a gap, creating a haven for fae among mortals, from which church bells could not drive them forth.

Their hands came to rest atop the London Stone. The light brightened once more; the moon continued along its course, and normalcy returned to the world.

They smiled at one another, exhausted, but exultant. "It is done."

PALACE OF PLACENTIA, GREENWICH
28–30 April 1590

Even the sprawling reaches of Hampton Court and Whitehall did not have room to house every courtier, merchant, and visiting dignitary that came seeking audience with the Queen and her nobles, especially not with their servants and train. Deven had asked for and received a leave of absence, with the result that when the court removed to Greenwich, he had no lodging assigned to him. He might have troubled Lord Hunsdon for one, especially as courtiers retired for the summer to their own residences, but it was simpler to take rooms at a nearby inn. From this staging point, closer to court than his London house but not in its midst, he tried to plan a course of action.

Judicious questions to the right people netted him a fuller story of Elizabeth's coronation, including those who had been involved. Deven could not rule out the possibility that the Queen was not, in fact, the other party to this rumoured pact; it might have been another. Lord Burghley leapt to mind. Sir William Cecil, as he was back then, had been a trusted adviser since the earliest days of the reign, and nothing short of the death he had put off for seventy years would make him retire. Moreover, he had taught Walsingham much of what the man knew about how to build an intelligence service.

Burghley was a good candidate. He might do a great deal to ensure his Queen stayed on her throne. But the question remained of how to approach him—or indeed, anyone else—about the matter.

I most humbly beg your pardon. But did you by any chance form a pact with a faerie queen thirty-one years ago?

He could not ask that question of anyone.

Deven supposed he had at least one advantage. Lune's bleak eyes had stayed with him, her resigned expression as she spoke so plainly of her Queen's murderous entertainments. Whatever other obstacles he faced—however much Elizabeth might rage and occasionally threaten to chop off someone's head—at least he did not fear for his life when in the presence of his sovereign.

How to do it? For all his fine words about ferreting out hidden information, Deven could not fathom how to begin. He was half-tempted to ask Lune to return to court as Anne Montrose, and let her handle the matter; if people imagined her to be mad, she lost nothing. Deven, on the other hand... he would be lucky if they simply thought him mad. Faeries were plausible; faeries beneath London, less so.

But the true danger would be if they believed him. It was a short step from faeries to devils, from lunacy to heresy. And even a gentleman could be executed for that.

If only he could have discovered this all before Walsingham died! Deven did not know how the Principal Secretary would have reacted, but at least then he could have shared it with someone. Walsingham, he was sure, would have believed, if shown the evidence. But Deven had been too slow; he had not completed the task his master set him until it was too late.

The thought came to him as he walked the bank of the Thames, the river wind blowing his hair back until it stood up in ruffling crests. He had done all of this because Walsingham asked it of him.

And therein lay the opening he needed.

He went to Lord Hunsdon for help. Beale could have done it, no doubt, using his influence as a secretary to the privy council, but Beale knew too much of what he was about, and would have asked too many questions. Hunsdon's aid was more easily obtained, though he was manifestly curious about Deven's purpose, and his recent absence from court.

A gift for Hunsdon; a gift for the Countess of Warwick; a gift for the Queen. Deven wondered about faerie gold, and whether the Goodemeades could not somehow fund the expense of this work. He was not at all certain he wanted to know.

His opportunity came on a crisp, bright Thursday, when the wind sent clouds scudding across the sun and the Queen rode out to hunt. She was accompanied, as always, by a selection of her ladies, several other courtiers, and servants to care for the hounds and hawks and other accoutrements that attended upon her Majesty; it seemed a great menagerie, when he thought about watching eyes, listening ears. But it was the best he could hope for.

"How stands the Queen's mood?" he asked Lady Warwick, as he rode out with the others into the unreliable brilliance of the morning.

The countess no doubt thought his question had something to do with Anne. "As changeable as the weather," she said, casting one eye skyward, at the racing clouds. "Whatever suit you wish to press, you might consider waiting."

"I cannot," Deven muttered. Even if their situation could wait—which he was not certain it could—his nerve could not withstand delay. "You and Lord Hunsdon have been most generous in arranging this private conference for me. If I do not take my opportunity today, who knows when it will come again?"

"Then I wish you good fortune, Master Deven."

With those reassuring words, the hunt began. Deven did not devote more than a sliver of his attention to its activity, instead rehearsing in his mind the words he would say. At length the hunt dismounted for a rest, and servants began to erect a pavilion in which the Queen would dine with the Earl of Essex. He saw Lady Warwick approach her, bearing in her hands the small book Deven had obtained from his father, and present it to the Queen. A murmured conversation, and then Elizabeth turned a sharp eye on him, across the meadow in which they rested.

The long-fingered hand beckoned, jewels flashing in the light; he crossed to where she stood and knelt in the grass before her. "Your Grace."

"Walk with me, Master Deven."

The beginnings of a headache were pulsing in his temples, keeping time with his thunderous heartbeat. Deven rose and followed the Queen, one respectful pace behind her, as she wandered the edge of the meadow. There were too many people around, passing into and out of earshot, but he could hardly ask her to withdraw farther; it was favour enough that she was granting him this semi-private audience.

"Lady Warwick tells me you bear a message of some importance," Elizabeth said.

"I do, your Majesty." Deven swallowed, then launched into the words he had rehearsed all morning, and half the day before.

"Prior to his death, Sir Francis set me a task. Were I a cleverer or more talented man, I might have completed it in time to share my discoveries with him, but I am come to my conclusions too late; only in the last few days have I uncovered the information he wished me to find. And in his absence, I have no master to whom 'tis fitting to report such matters. But I swore an oath not to conceal any matters prejudicial to your Grace's person, and with the loss of the Principal Secretary, my allegiance is, by that oath, to you alone." He wet his lips and went on. "Though it be presumptuous of me, I believe this issue of sufficient import as to be worth your Grace's time and attention, and your wisdom more than sufficient to judge how best to proceed."

Walking a pace behind Elizabeth, he could only see the edge of her face, but beneath the cosmetics he thought he discerned a lively interest. Walsingham to an extent, and Burghley even more, made a practice of trying to keep intelligence from the Queen; they preferred to control the information that reached her, so as to encourage her decisions in directions they favoured. But Elizabeth disliked being managed, and had a great fondness for surprising them with knowledge they did not expect her to have.

"Say on," she replied, her tone now more on the pleasant side of neutral.

Another deep breath. "The task the Master Secretary set me was this. He believed he had discerned, within the workings of your Grace's government, the hand of some unseen player. He wished me to discover who it is."

She was too experienced a politician to show surprise. Elizabeth's energetic stride did not falter, nor did she turn to look at him. But Deven noticed that their seemingly aimless wanderings now drifted, ever so slightly, toward a stand of birches that bordered the meadow. Away from those who might listen in.

"And you believe," the Queen said, "that you have discovered some such player?"

"I have indeed, madam. And having done so, I thought it all the more crucial that I convey this information to you alone."

They were far enough away; no one would overhear them. Elizabeth stopped and turned to face him, her back to the white trunks of the trees. Her aged face was set in unreadable lines. A cloud covered the sun, then blew away again, and Deven thought uneasily that perhaps he should have waited to find her in a fairer mood, after all.

"Say on," she commanded him again.

Too late to back out. Deven said, "Her name is Invidiana." He should have knelt to deliver the information; it would have been respectful. But he had to stand, because he had to be looking her in the face as he said it. This was his one chance to see her reaction, the one time she might betray some hint that would tell him what he needed to know. And even then, he almost missed it. Elizabeth had played this game for decades; she was more talented an actor than most who made their living from it. Only the tiniest flicker of tension at the corners of her eyes showed when he spoke the name: there for an instant, and then gone.

But it was there, however briefly.

Now Deven dropped to his knees, his heart fluttering so wildly it made his hands shake. "Your Majesty," he said, heedless of whether he might be cutting her off, desperate to get the words out before she could say anything, deny anything. "For days now I have thought myself a madman. I have met—people—spoken to them, heard stories that would be incredible were they played upon a stage. But I know them to be true. I have come to you today, risked speaking of this so openly, because events are in motion which could bring an upheaval as great as that threatened by Spain. Consider me a messenger, if you will."

And with that he halted; he could think of nothing more to say. The light shifted around him, and the wind blew more strongly, as if a storm might be on its way.

From above him, Elizabeth's measured, controlled voice. "She sent you to me?"

He swallowed. "No. I represent… others."

Footsteps approached; a rustle of satin, as Elizabeth gestured whomever it was away. When they were alone again, she said, "Explain yourself."

Those two words were very, very cold. Deven curled his gloved hands into fists. "I have come into contact with a group of… these people, who believe that a pact exists between their Queen and someone in your Majesty's own court—perhaps you yourself. The man who told them of this pact was of our own kind, and had long dwelled among them, but he died in the course of confessing this information. He claimed the pact was detrimental to both sides. They wish it to be broken, and have asked me to discover its nature and terms."

How he wished he could see her face! But Elizabeth had not told him to rise, nor did she interrupt his explanation. He had no choice but to continue. "Madam, I know not what to think. They say she is not their rightful Queen, that she deposed many others across England when she ascended to her throne. They say she is cold and cruel—that, at least, I most sincerely believe, for I do not think they could counterfeit such fear. They say their aid has helped maintain your Grace's own safety and security, and perhaps this is true. But if so..." His heart was hammering so loudly, the entire camp must be able to hear it. "I do not know if this pact *should* be broken. Even if I knew its terms, that is not a decision for me to make. All I can do, in good conscience, is lay what I know at your feet, and beg your good wisdom and counsel."

The long speech left his mouth bone dry. How many people were watching them discreetly, wondering what private suit drove him to his knees, with his face so pale? Did Elizabeth show anger, confusion, fear?

He might have just ended his career at court, in one disastrous afternoon.

Deven whispered, "If your Majesty is caught in some bargain from which you would escape, you have but to say so, and I will do everything I may to end it. But if these creatures are your enemies—if they threaten the security of your throne—then bid me stop them, and I will."

The sunlight flickered, then shone down with renewed strength. His linen undershirt was soaked with sweat.

Elizabeth said in courteous, impassive tones, "We thank you, Master Deven, and will take this information under advisement. Speak of this to no other."

"Yes, madam."

"Luncheon is served, it seems. Go you and eat, and send Lord Essex to me."

"I humbly take my leave." Deven rose, not looking at

her, backed away three steps, and bowed deeply. Then he fled, wishing it would not be an insult to quit the hunt early, before anyone asked him questions he could not answer.

MOOR FIELDS, LONDON

1 May 1590

The celebrations began in the hours before dawn, and would fade away with the morning light. To dance out here—in the open, under the stars, yet just outside the city walls—was an act of mad defiance, a fleeting laugh at the masses of humanity from which they ordinarily hid, holding their revels underground, or in wilder places. It also required a tremendous outlay of effort.

The laundresses' pegs and the archers' marks that normally dotted the open places of Moor Fields had been cleared away. The grass, trodden into dusty brownness and hard-packed dirt, was briefly, verdantly green, growing in a thick carpet that cushioned the bare feet of the dancers. The dark, sombre tones that predominated in the Onyx Hall had given way to riotous colour: pink and red and spring green, yellow and blue and one doublet of violent purple. Flower petals, fresh leaves, feathers whose edges gleamed with iridescent light; the garb tonight was all of living things, growing things, in honour of the first of May.

And the fae of the Onyx Court danced. Musicians wove competing tapestries in the air, flutes and hautbois and tabors sending forth sound and light and illusions that ornamented the dance. Orpheus wandered the edges, serenading the many lovers. Blossoms sprang up where he walked. Great bonfires burned at the four corners of their field, serving more than one purpose; they provided heat, light, fire for the festival, and foundation points for the immense web of charms that concealed all this revelry from watching eyes.

When the sun rose, mortals would go forth for their own May Day celebrations. They would pick flowers in the woods, dance around maypoles, and enjoy the onset of benevolent weather. But a few had started early: here and there, a human strayed near enough to the fires to pierce the veils that concealed them, and become aware of the crowds that had overtaken Moor Fields. A young man lay with his head in Lady Carline's lap, eating grapes from her fingers. Another scrambled on the ground in front of her, rump in the air, behaving for all the world like a dog in human form—but for once, those who laughed at him did so without the edge of cold malice their voices would ordinarily have borne. Maidens whirled about the dancing ground with faerie gentlemen who wove blossoms into their hair and whispered sweet nothings into their ears. Nor was everyone young: a stout peasant woman had wandered from her house on Bishopsgate Street to chase a dog not long after sundown on May Eve, and now stamped a merry measure with the best of them, her face red and shining with effort.

Amidst all this splendour, one figure was conspicuous

by her absence: conspicuous, but not missed. The Wild Hunt could more easily strike at this open field than at the subterranean confines of the Onyx Hall, and so Invidiana stayed below.

They had more fun without her.

The Queen's absence helped Lune breathe more easily. With wine flowing like water, everyone was merry, and many of them forgot to snub those who deserved snubbing. Nor did the snared mortals have any notion of politics. Shortly after midnight, a young man stumbled up to her, wine cup in hand, mouth languorous and searching for a kiss. He had brown hair and blue eyes, and Lune pushed him away, then regretted the violence of her action. But she did not need the reminder of Michael Deven, and the celebrations Elizabeth's court would engage in today.

Even on a night such as this, politics did not entirely cease. Everyone knew Tiresias was dead; everyone knew the Queen had been little seen by anyone since his body was discovered. A few thought she mourned him. Remembering the Goodemeades' tale, Lune felt cold. Invidiana mourned no one.

But his death created opportunity for those who needed it. Some who questioned Lune thought themselves subtle about it; others did not even try for subtlety. Certain mortals claimed the ability to foretell the future. Were any of them truly so gifted? Lune had lived among the mortal court; she might know something. They pestered her for information. Had she met Simon Forman? What of Doctor Dee? Did she perhaps know of any persuasive charlatans, who might be put forth as bait to trip up a political rival?

Lune joined the dancing to escape the questions, then abandoned dancing when it turned her mood fouler. There was no surcease for her here. But where would she go? Back down into the Onyx Hall? Its confines were unbearable to her now—and the Queen waited below. To the Angel Inn? She did not dare spend too much time there, and besides, the Goodemeades were here, along with every other fae from miles around. Lune knew the Goodemeades watched her, but she kept her distance.

A golden-haired elf lady she knew by sight but not name waylaid her. Was she familiar with John Dee? Where did he live? Was he old enough that it might not seem suspicious if he died in his sleep?

Lune fled her questioner, heading for one of the bonfires. Arriving at its edge, where the heat scorched her face with welcome force, she found there was one other person gazing into its depths.

From the far side of the bonfire, the hollow-cheeked, wasted face of Eurydice stared at her.

The mortal pet's presence at the May Day celebrations was like a splash of cold water from the Thames. Her black, sunken eyes saw what few others did: the spirits of the dead, those restless souls who had not passed on to their punishment or reward. And this was not All Hallows' Eve, not the time for such things.

But she did more than see. Few fae realised Eurydice was not just a curiosity to Invidiana; she was a tool. She not only saw ghosts: she could bind them to her will. Or rather, the Queen's will.

Lune knew it all too well. Invidiana had formed plans that depended closely on Eurydice's special skill, plans that Lune's disastrous embassy had undone. The folk of the sea wanted for little, and so the things she had gone there to offer them went unremarked. What they had wanted were the spoils of their storms: the souls of those sailors who drowned.

To what use Invidiana would have put such a ghost army, Lune did not know. Had she been aware that her Queen planned to create one, she might have bargained harder; the folk of the sea had no way to bind ghosts to their service. But she thought it a harmless thing, and so she agreed that Eurydice would come among them for a time, provided the ghosts were not turned against the Onyx Court. As long as the ships never reached England's shores, what did it matter?

Invidiana had seen it differently.

Eurydice's mouth gaped open in a broken-toothed, hungry grin. And suddenly, despite the blazing bonfire just feet away, Lune felt cold.

Ghosts.

Those who died in the thrall of faerie magic often lingered on as ghosts.

Francis.

Somehow, she kept herself from running. Lune met

Eurydice's gaze, as if she had no reason to fear. That hungry grin was often on the woman's face; it meant nothing. She had no assurance that Francis Merriman had lingered. After so long trapped in the Onyx Hall, his soul might well have fled with all speed to freedom and judgment.

Or not.

What did Invidiana know?

A chain of dancing fae went past, and someone caught Lune by the hand. She let herself be dragged away, following the line of bodies as they weaved in and out of the crowds of revellers, and did not extricate herself until she was at the far side of the field, safely distant from Eurydice's ghost-haunted eyes.

She should run now, while she could.

No. Running would bring her no safety; Invidiana ruled all of England. And there might be nothing to run from. But she must assume the worst: that the Queen had Francis's ghost, and knew from him what had transpired.

Why, then, would Lune still be alive?

Her mind answered that question with an image: a snake, lying with its jaws open and a mouse in its mouth, waiting. 'Tis safe, come in, come in. Why eat only one mouse when you might lure several? And that meant she could not follow her instinct, to run to the safety the Goodemeades offered. Invidiana could act on suspicion as well as proof, but would want to be sure she caught the true conspirators, and caught all of them. As long as she was not certain…

Lune stayed at the May Day celebrations, though it took

all her will. And in the remaining hours of dancing, and drinking, and fielding the questions of those who sought a new human seer, she caught one moment of relative privacy, while Rosamund dipped her a mug of mead.

"He may be a ghost," Lune whispered. It was all the warning she dared give.

PALACE OF PLACENTIA, GREENWICH
2 May 1590

In the days following his audience with the Queen, Deven considered abandoning his lodgings and returning to where Ranwell waited at his house in London. What stopped him was the thought that there, he would be sitting atop a faerie palace.

So he was still at Greenwich, though not at court, when the messenger found him.

He threw Colsey into a frenzy, demanding without warning that his best green satin doublet be brushed off and made ready, that his face needed shaving again already, that his boots be cleaned of infinitesimal specks of mud. But one did not show up looking slovenly when invited to go riding with the Queen.

Somehow his manservant got him out the door with good speed. Deven traversed the short distance to the palace, then found himself waiting; something had intervened, and her Majesty was occupied. He paced in a courtyard, his stomach twisting. Had he eaten anything that day, it might have come back up.

Nearly an hour later, word came that Elizabeth was ready at last.

She was resplendent in black and white satin embroidered with seed pearls, her made-up face and hair white and red above it. They did not ride out alone, of course; Deven might be one of her Gentlemen Pensioners, and therefore a worthy bodyguard, but one man was not sufficient for either her dignity or well-being. But the others who came kept their distance, maintaining the illusion that this was a private outing, and not a matter of state.

Everyone at court, from the jealous Earl of Essex down to the lowliest gentlewoman, and probably even the servants, would wonder at the outing, and speculate over the favour Elizabeth was suddenly showing a minor courtier. For once, though, their gossip was the least of Deven's concerns.

They rode in silence to begin with. Only when they were well away from the palace did Elizabeth say abruptly, "Have you met her?"

He had expected some preface to their discussion; her sudden question took him by surprise. "If you mean Invidiana, your Grace, I have not."

"Consider yourself fortunate, Master Deven." The line of her jaw was sagging with age, but steel yet underlay it. "What do you know of this pact?"

Deven chose his words with care. "Little to nothing, I fear. Only that on your Majesty's coronation day,

Invidiana claimed her own throne."

Elizabeth shook her head. "It began well before that."

The assertion startled him, but he held back his instinctive questions, letting the Queen tell it in her own time.

"She came to me," Elizabeth said softly, "when I was in the Tower." Her eyes were focused on something in the distance, and she controlled her horse with unconscious ease. Deven, watching her out of the corner of his eye, saw grimness in her expression. "My sister might have executed me. Then a stranger came, and offered me aid."

The Queen fell silent. Deven wanted to speak, to tell her that anyone might have made the same choice. Years later, there was still doubt in her, uncertainty about her actions. But he dared not presume to offer her forgiveness.

Elizabeth pressed her lips together, then went on. "She arranged my release from the Tower, and a variety of events that helped secure my accession. I do not know how much of that was her doing. Not all, certainly—even now, she does not have that much control. But some of it was hers. And in exchange, when I was crowned, I aided her. My coronation was hers as well." The Queen paused. "I did not know that it deposed others. But I would be false if I said that surprised me."

She hesitated again. At last, Deven prodded her onward. "And since then, your Grace?"

"Since then… it has continued. She has helped remove threats to my person, my throne, my people." Elizabeth's hands, encased in grey doeskin, tightened on her reins. "And in exchange, she has received concessions from me. Political decisions that suit some purpose of hers. The assistance of—mortals, to manipulate something of importance to her." Her stumble over the word was barely perceptible.

Deven ventured a reminder. "The man who spoke of this claimed, before he died, that it was causing harm to both sides."

For the first time since they rode out, Elizabeth turned her head to face him. The strength of her gaze shook him. It was easy to forget, when one saw her laughing with her courtiers, or smiling coquettishly at some outrageous compliment, that she was her father's daughter. But in that gaze lay all the fabled personality and will of Henry, eighth of that name, King of England. They had stores of rage within them, the Tudors did, and Elizabeth's was closer to the surface than he had realised.

"I do not know," Elizabeth said, "what this pact has cost her side. I do not care. She has often manipulated me, managed me, coerced me into positions I would not otherwise have occupied. Even that, I might have endured, if it meant the wellbeing of my people. But she went too far with our cousin Mary. I do not know how far back her interference extended, but I know this: were it not for that interference, I might never have been forced to sign that order of execution."

Deven saw, in his mind's eye, the chess pieces with which Walsingham had led him through the story of the Queen of Scots—and the white queen, standing on her own, caught halfway between the two sides.

"Then tell me the terms of your pact," he said quietly, "and I will see it ended."

She turned her gaze back to the landscape ahead, where the ground rose upward in a rocky slope. The men-at-arms were still all around, maintaining a respectful distance, and Deven was glad for them. He could not both navigate this conversation and keep watch for threats. How easy must it be, for fae to conceal themselves among the green?

"'Twas simple enough," the Queen said. "Do you know the London Stone?"

"On Candlewick Street?"

"The same. 'Tis an ancient symbol of the city, and a stone of oaths; the rebel Jack Cade once struck it to declare himself master of London. At the moment I was crowned, she thrust a sword into that stone, to claim her own sovereignty."

That was most promising; it gave him a physical target to attack. "Will it threaten your own position, if…?"

Elizabeth shook her head. "My throne came to me by politics, and the blessing of a bishop, speaking in God's name. What she has stolen is mine by right."

She spoke with certainty, but he had heard her at court, declaring with swaggering confidence that Spain would not dare attempt another invasion, or that some lord or other would never defy her will. She could feign confidence she did not feel. It felt like a sharp rock had lodged in Deven's throat when he swallowed. Would he help the fae depose Invidiana, only to find his own Queen overthrown?

Elizabeth was willing to risk it, to free herself from the

snare that trapped her. His was not to question it.

"If you bring her low," Elizabeth said in a hard, blazing voice, "then I will reward you well for it. She is a cold thing, and cruel in her pleasure. Princes must often be ruthless; this I knew, before I even ascended to my throne. But she has forced matters too far, more than once. There is no warmth in her, no love. And I despise her for it."

Deven thought of the Goodemeades—of Rosamund's story, and the conversation he had with Gertrude—and responded gently. "They tell me she was different once. Before her coronation. When she was still known as Suspiria."

Elizabeth spat, not caring if the gesture was coarse. "I would not know. I never knew this Suspiria."

They had ridden on several paces farther before Deven's hands jerked convulsively on the reins. His gelding short-stepped, then recovered. "Not even when first you met? Not even in the Tower?"

He had drawn level with the Queen again, and she was studying him in wary confusion. "The name she gave me was Invidiana. And I have never known her to show any kindness or human warmth, not since the moment she appeared."

"But—" Deven realised belatedly that he was forgetting to use titles, polite address, anything befitting a gentleman speaking to his Queen. "By the story I was told, madam, she was known as Suspiria until the moment of her coronation, and that while she bore that name, she was not so cold and cruel."

"Your friends are mistaken, or they have lied to you.

Although…" Elizabeth's dark eyes went distant, seeing once more into the past. "When I asked her name, she told me 'twas Invidiana. But the manner in which she said it…" The Queen focused on him once more. "It might have been the first time she claimed that name."

Deven was silent, trying to work through the implications of this. His mind felt overfull, too many fragments of information jostling each other, too few of them fitting together.

"I will bear this news to those who work against her," he said at last. If the Goodemeades had lied to him—trustworthy as they seemed, he had to consider it—then perhaps he could provoke some sign out of them. And if not…

If not, then nothing was quite what they had thought.

"You will keep us apprised of your work," Elizabeth said. The familiarity that had overtaken her during the ride, while she spoke of things he was certain she had divulged to no other, was gone without a trace, and in its place was the Queen of England.

Deven bowed in his saddle. "I will, your Majesty, and with all speed."

MEMORY

31 January 1587

The chamber was dim and quiet, all those who normally attended within it having been banished to other tasks.

Guards still stood outside the door—in times as parlous as these, dismissing them was out of the question—but the woman inside was as alone as she could ever be.

The cosmetics that normally armoured her face were gone, exposing the ravages wrought by fifty-three years of fear and anger, care and concern, and the simple burden of life. Her beauty had been an ephemeral thing, gone as her youth faded; what remained was character, that would bow but never break, under even such pressure as she struggled with tonight.

Her eyes shut and her jaw clenched as the fire flickered and she heard a voice speak out from behind her. Unannounced, but not unexpected.

"You know that you must execute her."

Elizabeth did not ask how her visitor had penetrated the defences that ringed her chamber. How had it happened the first time? Asking would but waste breath. She gathered her composure, then turned to face the woman who stood on the far side of the room.

Frustrated rage welled within her at the sight. Elaborate gowns, brilliant jewels, and a mask of cosmetics could create the illusion of unchanging beauty, but it was an illusion, nothing more, and one that failed worse with every passing year. The creature that stood before her was truly ageless. Invidiana's face and figure were as perfect now as they had been in the Tower, untouched by the scarring hand of time.

Elizabeth had many reasons to hate her, but this one was never far from her mind.

"Do not," she said in frigid reply, "presume to instruct me on what I must do."

Invidiana glittered, as always, in silver and black gems. "Would you rather be seen as weak? Her guilt cannot be denied—"

"She was *lured into it*!"

The faerie woman met her rage without flinching. "By your own secretary."

"With aid." Elizabeth spat the words. No one ever seemed to hear them, on the infrequent occasions that the two queens came face-to-face; she could shout all she wanted. "How much assistance did you provide? How much rope, that my cousin might hang herself? Or perhaps that was too inconvenient; perhaps 'twas simpler to falsify the letters directly. You have done it before, implicating her in her husband's murder. Had matters gone your way, she would have been dead ere she ever left Scotland."

The black eyes glimmered with cold amusement. "Or dead in the leaving, save that the nucklavee showed unexpected loyalty. I would the monster had drowned her; 'twould have saved much tedious effort on my part. And then your precious hands would be clean."

Words hovered behind Elizabeth's lips, all her customary oaths, swearing by God's death and his body and countless other religious terms. How fitting it would be, to hurl them now: proof that although Protestant rites might lack the power of Catholic tradition, words of faith yet held some force.

But again, what purpose would it serve? Nothing

she said now would save Mary. The Queen of Scots had been proven complicit in a scheme against Elizabeth and England; there was no concealing it. Invidiana had seen to that. Elizabeth's councillors, her parliament, her people—all wished to see Mary gone. Even James of Scotland had bowed to circumstances. His last letter, sitting open on a table nearby, offered no more trouble than the weak protest that his subjects would think less of him if he made no reprisal for his mother's execution.

"And what if I will not do it?" Elizabeth said. "'Tis plain you wish her gone for your own purposes. What if I refuse you? What if, this once, I refused to play a puppet's part?"

Invidiana's lips thinned in icy displeasure. "Would it please you more if I removed my hand from your affairs? Your end would surely then be swift."

Elizabeth almost told her to do it and be damned. The threats to English sovereignty were manifold—they were at war with Spain, and Leicester had bungled the campaign in the Low Countries—but she refused to believe herself dependent upon the faerie queen for her survival. *She* was Queen of England, by God, and needed no shadowy puppeteer to pull her strings.

Yet she could not deny the strings existed. Some of the demands Invidiana made of her seemed innocuous; some were not. The faerie woman had required no devilish rites, no documents signed in blood, but she had imposed a real cost—if a subtle one. A certain ruthless cast to particular affairs, colder and harder than it would otherwise have been. The persistent reminder of her own mortality, more unbearable because of its contrast with the faerie's eternal youth. And, in a blending of the personal and political, solitude.

Once, there had been many suitors for her hand. Leicester, Alençon, even the King of Sweden. None without complications of religion or faction, none without the threat of losing her independence as a ruling queen... but there might have been happiness with one of them. There might have been hope of marriage.

None of it had come to anything. And that, Elizabeth was certain, she could lay at the feet of her dark twin, the loveless, heartless, solitary faerie Queen.

She did not ordinarily resent the price she had been forced to pay, for security on her throne. What Elizabeth resented was the creature to whom she had been forced to pay it.

"You must execute her," Invidiana said again. "However you have come to this pass, no other road lies before you."

True, and inescapable. Elizabeth hated the elfin woman for it.

"Leave me be," she snarled. Invidiana smiled—beautiful, and ever so faintly mocking—and faded back into the shadows, returning whence she had come.

Alone in her bedchamber, Elizabeth closed her eyes and prayed. On the morrow, she would sign the order, and execute her cousin and fellow Queen.

BEER HOUSE, SOUTHWARK
5 May 1590

"The thing to remember," Rosamund said, "is that she's not all-knowing or all-powerful."

The words hardly reassured Lune. All around them the alehouse was bustling, with voices clamouring in half a dozen languages; the river thronged with travellers, merchants, and sailors from all over Europe, and the Beer House on the south bank attracted its fair share. The noise served as cover, but also made her nervous. Who might come upon them, without her ever knowing?

Rosamund clicked her tongue in exasperation. "She cannot have eyes and ears everywhere, my lady. Even if she has somehow trapped his ghost..." The prospect shadowed her face. "I know we haven't the rose here to protect us, but this will serve just as well. Her attention is bent where 'twill matter, and that is elsewhere."

The brownie was probably right. The greatest threat they faced here was from uncouth men who targeted them with bawdy jests. Lune and the Goodemeades had made certain they were not followed, and with glamours covering their true appearances, there was nothing to draw Invidiana's attention here.

They might as well meet; if Francis were in her clutches, Lune's only hope lay in following this matter through.

Her nerves wound a notch tighter when she saw a familiar head weaving through the noontime crowd. Deven wore a plain woollen cap and clothing more befitting a respectable clerk than a gentleman; Gertrude, who came into view before him, might have been any goodwife of the city. The brownie squeezed herself in next to her sister, leaving Deven no choice but to take the remaining place beside Lune. Rosamund passed them both jacks of ale.

Deven cast a glance around, then said in a voice barely audible through the racket, "Have a care what you say. Walsingham often picked up information from the docks."

Lune gave Rosamund a meaningful look.

He saw it, and an ironic smile touched his lips. For a moment the two of them were in accord; Gertrude, curse her, looked smug. "Escaping both sides at once takes more doing than this, I see. A moment." He vanished into the crowd, leaving behind his untouched ale and a fading warmth along Lune's side, where he'd pressed up against her.

He returned quickly, and gestured for them to follow. Soon they were upstairs, in a private room hardly big enough for the bed and table it held, but at least the noise faded. "Someone may try to listen at the door or through the wall," Deven said, "but 'tis better."

"I can help with that," Gertrude said, straightening up from where she crouched in the corner. A glossy rat sat on its hindquarters in her hands, and listened with a bright, inquisitive manner as she explained what she wanted. Deven watched this entire conference with a bemused air, but said nothing.

When the rat was dispatched to protect them from eavesdroppers, Deven gestured for the women to take the available seats. Lune perched on the edge of the bed—trying not to think about the uses to which it was put, nor what the Beer House's owner thought of the four of them—while the Goodemeades took the two stools.

Deven outlined for them in brief strokes what Elizabeth had said about the London Stone. "But I rode by it coming here," he said, "and saw no sign of a sword."

The fae all exchanged looks. "Have you ever seen it?" Gertrude asked, and Rosamund shook her head.

Lune followed their thoughts well enough. "But who knows every corner of the Onyx Hall? It might be there." Taking pity on Deven's confusion, she said, "The London Stone is half-buried, is it not? The lower end might extend into the palace below. But if it does, I know not where."

"She might well keep it hidden," Rosamund said.

Deven seemed less interested in this than he might have been. His face was drawn into surprisingly grim lines. "There's another problem."

Their speculation halted suddenly.

He looked straight at the Goodemeades. "You spun me a good tale the other day, of curses and lost loves. My Queen tells a different one. She met this Invidiana nearly five years before they were crowned, and says she was no kinder then than she is now, nor did she bear any other name. Have you any way to explain this?"

Lune was as startled as the brownies were. Had the

sisters lied? No, she could not believe it. Even knowing they could and did lie with great skill, she did not believe they were feigning their confusion now. Was this some game of Deven's? Or Elizabeth's?

"We do not," Gertrude whispered, shaking her head. "I—that is—"

The unexpected hostility of Deven's tone had distracted Lune, but now she thought about his words. Five years. Her grasp of mortal history was weak, but she thought she remembered this much. "Mary would have been on the throne then. Was that not when Suspiria lifted her curse?"

Rosamund's brow was still furrowed. "I suppose so, near enough. But I do not see—"

Lune rose to her feet. Deven was watching her, with his eyes that kept reminding her of Francis—more so since she learned what Francis had once been. "Not what *Invidiana* did—what *Suspiria* did. That is what he knew. That is what he was trying to say!"

"What?" Now everyone was staring at her.

She pressed one hand to the stiff front of her bodice, feeling sick. "He was dying, he could barely speak, but he tried to tell me—he could not get the words out—" Her fingers remembered the uncontrollable shaking of his body. Something hot splashed onto her hand. "The last thing he said. 'She is still c—'"

Lune looked down at the Goodemeades' pale faces. *"She is still cursed."*

"But that's impossible," Rosamund breathed. "'Tisn't

a glamour we see now; she is as she appears. Young and beautiful. She *must* have lifted the curse."

"Lifted?" Deven asked, from the other side of the table. "Or changed it somehow? Traded it for some other condition, escaped its terms?" He shrugged when Lune transferred her attention to him. "I know little of these things; you tell me if it is impossible."

"But did it happen before she met Elizabeth?" Rosamund twisted in her seat. "Or after?"

"Before, I think—but not long before. Elizabeth believes their meeting was the first time she claimed the name Invidiana."

Gertrude seized her sister's hand. "Rose, think. 'Twas after that she began gathering a court, was it not? No, she was not as we know her now—"

"But that might have been a mask." All the blood had drained from Rosamund's face; she looked dizzy. "She could have pretended to be the same. Ash and Thorn—that was when Francis began to lose his name. Do you remember? She always called him Tiresias, after that. And he said things had changed between them."

Lune said, "Then it was *not* Elizabeth's doing." Everything she had thought clear was fading away, leaving her grasping at mist. "But he said her pact..."

Into the ensuing silence, Deven said, "Perhaps this is a foolish question. But what certainty have we that she formed only one pact?"

No one seemed to be breathing. They had all leapt so

quickly to the thought of Elizabeth and the mortal court—and they had not been wrong. There *was* a pact there. But was that what Francis had meant? Or did he know something they had never so much as suspected?

Lune whispered, "Where do we *begin*?"

"With the curse," Deven said. "Everything seems to have spun out of whatever she did to escape it. Creating the Onyx Hall did not free her, you said. What did?"

"Something Francis saw," Gertrude said. "At least, we think so."

Lune lowered herself slowly back onto the bed. Briefly she prayed that the rats were doing their jobs, and no one was listening to this mad and treasonous conversation. "He said she misinterpreted it. But we cannot know what she did until we know what she was escaping. What crime did she commit, to be cursed in such fashion?"

"We never knew," Rosamund replied, clenching her small hands in frustration. "Even once she knew, she would not tell us. Or even Francis, I think."

"But where did she learn it herself?"

Gertrude answered Deven far more casually than her words deserved. "From Father Thames."

His shoulders jerked. "From *who*?"

"The river," the brownie replied.

"The river." Coming from him, it was an expression of doubt, and he turned to Lune for a saner answer, as if she would be his ally in disbelief.

"The spirit of it," she said; his jaw came just the slightest bit unhinged. Hers felt like doing the same. "She spoke to him? Truly?"

Rosamund shrugged. "She must have done. We were not in London when the curse was laid; 'twas long ago, when we lived in the North. Gertrude told her she must find someone who was here long ago. Who else could she turn to, save Old Father Thames himself?"

Who else, indeed. Lune felt dizzy. Father Thames spoke but rarely, and then to other creatures of the water. She did not know what could possibly induce him to speak to a fae of the land.

But she would have to find out, because they had no one else to question.

"I will try tonight, then," she said, and the Goodemeades nodded as if they had expected nothing else. She met Deven's gaze, briefly, and looked away. This was a faerie matter; he would want none of it.

"We should arrange to meet again," he said into the silence. "Your pigeon was most helpful, Mistress Goodemeade, but I pray you pardon me if I find communicating in such a manner to be... disconcerting." When the sisters smiled understandingly, he said, "There is a tavern along Fleet Street, outside the city's western wall. The Checkers. Shall we find one another there, three days hence?"

Three days. Giving Lune extra time, in case she failed the first night. Did he have so little confidence in her?

The brownies agreed, and they all dispersed, the

Goodemeades leaving first. Alone with Deven, Lune found herself without anything to say.

"Good luck," he murmured at last. His hand twitched at his side, as if he might have laid it briefly on her shoulder.

That simple note of friendship struck an unexpected chord. "Thank you," Lune whispered in response. Perhaps this alliance of theirs was leading him to forgive her—at least a little—for the harm she had done him before.

He stood a moment longer, looking at her, then followed the Goodemeades out the door.

Standing by herself in the centre of the room, Lune took a slow, deep breath. Father Thames. She did not know how to reach him, let alone gain his aid… but she had three days to find out.

RIVER THAMES, LONDON
5 May 1590

She had changed her appearance again, but Deven still recognised her. There were certain mannerisms—the way she walked, or held her head—that echoed his memories so powerfully it made him ache inside.

He followed the disguised Lune at a safe distance as she left the Beer House. Doing so required care; she was wary and alert, as if she might be observed or attacked. It was a tension that had not left her since he found her on Cloak Lane, wearing a bad illusion of Anne Montrose. The mere thought of living in such constant fear exhausted him. In comparison with life in her own court, masquerade as a human woman must have seemed a holiday for her.

Not that it excused a year of unending lies.

The crowds on London Bridge helped conceal him from her searching eyes. Disguised as he was, he blended in fairly well. So he followed her through the afternoon as she walked back and forth along the river's bank: first to the Tower of London, with its water port of the Traitor's Gate, then back westward to Billingsgate, the bridge, Queenhithe, Broken Wharf, pausing each time she passed a river stair, occasionally watching the watermen who rowed passengers from one bank to the other. Her feet at last took her to Blackfriars, on the far side of which the noisome waters of the Fleet poured out into the Thames. Deven, who had been wondering what manner of creature the spirit of the Thames would be, thought he would not want to meet anything that embodied the Fleet.

It seemed that Lune could not make up her mind where to make her attempt. Did it matter so much? Deven could imagine the Thames at its headwaters in the west might be a different being than the Thames where it passed London, but what might distinguish the Blackfriars Thames from the bridge Thames, he had no idea.

That was part of why he followed. Lune had not asked for aid, but his curiosity could not be suppressed. Though it was being sorely tested by all this walking; he had grown far too accustomed to riding.

Night fell, and still Lune delayed. Curfew had long since rung, and for the first time it occurred to him that his sober disguise might pose a problem; with neither horse, nor sword, nor finery, nor anything else save his word that might identify him as a gentleman, he had no excuse for why he might be on the street. The same was true of Lune, but remembering the befuddled guards at Aldersgate the night she fled the city, he was not concerned for her.

When the moon rose into the sky, she made her way back eastward, and Deven at last understood what she had been waiting for.

The tidal waters of the Thames, answering the call of the gibbous moon.

As the river's level rose, he trailed her through the darkness, and mentally rewarded himself the groat he had wagered. Lune was heading for the bridge.

For it, and onto it. The Great Stone Gate on the Southwark end would be closed for the night, but the north end was open. Deven wondered what she was doing, then cursed himself for distraction; he had lost her among the houses, chapels, and shops built along the bridge's length.

Only the scuff of a shoe alerted him. Peering cautiously over the edge in one of the few places it was accessible, he saw a dark shadow moving downward. The madwoman could have hired a wherry to take her there by water—well, perhaps not. Shooting the bridge, passing through the clogged, narrow races between the piers of the arches, was hazardous at the best of times; even the hardened nerves of a London waterman would be tested by a request to drop a passenger off along the way. But that might still have been better than climbing down the side of the bridge.

Lune reached safety below, on one of the wooden starlings that protected the stone piers from collision with debris or unlucky wherries. There was no way Deven could follow her without being heard or seen. He should give up, and he knew it, yet somehow his feet did not move homeward; instead they carried him to the other side of the bridge, one arch farther north, and then his hands were feeling the roughened stone as if this were not the worst impulse he'd had since the night he followed a faerie woman out of the city.

The first part was easy enough, where the pier sloped outward to a triangular point. The second, vertical part was the stuff of nightmares, clinging to crevices where the mortar had worn away, praying he did not fall to the starling below and alert Lune, praying he did not tumble into the Thames and ignominiously drown. But by then it was far too late to turn back.

And then he was safe, and tried not to think about how he would get off the starling again when this was done.

His cap had blown off in the river wind and was lost to the dark water. Shivering a little, though he was not cold, Deven crouched and peeked cautiously around the edge of the pier, looking across the intervening space to where Lune stood on the next platform over. No, not stood; knelt. The sound of the river had faded enough that her voice carried clearly to him.

"Father Thames," she said, respectful and solemn. The glamour that had disguised her all day was gone, but the shadow beneath the arch protected her from prying eyes on the riverbank. Only Deven could see her, a silver figure with her head bowed. "As the moon calls to your waters, so I, a daughter of the moon, call to you. I humbly beseech the gift of your presence and counsel. Secrets lie within your waters, the wisdom of ancient times; I beg you to relate to me the tale of Suspiria, and the curse laid upon her. I ask this, not for myself, but for my people; the good of faerie kind may hang upon this tale. For their sake, I pray you hear my words."

Deven hardly breathed, both from anticipation, and from fear of being overheard. The river licked the planks of the starlings, within arm's reach of the top edges. Every flicker of motion caught his eye—what sign would be returned? A face? A voice?—but it was never more than debris, floating through the narrow gaps of the races.

He waited, and Lune waited, and nothing came.

Then another sound laid itself over the quiet murmur of the water. Only when it recurred could he identify it: not speech, but choked-off breath, the ragged edge of fading control.

"Please," Lune whispered. Formality had failed; now she spoke familiarly. "Please, I beg you, answer me." The river made no reply. "Father Thames... do you wish her power to endure? Or do our acts mean nothing to you? She has warped her court. I can scarce remember where I was before I came here, but I know it was not this cold. I served her faithfully, beneath the sea, in Elizabeth's court, anywhere she has bid me, and now I am hounded to the edge of my life. There is no safety for me now, except in her overthrow. Without your aid, I have nothing. I..."

The words trailed off into another ragged breath. Her shoulders slumped with weariness, abandoning the armour of purpose and drive that ordinarily held her together. Her hands clutched the edge of the starling, white-knuckled in the night.

He should not have followed her. Deven was watching something private, that she would not have shown if she knew he was there. And it stirred something uncomfortable within him, where resentment had lodged itself when first he saw her true face.

That was, after all, the crux of it. Her *true* face. The other was a lie. He knew it, and yet some part of him had still grieved, still resented her, as if she had somehow stolen Anne Montrose from him—as if Anne were a real person, kidnapped away by the faerie woman.

But Anne had never existed. There was only ever Lune, playing a part, as so many did when they came to court.

Yet the part she played was a part of her, too. There had been more truth than he realised to the words she said back then: she could be at ease in his presence, as she could not elsewhere. Perhaps the Lune who existed before the Onyx Court had been more like Anne.

Or perhaps not. He had no way of knowing. But one thing he did know: Lune *was* Anne. He had loved this faerie woman before he knew the truth, and now that he did...

His feelings had not vanished when her mask did.

It might be foolish of him—no doubt it was—but also true.

"You do not have nothing," he murmured, mouthing the words soundlessly to himself. "You have your own strength. And the aid of the Goodemeades. And... you have me."

Slack water had come, the turn of the tide; the river was never more quiet than now. Why, then, did he hear a sound, as if something disturbed its tranquillity?

His first thought was that a boat approached; one hand went for his sword, remembered he did not have it, and groped instead for his knife. How would he explain their presence here? But no boat was near, and his fingers released the hilt, suddenly weak with shock.

The water between the two piers was swirling against all nature. The surface mounded, rose upward, then broke, and standing upon the Thames was an old man, broad-shouldered and tall, grey-bearded but hale, with centuries of wisdom graven upon his face.

"Rarely do I speak, in these times that so choke my waters with the passage of ordinary life," the spirit of the river said. His voice was deep and slow, rising and falling in steady rhythm. The murky grey fabric of his robe shimmered with hints of silver in its folds. "But rarely do two call me forth together, mortal and fae. Thus do I come, for the children of both worlds."

Deven froze. Lune's head came up like a doe's when

it hears the hunter's step. Could she see him, concealed behind the edge of the pier?

Father Thames was not looking at him, but still he felt shamed. He could not hide from the venerable spirit.

Stepping around the edge, onto the nearer half of the starling, Deven made his most respectful bow, as if he approached the Queen herself. "We are most grateful for your presence." A back corner of his mind worried, *What form of address does one use for a river god?*

Lune rose slowly to her feet, staring at him. She seemed to speak out of reflex. "As he says, Old Father. You honour us by rising tonight."

Deven moved far enough that they both stood before the spirit, on opposite sides of the arch. The fathomless eyes of Father Thames weighed them each in turn. "Daughter of the moon, you spoke the name of Suspiria." Lune nodded, as if she did not trust her voice. "An old name. A forgotten name."

"Forgotten not by all," she whispered. "We seek knowledge of her—this mortal man and I. Can you tell us of her? What wrong did she commit, that she was cursed to suffer as if human?"

The spirit's gaze fixed inexplicably on Deven, who tried not to shiver. "She came to me for this tale, begging every night for a year and a day until I took pity on her and spoke. Her mind was clouded by her suffering. She did not remember.

"'Twas long and long ago. A town stood upon my banks, little more than a village, save that the chieftain of the mortal people dwelt within its palisade, and thereby lent it dignity beyond its size. Within the hollow hills lived the faerie race, and there was often conflict between the two.

"And so a treaty was struck, a bargain to bring peace for both peoples. Faerie kind would walk in freedom beneath the sun, and mortals go in safety beneath the earth. But 'twas not enough simply to agree; the bargain must be sealed, some ritual enacted to bind both sides to honour its terms. The son of the chieftain had gone more than was wise among the faerie people, and seen many wonders there, but one stood high in his mind: the beauty of an elfin lady, who of all things seemed to him most fair.

"Thus was it proposed: that the treaty be sealed by marriage, joining a son of mortality to a daughter of faerie."

The measured, flowing cadence of Father Thames's words carried the rhythm of simpler times. Not the crowded, filthy bustle of London as it was now: the green banks of the Thames, a village standing upon them, a young man dreaming of love.

The river god's eyes weighed Deven, seeing deep into his thoughts, and the admission he had not spoken aloud. Then the spirit continued on.

"But the lady refused her part.

"The dream that might have been was broken. The peace that would have been faded ere it took hold. Spurned, the man cursed her. If she held mortality in such disdain, then he condemned her to suffer its pangs, to feel age and sickness and debility, until she understood and atoned for her error."

Lune whispered, "The Onyx Hall."

At last Father Thames shifted his attention to her. "The time for treaties between the two peoples has passed. The beliefs of mortals are anathema now, and drive fae kind ever farther into the wilderness, where faith and iron do not yet reach. Only here, in this one place, do faeries live so closely with human kind."

"But 'twas not enough, Old Father," Deven said. "Was it? She created the Onyx Hall, and still was cursed. How did she escape it in the end?"

Water rippled around the hem of the spirit's robe. "I know not," he said simply. "That which occurs upon my waters, along my banks, from the dawn of time until now: all that lies within my ken. But that which is done beyond my sight is hidden to me."

In Lune's gaze, Deven read the thoughts that filled his own mind. They still did not have the answer they needed: what pact Suspiria had formed. But this tale mattered, if only because it helped them understand the being that had once stood where Invidiana did now.

"I will bear you safe to shore," Father Thames said, and held out his broad hands.

Without thinking, Deven stepped forward to accept. Only when it was too late did he realise his feet had left the starling. But he did not fall: the surface of the water bore him like a slightly yielding carpet, against all the custom of nature.

Lune took Father Thames's left hand, and then the river flowed beneath them. With gentle motion it carried them out from under the bridge, slantwise across the breadth of the water, until they came to the base of the Lyon Key stair, within sight of the Tower wall. When his feet were securely on the stone, Deven turned back.

Father Thames was gone.

Then he looked at the rippling surface, and understood. The river god was never gone.

"Thank you," he murmured, and Lune echoed him, her own words no louder than his.

BRIDGE AND CASTLE
BAYNARD WARD, LONDON
6 May 1590

Lune realised, as if through a great fog, that she stood openly on the bank of the Thames, her elfin form undisguised, Michael Deven at her side.

Summoning a glamour took tremendous effort. She should not have let her guard down, out there on the starling; not only had Deven been listening—why had he followed her?—but allowing herself to relax her control had been a mistake. Weariness dragged at her like leaden chains, and she could not focus.

What face could she wear? Not Anne Montrose. Not Margaret Rolford. Her first attempt failed and slipped, without even being tested. She took a deep breath and tried again. The illusion she created was a poor one, unnaturally generic; it would seem strange, like a badly crafted doll, if anyone looked at her closely. But it was the best she could do.

She surfaced from her concentration to find Deven watching her with an odd expression.

"Have you somewhere safe to go?" he asked.

Lune forced herself to nod. "There's a chamber in the Onyx Hall I have been permitted to claim as my own."

He bit his lower lip, apparently unconscious of the gesture. "But will you be safe there?"

"As safe as I may be anywhere at court." It sounded stiff even to her; she did not want to appear like she sought pity. "We should part. 'Tis not safe for you to be seen with me."

They still stood a little below the surface of the wharf, not that it would protect them much. Deven gave her a frank appraisal. "And if you go there now, tired as you are, will I be any safer? Weariness can drive any man to error." His mouth quirked wryly. "Or any woman. Or faerie."

Lune did not want to hear of the risk. She had measured it herself, time and again, even before she fell out of favour. It was the way of the Onyx Court. One mistake, one wrong word… she was so very tired of that world.

"Come," Deven said, and took her by the hand.

She followed him in a daze, too tired to ask questions. He led her through the streets, and it seemed like they walked forever before they arrived at a house. Lune knew she should protest—her absence might raise suspicion—but pathetic as it was, the thought of spending even one night outside the Onyx Hall was enough to make her weep with relief.

A single candle lit their way, kindled from one by the door; up the stairs they went, and then something made an appalling amount of noise as Deven dragged it from under the bed and out the chamber door. "Sleep here for tonight," he said, before he left her. "I'll keep near."

That was not safe. But Lune was well past the point of arguing. She collapsed onto the bed, barely pausing to pull a blanket over herself, and slept.

When morning came, she found herself in a small, moderately appointed bedchamber. On the floor beside her was a large, empty box; the noise must have been Deven dragging a mattress free of the truckle bed.

Her glamour had faded while she slept. It had been too long since she ate from that loaf of bread. Lune closed her eyes and rebuilt it, far better this time, making herself into an auburn-haired young woman with work-roughened hands, then pinched off another bite from the lump she carried with her. She dared not leave it behind in the Onyx Hall.

In the neighbouring chamber, the mattress lay on the floor; Deven she found downstairs. Pausing at the door, Lune looked around at the modest pewter plate on the sideboard, the cittern in one corner, with two of its strings broken. She had thought briefly last night that the place might belong to some former agent of Walsingham, and had been both right and wrong. "This is your house."

He'd glanced up at her approach. "Yes." After a pause, he added, "I know. I should not have brought you here. But I was tired, too, and did not know where else to go; it seemed unkind to put my father in danger. We shall not do it again."

It was too late to undo. Lune came forward a few steps, smoothing the apron over her skirt. She looked like a maidservant for the house. "I'll take my leave, then."

"They know about me, do they not?" The room was dim, even though it was morning; Deven had kept the windows shuttered, and only a few lights burned. They accentuated the hollows in his face. He had not been sleeping well.

"Yes," Lune said. Taking a deep breath, she added with sincerity, "I am sorry for it. I cannot play the part of a man, and Walsingham would not take a woman into his confidence. The only course open to me was to attach myself to someone in his employ." That skirted too close to the wound between them. "They knew you were my contact in his service."

He was dressed once again as a gentleman, though not completely; his servants were nowhere in evidence. The sleeves of his doublet lay across his knees, and the aglets of his points dangled loose from his shoulders. "Would I still be in danger, had it ended at that?"

"Yes." He deserved honesty. "They would kill you, to make themselves safe."

"Well." Deven's fingers brushed over the vines embroidered on one sleeve, then stilled. "My Queen has commanded me to break her pact with yours. Even if that is a separate thing from this other one we are chasing, we need each other's aid. But I do not know how one might escape a curse."

The door was behind Lune, a silent reminder that she should leave. Doing so would only slow their progress, though; the only true safety lay in completing their task. Hoping she was not making a terrible mistake, Lune sat.

"Nor I. A curse may only be ended on its own terms. But Suspiria tried that, and failed. I do not think anyone could absolve her of it. The man who laid it might have lifted it, but he is long dead. And Tiresias—Francis—believed it still bound her."

"He had some vision, the Goodemeades said. Did he never speak of it to anyone else?"

Lune shook her head, more in bafflement than confident denial. "If 'twere part of the binding she laid on him—she has this jewel, that she can use to place commands on others, so they must obey or die. She might have bound him not to speak of that vision. But even if she did not... he died before he could tell me."

"And he never mentioned it at other times."

"How could I know?" Frustration welled up; Lune forced herself not to turn it on him. "You must understand. Dwell among fae for long enough, drink our wine, eat our food... it changes a man. And he had been there for years. He raved, he lived in dreams; nine-tenths of what he said was madness, and the other tenth too obscure to understand. He might have told a dozen people the content of his vision, and we would never know."

"Did he never say anything else?" Deven leaned forward, elbows on knees, face earnest and alert. "Not of the vision specifically. Anything touching on Suspiria, or curses, or the Onyx Hall… he would not speak entirely at random. Even madmen follow a logic of their own."

It might be true of ordinary madmen, but Tiresias? Under Deven's patience gaze, Lune disciplined her mind. When else had he spoken to her of the past?

When he told her to find him.

"He remembered his name," she murmured, recalling it. "Before I came to Elizabeth's court. He bade me find Francis Merriman; not until later did I realise he had forgotten who he was."

"Begin with that," Deven said. They had not worked together like this, piecing together an image from fragments, in months. And this time she was on his side. "He wanted you to find him. When you did…"

"He died." Deven had never known Francis; she could tell him what she could not tell the Goodemeades, who had loved the mad seer. "I forced him. He was afraid to speak, but I would not let him back away; I thought finding him would better my position in the Onyx Court." Quite the opposite, and the memory left a bitter taste in her mouth. "I demanded he tell me what he knew. And so he died."

She could not look away from his intent blue eyes. Speaking softly, Deven said, "Did you rack him? Put him to the question? Of course not. You kept him from fleeing at the last, perhaps, but unless there is something you have not told me, he chose to speak."

She would not cry. Lune turned her head away, by sheer

force of will, and studied the linen-fold panelling of the wall until she had her composure again.

Deven granted her that space, then spoke again. "Go back to when he bade you find him. What precisely did he say?"

What *had* he said? Lune tried to think back. She had been wondering how to regain Invidiana's favour. Tiresias had been in her chamber. She had mortal bread…

"He spoke of Lyonesse," she said. "The lost kingdom. Rather, he looked at my tapestry of Lyonesse, and spoke of errors made after it sank." Or had he been speaking of other things? The Onyx Hall, perhaps? She could see him in her mind's eye now, a slender, trembling ghost. "He did not want to dream. I know he thought of his visions as dreams… then he said something about time having stopped."

This made both of them sit more sharply upright. "So it had, for Suspiria," Deven said. Excitement hummed in his voice, held carefully in check.

But there had been something before that. *If I should find this Francis Merriman, what then?*

He had sounded so lucid, yet spoken so strangely. Lune echoed his words. "'Stand still, you ever-moving spheres of Heaven…'"

Deven's breath caught. "What?"

She had not expected such a reaction. "'Tis what he said, when I asked him what I should do. '*Stand still, you ever-moving spheres of Heaven—*'"

"*—that time may cease, and midnight never come.*"

"He did not use those words. But 'twas then he said time had stopped." Deven's expression baffled her. "What is it?"

He answered with half a laugh. "You never go to the theatre, do you."

"Not often," she said, feeling obscurely defensive. "Why? What are those words?"

"They come from a play." Deven rose and went to the cold fireplace, laying one hand on the mantel, chin tucked nearly to his chest. "The man who wrote them... he has served Burghley in the past, but he's more a poet than a spy. Sir Francis's cousin is a friend of his. I met him once, at dinner." He turned back to face her. "Is the name Christopher Marlowe familiar to you?"

Lune's brow furrowed. "I have heard it. But I do not know him."

"Nor his work, 'twould seem. The lines are from his play *Doctor Faustus.*"

She shook her head; the title meant nothing to her.

Deven's jaw tensed, and he said, "The story is of a man who makes a pact with the devil."

Lune stared up at him, utterly still.

"Tell me," Deven said. "Have your kind any dealings with Hell?"

She spoke through numb lips, as if someone else answered for her. "The Court of Thistle in Scotland tithes to them every seven years. A mortal, not one of their own. I do not know how they were bound into such obligation. But Invidiana—Suspiria—surely she..."

"Would not have done so?" Deven's voice was tight with something: anger, fear, perhaps both. "By what you said, your own people would have no way to restore her beauty without lifting her curse. And she is still cursed. Some other power *must* have aided her."

Not a celestial power; an infernal one. What had she offered them in return? Any kindness she had once possessed, it seemed. And to fill that void, she craved power, dominion, control. She made of the Onyx Court a miniature Hell on earth; Tiresias had said often enough that it was so.

And he had told her what to do.

"Sun and Moon," Lune breathed. "We must break her pact with Hell."

ACT FIVE

Ah, Faustus,
Now hast thou but one bare hower to liue,
And then thou must be damnd perpetually:
Stand stil you euer mouing spheres of heauen,
That time may cease, and midnight neuer come!

CHRISTOPHER MARLOWE
THE TRAGICALL HISTORY OF D. FAUSTUS

The long gallery is lined from end to end with tapestries, each one a marvel of rich silk and intricate detail, limned in gold and silver thread. The figures in them seem to watch, unblinkingly pitiless, as he stumbles by them, barefoot, without his doublet, his torn shirt pulled askew. His lips ache cruelly. There is no one present to witness his suffering, but the embroidered eyes weigh on him, a silent and judgmental audience.

He spins without warning, shoulders thrown back, to tell the figures in the tapestries they must leave him be—but the words never leave his mouth.

The scene that has arrested his attention might depict anyone. Some faerie legend, some ancient lord whose name has escaped his mind, slipping through the cracks and holes like so much else. But his eye is transfixed by the two central images: a lone swordsman in a field, gazing at the moon high above.

His bruised lips part as he stares at those two. The broken spaces of his mind fill suddenly with a barrage of other pictures.

He sees another Queen. A canopy of roses. A winter garden. A stool, alone in a room. Lightning, splitting the sky. A loaf of bread. A sword, clutched in a pale hand. Two figures on a horse.

Shattered crystal, littering the floor, and an empty throne.

He presses one hand to his mouth, trembling.

He has seen it before. Not these same images, but other possibilities, other people. They have not come to pass. But who knows how far in the future a vision may lie? Who is to say whether one might not yet become true?

Some of those he has seen lie dead now. Or so he thinks. He has lost all grip on time; past, present, and future long since ceased to hold any meaning. He does not age, and neither does she, and it is always night below. There is no anchor for his mind, to make events proceed in their natural order, first cause, then effect.

It may be nothing more than the desperate hope of his heart. But he clings to it, for he has so little else. And he will bury this new one with the others, so deep that even he will not recall it, for that is the only way to keep such things from her.

She has gotten some of them. Or will get them. That is why those people are dead, or will be.

But not him. Never him. She will never let him go.

He tugs the tattered remains of his shirt about himself and hurries away from the tapestries. Must not be seen looking at them. Must not give her that hint.

Someday, perhaps, he will see one of these visions come to pass.

ST. PAUL'S CATHEDRAL, LONDON
6 May 1590

Deven thought, a little wildly, *God have mercy. I'm bringing a faerie woman to church.*

By her expression, Lune might have been thinking the same thing. Mindful of how the stone would carry her voice, she murmured acidly, "Do you expect some priest here to stop her?"

"No. But at the very least, we are less likely to be overheard within these walls." One hand on her elbow, he pulled her farther down the nave. Outside, the churchyard echoed with its usual clamour, booksellers and bookbuyers and men looking for work. The vaulting interior of the cathedral somehow remained untouched by it all, a small island of sanctity in the midst of commerce.

"We can still come inside, when prepared; you yourself have seen me at chapel."

"True. But I doubt your kind wander in idly." Deven broke off as one of the cathedral canons passed by, giving him an odd look.

Without realising it, he had led them to an all-too-familiar spot. Lune was not paying attention; she did not seem to notice that the magnificent tomb nearby contained not just Sir Philip Sidney, but his erstwhile father-in-law Walsingham.

Deven pulled up short, turning her to face him. "Now, tell me true. Do you think it mere chance, that your seer spoke that line? If you have *any* doubts…"

Lune shook her head. She still appeared a common maidservant, but he no longer had any difficulty imagining her true face behind it, silver hair and all. "I do not. I even thought, at the time, that he sounded sane… I simply could not make sense of the words. And I know of no power our kind possess to effect such a change, against the force of the curse."

He had hoped she would say it was a lunatic idea. Hoped for it, but not expected it.

"How do we break such a pact?" she whispered. She looked lost, stumbled without warning into a realm alien to her faerie nature. "Mere prayer will not do it. And I doubt she would stand still for an exorcism—if such would even touch her."

Despite his resolution not to draw her attention to his dead master, Deven found himself looking at the tomb that held Walsingham's body. Puritan belief was strong against them, Lune had said; Puritan, and Catholic. He was not on good terms with any Catholics. And he could not possibly ask Beale for help with this.

No, not Beale. The realisation came upon him like a blessing from God.

"Angels," Deven said. "To break a pact with the devil… one would need an angel."

Lune's face paled as she followed his logic. "Dee."

The old astrologer, the Queen's philosopher. How many of the stories were true? "They say he speaks with angels."

"Or devils."

"I do not think so," Deven replied, soberly. "Not from what I saw of him… he might have feigned piety, of course. But can you think of one better?"

She wanted to; he could see her trying, calling up and then discarding names, one by one. "No."

Now Deven regretted his contrived visit of a few weeks before; how would he look, a supposedly lovestruck fool, coming back and asking for aid against a faerie queen? His audience with Elizabeth would seem simple by comparison. But Dee had been a faithful supporter of Elizabeth since even before her accession; it should be possible to convince him to act against her enemy, however strange that enemy might be. And Walsingham had set him on this road—though the Principal Secretary could not have guessed where it would lead.

"I'll go to him," Deven promised. "Without delay. You…"

"Will warn the Goodemeades."

He could not quite suppress his ironic smile. "They have set a few pigeons to shadow me; one should be at my house. 'Twill carry a message, if you can find paper—"

Her own mouth quirked, and he remembered what lay outside the cathedral doors.

"'Twill carry a message to them," he finished lamely.

Then they stood in awkward silence, the shared tomb of Sidney and Walsingham a mute presence beside them.

At last Lune said, the words coming out stiffly, "Be careful as you ride. They know who you are."

"I know," he replied. They stood only a step apart; the intervening space was both a yawning gulf, and intimately close. He would have taken Anne's hands, but what would Lune make of such a gesture? "Have a care for yourself. 'Tis you who must go into the viper's den, not I."

Lune smiled grimly and moved past him, heading for the cathedral doors. "I have lived with the viper for years. And I am not without my own sting."

QUEENHITHE WARD, LONDON

6 May 1590

Only after Lune was gone did Deven realise he had left his sleeves behind at the house. No wonder the canon had stared.

He needed to put himself together properly if he was to visit Dee. He needed Colsey; he needed his horse. The previous day had left the pieces of his ordinary life scattered around London like debris after a storm.

Assembling himself again took until the afternoon. Colsey was mutinously silent while tending his master, no doubt anticipating what would come; he did not even blink when Deven said, "I must go alone."

"Again."

"Yes." Deven hesitated. How much could he say? Not much. He laid one hand on Colsey's shoulder and promised, "This will be over soon."

Ranwell had readied his black stallion for some reason; the warhorse stood rock still as he mounted. The day was half-spent. He would spend the other half getting to Mortlake, and hope Dee granted him an audience at the end of it. At least he would be out of London, with no faerie palace lurking beneath his feet.

The congestion of the city's streets had never irritated him so much. He should have gone west, made for the horse ferry at Fulham, but by the time he thought of it he was halfway to the bridge, with no point in backtracking. A cart in the process of unloading had mostly blocked Fish Street ahead of him; standing briefly in his stirrups, Deven scowled at the ensuing knot, as people tried to edge by. Then he cast a sideways glance at a narrow, lamp-lit lane whose name he did not recall. If memory served, it ran through to Thames Street.

Turning the black stallion's head, he edged behind a heavily laden porter and into the lane.

Lamplight marked his way through the shadows. The lane brought him into a small courtyard, not Thames Street, but on the far side there was an archway, and his horse paced toward it without needing to be nudged. The lamp hovered above that arch, but did nothing to touch the darkness within...

"Master! *Don't follow the light!*"

Irritation seized him. What was Colsey doing, following against his orders? He turned in his saddle to reprimand the servant, and found himself crying out instead. "*Ware!* "

Colsey leapt to the side just in time to dodge the grasping hands of the man behind him. A strange man, clad in nothing more than a brief loincloth and sandals, but broad-shouldered and muscled like a wrestler. He was unarmed, though, and in the close confines of the courtyard, he would be easy enough to ride down.

Except that Deven's horse stood like a rock when he jerked at the reins, heedless of his master's command.

And when he tried to swing his leg over the saddle, to go to Colsey's aid, he found himself rooted as if his feet were tied to the stirrups.

The strange, eldritch light hovered and pulsed as he fought to free himself. Across the way, Colsey slashed out with his knife at the half-naked stranger, who parried his blows and stalked him with hands spread wide. Christ above, the horse wasn't his; how had he ever mistaken it for his own stallion?

Christ. "In the name of the Lord God," Deven snarled, "release me!"

The animal bucked with apocalyptic force, hurling him through the air and into the wall of a neighbouring house. All the air was driven from Deven's lungs, and he crashed heavily to the dirt below. But he untangled himself and lurched to his feet in time to see the creature shuddering and writhing into a two-legged shape, a man with a shock of black hair and large, crushing teeth.

The stranger fighting Colsey was blocking the exit to Fish Street—Deven no longer felt the slightest urge to go through the black archway at the other end—and as he looked, the man seized Colsey's knife hand and twisted it cruelly. The servant cried out and dropped his dagger.

Deven charged toward them, but his sword was only half-clear of its sheath when something cannoned into him from the side. The horse thing knocked him into the wall again, and Deven gasped for air. Reflex saved him; he kept drawing and now had three feet of steel to keep the creature from him. It danced back, suddenly wary.

Colsey had broken free, but now he was unarmed. "Get to the street!" Deven shouted, or tried to; the words rasped painfully out of him. If Colsey could rouse some kind of aid—

Except that Colsey shook his head and backed up two steps, retreating toward Deven's side. "Damn your eyes," Deven snarled, "do as I say!"

"And leave you with yonder two? With the greatest respect, master, shove it." Colsey made a swift lunge, but not toward their opponents. Deven's own knife whisked clear of its sheath, into the servant's hand.

They had another weapon, though, better than steel. "By the most Holy Trinity," Deven said, advancing a step, "by the Father, the Son, and the Holy Spirit—"

He got no farther. Because although the horse-thing shrank back and the hovering light snuffed out as if it had never been, the strange man charged in without flinching.

His bulk bowled Colsey away from Deven's side, dividing them again. Deven lunged, but retracted it as the stranger whirled to grab for his arm; he dared not lose hold of his sword. Then the horse-thing was there again, kicking out and getting stabbed for his pains, and Colsey circled with his opponent, slashing with the knife to keep him at bay.

But not well enough. The stranger stepped in behind a slash, closing with the servant. A swift kick to the back of the leg dropped Colsey to one knee, and then the broad, hard hands closed around his head.

The crack echoed from the walls of the small courtyard. Deven crossed the intervening space in an eyeblink, but too late; Colsey's limp body dropped to the ground even as his master's blade scored a line across the back of his murderer. And the stranger did not seem to care. He turned with a feral grin and said, "Come on, then," and spread his killing hands wide.

The horse-thing faded back, clutching his wounded side and seeming glad to leave this fight to its partner. Deven focused on the man before him. The tip of his blade flickered out, once, twice, a third time, but the stranger dodged with breath-taking speed, more than a fellow of his size should possess. "Drop the sword," the stranger suggested, with a grin of feral pleasure. "Face me like a proper man."

Deven had no interest in playing games. He advanced rapidly, trying to pin the man against a wall where he could not dodge, but his opponent sidestepped and moved to grab his arm again. Deven slammed his elbow into the other man's cheek, but the stranger barely blinked. Then they were moving, back across the courtyard, not so much advancing or retreating as whirling around in a constantly shifting spiral, the stranger trying to close and get a hold on him, Deven trying to keep him at range. He wounded the man a second time, a third, but nothing seemed to do more than bleed him; the grin got wilder, the movements faster. Jesu, what was he?

They were almost to the courtyard entrance. Then Deven's footing betrayed him, his ankle turning on an uneven patch of ground, and what should have been a lunge became a stagger, his sword point dropping to strike the dirt.

And a sandaled foot descended on it from above, snapping the steel just above the hilt.

A calloused hand smashed into his jaw, knocking him backward. Deven punched out with the useless hilt and connected with ribs, but he had lost the advantage of reach; an instant later, the man was behind him, locking him into a choke hold. Gasping, Deven reversed his grip and stabbed blindly backward, gouging the broken tip into flesh.

The stranger ignored that wound, as he had ignored all others.

The world was fading, bright lights dancing with blackness. The hilt fell from his nerveless fingers. Deven reached up, trying to find something to claw, but there was no strength in his arms. The last thing he heard was a faint, mocking laugh in his ear.

TURNAGAIN LANE, BY THE RIVER FLEET
6 May 1590

The sluggish waters of the Fleet reeked, even up here by Holborn Bridge, before it passed the prison and the workhouse of Bridewell and so on down to the Thames. It was an ill-aspected river, and always had been; again and again the mortals tried to cleanse it and make its course wholesome once more, and always it reverted to filth. Lune had once been unfortunate enough to see the hag of the Fleet. Ever since then, she kept her distance.

Except when she had no choice.

The alehouse her instructions had told her to find was a dubious place in Turnagain Lane, frequented by the kind of human refuse that clustered around the feet of London, begging for scraps. She had disguised herself as an older woman, and was glad of her choice; a maiden wouldn't have made it through the door.

She had been given no description, but the man she sought was easy enough to find; he was the one with the wooden posture and the disdainful sneer on his face.

Lune slipped into a seat across from him, and wasted no time with preliminaries. "What do you want from me?"

The glamoured Vidar tsked at her. "No patience, and no manners, I see."

She had barely sent word off to the Goodemeades when Vidar's own messenger found her. The added delay worried her, and for more than one reason: not only might Invidiana wonder at her absence, but the secrecy of this meeting with Vidar meant he had not called her for official business.

She had not forgotten what she owed him.

But she could use that to her advantage, if only a little. "Do you want the Queen to know of this conference? 'Tis best for us both that we be quick about it."

How had he ever managed his extended masquerade as Gilbert Gifford? Vidar sat stiffly, like a man dressed up in doublet and hose that did not fit him, and were soiled besides. Lune supposed the preferment he got from it had been motive enough to endure. Though he had been squandering that preferment of late; she had not seen him at court in days.

Vidar's discomfort underscored the mystery of his absence. "Very well," he said, dropping his guise of carelessness. What lay beneath was ugly. "The time has come for you to repay that which you owe."

"You amaze me," Lune said dryly. She had made no oath to be polite about it.

He leaned in closer. The face he had chosen to wear was sallow and ill shaven, in keeping with the tenor of the alehouse; he had forgotten, however, to make it smell. "You will keep silent," Vidar growled, "regarding any other agents of the Wild Hunt you may uncover at court."

Lune stared at him, momentarily forgetting to breathe.

"As I kept silent for you," he said, spitting the words out one by one, "so you shall for me. Nor, by the vow you swore, will you let any hint of this matter leak to the Queen—by *any* route. Do you understand me?"

Corr. No wonder Vidar had been so absent of late; he must have feared what Invidiana would uncover about the dead knight... and about him.

Sun and Moon—what was he planning?

Lune swallowed the question, and her rudeness. "I understand you very well, my lord."

"Good." Vidar leaned back and scowled at her. "Then get you gone. I relish your company no more than you relish mine."

That command, she was glad to obey.

FARRINGDON WARD WITHIN, LONDON
6 May 1590

Her quickest path back to the Onyx Hall led through Newgate, and she walked it with her mind not more than a tenth on her surroundings, working through the implications of Vidar's demand.

He must have formed an alliance with the Hunt. But *why*? Had he given up all hope of claiming Invidiana's throne for himself? Knowing what she did now, Lune could not conceive of those exiled kings permitting someone to take the usurper's place. If he thought he could double-cross them...

She was not more than ten feet from the Hall entrance in the St. Nicholas Shambles when screeching diverted her attention.

Fear made her heart stutter. In her preoccupation, someone might have crept up on her with ease, and now her nerves all leapt into readiness. No one did more than eye her warily, though, wondering why she had started in the middle of the street.

The noise didn't come from a person. It came from a jay perched on the eave of a building just in front of the concealed entrance. And it was staring straight at her.

Watching it, Lune came forward a few careful steps.

Wings flapped wildly as the jay launched itself at her face, screaming its rasping cry. She flinched back, hands coming up to ward her eyes, but it wasn't attacking; it just battered about her head, all feathers and noise.

She had not the gift of speaking with birds. It could have been saying anything, or nothing.

But it seemed very determined to keep her from the entrance to the Onyx Hall—and she did know someone who might have sent it.

Lune retreated a few steps, ignoring the staring butchers that lined both sides of the shambles, and held up one hand. Now that she had backed away, the jay quieted, landing on her outstretched finger.

Something in her message must have panicked the Goodemeades. But what?

She dared not go to them to ask. She had to hide herself, and then get word to the sisters. Not caring how it seemed to onlookers, Lune cupped the bird in her hands, closing her fingers around its wings, and hurried back out through Newgate, wondering where—if anywhere—would be safe.

THE ONYX HALL, LONDON
7 May 1590

Instinct stopped him just before he would have moved.

He could feel ropes binding his ankles together, his arms behind him. The stone beneath him was cold and smooth. In the instant when he awoke, before he shut his eyes again, he saw a floor of polished black and white and grey. The air on his skin, ghosting through the rents in his clothing, was cool and dry.

He knew where he was. But he needed to know more.

Footsteps tapped a measured beat on the stone behind him. Deven kept his body limp and his eyes shut. Let them think him still unconscious.

Then he began to move, without a single hand touching him.

Deven felt his body float up into the air and pivot so that he hung upright, facing the other direction. His arms ached at the change in position, cold and cramped from the ropes and the stone. Then a voice spoke, as cool and dangerous as silk over steel. "Cease your feigning, and look at me."

For a moment he considered disobeying. But what would it gain him?

Deven opened his eyes.

The breath rushed out of him in a sigh. *Oh, Heaven save me...* They had spoken of her beauty, but words could not frame it. All the poetry devoted to Elizabeth, all the soaring, extravagant compliments, comparing her to the most glorious goddesses of paganism—every shred of it should have been directed here, to this woman. Not the slightest imperfection or mark interrupted the alabaster smoothness of her skin. Her eyes were like black diamonds, her hair like ink. High cheekbones, delicately arched brows, lips of a crimson hue both forbidding and inviting...

The words tore their way free of him, driven by some dying instinct of self-preservation. "God in Heaven..."

But she did not flinch back. Those red lips parted in an arrogant laugh. "Do you think me so weak? I do not fear your God, Master Deven."

If she did not fear the Almighty, still His name had given Deven strength. He wrenched his gaze away, sweating. They had spoken of Invidiana's beauty, but he had imagined her to be like Lune.

She was nothing like Lune.

"You are not surprised," Invidiana said, musingly. "Few men would awake in a faerie palace and be unamazed. I took you for bait, but you are more than that, are you not, Master Deven? You are the accomplice of that traitor, Lune."

How much did she know?

How much could he keep from her?

"Say rather her thrall," Deven spat, still not looking at her.

"I care nothing for your politics. Free me from her, and I will trouble you no more."

Another laugh, this one bidding fair to draw blood by sound alone. "Oh, indeed. 'Tis a pity, Master Deven, that I did not have Achilles steal you sooner. A man who so readily resorts to lies and deception, manipulation and bluff, could well deserve a place in my court. I might have made a pet of you.

"But the time for such things has passed." The idle amusement of her voice hardened. "I have a use for you. And if that use should fail... you will provide me with other entertainment."

Deven shuddered uncontrollably, hearing the promise in those words.

"You are my guest, Master Deven." Now it was mock courtesy, as disturbing as everything else. "I would give you free run of my domain, but I fear some of my courtiers do not always distinguish guests from playthings. For your own safety, I must take precautions."

The force that held him suspended now lowered him. The toes of his boots touched the floor; then she pushed him farther, until he knelt on the stone, arms still bound behind his back.

His head was dragged forward again; he could not help but look.

Invidiana was lifting a jewel free of her bodice. He had a glimpse of a black diamond housed in silver, edged with smaller gems; then he tried to flinch back and failed as her hand came toward his face.

The metal was cool against his skin, and did not warm at the contact. An instant later Deven shuddered again, as six sharp points dug into his skin, just short of drawing blood.

"This ban I lay upon thee, Michael Deven," Invidiana murmured, the melody of her voice lending horror to her words. "Thou wilt not depart from this chamber by any portal that exists or might be made, nor send messages out by any means; nor wilt move in violence against me, lest thou die."

Every vein in his body ran with ice. Deven's teeth clenched shut, his jaw aching with sudden strain, while six points of fire fixed into the skin of his brow.

Then it was gone.

Invidiana replaced the gem, smiling, and the bonds holding him fell away.

"Welcome, Master Deven, to the Onyx Hall."

* * *

DEAD MAN'S PLACE, SOUTHWARK

7 May 1590

There was something grimly appropriate, Lune thought, about hiding a stone's throw from an Episcopal prison full of heretics. But Southwark was a good place for hiding; with its stews and bear-baiting, its prisons and general licentiousness, a woman on her own, renting out a room for a short and indefinite period of time, was nothing out of the ordinary way. Lune would simply have to be gone before her faerie gold—or rather, silver—turned back to leaves.

Had the jay in truth belonged to the Goodemeades? Or had it taken her message to another? Would the Goodemeades come? What had happened, that they were so determined to keep her from the Onyx Hall?

Footsteps on the stair; she tensed, hands reaching for weapons she did not have or know how to use. Then a soft voice outside: "My lady? Let us in."

Trying not to shake with relief, Lune unbarred the door.

The Goodemeades slipped inside and shut it behind them. "Oh, my lady," Gertrude said, rushing forward to clasp her hands, "I am so sorry. We did not know until too late!"

"About the pact?" Lune asked. She knew even as she said the words that wasn't it, but her mind had so fixated on it, she could not think what Gertrude meant.

Rosamund laid a gentle hand on her arm. The touch alone said too much. "Master Deven," the brownie said. "She has taken him."

There was no refuge in confusion, no stay of understanding while Lune asked what she meant. Fury began instantly, a slow boil in her heart. "I trusted you to warn him. He's as much in danger as I; why did you warn only me?"

The sisters exchanged confused looks. Then Rosamund said, "My lady… the birds stopped you of their own accord. Her people ambushed him on the street yesterday. We did not even know of it until later. We sent birds some time ago, to watch you both. They had lost you, but when one saw him taken, they chose to watch the entrances and stop you if they could."

Lost her. Because she had tried so very hard to keep anyone from following her when she went to meet Vidar. Where had she been, when they attacked him? Had Vidar distracted her on purpose?

"Tell me," Lune said, harsh and cold.

Gertrude described it softly, as if that lessened the dreadfulness of what she said. "A will-o'-the-wisp to lead him astray. A tatterfoal, to replace his own horse and carry him into the trap." She hesitated before supplying the last part. "And Achilles, to bring him down."

One tiny comfort Lune could take from that: Invidiana must not mean to have Deven battle to the death, or she would have saved Achilles for later, and sent Kentigern instead.

"There's more," Rosamund said. "His manservant Colsey was following him, it seems. I do not know why, or what happened… but he's dead."

Colsey. Lune had met him, back when they were all

at court, and her greatest concern had been how to evade Deven's offer of marriage without losing his usefulness to her. She had liked him, and his close-mouthed loyalty to his master.

Gone, that easily. And Deven...

Lune turned away and walked two paces. She could go no farther; the room she had rented was scarcely larger than a horse's stall.

The lure was plain. The question was whether she would take it.

It hardly mattered whether Invidiana had Francis Merriman's ghost. The Queen knew enough. Would Lune now walk into her trap?

Without thinking, one hand dropped to touch the purse that held the last of the loaf Deven had given her. Mortal bread. She had consumed so much of it, since she met him. Not enough to make her human, but enough to change her.

Michael Deven loved her. Not Anne Montrose, but *Lune*. She knew it the night he led her to his house. What did that love mean to her?

Would she spurn it, and flee to save herself?

Or would she accept it—return it—despite the cost?

She had never felt that choice within her before. Too much mortal bread; it brought her to an unfamiliar precipice. Her mind moved in strange ways, wavering, uncertain.

"My lady?" Gertrude whispered from behind her.

Lune's hands stilled on her skirt. She turned to find the two brownies watching her with hesitant expressions.

It was the first time she had seen them show fear. They had spent years opposing Invidiana; now, at long last, their game might be at an end.

"The London Stone lies within the Onyx Hall," Lune said. "So does Invidiana, who made a pact with Hell. And so does Michael Deven.

"I will do what we had intended. I will seek out Doctor Dee."

MEMORY

Long and long ago...

There was a beauty of night, pale as the moon, dark as her shadow, slender and graceful as running water. A young man saw her dancing under the stars, and loved her; he pined and sighed for her, until his mother feared he would waste away, lost in dreams of love. For that happened at times, that folk should die for love of the strangers under the hills.

Such was not this young man's lot. A plan was formed, wherein he would have the beautiful stranger to wife. Great preparations were made by his people and by hers, a glorious midsummer wedding on the banks of the river, a little distance from the village where the young man's father ruled. There would be music and dancing, good food and drink, and if the maidens and youths of the village fell in love with their guests from the other side, perhaps this wedding would be only the first of many. And when it was done, the young man would have a fine house to share with his wife, in time succeeding his father as chieftain and ruling in his place.

So it was planned. But it did not come to pass.

The guests gathered beneath the twilit summer sky. On the one side, the weathered faces of the villagers, tanned by the sun in their labours, the old ones wrinkled, the young ones round-cheeked and staring at the folk across the field. There stood creatures tall and tiny, wide-shouldered and slender, some with feathers, hooves, tails, wings.

The one the young man loved looked at her people, in all their wild glory, and even their ugliness was more beautiful to her, because it was what they were and always would be.

Then she looked at the people of the village, and she saw how accidents marked their bodies, how they soon crumbled and fell, how their houses stood on bare dirt and they scratched out their living with toil.

And she asked herself: *Am I to go from this to that?*

So she fled, leaving the young man alone beneath the rising moon, with his heart broken into pieces.

He sickened and died, but not for love. Yet he took strange pride in his illness, laughing a mad laugh that grieved his mother unbearably. *You see, we prove her right. We die so soon, so easily; she will remain long after I am gone. I do not mourn the mayfly, nor yoke my heart to its; why should it be different with her?*

Bitterness poisoned the words, the terrible knowledge that his love was as nothing to the immortal creature upon whom it had fixed.

The moon waned and waxed, and when it was full once more, the young man died. On his deathbed he spoke his last words, not to his family, but to the absent creature that had been the end of him. *May you suffer as we suffer, in sickness and age, so that you find no escape from that which you fled. May you feel all the weight of mortality, and cry out beneath your burden, until you atone for the harm you have done and understand what you have spurned.*

Then he died, and was buried, and never more did the villagers gather in harmony with the strangers under the hills.

MORTLAKE, SURREY

7 May 1590

The house, with all its additions and extensions, was like an old man dreaming in the afternoon sunlight, relaxed into a sprawling doze. Yet to Lune it seemed more foreboding than the Onyx Hall: a lair of unknown dangers.

Be it angels or devils he summoned inside, it was not a place a faerie should go.

Lune put her shoulders back and approached the door with a stride more resolute than she felt.

She was a woman on her own, with no letter of introduction to smooth her way. But the maidservant was easy enough to charm, and Dee's wife proved sympathetic. "He's at his studies," the woman said, shifting the infant she held onto her other hip. The small creature stared frankly at Lune, as if it could see through the glamour. "But if 'tis urgent..."

"I would be most grateful," Lune said.

Her reception was warmer than expected. "You will forgive my frankness in asking," Dee said, once the formalities were dispensed with, "but has this anything to do with Michael Deven?"

This was not in the mental script Lune had prepared on her journey to Mortlake. "I beg your pardon?"

A surprising twinkle lightened the astrologer's tired eyes. "I am not unaware of you, Mistress Montrose. Your lady the Countess of Warwick has been kind to me since my return, and I had the honour of friendship with Sir Francis Walsingham. When Master Deven came to my door, asking for aid in the matter of a young gentlewoman, 'twas not difficult to surmise whom he meant."

No magic, just an observant mind. Lune began to breathe again. "Indeed, Doctor Dee—it has everything to do with him. Will you aid me?"

"If I can," Dee said. "But some things are beyond my influence. If he is in some political difficulty—"

Not of the sort he thought. Lune clasped her hands in her lap and met the old man's gaze, putting all the sincerity she could muster into it. "He is in great peril, and for reasons I fear must be laid at my feet. And it may be, Doctor Dee, that you are the only man in England who could help us."

His face stilled behind its snowy beard. "And why would that be?"

"They say you speak with angels."

All pleasantness fell away, but his eyes were as bright and unblinking as a hawk's. "I fear, Mistress Montrose, that you may have an overly dramatic sense of his danger, my abilities, or both. Angels—"

"I am not overly dramatic," she snapped, forgetting in her distress to be polite. "I assure you. The tale is a complex one, Doctor Dee, and I have not the time to waste on it if at the end you will tell me you can be of no aid. Do you hold conference with angels, or not?"

Dee rose from his seat, ink-stained fingers twitching his long robe straight. Turning away to pace across the room, he spoke very deliberately. "I see that you are distraught, Mistress Montrose, and so I will lay two things before you. The first is that angelic actions are no trivial matter, no miracle that can be summoned at a whim to solve worldly ills.

"The second..." He paused for a long time, and his hands, clasped behind his back, tightened. Something hardened his voice, lending it an edge. "The second is that such efforts require assistance—namely, the services of a scryer, one who can see the presences when they come. My former companion and I have parted ways, and I have found no suitable replacement for him."

The first point did not worry her; the second did. "Can you not work without such assistance?"

"No." Dee turned back to face her. His jaw was set, as if against some unhappy truth. "And I will be honest with you, Mistress Montrose. At times I doubt whether I have *ever* spoken with an angel, or whether, as they accuse me, I have done naught but summon devils, who play with me for their own amusement."

Her mouth was dry. All her hope crumbled. If not Dee, then who? A priest? Invidiana had destroyed priests before. And Lune did not think a saint would answer the call of a faerie.

"Mistress Montrose," Dee said softly. Despite the lines that had sobered his face, his manner was compassionate. "Will you not tell me what has happened?"

A simple question, with a dangerous answer. Yet some corner of Lune's mind was already calculating. If he were not the sorcerer she had expected, then a charm might bedazzle him long enough for her to escape, should all go poorly. She would be destroying Anne Montrose, but no life remained for that woman regardless...

She truly was thinking of doing it.

"Can I trust you?" Lune whispered.

He crouched in front of her, keeping space between them, so as not to crowd her. "If it means no harm to England or the Queen," Dee said, "then I will do my best to aid you in good faith."

The door was closed. They were private.

Lune said, "I am not as I seem to be." And, rising to her feet, she cast aside her glamour.

Dee rose an instant later, staring.

"The Queen of faerie England," she said, every muscle tensed to flee, "has formed a pact with Hell. I need the aid of Heaven to break it. On this matter rests not only the safety of Michael Deven, but the well-being of your own kingdom and Queen."

He did not shout. He did not fling the name of God up as defence. He did nothing but stare, his eyes opaque, as if overtaken by his thoughts.

"So if you cannot summon angels," Lune said, "then tell me, Doctor Dee, what I should do. For I do not know."

Within the mask of his beard, his mouth was twitching; now she read it as a kind of bitterness, surprising to her. "Did you send him?" he asked abruptly.

"Michael Deven?"

"Edward Kelley."

The name ground out like a curse. Where did she know it from? She had heard it somewhere...

"When he came to me," Dee said coldly, "he offered to further my knowledge in magic with faeries."

Memory came. A human man with mangled ears; she had seen him once or twice at court—her own court—and heard his name. She had never known more. "I did not send him," Lune said. "But someone may have. Who was he?"

"My scryer," Dee replied. "Whom I have long suspected of deception. He came to me so suddenly, and seemed to have great skill, but we so often fought..." Now she recognised the note in his voice; it was the sound of affection betrayed. This Kelley had been dear to him once.

"He is gone now?" Lune asked.

Dee made a cut-off gesture with one hand. "We parted ways in Trebon. He is now court alchemist to the Holy Roman Emperor." Then he truly was out of reach. Lune said, "Please, Doctor Dee. I beg you." Never in all the ages she could remember had she knelt, as a fae, to a mortal, but she did it now. "I know I am no Christian soul, but Michael Deven is, and he will die if I cannot stop this. And does not your God oppose the devil, wherever he may work? Help me, I beg. I do not know who else to ask." Dee gazed blindly down at her, distracted once more. "I have no scryer. Even Kelley may have given me nothing but falsehoods, and I myself have no gift for seeing. It may be that I have no more power to summon angels than any other man."

"Will you not try?" Lune whispered.

With her eyes fixed on him, she saw the change. Some thought came to him, awakening all the curiosity of his formidable mind. The expression that flickered at the edge of his mouth was not quite a smile, but it held some hope in it. "Yes," Dee said. "We will try."

THE ONYX HALL, LONDON

8 May 1590

Thirst was the greatest threat.

Deven tried to distract himself. The room, he came to

realise, was Invidiana's presence chamber. Larger by far than Elizabeth's, it had an alien grandeur a mortal queen could only dream of, for in this place, fancies of architecture could truly take flight. The pillars and ribs that supported the arching ceiling were no more than a decoration born from some medieval fever dream; they were not needed for strength. The spaces between them were filled with filigree and panes of crystal, suspended like so many fragile swords of Damocles.

Beneath and among these structures wandered fae whom he presumed to be the favoured courtiers of this Queen. They were a dizzying lot: some human-looking, others supernaturally fair, others bestial, and clad in finery that was to mortal courtiers' garb as the chamber was to mortal space. They all watched him, but none came near him; clearly word had gone around that he was not to be touched. How much did they know of who he was, and why he was there?

Lacking an answer to that question, Deven decided to test his boundaries. He tried to speak to others; they shied away. He followed them around, eavesdropping on their conversations; they fell silent when he drew near, or forwent the benefit of being so near the Queen and left the chamber entirely. The fragments he overheard were meaningless to him anyway.

He spoke of God to them, and they flinched, while Invidiana looked on in malicious amusement.

She was less amused when he decided to push harder.

Deven took up a position in the centre of the chamber, facing the throne, and crossed himself. Swallowing against the dryness of his mouth, he began to recite.

"Our father, which art in Heaven, hallowed be Thy Name. Thy kingdom come. Thy will be done in earth, as it is in Heaven. Give us this day our daily bread, and forgive us our trespasses, as we forgive them that trespass against us. And lead us not into temptation, but deliver us from evil. Amen."

The chamber was half-empty before he finished; most of those who remained were bent over or sagged against the walls, looking sick. Only a few remained untouched; those, he surmised, had eaten of mortal food recently. But even they did not look happy.

Nor did Invidiana. She, for the first time, was angry.

He tried again, this time in a different vein, dredging up faded memories of prayers heard from prisoners and recusants. *"Pater noster, qui es in caelis: sanctificetur Nomen Tuum..."*

This time even he felt its force. The hall trembled around him; its splendour dimmed, as if he could see through the marble and onyx and crystal to plain rock and wood and dirt, and all the fae stood clad in rags.

Then something slammed into him from behind, knocking him to the floor and driving all breath from him. His Catholic prayer ended in a grunt. A voice spoke above him, one he knew too well, even though he had heard scarcely a dozen words from it. "Should I cut out his tongue?" Achilles asked.

"No." If the Latin form had shaken Invidiana, she

gave no sign. "We may yet need him to speak. But stop his mouth, so he may utter no more blasphemies."

A wad of fabric was shoved into Deven's mouth and bound into place. His thirst increased instantly as every remaining bit of moisture went into the cloth.

But his mind was hardly on that. Instead he was thinking of what he had seen, in that instant before Achilles took him down.

Invidiana's throne sat beneath a canopy of estate, against the far wall. Under the force of his prayer, it seemed for a moment that it masked an opening, and that something lay in the recess behind it.

What use he could make of that knowledge, he did not know. But with his voice taken away, knowledge was his only remaining weapon.

MORTLAKE, SURREY

8 May 1590

"You are mad," Lune said.

"Perhaps." Dee seemed undisturbed by the possibility; no doubt he had been accused of it often enough. "But children are ideal for scrying; children, and those who suffer some affliction of the mind. Kelley was an unstable man—well, perhaps that is no recommendation, if in truth he did naught but deceive me. Nonetheless. The best scryers are those whose minds are not too shackled by notions of possibility and impossibility."

"You yourself, then."

He shook his head. "I am too old, too settled in my ways. My son has shown no aptitude for it, and we have no time to find another."

Lune took a slow breath, as if it would banish her feeling that all this had taken a wrong turn somewhere. "But if you question whether you have ever spoken with an angel before, what under the sun and moon makes you believe one will answer to a *faerie*?"

They were in his most private workroom, with strict orders to his surprisingly large family that under no circumstances were they to be disturbed. Lune hoped it would be so; at Dee's command, she had eaten no food of any kind since the previous day—which meant no mortal bread.

He knew quite well what that meant, for she had told him. At great length, when she began to understand what he had in mind. And that was *before* he voiced his decision to use her as his scryer.

The philosopher shook his head again. "You misunderstand the operation of this work. Though you will be a part of it, certainly, your role will be to perceive, and to tell me what you see and hear. The calling is mine to perform. I have been in fasting and prayer these three days, for I intended to try again with my son; I have purified myself, so that I might be fit for such action. The angel—if indeed one comes—will come at my call."

Now she understood the fasting. But prayer? "I have

not made such thorough preparations."

The reminder dimmed his enthusiasm. "Indeed. And if this fails, then we will try again, three days from now. But you believe time to be of the essence."

Invidiana had the patience of a spider; she would wait three years if it served her purpose. But the longer Deven remained in the Onyx Hall, the greater the likelihood that the Queen would kill him—or worse.

Worse could take many forms. Some of them were the mirror image of what Lune risked now. Baptism destroyed a fae spirit, rendering it no more than mortal henceforth. Dee had not suggested that rite, but who knew what effect this "angelic action" would have?

That frightened her more than anything. Fae could be slain; they warred directly with one another so rarely because children were even more rare. But death could happen. Nor did anyone know what if anything lay beyond it, though faerie philosophers debated the question even as their human counterparts did. The uncertainty frightened Lune less than the certainty of human transformation. 'Twas one thing to draw close to them, to bask in the warmth of their mortal light. To *be* one…

She had already made her choice. She could not unmake it now.

Lune said, "Then tell me what I must do."

Dee took her by the hand and led her into a tiny chapel that adjoined his workroom. "Kneel with me," he said, "and pray."

Her exposed faerie nature felt terrifyingly vulnerable. With mortal bread shielding her, she could mouth words of piety like any human. But now?

He offered her a kindly smile. If her alien appearance disturbed him, he had long since ceased to show it. "You need not fear. Disregard the words you have heard others say—Catholic and Protestant alike. The Almighty hears the sentiment, not the form."

"What kind of Christian are you?" Lune asked, half in astonishment, half to stall for time.

"One who believes charity and love to be the foremost Christian virtues, and the foundation of the true Church, that lies beyond even the deepest schism of doctrine." His knobbled hand pressed gently on her shoulder, guiding her to her knees. "Speak in love and charity, and you will be heard."

Lune gazed up at the cross that stood on the chapel's wall. It was a simple cross, no crucifix with a tormented Christ upon it; that made it easier. And the symbol itself did not disturb her—not here, not now. Dee believed what he said, with all his heart. Without a will to guide it against her, the cross was no threat.

Speak in love and charity, he had said.

Lune clasped her hands, bent her head, and prayed.

The words came out hesitantly at first, then more fluidly. She wasn't sure whether she spoke them aloud, or only in her mind.

Some seemed not even to be words: just thoughts,

concepts, inarticulate fears, and longings, set out first in the manner of a bargain—*help me, and I will work on your behalf*—then as justifications, defences, an apology for her faerie nature. *I know not what I am, in the greater scope of this world; whether I be fallen angel, ancient race, unwitting devil, or something mortals dream not of. I do not call myself Christian, nor do I promise myself to you. But would you let this evil persist, simply because* I *am the one who works against it? Does a good deed cease to be good, when done by a heathen spirit?*

At the last, a wordless plea. Invidiana—Suspiria—had taken this battle into territory foreign to Lune. Adrift, lost in a world more alien than the undersea realm, she could not persevere without aid.

So far did she pour herself into it, she forgot this was preparation only. She jerked in surprise when Dee touched her shoulder again. "Come," he said, rising. "Now we make our attempt."

The workroom held little: a shelf with a few battered, much-used books. A covered mirror. A table in the centre, whose legs, Lune saw, rested upon wax rondels intricately carved with symbols. A drape of red silk covered the tabletop and something else, round and flat.

Upon that concealed object, Dee placed a crystalline sphere, then stepped back. "Please, be seated."

Lune settled herself gingerly on the edge of a chair he set facing the sphere.

"I will speak the invocation," he said, picking up one

of the books. Another bound volume sat nearby, open to a blank page; she glimpsed scrawled handwriting on the opposite leaf, that was evidently his notes, for he had ink and a quill set out as well.

She wet her lips. "And I?"

"Gaze into the stone," he said. "Focus your mind, as you did when you prayed. Let your breathing become easy. If you see aught, tell me; if any being speaks to you, relate its words." He smiled at her once more. "Do not fear evil spirits. Purity of purpose, and the formulas I speak, will protect us."

He did not sound as certain as he might have, and his hand tightened over the book he held, as if it were a talisman. But Lune was past the point of protest; she simply nodded, and turned her attention to the crystal.

John Dee began to speak.

The first syllables sent a shiver down her spine. She had expected English, or Latin; perhaps Hebrew. The words he spoke were none of these, nor any language she had ever heard. Strange as they were, yet they reverberated in her bones, as if the sense of them hovered just at the edge of her grasp. Did she but concentrate, she might understand them, though she had never heard them before.

The words rolled on and on, in a sonorous, ceaseless chant. He supplicated the Creator, Lune sensed, extolling the glory of Heaven and its Lord, describing the intricate structure of the world, from the pure realms of God down to the lowliest part of nature. And for a brief span she perceived it as if through his eyes: a beautifully mathematical cosmos, filled with pattern, correspondence, connection, like the most finely made mechanical device, beyond the power of any mind save God's to apprehend in its entirety, but appreciable through the study of its parts.

To this, he had devoted his life. To understanding the greatest work of God.

In that moment, all the aimless, immortal ages of her life seemed by comparison to be flat and without purpose.

And then she felt suffused by a radiance like that of the moon, and her lips parted; she spoke without thinking. "Some thing comes."

Dee's invocation had finished, she realised, but how much time had passed, she did not know. A soft scratching reached her ears: his quill upon paper. "What do you see?"

"Nothing." The sphere filled her vision; how long since she had last blinked?

"Speak to it."

What should she say? Her mind was roaringly empty of words. Lune groped for something, anything. "We—I—most humbly beseech your power, your aid. The Queen of the Onyx Court has formed a pact with Hell. Only with your power may it be broken. Will you not help us?"

Then she gasped, for the crystal vanished; she saw instead a figure, its form both perfect and indefinable. The table was gone, the chair was gone; she stood in an empty space before the terrible glory of the angel, and sank to her knees without thinking, in respect and supplication.

As if from a great distance, she heard Dee utter one

word, his own voice trembling in awe. *"Anael."*

Her spirit lay exposed, helpless, before the angel's shining might. With but a thought, it could destroy her, strip all faerie enchantment from her being, leave her nothing more than a mortal remnant, forever parted from the world that had been hers. She was no great legend of Faerie to defend herself against such, and she had laid herself open to this power of her own free will.

All that defended her now was, as Dee had said, charity and love.

She trembled as the figure drew closer. The strength might have crushed her, but instead it held her, like a fragile bird, in the palm of its hand. Lune felt lips press against hers, and the cool radiance flooded her body; then they were gone.

"Bear thou this kiss to him thou lovest," the angel Anael said, its words the true and pure form of the language Dee had spoken, a force of beauty almost too much to bear.

Then the light receded. She was in her chair; the crystal was before her; they were alone once more in the room.

Dee murmured a closing benediction, and sank back into his own chair, from which he had risen without her seeing. The notebook sat next to him, hardly touched.

Lune's eyes met the philosopher's, and saw her own shock echoed there.

He, who had no gift for seeing, had seen something. And he knew, as she did, that it was a true angelic presence, and it had answered her plea.

Bear thou this kiss to him thou lovest.

She had made that choice. What it meant, she did not know; she had never given her heart before. How a kiss would aid her, she could not imagine. It seemed a weak weapon against Invidiana.

But it was Heaven's response to her plea. For Michael Deven's sake, she would go into the Onyx Hall, and somehow win her way through to him. She would bring him Anael's kiss.

What happened after that was in God's hands.

THE ANGEL INN, ISLINGTON
8 May 1590

"She must be distracted," Lune said. "Else she will place all her knights and guardsmen and other resources between me and Deven, and I will stand no hope of reaching him. They will kill me, or they will bind me and drag me before her; either way, I will not be able to do what I must."

The Goodemeades did not question that part of it. Lune had told them in brief terms of what had passed in Mortlake—brief not because she wished to hide anything from them, but because she had few words to describe it. Their eyes had gone round with awe, and they treated her now with a reverent and slightly fearful respect that unnerved her.

Not so much respect, though, that they didn't question

certain things. "My lady," Gertrude said, "she will be expecting you to do exactly that. You have not come back, which means you know of your peril. If you are not simply to walk into her claws, then you must try to draw her attention away. But she will recognise any diversion as just that—and ignore it."

From across the rose-guarded room, Rosamund, who had been silent for several minutes, spoke up. "Unless the diversion is something she cannot ignore."

"The only thing she could not ignore would be—"

"A real threat," Lune said.

Something Invidiana truly did have to fear. A war on her very doorstep, that she must send her soldiers to meet, or risk losing her throne.

The list of things that fit that name was short indeed.

Gertrude's face had gone white, and she stared at her sister. Grimness sat like a stranger on Rosamund's countenance, but if a brownie could look militant, she did. "We could do it," she said. "But, my lady, once such a force is unleashed, it cannot be easily stopped. We all might lose a great deal in the end."

Lune knew it very well. "Could anything stop them?"

"If she were to draw the sword out again—perhaps. That, more than anything, is what angers them. They might be satisfied, if she renounced it."

"But Invidiana would never do it," Gertrude said. "Only Suspiria, and perhaps not even her." She stared up at Lune, her eyes trembling with tears. "Will we have her back, when you are done?"

The unspoken question: *Or do you go to kill her?*

Lune wished she could answer the brownie's question, but she was as blind as they. The angel's power waited within, alien and light, but she did not know what it would do. Could a faerie spirit be damned to Hell?

Her reply came out a whisper. "I can make no promises."

Rosamund said heavily, "With that, we must be content. We have no other choice."

"You must move with haste." The knot of tension in Lune's stomach never loosened, except for a few timeless moments, in the angel's presence. "Use Vidar."

"Vidar?"

"Corr was his agent, or at least an ally. He bade me be silent about any others I might find at court. I do not know his scheme, but there must be one; we can make use of it." Her vow did not prevent her from telling the Goodemeades; the last person in creation they would share the information with was Invidiana. But she had never expected to use such a loophole.

Rosamund came forward, smoothing her apron with careful hands, and put an arm around her white-faced sister. "Make your preparations, my lady. Gertrude and I will raise the Wild Hunt."

* * *

THE ONYX HALL, LONDON

9 May 1590

The sun's heat baked his shoulders and uncovered head. His ride had been a long one, and he was tired; he swung his leg over the saddle and dropped to the ground, handing off his reins to a servant. They were gathered by the riverbank, an elegant, laughing crowd, playing music, reciting poetry, wagering at cards. He longed to join them, but ah! He was so thirsty.

A smiling, flirtatious lady approached him, a cup of wine in each hand. "My lord. Will you drink?"

The chased silver was cool in his fingers. He looked down into the rich depths of the wine, smelling its delicate bouquet. It would taste good, after that long ride.

With the cup halfway to his lips, he paused. Something…

"My lord." The lady rested one hand gently on his arm, standing closely enough that her breasts just touched his elbow. "Do you not like the wine?"

"No," he murmured, staring at the cup. "That is…"

"Drink," she invited him. "And then come with me."

He was so thirsty. The sun was hot, and the wine had been cooled in the stream. He had not eaten recently; it would go to his head. But surely that did not matter—not in this gay, careless crowd. They were watching him, waiting for him to join them.

He brought the cup to his lips and drank.

The liquid slid down his throat and into his belly, chilling him, making all his nerves sing. No wine he had ever drunk tasted thus. He gulped at it, greedy and insatiable; the more he drank, the more he wanted, until he was tipping the cup back and draining out the last drops, and shaking because there was no more—

There was no sunlight. There was no meadow by the stream. There were courtiers, but the faces that watched were wild and inhuman, and all around him was darkness.

The lush faerie lady stepped back from him, her face avid with delight, and from some distance away Invidiana gave sardonic applause. "Well done, Lady Carline. Achilles, you need not restore his gag." The Queen smiled across the chamber at Deven, letting all her predatory pleasure show. "He will speak no names against us now."

The cup fell from Deven's hand and clanked against the stone, empty to the dregs. Faerie wine. He had refused all food, all drink, knowing the danger, but in the end his body had betrayed him, its mortal needs and drives making it an easy target for a charm.

Even if Lune came for him now, it was too late.

He reached for the names that had been his defence, and found nothing. A mist clouded his mind, obscuring the face of… what? There had been something, he knew it; he had gone to church, and prayed…

But the prayers were gone. Those powers were no longer within his reach.

Laughter pursued him as he stumbled away, seeking refuge in a corner of the chamber. Now, at last, the stoicism he had clung to since his capture failed him. He wanted more; his body ached with the desire to beg. Another cup—a sip, even—

He clenched his hands until his knuckles creaked, and waited, trembling, for the next move.

LONDON

9 May 1590

The moon rose as the sun set, its silver disc climbing steadily into the sky.

The curfew bells had rung. London was abed—or ought to be; those who were out late, the drunken gentlemen and the scoundrels who waited to prey on them, deserved, some would say, whatever happened to them.

On the northern horizon, without warning, storm clouds began to build.

They moved from north to south, against the wind, as clouds should not have done. In their depths, a thunder like the pounding of hoofbeats against the earth, up where no earth was. A terrible yelping came from the clouds, that more sceptical minds would dismiss as wild geese. Those who knew its true source, hid.

Brief flashes of lightning revealed what lay within the clouds.

The hounds ran alongside, leaping, darting, weaving in and out of the pack. Black hounds with red eyes; white hounds with red ears; all of them giving that terrible, belling cry, unlike any dog that ever mortal bred.

Horses, shod with silver and gold, flaring with spectral light. Formed from mist, from straw, from fae who chose to run in such shape, their headlong gallop brought them on with frightening speed. And astride their backs rode figures both awful and beautiful.

Stags' horns spiked the sky like a great, spreading crown. Feathered wings cupped the air, pinions whistling in the storm wind. Their hair was yellow as gold, red as blood, black as night; their eyes burned with fury, and in their hands were swords and spears out of legend.

The forgotten kings of faerie England rode to war.

It went by many names. Wisht Hounds, Yeth Hounds, Gabriel Rachets, Dando and His Dogs. A dozen faces and a dozen names for the Wild Hunt, united now in a single purpose.

They would not involve mortals in their war, and for decades their enemy had lain safe behind that shield. But something else was vulnerable, could not be hidden entirely away; to do so would negate its very purpose, and break the enchantment it held in trust. And so it stood in the open, unprotected, on Candlewick Street.

The Wild Hunt rode to destroy the London Stone.

* * *

ST. PAUL'S CATHEDRAL, LONDON

9 May 1590

The wind was already stirring, fleeing before the oncoming storm, when Lune reached the western porch of St. Paul's Cathedral.

"The entrances will be watched, my lady," Gertrude had said, when word came that the irrevocable move was made, the Wild Hunt was alerted to the secret of the London Stone, and the battle would take place under the full moon. "But there's one she cannot guard against you."

St. Paul's and the White Tower. The two original entrances to the Onyx Hall, created in the light of the eclipse. The latter lay within the confines of a royal fortress, and would have its own protection below.

But the former lay on Christian ground. No faerie guard could stay there long, however fortified with mortal bread he might be. None had passed through it since Invidiana had confined Francis Merriman to the chambers below.

The only question was whether it would open for Lune.

She passed the booksellers' stalls, closed up for the night. The wind sent refuse rattling against their walls. A snarl split the air, and she halted in her tracks. Light flashed across the city, and then from the sky above, a roar.

She glimpsed them briefly, past the cathedral's spire. Dame Halgresta Nellt, towering to a height she could never reach in the Onyx Hall. Sir Kentigern, at his sister's right hand, howling a challenge at the oncoming storm.

Sir Prigurd, at the left, his blunt features composed in an expression of dutiful resolution. She had always liked Prigurd the best. He was not as brutal as his siblings, and he was that rarity in the Onyx Hall: a courtier who served out of loyalty, however misplaced.

They stood at the head of the Onyx Guard, whose elf knights blazed in martial glory. Their armour gleamed silver and black and emerald, and their horses danced beneath them, tatterfoals and brags and grants eager to leap into battle. Behind stood the massed ranks of the infantry, boggarts and barguests, hobyahs and gnomes, all the goblins and pucks and even homely little hobs who could be mustered to fight in defence of their home.

The Onyx Hall. It *was* their home. A dark one, and twisted by its malevolent Queen, but home nonetheless.

Before the night was done, the Wild Hunt might reduce it to rubble.

But if Lune let herself question that price, she would be lost before she ever started.

The great doors of the western porch swung open at her approach. Stepping within, she felt holiness pressing against her skin, weirdly close and yet distant; the waiting tension of the angel's kiss thrummed within her. Like a sign shown to sentries, it allowed her passage.

She did not know what she sought, but the angel's power resonated with it, like a string coming into tune. *There.* A patch of floor like any other in the nave; when she stepped on it, the shock ran up her bones.

Here, faerie magic erupted upward. Here, holy rites saturated the ground. Here, London opened downward, into its dark reflection.

Lune knelt and laid one hand against the stone of the floor. The charm that governed the entrance spoke to her fingers. Francis had prayed, the words of God bringing him from one world to the other without any eyes seeing him. For her, the angelic touch sufficed.

Had any observer been there to watch, the floor would have remained unchanged. But to Lune's eyes, the slabs of stone folded away, revealing a staircase that led downward.

She had no time to waste. Gathering her courage, Lune hurried below—and prayed the threat of the Hunt had done its job.

THE ONYX HALL, LONDON

9 May 1590

The marble walls resonated with the thunder above, trembling, but holding strong.

Seated upon her throne, Invidiana might have been a statue. Her face betrayed no tension—had been nothing but a frozen mask since a hideous female giant brought word that the Wild Hunt rode against London.

Whatever he might say against her, Deven had to grant Invidiana this: she was indeed a Queen. She gave orders crisply, sending her minions running, and in less time than he would have believed possible, the defence of the Onyx Hall was mustered.

The presence chamber was all but empty. Those who had not gone to the battle had departed, hiding in their chambers, or fleeing entirely, in the hope of finding some safety.

Most, but not all. Invidiana, motionless upon her throne, was flanked by two elf knights, black-haired twin brothers. They stood with swords unsheathed, prepared to defend her with their lives. A human woman with a wasted, sunken face and dead eyes crouched at the foot of the dais.

And Achilles stood near Deven, clad only in sandals and a loincloth, his body tense with desire to join in the slaughter.

The thunder grew stronger, until the entire chamber shook. A crashing sound: some of the filigree had detached from between the arches, and plummeted to the floor. Deven glanced up, then rolled out of the way just in time to save himself as an entire pane of crystal shattered upon the stones.

Achilles laughed at him, fingers caressing the hilt of the archaic Greek sword he wore.

Where was Lune, in it all? Up in the sky, riding with the Hunt to save him? Battling at some entrance against guards that would keep her from the Onyx Hall?

Would she bring the miracle he needed?

He hoped so. But a miracle would not be enough; when she arrived, Achilles and the two elf knights would destroy her.

His sword was gone, broken in the battle against Achilles; he had not even a knife with which to defend himself. And he had no chance of simply snatching a weapon from the mortal or the knights. While he struggled with one, the others would get him from behind.

His eye fell upon the debris that now littered the floor, and a thought came to him.

They said Suspiria had called her lover Tiresias, for his gift. She had clearly continued the practice, naming Achilles for the great warrior of Greek legend.

Deven glanced upward. More elements of the structure were creaking, cracking; he dove suddenly to one side, as if fearing another would fall on him. The movement brought him closer to Achilles, and when he rose to a kneeling position, a piece of crystal was cold in his palm, its razor edges drawing blood.

He lashed out, and slashed the crystal across the backs of Achilles's vulnerable heels.

The man screamed and collapsed to the floor. Downed, but not dead, and Deven could take no chances. He seized a fragment of silver filigree and slammed it down onto his enemy's head, smashing his face to bloody ruin and sending the muscled body limp.

He got the man's sword into his hand just in time to meet the rush of the knights.

* * *

The palace groaned and shook under the assault of the battle above. How long would the Nellt siblings and their army hold off the Wild Hunt?

She ran flat out for the presence chamber. The rooms and galleries were deserted; everyone had gone to fight, or fled. Everywhere was debris, decorations knocked to the floor by the rattling blasts. And then the doors of the presence chamber were before her, closed tight, but without their usual guard. She should pause, listen at the crack, try to discover who was inside, but she could not stop; she lacked both the time and the courage.

Lune hit the doors and flung herself into the room beyond.

A wiry arm locked around her throat the instant she came through, and someone dragged her backward. Lune clawed behind herself, arms flailing. Fingers caught in matted hair. *Eurydice. Sun and Moon, she knows...*

Achilles lay in a pool of his own blood along one wall. Sir Cunobel of the Onyx Guard groaned on the floor not far away, struggling and failing to rise. But his twin Cerenel was still on his feet, and at the point of his sword, pinned with his back to a column, Michael Deven.

"So," Invidiana said, from the distant height of her throne. "You have betrayed me most thoroughly, it seems. And all for *this*?"

Deven was bruised and battered, his right hand bleeding; great tears showed in his doublet, where his opponents had nearly skewered him. His eyes met hers.

They were not so very far apart. If only she could get to him, just for an instant—

One kiss. But was it worth them both dying, to deliver it? What would happen, once their lips met?

Lune forced herself to look at Invidiana. "You mean to execute us both."

The Queen's beauty was all the more terrible, now that Lune knew from whence it came. Invidiana smiled, exulting. "Both? Perhaps, and perhaps not… he has drunk of faerie wine, you see. Already he is becoming ours. Once they take the first step, 'tis so easy to draw them in further. And you have deprived me of two of my pets. It seems only fitting that one, at least, should be replaced."

She saw the signs of it now, in the glittering of his eyes, the hectic flush of his cheeks against his pale skin. How much had he drunk? How far had he fallen into Faerie's thrall?

Some. But not, perhaps, enough.

Lune faced the Queen again. "He is stubborn. 'Tis a testament to your power that he drank even one sip. But a man with strong enough will can cast that off; he may refuse more. I know this man, and I tell you now: you will lose him. He will starve before he takes more from your hand, or from any of your courtiers."

Invidiana's lip curled. "Tell me now what you think to offer, traitor, before I lose patience with you."

Eurydice's bony arm threatened to choke her. Lune rasped out, "Promise me that you will keep him alive, and I will convince him to accept more food."

"I make no *promises*," Invidiana spat, her rage suddenly breaking through. "You are not here to bargain, traitor. I need do nothing you ask of me."

"I understand that." Lune let her weight drop; Eurydice was not strong enough to keep her upright, and so she sagged to her knees on the floor, the mortal now clinging to her back. Bowing her head against the restricting arm around her throat, Lune said, "With nothing left to lose, I can only beg, and offer my assistance—in hopes of buying this small mercy for him."

Invidiana considered this for several nerve-racking moments. "Why would you wish for that?"

Lune closed her eyes. "Because I love him, and would not see him die."

Soft, contemptuous laughter. Invidiana must have guessed it, but the admission amused her. "And why would he accept from you what he would not take from us?"

Her fingernails carved crescents into her palms. "Because I placed a charm on him, when I went to the mortal court, that made his heart mine. He will do anything I ask of him."

The battle still shook the walls of the presence chamber. Most of what could fall, had fallen; the next thing to go would be the Hall itself.

Eurydice's arm vanished from her throat.

"Prove your words true," Invidiana said. "Show me this mortal is your puppet. Damn him with your love. And perhaps I will hear your plea."

Lune pressed one trembling hand to the cold floor, pushed herself to her feet. She found Eurydice offering her a dented cup half-filled with wine. She took it, made a deep curtsy to the Queen, and only then turned to face Michael.

His blue eyes stared at her unreadably. There was no way to tell him what she intended, no way to tell him her words were a lie, that she had placed no charm upon him, that she would see him dead before she left him to be tormented by Invidiana, as Francis had been. All that would have to come later—if there was a later.

All that mattered now was to get close to him, for just one heartbeat.

Sir Cerenel sidestepped as she approached, but kept his blade at Deven's throat, and now a dagger flickered out, its point trained on her. Lune drew close, raised the cup, and leaned in just a fraction closer, so she could smile into his eyes, as if drawing upon a charm. "Drink for me, Master Deven."

His hand dashed the cup to the floor, and the instant it was gone from between them, she threw herself forward and kissed him.

As their lips met—as Lune kissed him as herself for the first time, with no masks between them—a voice rang out in the Onyx Hall, high and pure, speaking the language that lay beyond language.

"Be now freed all those whose love hath led them into chains."

* * *

Fire burned again on Deven's brow, six points in a ring, and he cried out against Lune's mouth, thinking himself about to die.

But it was a clean fire, a white heat that burned away whatever Invidiana had left there, and it caused him no pain; when it ended, he knew himself to be free.

Nor was he the only one.

The elf knight staggered away, dropping his weapons, hands outstretched, as if the power of that angelic presence had blinded him. The mortal woman collapsed on the floor, mouth open in a silent scream.

And in the centre of the chamber, in the very place Deven had stood to pray, he saw a slender, dark-haired man with sapphire eyes.

Francis Merriman stood loose and straight, his shoulders unbowed, his chin high, his eyes clear. Deven could see a shadow falling away from him, the last remnant of Tiresias, the maddened reflection that wandered lost in these halls for so many years. But it was a shadow only: death had freed him from the grip of dreams, and restored the man Suspiria once loved.

And Invidiana's icy calm shattered beneath his gaze.

"Control him!" she screamed at the woman on the floor, her fingers clutching the arms of her throne. "I did not summon him—"

"Yet I am come," Francis Merriman said. His voice was a light tenor, clear and distinct. "I have never left your side, Suspiria. You thought you bound me, first with your jewel, then by Margaret's arts—" The mortal woman gasped at the name. "But the first and truest chains that bound me were ones I forged myself. They are my prison, and my shield. They protected me against you after my death, so that I told you nothing I did not wish you to know. And they bring me to you now."

"Then I will banish you," Invidiana spat. Rage distorted the melody of her voice. "You are a ghost, and nothing more. What Hell waits for your unshriven soul?"

She should not have mentioned Hell. Francis's face darkened with sorrow. "You need not have made that pact, Suspiria. Nor need you have hidden from me. Did you think me, with my gift, blind to what you were? What you suffered? I stayed with you, knowing, and would have continued so."

"Stayed with me? With what? A shrivelling, rotting husk—you speak of prisons, and you know nothing of them. To be trapped in one's own flesh, every day bringing you closer to worms—a fitting fate for you, perhaps, but not for me. I did what he demanded, and yet to no avail. Why should I go on trying? I would endure his punishment *no more.*"

Then her voice dropped from its heightened pitch, growing cold again. "Nor will I endure you."

She raised her long-fingered hands, like two white spiders in the gloom. Deven's entire body tensed. A darkness hovered at the edge of his vision, deeper than the shadows of the Onyx Hall, and more foul. A corruption to match the purity that had touched him with Lune's kiss. It but waited for someone to invite it in.

Francis stopped her. He came forward with measured strides, approaching the throne, and despite herself Invidiana shrank back, hands faltering. "You did not give them your soul. You were never such a fool. No, you sold something else, did you not?" His voice was full of sorrow. "I saw it, that day in the garden. A heart, traded for what you had lost."

Her mouth twisted in fury: an open admission of guilt.

The man who had been her lover watched her with grieving eyes. "You bartered away your heart. All the warmth and kindness you could feel. All the love. Hell gained the evil you would wreak, and you gained a mask of ageless, immortal beauty.

"But I knew you without that mask, Suspiria. And I know what you have forgotten."

He mounted the steps of the dais. Invidiana seemed paralyzed, her black eyes fixed unblinking upon him.

"You gave your heart years before you sold it to the devil," Francis said. "You gave it to me. And so I return it to you."

The ghost of her love bent and kissed her, as Lune had kissed Deven moments before.

A scream echoed through the Onyx Hall, a sound of pure despair. The flawless, aching beauty of Invidiana shrivelled and decayed, folding in upon itself; the woman herself shrank, losing her imposing height, until what sat upon the throne seemed like a girl, not yet at her full growth, sitting upon a chair too large for her. But no girl would ever have looked so old.

Deven flinched in revulsion from the ancient, haggard thing Invidiana had become.

As the pact with Hell snapped, as the Queen of the Onyx Court dwindled, so, too, did the ghost of Francis Merriman fade. He grew fainter and fainter, and his last words whispered through the chamber.

"I will wait for you, Suspiria. I will never leave you."

The last wisp of him disappeared from view.

"Please—do not leave me."

Deven and Lune were left, the only two still standing, before the throne of the Onyx Hall.

A sound pierced the air, faint but passionate: part snarl, part shriek. The creature before them should not have been able to move, but she shifted forward, rising to her feet, and she had not lost the force of her presence; hatred beat outward like heat from a forge. Her voice was a shredded remnant of itself, grinding out the accusation. *"You brought this upon me!"*

Lune opened her mouth, her eyes full of urgency. But Deven stepped forward, interposing himself between his lady and the maddened shell of the Queen. He recognised what he saw in her eyes. Fury, yes, but fury to cover what lay beneath: a bottomless well of pain. She had her heart again; with it must have come all the emotions she had lost. Including remorse, for what she had done to the man she loved.

He had to say it now, before it was too late; the chance would not come again.

"Suspiria." It was important he use that name. The pieces had fallen together in the depths of his mind; he spoke from instinct. "Suspiria—*I know why you are still cursed*."

The withered hag twitched at his words.

"You had so much of it right," he said. Lune came forward a step, moving to stand at his side. "You atoned for your error. The Onyx Hall was a creation worthy of legend—a place for fae to live among mortals in safety, a place where the two could come together. You had so much of it right. But you did not *understand*.

"The chieftain's son loved you. But you disdained mortality, did you not? You could not bear to join yourself to it. And so you cast him aside, cast his love aside, as a thing without value, for what can it be worth, when it dies so soon? But the ages you endured after that must have taught you something, as they were intended to do; else you would not have made this great hall. And you would not have loved Francis Merriman."

He could feel the presence still. The ghost was gone, but Francis was not. The man had said it himself. He would never leave her. The love he felt joined them still.

And he had restored *her* ability to love.

"You did everything right," Deven said. "Your mistake came when you did not trust it. Faced with a future alongside the man you loved—suffering a sort of mortality, yes, aging while you watched him stay eternally young—you let your fear, your disdain, triumph again. You cast aside his love, and the love you felt for him. You failed to understand its worth."

A heart, traded for what she had lost. Youth. Beauty. Immortality. The answer had been in her hands, had she but accepted it.

Do not leave me, Francis had said.

"You face that decision again," Deven whispered. "Your true love waits for you. Honour that love as it deserves. Do not cast it aside a third time." This world operated by certain rules he did not have to explain to her or Lune. What was done a third time, was done forever.

For the first time since she bargained with Invidiana, Lune spoke. "Once we love, we cannot revoke it," she said. "We can only glory in what it brings—pain as well as joy, grief as well as hope. He is as much a fae creature now as a mortal. Where you will go, I do not know. But you can go with him."

Suspiria lifted her wasted face, lowering the claw-like hands that had risen to hide it. Only after a moment did Deven realise she was crying, the tears running down the deep gullies of her wrinkles, almost hidden from sight.

Invidiana had been evil. Suspiria was not. His heart gave a sharp ache, and a moment later, he felt Lune's hand slip into his own.

The change happened too subtly to watch. Without him ever seeing how, the wrinkles grew shallower, the liver spots began to fade. As age had shrivelled her a moment ago, now it acted in reverse, all the years lifting away, revealing the face of the woman Francis had loved.

She had the pale skin, the inky hair, the black eyes and

red lips. But what had been unnerving in its perfection was now mere faerie beauty: a step sideways from mortality, enough to take the breath away, but bearable. And *right*.

A last, a crystalline tear hovered at the edge of her lashes, then fell.

"Thank you," Suspiria whispered.

Then, like Francis Merriman, she faded from view, and when the throne was empty Deven knew they were both gone forever.

For a moment they stood silently in the presence chamber, with the corpse of Achilles, the huddled forms of Eurydice and the two elf knights, while Lune absorbed what she had just seen and done.

Then a pillar cracked and split in two, and Lune realised the thunder had not stopped. It had drawn nearer.

And Suspiria was gone.

Deven saw the sudden panic in her face. "What is it?"

"The Hunt," she said, unnecessarily. "I was to ask Suspiria—the Stone—they think the kings might relent, if she relinquished her sovereignty—but what will happen, now that she is gone?"

He took off before she even finished speaking, flying the length of the presence chamber at a dead run, heading directly for the throne. No, not directly; he went to one side of it, and laid hold of the edge of the great silver arch. "Help me!"

"With *what*?" She came forward regardless. "The throne

does not matter; we have to find the London Stone—"

"'Tis here!" Tendons ridged the backs of his hands as he dragged ineffectually at the throne. "A hidden chamber—I saw it before—"

Lune stood frozen for only a moment; then she threw herself forward and began to pull at the other side of the seat.

It moved reluctantly, protecting its treasure. "Help us!" Lune snapped, and whether out of reflexive obedience or a simple desire not to die at the hands of the Hunt, first Sir Cerenel and then Eurydice picked themselves up and came to lend their aid. Together the four of them forced it away from the wall, until there was a gap just wide enough for Lune and Deven to slip through.

The chamber beyond was no more than an alcove, scarcely large enough for the two of them and the stone that projected from the ceiling. A sword was buried halfway to the hilt in the pitted surface of the limestone, its grip just where an extremely tall woman's hand might reach.

Lune did not know what effect the sword had, now that one half of its pact had passed out of the world, but if they could take it to the Hunt, as proof of Invidiana's downfall… a slim hope, but she could not think of anything else to try.

Her own fingers came well short of the hilt. She looked at Deven, and he shook his head; Invidiana had been even taller than he, and he looked reluctant to touch a faerie sword regardless.

"Lift me," Lune said. Deven wrapped his bloodstained hands about her waist, gathered his strength, and sent her into the air, as high as he could.

Her hand closed around the hilt, but the sword did not pull free.

Instead, it pulled her upward, with Deven at her side.

CANDLEWICK STREET, LONDON

9 May 1590

She understood the truth, as they passed with a stomach-twisting surge from the alcove to the street above. The London Stone, half-buried, did not extend downward into the Onyx Hall. The Stone below was simply a reflection of the Stone above, the central axis of the entire edifice Suspiria and Francis had constructed. In that brief, wrenching instant, she felt herself not only to be at the London Stone, but at St. Paul's and the Tower, at the city wall and the bank of the Thames.

Then she stood on Candlewick Street, with Deven at her side, the sword still in her hand.

All around them was war. Some still fought in the sky; others had dragged the battle down into the streets, so that the clash of weapons came from Bush Lane and St. Mary Botolph and St. Swithins, converging on where they stood. Hounds yelped, a sound that made her skin crawl, and someone was winding a horn, its call echoing over the city rooftops. But she had eyes only for a set of figures mounted on horseback that stood scant paces from the two of them.

She thrust the sword skyward and screamed, "*Enough!*"

And her voice, which should not have begun to cut through the roar of battle, rang out louder than the horn, and brought near-instant silence.

They stared at her, from all around where the fighting had raged. She did not see Sir Kentigern, but Prigurd stood astraddle the unmoving body of their sister, a bloody two-handed blade in his grip. Vidar was missing, too. Which side did he fight on? Or had he fled?

It was a question to answer later. In the sudden hush, she lowered the tip of the sword until it pointed at the riders—the ancient kings of Faerie England.

"You have brought war to my city," Lune said in a forbidding voice, a muted echo of the command that had halted the fighting. "You *will* take it away again."

Their faces and forms were dimly familiar, half-remembered shades from scarcely forty years before. Had one of them once been her own king? Perhaps the one who moved forward now, a stag-horned man with eyes as cruel as the wild. "Who are you, to thus command us?"

"I am the Queen of the Onyx Court," Lune said.

The words came by unthinking reflex. At her side, Deven stiffened. The sword would have trembled in her grasp, but she dared not show her own surprise.

The elfin king scowled. "That title is a usurped one. We will reclaim what is ours, and let no pretender stand in our way."

Hands tensed on spears; the fighting might resume at any moment.

"I am the Queen of the Onyx Court," Lune repeated. Then she went on, following the same instinct that had made her declare it. "But not the Queen of faerie England."

The stag-horned rider's scowl deepened. "Explain yourself."

"Invidiana is gone. The pact by which she deprived you of your sovereignty is broken. I have drawn her sword from the London Stone; therefore the sovereignty of this city is mine. To you are restored those crowns she stole years ago."

A redheaded king spoke up, less hostile than his companion. "But London remains yours."

Lune relaxed her blade, letting the point dip to the ground, and met his gaze as an equal. "A place disregarded until the Hall was created, for fae live in glens and hollow hills, far from mortal eyes—except here, in the Onyx Hall. 'Twas never any kingdom of yours. Invidiana had no claim to England, but here, in this place, she created a realm for herself, and now 'tis mine by right."

She had not planned it. Her only thought had been to bear the sword to these kings, as proof of Invidiana's downfall, and hope she could sue for peace. But she felt the city beneath her feet, as she never had before. London was *hers*. And kings though they might be, they had no right to challenge her here.

She softened her voice, though not its authority. "Each side has dead to mourn tonight. But we shall meet in peace anon, all the kings and queens of faerie England, and when our treaty is struck, you will be welcome within my realm."

The red-haired king was the first to go. He wheeled his horse, its front hooves striking the air, and gave a loud cry; here and there, bands of warriors followed his lead, vaulting skyward once more and vanishing from sight. One by one, the other kings followed, each taking with them some portion of the Wild Hunt, until the only fae who remained in the streets were Lune's subjects.

One by one, they knelt to her.

Looking out at them, she saw too many motionless bodies. Some might yet be saved, but not all. They had paid a bloody price for her crown, and they did not even know why.

This would not be simple. Sir Kentigern and Dame Halgresta, if they lived—Lady Nianna—Vidar, if she could find him. And countless others who were used to clawing and biting their way to the top, and fearing the Queen who stood above them.

Changing that would be slow. But it could begin tonight.

To her newfound subjects, Lune said, "Return to the Onyx Hall. We will speak in the night garden, and I will explain all that has passed here."

They disappeared into the shadows, leaving Lune and Deven alone in Candlewick Street, with the sky rapidly clearing above them.

Deven let out his breath slowly, finally realising they might—at last—be safe. He ached all over, and he was light-headed from lack of food, but the euphoria that followed a battle was beginning to settle in. He found himself grinning wryly at Lune, wondering where to start with the things they needed to say. She was a *queen* now. He hardly knew what to think of that.

She began to return his smile—and then froze.

He heard it, too. A distant sound—somewhere in Cripplegate, he thought. A solitary bell, tolling.

Midnight had come. Soon all the bells in the city would be ringing, from the smallest parish tower to St. Paul's Cathedral itself. And Lune stood out in the open, unprotected; the angel's power had gone from her. The sound would hurt her.

But it would destroy something else.

He had felt it as they passed through the London Stone. St. Paul's Cathedral, one of the two original entrances to the Onyx Hall. The pit still gaped in the nave, a direct conduit from the mortal world to the fae, open and unprotected.

In twelve strokes of the great bell, every enchantment that bound the Onyx Hall into being would come undone, shredded by the holy sound.

"Give me your hand." Deven seized it before she could even move, taking her left hand in his left, dragging her two steps sideways to the London Stone.

"We will not be safe within," Lune cried. Her body shook like a leaf in the wind, as more bells began to ring.

Deven slapped his right hand onto the rough limestone surface. "We are not going within."

It was the axis of London and its dark reflection, the

linchpin that held the two together. Suspiria had not made the palace alone, because she *could* not; such a thing could only be crafted by hands both mortal and fae. Deven would have staked his life that Francis Merriman was a true Londoner, born within hearing of the city bells.

As Deven himself was.

With his hand upon the city's heart, Deven reached out blindly, calling on forces laid there by another pair before them. He had drunk of faerie wine. Lune had borne an angel's power. They had each been changed; they were each a little of both worlds, and the Onyx Hall answered to them.

The Thames. The wall. The Tower. The cathedral.

As the first stroke of the great bell rang out across the city, he felt the sound wash over and through him. Like a seawall protecting a harbour in a storm, he took the brunt of that force, and bid the entrance close.

A fourth stroke; an eighth; a twelfth. The last echoes of the bell of St. Paul's faded, and trailing out after it, the other bells of London. Deven waited until the city was utterly silent before he lifted his hand from the Stone.

He looked up slowly, carefully, half-terrified that he was wrong, that he had saved the Hall but left Lune vulnerable, and now she would shatter into nothingness.

Lune's silver eyes smiled into his, and she used their clasped hands to draw him toward her, so she might lay a kiss on his lips. "I will make you the first of my knights—if you will have me as your lady."

MEMORY

9 January 1547

The man walked down a long, colonnaded gallery, listening to his boot heels click on the stone, trailing his fingers in wonder across the pillars as he passed them by. It was impossible that this should all be here, that it should have come into being in the course of mere minutes, and yet he had seen it with his own eyes. Indeed, it was partly his doing.

The thought still dizzied him.

The place was enormous, far larger than he had expected, and so far almost entirely deserted. The sisters had chosen to stay in their own home, though they visited from time to time. Others would come, they assured him, once word spread farther, once folk believed.

Until then, it was just him, and the woman he sought.

He found her in the garden. They called it so, even though it was barely begun: a few brave clusters of flowers—a gift from the sisters—grouped around a bench that sat on the bank of the Walbrook. She was not seated on the bench, but on the ground, trailing her fingers in the water, a distant expression on her face. The air in the garden was pleasantly cool, a gentle contrast to the winter-locked world outside.

She did not move as he seated himself on the ground next to her. "I have brought seeds," he said. "I have no gift for planting, but I am sure we can convince Gertrude—since they are not roses." She did not respond, and his expression softened. He reached for her nearer hand and took it in his own. "Suspiria, look at me."

Her eyes glimmered with the tears she was too proud to shed. "It has accomplished nothing," she said, her low, melodic voice trembling.

"Did you hear that sound, half an hour ago?"

"What sound?"

He smiled at her. "Precisely. All the church bells of the city rang, and you did not hear a thing. This is a haven the likes of which has *never* existed, not even in legend. In time many fae will come, all of them dwelling in perfect safety beneath a mortal city, and you say it has accomplished nothing?"

She pulled her hand from his and looked away again. "It has not lifted the curse."

Of course. Francis had known Suspiria far longer than his appearance would suggest; he had not dwelt among mortals for many a year now. This hall had been an undertaking in its own right, a challenge that fascinated them both, and they had many grand dreams of what could be done with it, now that it was built. But it was born for another purpose, one never far from Suspiria's mind.

In that respect, it had failed.

He shifted closer and put gentle pressure on her shoulder, until she yielded and lay down, her head in his lap. With careful fingers he brushed her hair back, wondering if he should tell her what he knew: that the face he saw was an illusion, crafted to hide the age and degeneration beneath. The truth did not repel him—but he feared it would repel her, to know that he knew.

So he kept silent as always, and closed his eyes, losing himself in the silky touch of her hair, the quiet rippling of the Walbrook.

The gentle sound lifted him free of the confines of his mind, floating him into that space where time's grip slackened and fell away. And in that space, an image formed.

Suspiria felt his body change. She sat up, escaping his suddenly stilled arms, and took his face in her delicate hands. "A vision?"

He nodded, not yet capable of speech.

The wistful, loving smile he knew so well softened her face. He had not seen it often of late. "My Tiresias," she said, stroking his cheekbone with one finger. "What did you see?"

"A heart," he whispered.

"Whose heart?"

Francis shook his head. Too often it was thus, that he saw without understanding. "The heart was exchanged for an apple of incorruptible gold. I do not know what it means."

"Nor I," Suspiria admitted. "But this is not the first time such meaning has eluded us—nor, I think, will it be the last."

He managed a smile again. "A poor seer I am. Perhaps I have been too long among your kind, and can no longer tell the difference between true visions and my own fancy."

She laughed, which he counted a victory. "Such games we could play with that; most fae would believe even the strangest things to be honest prophecy. We could go to Herne's court and spread great confusion there."

If it would lighten her heart, he would have gladly done it, and risked the great stag-horned king's wrath. But sound distracted him, something more than the gentle noise of the brook. Someone was coming, along the passage that led to the garden.

Suspiria heard it, too, and they rose in time to see the plump figure of Rosamund Goodemeade appear in an archway. Nor was she alone: behind her stood a fae he did not recognise, travel-stained and weary, with a great pack upon his back.

Francis took Suspiria's hand, and she raised her eyebrows at him. "It seems another has come to join us. Come, let us welcome him together."

WINDSOR GREAT PARK, BERKSHIRE
11 June 1590

The oak tree might have stood there from the beginning of time, so ancient and huge had it grown, and its spreading branches extended like mighty sheltering arms, casting emerald shadows on the ground below.

Beneath this canopy stood more than two score people, the greatest gathering of faerie royalty England had ever seen. From Cumberland and Northumberland to Cornwall and Kent they came, and all the lands in between: kings and queens, lords and ladies, a breathtaking array of great and noble persons, with their attendants watching from a distance.

They met here because it was neutral ground, safely removed from the territory in dispute and the faerie palace many still thought of as an unnatural creation, an emblem of the Queen they despised. Under the watchful aegis of the oak, the ancient tree of kings, they gathered to discuss the matter—and, ultimately, to recognise the sovereignty of a new Queen.

It was a formality, Lune knew. They acknowledged her right to London the moment they obeyed her command to leave. Her fingers stroked the hilt of the sword as one of the kings rolled out a sonorous, intricate speech about the traditional rights of a faerie monarch. She did not want to inherit Invidiana's throne. It carried with it too many dark memories; the stones of the Onyx Hall would never be free of blood.

But that choice, like others, could not be unmade.

The orations had gone on for quite some time. Lune suspected her fellow monarchs were luxuriating in the restoration of their dignity and authority. But in time she grew impatient; she was glad when her own opportunity came.

She stood and faced the circle of sovereigns, the London Sword sheathed in her hands. The gown she wore, midnight-blue silk resplendent with moonlight and diamonds, felt oddly conspicuous; she still remembered her time out of favour, hiding in the corners of the Onyx Hall, dressed in the rags of her own finery.

But the choice was deliberate: many of those gathered about her wore leather or leaves, clothing that less closely mirrored that of mortals.

Lune had a point to make. And to that end, she lifted her gaze past those gathered immediately beneath the oak, looking to the attendant knights and ladies that waited beyond.

Lifting one hand, she beckoned him to approach.

Standing between the Goodemeade sisters, Michael Deven hesitated, as well he might. But Lune raised one eyebrow at him, and so he came forward and stood a pace behind her left shoulder, hands clasped behind his back. He, too, was dressed in great finery, faerie-made for him on this day.

"Those of you gathered here today," Lune said, "remember Invidiana, and not fondly. I myself bear painful memories of my life under her rule. But today I ask you to remember someone else: a woman named Suspiria.

"What she attempted, some would say is beyond our reach. Others might say we *should not* reach for it, that mortal and faerie worlds are separate, and ever should stay so.

"But we dwell here, in the glens and the hollow hills, because we do not believe in that separation. Because we seek out lovers from among their kind, and midwives for our children, poets for our halls, herdsmen for our cattle. Because we aid them with enchantments of protection, banners for battle, even the homely tasks of crafting and cleaning. Our lives are intertwined with theirs, to one degree or another—sometimes for good, other times for ill, but never entirely separate.

"Suspiria came to believe in the possibility of harmony between these two worlds, and created the Onyx Hall in pursuit of that belief. But we do wrong if we speak only of her, for that misses half the heart of the matter: the Hall was created by a faerie and a mortal, by Suspiria and Francis Merriman."

Reaching out, Lune took Deven by the hand, bringing him forward until he stood next to her. His fingers tightened on hers, but he cooperated without hesitation.

"I would not claim the Onyx Hall if I did not share in their belief. And I will continue to be its champion. So long as I reign, I will have a mortal at my side. Look upon us, and know that you look upon the true heart of the Onyx Court. All those who agree will ever be welcome in our halls."

Her words carried clearly through the still summer air. Lune saw frowns of disagreement here and there, among the kings, among their attendants. She expected it. But not everyone frowned. And she had established her own stance as Queen—her similarity to Suspiria, her difference from Invidiana—and that, more than anything, was her purpose here today.

The day did not end with speeches. There would be celebrations that night, and she would take part, as a Queen must. But two things would happen before then.

She walked with Deven at twilight along the bank of a nearby stream, once again hand in hand. They had said many things to one another in the month since the battle, clearing away the last of the lies, sharing the stories of what had happened while they were apart. And the stories of what had happened while they were together—truths they had never admitted before.

"Always a mortal at your side," Deven said. "But not always me."

"I would not do that to you," Lune responded, quietly serious. "'Twas not just Invidiana's cruelty that warped Francis. Living too long among fae will bring you to grief, sooner or later. I love the man you are, Michael. I'll not make you into a broken shell."

He could never leave her world entirely. The faerie wine he drank had left its mark, as Anael's power had done to her. But it did not have to swallow him whole.

He sighed and squeezed her hand. "I know. And I am thankful for it. But 'tis easy to understand how Suspiria came to despair. Immortality all around, and none for her."

Lune stopped and turned him to face her, taking his other hand. "See it through my eyes," she said. "All the passion of humanity, all the fire, and I can do no more than warm myself at its edge." A presentiment of sorrow roughened her voice. "And when you are gone, I will not grieve and recover, as a human might. I may someday come to love another—perhaps—but this love will never fade, nor the pain of its loss. Once my heart is given, I may never take it back."

He managed a smile. "Francis gave Suspiria's heart back."

Lune shook her head. "No. He shared it with her, and

reminded her that she loved him, still and forever."

Deven closed his eyes, and Lune knew he, like her, was remembering those moments in the Onyx Hall. But then an owl hooted, and he straightened with a sigh. "We are due elsewhere. Come—she does not like to be kept waiting."

WINDSOR CASTLE, BERKSHIRE
11 June 1590

When all the attendants and ladies-in-waiting had been dismissed, when the room was empty except for the three of them, Elizabeth said, "I think 'tis time you showed me your true face, Mistress Montrose."

Deven watched Lune out of his peripheral vision. She must have been half-expecting the request, for she did not hesitate. The golden hair and creamy skin faded away, leaving in their place the alien beauty of a faerie queen.

Elizabeth's mouth pressed briefly into a thin, hard line. "So. You are her successor."

"Yes." Deven winced at Lune's lack of deferential address, but she was right to do it; Elizabeth must see her as a fellow queen, an equal. "And on behalf of my people, I offer you a sincere apology for the wrongs your kingdom suffered at the hands of Invidiana."

"Is that so." Elizabeth fingered her silken fan, studying Lune. "She did much that was ill, 'tis true."

Deven could not make up his mind which queen to

watch, but something in Elizabeth's manner sparked a notion deep within his brain. "Your Majesty," he asked, directing the words at the aging mortal woman, "how long did you know Anne Montrose was not what she seemed to be?"

Elizabeth's dark gaze showed unexpected amusement, and a smile lurked around the corners of her mouth, proud and a little smug. "My lords of the privy council take great care to watch the actions of my royal cousins in other lands," she said. "Someone had to keep an eye on the one that lived next door."

This *did* startle Lune. "Did you—"

"Know of others? Yes. Not all of them, to be sure; no doubt she sent temporary agents to manipulate my lords and knights, whom I never saw. But I knew of some." Now the pride was distinctly visible. "Margaret Rolford, for one."

Lune gaped briefly, then recovered her dignity and nodded her head in respectful admission. "Well spotted. I would be surprised you allowed me to remain at court—but then again, 'tis better to know your enemy's agents and control them, is it not?"

"Precisely." Elizabeth came forward, looking thoughtful. She stood a little taller than Lune, but not by much. "I cannot say I will like you. There is too much of bad blood, not so easily forgotten. But I hope for peaceful relations, at least."

Lune nodded. Looking at the two of them, Deven marked their choice of colour: Lune in midnight blue and silver, Elizabeth in russet brocade with gold and jewels.

Neither wore black, though Elizabeth often favoured it. For the striking contrast with her auburn hair and white skin, or out of some obscure connection to or competition with Invidiana? Either way, it seemed both were determined to separate themselves from that past, at least for today.

Elizabeth had turned away to pace again; now she spoke abruptly. "What are your intentions toward my court?"

This was the true purpose of the meeting, the reason why "Mistress Montrose" had made a visit to Windsor Castle. Deven and Lune had talked it over before coming, but neither could guess what answer Elizabeth wanted to hear. All they could offer was the truth.

"'Tis a delicate balance," Lune said. "Invidiana interfered too closely, appropriating your actions for her own ends, and treading upon your sovereign rights. I have no wish to imitate her in that respect. But we also have no interest in seeing England fall to a Catholic power. I do not speak for all the faerie kingdoms, but if there is need of defence, the Onyx Court will come to your aid."

Elizabeth nodded slowly, evaluating that. "I see. Well, I have had enough of pacts; I want no swords in stones to bind us to each other. If such a threat should arise, though, I may hold you to your word."

Then she turned, without warning, to Deven. "As for you, Master Deven—you offered to free me from that pact, and so you did. What would you have of me in return?"

His mind went utterly blank. How Colsey would have laughed to see him now, and Walsingham, too; he had come to court with every intention of advancing himself, and now that his great opportunity came, he could not think what to ask. His life had gone so very differently than he expected.

Kneeling, he said the first thing that came into his head. "Madam, nothing save your gracious leave to follow my heart." Elizabeth's response was cool and blunt. "You cannot marry her, you know. There's not a priest in England that would wed you." John Dee might do it, but Deven had not yet worked up the courage to ask. "I do not speak only of marriage."

"I know." Her tone softened. Deep within it, he heard the echo of a quiet sorrow, that never left her heart. "Well, it cannot be made official—I would not fancy explaining it to my lords of the council—but if our royal cousin here finds it acceptable, you shall be our ambassador to the Onyx Court."

He could almost hear Lune's smile. "That would be most pleasing to us."

"Thank you, madam." Deven bowed his head still further. "But there is one difficulty." Elizabeth came forward and put her white fingers under his chin, tilting his head up so he had no choice but to meet her dark, level gaze. "'Twould be an insult to send a simple gentleman to fill such a vital position." She pretended to consider it, and he saw the great pleasure she took in this, dispensing honours and rewards to those who had done her good service. "I believe we shall have to knight you. Do you accept?"

"With all my heart." Deven smiled up at one of his

queens, and out of the corner of his eye, saw his other queen echo the expression.

It was a divided loyalty, and if a day should come that Elizabeth turned against Lune, he would regret occupying such a position.

But he could not leave the faerie world, and he could not leave Lune. So together, they would ensure that day never came.

EPILOGUE

RICHMOND PALACE, RICHMOND

8 February 1603

The coughing never went entirely away anymore. They sent doctors to pester her; she mustered the energy to drive them out again, but every time it was harder. The rain beat ceaselessly against the windows, a long, dreary winter storm, and it was easy to believe that all the world had turned against her. She sat upon cushions before the fire, and spent many long hours staring into its depths.

Her mind drifted constantly now, forgetting what it was she had been doing. Cecil came occasionally with papers for her to sign; half the time she was surprised to see it was Robert, William's hunchbacked little son. The wrong Cecil. Burghley, her old, familiar Cecil, had died... how long ago now?

Too long. She had outlived them all, it seemed. Burghley, Walsingham, Leicester. Her old enemy Philip of Spain. Essex, executed on Tower Hill—oh, how he had gone wrong. She could have handled him differently, perhaps, but when all was said and done he would never forgive her for being an old woman, too proud to give in, too stubborn to die. She was approaching seventy. How many could boast reaching such a great age?

She could think of some, but her mind flinched away. Those thoughts were too painful, now that illness and the infirmity of old age were defeating her at last.

"But I have done well, have I not?" she whispered to the fire. "I have done well. 'Twas not all because of her."

She glanced behind her, half-expecting to see a tall figure in the shadows, but no one was there. Just two of her closest maids, keeping weary vigil over their crabbed old queen, periodically trying and failing to convince her to go to bed. She looked away again, quickly, before they could raise their incessant refrain again.

Sometimes she could almost believe she had imagined it all, from her visitor in the Tower onward. But no—it had been real. Invidiana, and all the rest.

So many regrets. So many questions: What would have been different, had she never formed that pact? Would the Armada have reached the shores of England, bearing Parma's great army to overrun and subjugate them beneath the yoke of Spain? Or would she never have gotten that far? Perhaps she would have died in the Tower, executed for her Protestant heresy, or simply permitted to perish from the damp cold there, as she was perishing now. Mary Stewart might have had her throne, one Catholic Mary to follow another.

Or not. She had survived thirteen years without Invidiana, through her own wits and will, and the aid of those who served her. She was the Queen of England, blessed by God, beloved of her people, and she could stand on her own.

"And I have," she whispered, her lips moving near-soundlessly. "I have been a good queen."

The rain drumming against the windows made no reply. But she heard in it the cheers of her subjects, the songs in her honour, the praise of her courtiers. She had not been perfect. But she had done her best, for as long as she could. Now the time had come to pass her burden to another, and pray he did well by her people.

Pray they remembered her, and fondly.

Gazing into the fire, Elizabeth of England sank into dreams of her glorious past, an old woman, wrinkled and ill, but in her mind's eye, now and forever the radiant Virgin Queen.

ACKNOWLEDGMENTS

I owe a great debt of gratitude to the many people who helped me research this novel. During my trip to England, I was assisted by the following wonderful volunteers: from the Shakespeare's Globe Library and Archives, Victoria Northwood; from the National Trust, Kate Wheeldon at Hardwick Hall; and from Historic Royal Palaces, Alison Heald, Susan Holmes at the Tower of London, and Alden Gregory at Hampton Court Palace. (The rooftop scene is his fault.)

I'm also grateful to Kevin Schmidt, for the astrology in Act Three, and to Dr. William Tighe, who taught me everything I know about the Gentlemen Pensioners, and mailed me his dissertation to boot. He is not to be blamed for anything I got wrong.

Finally, I have to thank Kate Walton, for needing someone to keep her awake on a late-night drive to the airport back in June of 2006. It was the first of many fruitful midnight conversations about this story, and it wouldn't have been the same without her.

ABOUT THE AUTHOR

American fantasy writer Marie Brennan habitually pillages her background in anthropology, archaeology and folklore for fictional purposes. In addition to the *Onyx Court* series, she is author of the *Doppelganger* duology of *Warrior* and *Witch*, the urban fantasy *Lies and Prophecy*, and the highly acclaimed *Natural History of Dragons* fantasy series, as well as more than forty short stories.

ALSO AVAILABLE FROM TITAN BOOKS

A NATURAL HISTORY OF DRAGONS
A MEMOIR BY LADY TRENT
by Marie Brennan

Everyone knows Isabella, Lady Trent, to be the world's preeminent dragon naturalist. Here at last, in her own words, is the true story of a pioneering spirit who risked her reputation, prospects, and her life to satisfy scientific curiosity; of how she sought true love despite her lamentable eccentricities; and of her thrilling expedition to the mountains of Vystrana, where she made discoveries that would change the world...

"Fans of fantasy, science, and history will adore this rich and absorbing tale of discovery."
Publishers Weekly, starred review

"Though dragons may not exist in our own world's history, it is certain that the struggles of women like Trent do, and Brennan captures both with equal elegance in this first volume in the memoirs of Lady Trent."
Shelf Awareness, starred review

"Courageous, intelligent and determined Isabella's account is colorful, vigorous and absorbing... "
Kirkus

ALSO AVAILABLE FROM TITAN BOOKS

THE TROPIC OF SERPENTS
A MEMOIR BY LADY TRENT

by Marie Brennan

Three years after her fateful journeys through Vystrana, the widowed Mrs. Camherst defies convention to embark on an expedition to the savage, war-torn continent of Eriga, home of the legendary swamp-wyrms of the tropics. Accompanied by an old associate and a runaway heiress, Isabella must brave heat, fevers and palace intrigues to satisfy her boundless fascination with all things draconian.

"Marie Brennan's *The Tropic of Serpents* brings dragons into the Victorian world, and flawlessly so… [It] gives fantasy readers another good reason to enter Brennan's fantastical world."
Washington Post

"As uncompromisingly honest and forthright as the first, narrated in Brennan's usual crisp, vivid style, with a heroine at once admirable, formidable and captivating. Reader, lose no time in making Isabella's acquaintance."
Kirkus, starred review

"A thoroughly enjoyable book. I recommend it without hesitation."
Tor.com

TITANBOOKS.COM